UNDERSTANDING GLIAL CELLS

UNDERSTANDING GLIAL CELLS

edited by

Bernardo Castellano, Ph.D.
Department of Cell Biology and Physiology
Universitat Autònoma Barcelona

Berta González, Ph.D.
Department of Cell Biology and Physiology
Universitat Autònoma Barcelona

Manuel Nieto-Sampedro, Ph.D.
Cajal Institute
C.S.I.C., Madrid

SPRINGER SCIENCE+BUSINESS MEDIA, LLC

Library of Congress Cataloging-in-Publication Data

Understanding glial cells / edited by Bernardo Castellano, Berta González, Manuel Nieto-Sampedro.
p. cm.
Includes bibliographical references and index.
ISBN 978-0-7923-8140-2 ISBN 978-1-4615-5737-1 (eBook)
DOI 10.1007/978-1-4615-5737-1
1. Neuroglia. I. Castellano, Bernardo, 1957- II. González, Berta, 1955- III. Nieto-Sampedro, Manuel, 1944-
QP363.2.U53 1998
573.8'636--dc21 98-12036
CIP

Printed on acid-free paper.

CONTENTS

II. MOLECULAR AND BIOCHEMICAL PROPERTIES

III. INVOLVEMENT IN DAMAGE AND REGENERATION

LIST OF CONTRIBUTORS

Laia Acarín
Unitat Histologia, Facultat de Medicina
Departament Biologia Cel.lular i Fisiologia
Universitat Autònoma Barcelona
08193 Bellaterra, Spain.

José Aijón
Departamento de Biología Celular y Patología
Universidad de Salamanca
37007 Salamanca, Spain.

Angela Alcalá
Departamento Ciencias Morfológicas
Facultad de Medicina
Universidad de Zaragoza
50009 Zaragoza, Spain.

Alberto Alcázar
Servicio de Bioquímica
Departamento de Investigación
Hospital Ramón y Cajal, INSALUD
28034 Madrid, Spain.

Ramón Anadón
Departmento de Biología Fundamental
Universidad Santiago de Compostela
15706 Santiago de Compostela, Spain.

Jean de Barry
Laboratoire de Neurobiologie Cellulaire
(UPR 9009 CNRS)
Centre de Neurochimie
Université Louis Pasteur
Strasbourg, France.

Eulalia Bazán
Servicio de Bioquímica
Departamento de Investigación
Hospital Ramón y Cajal
INSALUD
28034 Madrid, Spain.

Elena Becker
Departamento de Microbiología y Genética
C.S.I.C., Universidad de Salamanca
37007 Salamanca, Spain.

Elena Caminos
Departamento de Biología Celular
y Patología
Facultad de Medicina
Universidad de Salamanca
37007 Salamanca, Spain.

Maria José Casarejos
Departamento de Investigación
Hospital Ramón y Cajal
INSALUD
28034 Madrid, Spain.

Bernardo Castellano
Unitat Histologia,
Facultat de Medicina
Departament Biologia Cel.lular i
Fisiologia
Universitat Autònoma Barcelona
08193 Bellaterra, Spain.

Anthony J. Castro
Department of Cell Biology,
Neurobiology and Anatomy,
Stritch School of Medicine
Loyola University Chicago
Maywood (IL 60153, U.S.A.

Julie Ann Chowen
Instituto Cajal, C.S.I.C.
28002 Madrid, Spain.

Blanca Conde
Departamento Ciencias Morfológicas
Facultad de Medicina
Universidad de Zaragoza
50009 Zaragoza, Spain.

Ishar Dalmau
Unitat Histologia, Facultat de Medicina
Departament Biologia Cel.lular i
Fisiologia
Universitat Autònoma Barcelona
08193 Bellaterra, Spain.

Sofía M. Díaz-Regueira
Departmento de Biología Celular
y Molecular
Universidad de La Coruña
15071 La Coruña, Spain.

Francisco José Esteban
Departmento de Biología Celular
Facultad de Ciencias Experimentales
Universidad de Jaén
23071 Jaén, Spain.

María del Carmen Fernández-Galaz
Instituto Cajal, C.S.I.C.
28002 Madrid, Spain.

David J. Fogarty
Departamento de Neurociencias
Universidad del País Vasco
48940 Leioa, Vizcaya, Spain.

José María García-Barcina
Departamento de Neurociencias
Universidad del País Vasco
48940 Leioa, Vizcaya, Spain.

Justo García de Yébenes
Servicio de Neurología
Fundación Jiménez Díaz
Madrid, Spain.

Luis Miguel García-Segura
Instituto Cajal, C.S.I.C.
28002 Madrid, Spain.

Ana Gascón
Departamento Ciencias Morfológicas
Facultad de Medicina
Universidad de Zaragoza
50009 Zaragoza, Spain.

Giorgio Gombos
Laboratoire de Neurobiologie Cellulaire
(UPR 9009 CNRS)
Centre de Neurochimie
Université Louis Pasteur
Strasbourg, France.

Pilar Gómez-Ramos
Departamento de Morfología
Facultad de Medicina
Universidad Autónoma de Madrid
28029 Madrid, Spain.

Berta González
Unitat Histologia,
Facultat de Medicina
Departament Biologia Cel.lular i Fisiologia
Universitat Autònoma Barcelona
08193 Bellaterra, Spain.

Miroslav Gottlieb
Departamento de Neurociencias
Universidad del País Vasco
48940 Leioa, Vizcaya, Spain.

Carmen Guaza
Grupo de Psicobiología
Unidad de Plasticidad Neural
Instituto Cajal, C.S.I.C.
28002 Madrid, Spain

Consuelo Guerri
Instituto de Investigaciones Citológicas
(F.I.B.)
46010 Valencia, Spain.

Antonio S. Herranz
Servicio de Neurobiología
Departamento de Investigación
Hospital Ramón y Cajal, INSALUD
28034 Madrid, Spain.

Nieves Ibarrola
Instituto Investigaciones Biomédicas
C.S.I.C., Madrid, Spain.

David Jimeno
Dpto. de Biología Celular y Patología
Universidad de Salamanca
37007 Salamanca, Spain.

Juan M. Lara
Dpto. de Biología Celular y Patología
Universidad de Salamanca
37007 Salamanca, Spain.

Concepción Lillo
Dpto. de Biología Celular y Patología
Universidad de Salamanca
37007 Salamanca, Spain.

Alberto Lledó
Grupo de Psicobiología
Unidad de Plasticidad Neural
Instituto Cajal
C.S.I.C.
28002 Madrid, Spain

Miguel A. López-Toledano
Servicio de Neurobiología
Departamento de Investigación
Hospital Ramón y Cajal,
INSALUD
28034 Madrid, Spain.

Dionisio Martín-Zanca
Departamento de Microbiología
y Genética,
C.S.I.C.
Universidad de Salamanca
37007 Salamanca, Spain.

Manuel Martínez
Departmento de Biología Celular
Facultad de Ciencias Experimentales
Universidad de Jaén
23071 Jaén, Spain.

Carlos Matute
Departamento de Neurociencias
Universidad del País Vasco
48940 Leioa, Vizcaya, Spain.

Luis Megías
Departamento Ciencias Morfológicas
Universidad de Granada
18071 Granada, Spain.

María Angeles Mena
Servicio de Neurobiología
Departamento de Investigación
Hospital Ramón y Cajal, INSALUD
28034 Madrid, Spain.

Francisco Molina-Holgado
Grupo de Psicobiología
Unidad de Plasticidad Neural
Instituto Cajal, C.S.I.C.
28002 Madrid, Spain

Eduardo Molina-Holgado
Grupo de Psicobiología
Unidad de Plasticidad Neural,
Instituto Cajal
C.S.I.C., 28002 Madrid, Spain

Maximina Monzón-Mayor
Dept. de Morfología y Biología Celular
Facultad de Ciencias de la Salud
Universidad de Las Palmas
de Gran Canaria
35080 Islas Canarias, Spain.

María Luisa del Moral
Departmento de Biología Celular
Facultad de Ciencias Experimentales
Universidad de Jaén
23071 Jaén, Spain.

María Asunción Morán
Departamento de Morfología
Facultad de Medicina
Universidad Autónoma de Madrid
28029 Madrid, Spain.

María José Morán
Departamento de Neurociencias
Universidad del País Vasco
48940 Leioa, Vizcaya, Spain.

Xavier Navarro
Unitat Fisiologia
Facultat de Medicina
Departament Biologia Cel.lular i
Fisiologia
Universitat Autònoma Barcelona
08193 Bellaterra, Spain.

Manuel Nieto-Sampedro
Departamento de Plasticidad Neural
Instituto Cajal
C.S.I.C.
28002 Madrid, Spain.

Neville N. Osborne
Nuffield Laboratory of Ophthalmology
University of Oxford
Oxford, United Kingdom.

Carlos L. Paíno
Servicio de Neurobiología
Departamento de Investigación
Hospital Ramón y Cajal
INSALUD
28034 Madrid, Spain.

Juan Angel Pedrosa
Departmento de Biología Celular
Facultad de Ciencias Experimentales
Universidad de Jaén
23071 Jaén, Spain.

María Angeles Peinado
Departmento de Biología Celular
Facultad de Ciencias Experimentales
Universidad de Jaén, 23071 Jaén, Spain.

José María Peinado
Departmento de Bioquímica
Instituto de Neurociencias F. Oloriz
Universidad de Granada
18071 Granada, Spain.

Adoración Quesada
Departmento de Biología Celular
Facultad de Ciencias Experimentales
Universidad de Jaén
23071 Jaén, Spain.

María Jesús Ramírez
Departmento de Biología Celular
Facultad de Ciencias Experimentales
Universidad de Jaén
23071 Jaén, Spain.

José M. Ramírez
Instituto de Investigaciones
Oftalmológicas Ramón Castroviejo
Facultad de Medicina
Universidad Complutense de Madrid
28040 Madrid, Spain.

Ana I. Ramírez
Instituto de Investigaciones
Oftalmológicas Ramón Castroviejo
Facultad de Medicina
Universidad Complutense de Madrid
28040 Madrid, Spain.

Carolina Redondo
Servicio de Neurobiología
Departamento de Investigación
Hospital Ramón y Cajal
INSALUD
28034 Madrid, Spain.

Jaime Renau-Piqueras
Centro de Investigación
Sección de Biología y Patología Celular
Hospital La Fe
46009 Valencia, Spain.

José Rodrigo
Departmento de Neuroquímica
Instituto Cajal (C.S.I.C.)
28002 Madrid, Spain.

Angeles Rodríguez-Peña
Instituto Investigaciones Biomédicas
(C.S.I.C.)
Madrid, Spain.

María M. Romero-Alemán
Departamento de Morfología
Sección de Biología Celular
Facultad de Ciencias de la Salud
Universidad de Las Palmas
de Gran Canaria
35080 Islas Canarias, Spain.

Mercedes Ruidíaz
Departamento Ciencias Morfológicas
Facultad de Medicina
Universidad de Zaragoza
50009 Zaragoza, Spain.

Juan J. Salazar
Instituto de Investigaciones
Oftalmológicas Ramón Castroviejo,
Facultad de Medicina
Universidad Complutense de Madrid
28040 Madrid, Spain.

María Victoria Sánchez-Gómez
Departamento de Neurociencias
Universidad del País Vasco
48940 Leioa, Vizcaya, Spain.

Elena Sinués
Departamento Ciencias Morfológicas
Facultad de Medicina
Universidad de Zaragoza
50009 Zaragoza, Spain.

Robert R. Sturrock
Department of Anatomy
and Physiology
University of Dundee
Dundee, United Kingdom.

Alberto Triviño
Instituto de Investigaciones
Oftalmológicas Ramón Castroviejo,
Facultad de Medicina
Universidad Complutense de Madrid
28040 Madrid, Spain.

Elena Vecino
Departamento de Biología Celular
y Ciencias Morfológicas
Facultad de Medicina
Universidad del País Vasco, Spain.

Sonia Vega
Instituto Investigaciones Biomédicas
(C.S.I.C.)
Madrid, Spain.

José Miguel Vela
Unitat Histologia
Facultat de Medicina
Departament Biologia Cel.lular i
Fisiologia
Universitat Autònoma Barcelona
08193 Bellaterra, Spain.

Almudena Velasco
Dpto. de Biología Celular y Patología
Universidad de Salamanca
37007 Salamanca, Spain.

Enrique Verdú
Unitat Fisiologia, Facultat de Medicina
Departament Biologia Cel.lular i
Fisiologia
Universitat Autònoma Barcelona
08193 Bellaterra, Spain.

Carmen M. Yanes
Departamento de Biología Celular
Facultad de Biología
Universidad de La Laguna
Tenerife, Islas Canarias, Spain.

PREFACE

The studies by Santiago Ramón y Cajal on astrocytes, and the extense work developed by Pío del Río Hortega dealing with microglia and oligodendrocytes are with no doubt two important milestones of neurobiology, and specially of glial research.

Since these classical studies, the curiosity for understanding glial cells has continuously grown. During the last two decades, the development of new methods of study, including cell culture techniques and the use of histochemical and immunocytochemical markers, has allowed a remarkable advance in this field. This increasing interest is reflected by the creation of local and international glial networks, the publication of the journal *Glia*, the celebration of periodic international meetings and the publication of extensive books focused on glial cells.

In February 1997, with the establishment of the Spanish Glial Network, arose the early idea of publishing a compendium of selected work by Spanish glial research teams, which has finally materialised in *Understanding Glial Cells*, a book whose objective is to present topics of current interest in glial cell biology. The aim of *Understanding Glial Cells* is far from competing with other recently published books covering wider aspects, but to complement this information with actualised reviews on specific topics. Chapters that constitute the four main sections of this book are directed to readers with a pre-existing interest in the field, being specially indicated for teachers, basic-scientists and clinicians that, within neurobiology, neurology or neuropathology, want to actualize their knowledge on the topic. In addition, we hope that medical and graduate students will find this book a useful reference volume reviewing the current state of research dealing with some particular aspects of glial cell biology.

Spanish Glial Network
http://blues.uab.es/redglial/

ACKNOWLEDGMENTS

The publication of **Understanding Glial Cells** has been possible thanks to the effort of the Spanish Glial Network and to the economical support received from the Spanish Ministry of Education and Culture, Autonomous University of Barcelona, Basque Country University, Las Palmas de Gran Canaria University and Zaragoza University.

Universitat Autònoma de Barcelona

Universidad del País Vasco

Euskal Herriko Unibertsitatea

I. MORPHOLOGY AND ONTOGENY

1

CHARACTERIZATION OF THE GLIAL CELLS IN THE TELEOST VISUAL PATHWAY

Juan M. Lara, Almudena Velasco, Concepción Lillo, David Jimeno and José Aijón

Continual growth and regenerative capacity are characteristics of the Central Nervous System (CNS) of some teleosts, and both particularities are especially notable in the visual system. Within this system, the optic nerve is an ideal structure for the study of the glia of these animals and of their involvement in the processes of development and regeneration. The cyto-architecture of the optic nerve determines a precise temporal and spatial order between the retinofugal fibres that the neuroglial population establishes and maintains. In the distinct parts of the optic nerve (intraorbital, mediorbital, optic chiasm and anterior optic tract), astrocytes and oligodendrocytes are responsible for the fasciculation and the myelination, adapting their cytological characteristics and their arrangement to the precise necessities of each zone. Throughout the extension of the nerve the continuous integration of new axons also requires a continuous supply of neuroglial cells that myelinate and fasciculate the young fibres. This process is particularly evident in the mediorbital portion where the nerve is structured like a folded ribbon and in which its elements, axons and glia, are ordered from one edge to the other according to degrees of maturity. Thus, while at one edge the axons originate from the oldest portions of the retina and are strongly myelinated by completely mature oligodendrocytes and organised in fascicles by the prolongations of equally mature astrocytes; at the other germinal edge the axons are non-myelinated, the fasciculation is outlined by immature astrocytes and until these reach a certain maturity oligodendrocytes are not identified.

In numerous species of teleosts the retina grows throughout the life of the animal by the proliferation of a neuroepithelium, located at the margin of the retina, which determines the ordered addition of peripheral rings of new neural tissue which differentiates until it becomes morphologically and functionally mature retina (Johns, 1977, 1982). The addition of new ganglion cells to the retina involves an increase in the number of axons in the optic nerve and in the retinocipient encephalic centres. The optic nerve of the teleosts permits a precise order both in the disposition of the mature axons and in the addition of new elements: on their path along the optic nerve the axons of the ganglion cells are organised in fascicles whose elements originate from ganglion cells related both spatially (close in their location in the retina) and chronologically (differentiating at similar times).

Furthermore, the visual system of teleosts conserves a regenerative capacity which permits not only, after a traumatic lesion of the optic nerve such as crushing or sectioning, the ganglion cells of the retina to spontaneously regenerate their axons until they re-innervate the original target centres (Sperry, 1948; Gaze, 1970) and re-establish the function of the system but, what is more, after lesions in the retina, the proliferation of the retina cells to generate new neurons until the initial structure is re-established (Raymond et al., 1988; Hitchcock and Raymond, 1992; Braisted et al., 1994).

Therefore, the Central Nervous System (CNS) of some teleosts presents two characteristics, continual growth and regenerative capacity, which differentiate it from the CNS of amniotes. Both features are especially notable in the visual system.

In addition to the neuron populations which are adequate for continual growth and/or with regenerative capacity, so that these processes are successful, the visual system of the teleosts requires the establishment of neuron-glia interactions between the growing or re-growing axons and a glial stroma which permits and facilitates their development (Stuermer et al., 1992; Bastmeyer et al., 1993).

The optic nerve is a unique component of the CNS in the sense that it is physically "isolated" from the rest. Its nerve components present little variation since it contains, almost exclusively, axons of ganglion cells of the retina and glial elements. The former make it an ideal structure for the study of the glia of these animals and of their involvement in the processes of development and regeneration.

THE STRUCTURE OF THE OPTIC NERVE OF TELEOSTS

The optic nerve of fish possesses anatomic particularities of great interest. It is submitted to strong tensions which do not exist in other parts of the CNS, thus it must be elastic and resistant to stretching (Maggs and Scholes, 1986, 1990; Abbot, 1995). In contrast to the compact cylindrical structure of the optic nerve of mammals that of teleosts is a folded ribbon (Duke-Elder, 1958), in much of its extension, which favours the elasticity and resistance to stretching by notably increasing the quantity of connective tissue directly related with the optic nerve.

Moreover, it permits the ordering of its axons in accordance with their time of differentiation (Easter et al., 1981, 1984).

Taking as a base both the macroscopic structure and the histological organisation in the optic nerve of fish four parts can be considered, denominated: the intraorbital portion, the mediorbital portion, the optic chiasm and the optic tract (Easter et al.,1981).

The intraorbital portion, or head of the nerve, from the optical papilla to the zone of integration of the central artery of the optic nerve. The mediorbital portion is the zone between the intra-orbital portion and the chiasm. The optic chiasm is macroscopically differentiable because it is the zone where the optic nerves cross.

The optic tract, or, more precisely denominated, the anterior optic tract (Velasco et al., 1995, 1997), is the portion of the optic nerve between the chiasm and the integration in the diencephalon. In teleosts this portion is macroscopically differentiable (Ariëns Kapers et al., 1936).

The intraorbital portion, or head of the optic nerve. The retina/optic nerve transition zone

The head of the optic nerve is well differentiated from the layer of the fibres of the optic nerve, since in this zone the axons are arranged parallel to the vitreal surface while in the transition to the head of the nerve they change orientation and are arranged parallel to the longitudinal axis of the optic nerve. In diverse teleosts myelinated axons have been described in the layer of the fibres of the optic nerve, identified in some cases as retinopetals (Witowsky, 1971; Uchiyama, 1989). In tench (Lillo, 1996) and in goldfish (Wolgurg, 1981; Easter et al., 1984) the number of myelinated axons in this zone of the retina is excessive to consider them exclusively centrifugal. Near the zone of transition, still in the layer of fibres of the optic nerve, a series of cells can be distinguished disposed in rows and in close contact with one another, in goldfish they have been identified as astrocytes without prolongations (Skoff et al., 1986). In tench the electron microscope shows that these cells are astrocytes and oligodendrocytes, probably in processes of development. The astrocytes are orientated parallel to the axons. They have a very euchromatic nucleus and gliofilaments, although less abundant than in mature astrocytes; in contrast the cytoplasm is richer in organelles, most importantly short and dilated cisterns of rough endoplasmic reticulum. These characteristics coincide with those of immature astrocytes of the CNS in reptiles (Monzón Mayor et al., 1990a) and mammals (Peters et al., 1991) and with those of the optic nerve of rat (Peters and Vaughn, 1967; Skoff et al., 1976a,b), although the granules of glycogen are very scarce in the immature astrocytes found in this zone of the teleost visual pathway. In the layer of the fibres of the ganglion cells of the retina mature astrocytes are also located, with the characteristic components of this cell type (Lara et al., 1995).

The immature or active oligodendrocytes are characterised by possessing an irregularly shaped nucleus, with peripheral heterochromatin, moderately dense cytoplasmic matrix, numerous mitochondria, rough endoplasmic reticulum and the Golgi complex well developed, abundant free ribosomes and microtubules as the outstanding elements of the cytoskeleton. These characteristics coincide with those described for oligodendrocytes in development in other vertebrates (Monzón-Mayor et al., 1990b; Peters et al., 1991).

In this transition zone between the layer of the fibres of the ganglion cells of the retina and the optic nerve the rows of cells can be constituted by series of elements, homogeneous or heterogeneous, in different states of maturity (fig. 1-1A).

The head of the optic nerve

Once the axons are orientated parallel to the longitudinal axis of the optic nerve, its shape is rounded and in its centre is the central artery of the retina; surrounding and adjacent to this other vessels of lesser diameter are located. Surrounding the central artery is a totally non-myelinated zone, differentiable in *Carassius auratus* and in tench by its tenuous staining using Toluidine Blue, while the rest of the head of the nerve is composed of small and medium sized myelinated axons. In goldfish Bunt (1982) considers that the axons originating from the retina remain non-myelinated to 200 μm in the head of the nerve, and from here they all become myelinated except for a ring of them that remains non-myelinated around the central artery. Our results show that in the tench there are myelinated axons both in the retina (as mentioned) and in the head of the nerve, without our having been able to confirm the existence of a global myelinated exclusion plane, but we do not discard the possibility that for each one of the axons a non-myelinated zone exists in the head of the nerve.

Both in goldfish (Bunt, 1982; Easter et al., 1981) and in tench the ultrastructural analysis of this zone confirms that the pallid area around the central artery corresponds to large groups of non-myelinated axons of small size. The rest of the axons of the head of the nerve are myelinated.

The fasciculation begins in this zone, originated by the prolongations of the astrocytes (Skoff et al., 1986). The fasciculation is a process by which large packets of axons are formed, which in goldfish (Easter et al., 1981; Murray 1982) and tench are of different diameter and thickness with regard to the myelinated sheath, although other authors (Rusoff 1980; Stuermer and Easter 1984) consider that the axons of one fascicle have a similar diameter in each zone of the nerve. In semithin transversal sections of this zone of the optic nerve cells can be observed with relatively large somata, palely coloured with Toluidine Blue, with several radial ramifications, some directed to the surface of the nerve and others towards capillaries of the zone. These characteristics coincide with those of the astrocytes of teleosts (Tapp, 1973; Maggs and Scholes, 1986, 1990; Skoff et al., 1986;

Levine, 1989), amphibians (Stensaas, 1977), rat (Butt and Ransom, 1993), mouse (Butt et al., 1994), cat and monkey (Büssow, 1980) and mammals in general (Peters et al., 1991).

The astrocytes of the head of the nerve are of great size, particularly thick in the somata and the beginnings of the principal processes. Using electron microscopy it can be observed that they have a euchromatic nucleus of irregular or kidney shape; in the cytoplasm, which is not electrondense, the intermediate filaments are noteworthy, forming large bundles that are particularly conspicuous in the prolongations; the microtubules, interspersed among the filaments, are not numerous (Tapp, 1973; Battisti et al., 1995). The organelles are scarce. Although infrequent, in some astrocytes dense homogeneous or heterogeneous bodies are found. From the somata thick radial processes originate which are usually associated with others to form large partitions between fascicles of axons; the principal processes divide in secondary thinner processes which also participate in the fasciculation; some processes invade the groups of axons (Skoff et al., 1986; Maggs and Scholes, 1986, 1990; Levine, 1989; Dowding et al., 1991).

Some of the astrocyte processes reach the surface of the nerve, and associated, frequently through desmosomes, form the glia limitans (Peter et al., 1991). The external surface of the nerve, in contact with the connective tissue, is covered by a basal lamina. Some authors state that the processes of all the astrocytes form branches which end in the pial surface and in blood vessels, inasmuch as other branches remain free in the nerve and make contact with the nodules of Ranvier of the axons (Levine, 1989, Maggs and Scholes, 1990; Butt et al., 1994).

In the immature zone of the head of the nerve, where the axons are still non-myelinated, the glial cells in the process of differentiation, of similar morphology to that formerly described in astrocytes and immature oligodendrocytes of the layer of fibres of the retina, are located. In the rest of the head of the optic nerve, where the majority of the axons are myelinated, the oligodendrocytes are typically mature, located between the axons, within the fascicle (Peter et al., 1991; Ransom et al., 1991). The cellular body is usually spherical or polygonal and its nucleus rich in heterochromatic accumulations. They are cells whose cytoplasm is more electrondense than that of the astrocytes, with rough endoplasmic reticulum and well developed Golgi complex and many free ribosomes grouped in rosettes; the microtubules are more numerous in the processes than the perikaryon, while there are very rarely filaments. The presence of dense homogeneous and heterogeneous bodies is habitual (Tapp, 1973).

The mature oligodendrocytes have few observable processes using an electron microscope and these normally penetrate between the little-myelinated axons. As the myelinated sheath develops the prolongations become less evident (Peters et al., 1991; Lara et al., 1995). In the head of the tench optic nerve the cellular bodies of the oligodendrocytes are frequently found to be contiguous to the processes of the astrocytes, and from that position they emit their prolongations between the axons (fig. 1-1B).

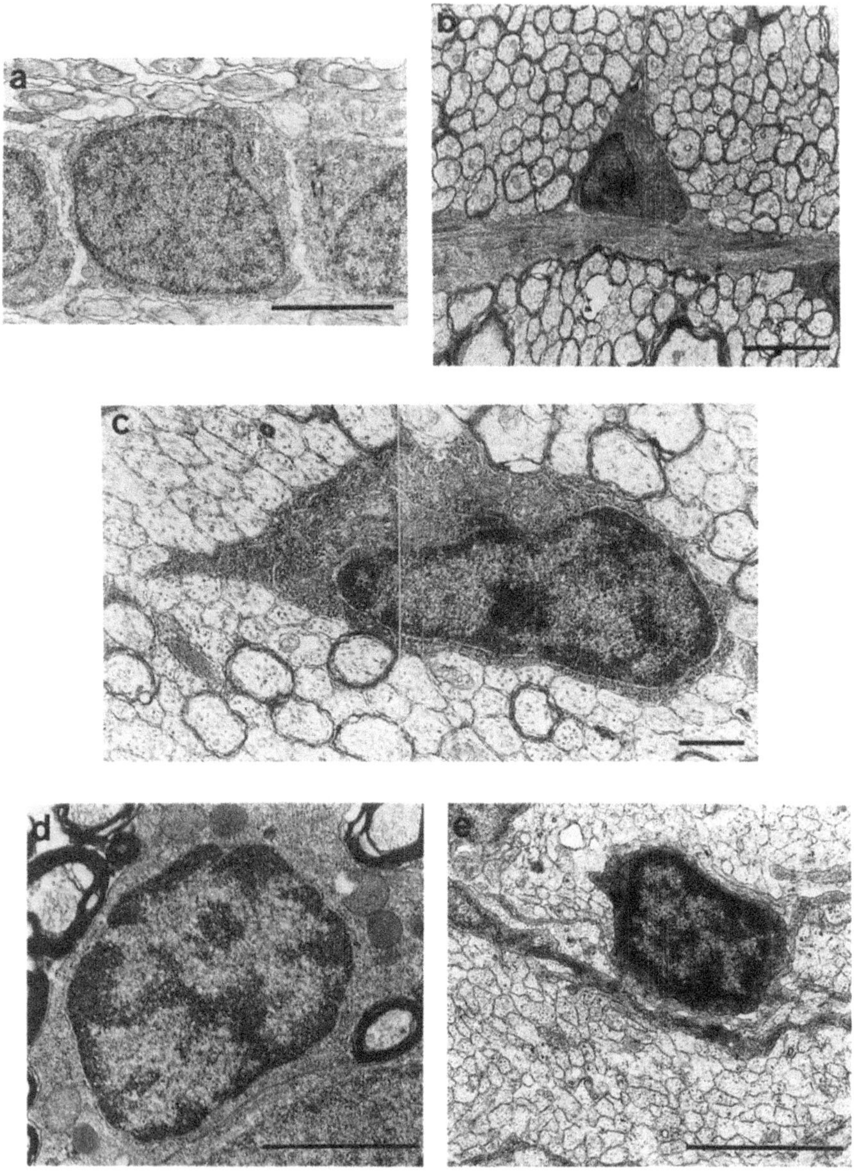

Figure 1-1. A, Immature oligodendrocytes surrounded by slightly myelinated axons. Bar = 2,4 μm. B, A large interfascicular astrocytic process full of gliofilaments. Oligodendrocyte cell bodies can be observed in close contact with astrocytic processes. Bar = 2,4 μm. C, Mature oligodendrocyte immersed in a packet of myelinated and non-myelinated axons. Bar = 600 nm. D, Microglial cell with dense body in its cytoplasm, mitochondrias and a moderately heterochromatic nucleus. Bar = 1,3 μm. E, Microglial cell body adjacent to a laminar astrocytic process of the germinal zone of the optic nerve. Bar = 2,4 μm.

The microglial cells, scarce in the tench optic nerve in the normal state (Velasco et al., 1995), are present in this zone of the optic nerve, although they are difficult to identify with the light microscope with purely morphological techniques. Using the electron microscope the microglial cells are small in comparison to the astrocytes. The nucleus is rich in heterochromatic accumulations. The cytoplasm is of a density intermediate between that of the astrocytes and oligodendrocytes, rich in organelles among which free ribosomes are outstanding and dilated cisterns of rough endoplasmic reticulum, clear vesicles, dense bodies and on occasions large myeloid bodies exist. The microtubules are scarce and filaments are not distinguishable (fig. 1-1D). These characteristics coincide with those of the microglia of different vertebrates (Tapp, 1973; Stensaas, 1977; Dowding et al., 1991; Peters et al., 1991; Battisti et al., 1995). In the tench optic nerve the microglia are located near the glia limitans and in the area of the blood vessels, where they can be next to the vascular prolongations of the astrocytes.

In the optic nerve the central artery is gradually displaced from its initial position in the optic papilla towards the ventral periphery, and the non-myelinated axons accompany it in its displacement. At approximately 1 mm. from the ocular globe the artery leaves the nerve through its ventral margin and the non-myelinated axons stop at the margin (Easter et al., 1984).

The mediorbital portion

The organisation of the optic nerve caudal to the entrance of the artery is completely different to that described earlier. In the mediorbital portion the optic nerve of teleosts adopts a structure of folded ribbon (Tapp, 1973; Easter et al., 1981; Bunt, 1982; Vanegas and Ito, 1983) where the organisation of the axons is such that, still maintaining spatial relationships, a gradient of maturity (of age) is established which maintains, on one side the axons of the oldest ganglion cells of the retina (central retina) while on the other side of the ribbon the youngest axons are arranged, originating from the recent peripheral rings of the retina.

The folded ribbon, in its entirety, is wrapped in a sheath of connective tissue and capillaries which is a continuation of the piamater which sheaths the encephalon (Tapp, 1973; Wolburg, 1981; Levine, 1989; Maggs and Scholes, 1990). The optic nerve in teleosts is more resistant to traction than other parts of the CNS and bears a certain similarity to a peripheral nerve which is reinforced externally by a matrix of fibrous collagen synthesised by the Schwann cells. The glia of the optic nerve are considered to increase its cytoskeleton and establish unions through desmosomes and other differentiations as an alternative strategy of reinforcement (Maggs and Scholes, 1990).

The folding of the ribbon that constitutes this zone of the optic nerve of teleosts allows the sheath of connective tissue to penetrate the folds and with it the immersed capillaries in this connective tissue. In the tench, as well as these capillaries, small vessels, situated endoneurally, are abundant in this zone of the nerve and in optic tract. This is not in accordance with the description of optic nerve of teleosts as non-vascular (Maggs and Scholes, 1986; Abbott, 1995).

Using light microscopy it is observed that the immense majority of the axons of the mediorbital portion are myelinated, very varied in their diameters and in the thickness of myelinated sheaths. The population of cellular somata is very abundant and from some of them processes emerge which limit the fascicles or are directed to the glia limitans or to a blood vessel. One of the borders of the folded ribbon that forms the nerve is completely different from the rest because of the absence of myelin and the arrangement of its cells and axons. The majority of the cell bodies are arranged on the periphery of the nerve and emit projections, but in this case not radial but parallel to each other (Maggs and Scholes, 1986, 1990; Abbott, 1995). In tench, due to the folding of the nerve, although this area is situated at one of its edges, it is surrounded by other folds, being situated more or less in the centre of the group of turns of the nerve, contrarily, in goldfish, this zone is situated on the margin of the entirety of the structure (Maggs and Scholes, 1986, 1990).

The ultrastructure of this part of the optic nerve shows that the fasciculation of the axons through the processes of the astrocytes is very pronounced and that the great majority of the axons are myelinated and tightly packed (figs. 1-2B, E). The number of fascicles is greater than in the head of the nerve and in each packet the axons are greater in diameter and in the thickness of their myelinated sheaths (figs. 1-2B, D, E). In tench, as occurs in goldfish (Murray, 1982; Easter et al., 1984) and in *Eugerres plumieri* (Tapp, 1973), the myelinated axons of each bundle vary in diameter and in the thickness of the myelinated sheath, in contrast to other teleosts (Rusoff, 1984).

It is noteworthy that in the zones close to the non-myelinated edges, in the same fascicle there may be myelinated and non-myelinated axons (figs. 1-2A, D).

Using electron microscopy it is evident that the cell bodies of the mature astrocytes are found between the fascicles and from them emerge the prolongations of average thickness (figs. 1-2B, E), full of filaments that form packets and are orientated parallel to the longitudinal axis of the astrocyte processes (figs. 1-2B, F). Near the plasmatic membrane of many astrocytes clear vesicles can be observed (fig. 1-2F), probably secretory, since this cell type is considered to be related, among other functions, with the formation of the extracellular matrix of the nerve (Baehr and Bunge, 1990; Smith et al., 1990). In the mediorbital portion, as in the nerve head, the partitions that separate the fascicles of axons are formed by the association of various astrocytic processes anchored by desmosomes (fig. 1-2D). In this zone of the optic nerve, the desmosomes, as well as anchoring astrocytic prolongations to each other, can also join these prolongations to astrocytic somata, or be found between two astrocyte cell bodies.

The oligodendrocytes are rather more numerous in the mediorbital portion than in the head of the nerve, probably due to the large number of myelinated axons. The cell bodies here are also usually found within the fascicles of axons. In the completely mature zone of the optic nerve, the oligodendrocytes also present the characteristic appearance of fully-developed cells (figs. 1-1B, C).

The edge of the ribbon that composes the optic nerve, which using light microscopy presents a paler staining, is seen ultrastructurally as the extreme of growth of the nerve, where the vast majority of axons are non-myelinated. This edge of growth of the nerve is considered as a germinal zone that only contains non-myelinated axons, glial progenitor cells and immature glia (Maggs and Scholes, 1986, 1990; Dowding et al., 1991). In tench this growth zone is very limited and there are very few fascicles that contain only non-myelinated axons (figs. 1-2A, C); the mixed fascicles of myelinated and non-myelinated axons are more frequent (figs. 1-2A, D). The analysis of the entirety suggests that, in fact, some completely non-myelinated fascicles exist on the margin of the germinal portion, in direct contact with the immature elements which, in the future, will produce the glial support for these axons; as distance from the germinal edge increases the fascicles become defined by prolongations of ever more mature astrocytes and myelinated sheaths begin to appear and synchronised to these the first more or less mature oligodendrocytes are identified. In more internal portions of the ribbon (more mature fascicles), the percentage of myelinated axons increases notably at the same time as the glia (both astrocytes and oligodendrocytes) acquire characteristics of full maturity. The sequence described is produced in the course of a few hundreds of μm; nevertheless, even in the fully mature zones close to this transition zone, fascicles are located with some isolated non-myelinated elements (Easter et al., 1981).

Some authors consider all the cells of the germinal zone immature, situated only on the periphery of the nerve.These cells have fine processes between the myelinated axons (Easter et al.,1981; Maggs and Scholes, 1986). These authors consider an indicator of the end of differentiation to be the appearance of desmosomes between prolongations and of some myelin sheaths. However, in the optic nerve of tench, as well as cells in the process of differentiation, in this zone in growth it is easy to distinguish mature astrocytes which emit processes that form the glia limitans of the extreme of growth, full of filaments and joined by desmosomes (figs. 1-2A, C). As some other authors have observed in *Carassius auratus* (Maggs and Scholes, 1986), mature oligodendrocytes are not found in the germinal zone of the nerve, though, nearby, there are active oligodendrocytes, as an indicator that in this zone the myelination begins very early. As in the rest of the nerve, in this zone we also find microglial cells, close to the glial limitant, next to the astrocytic processes (fig. 1-1E).

Optic chiasm

The structure in folded ribbon of the nerve reaches this portion (Tapp, 1973), where the axons change orientation. At this point, the two optic nerves cross to later reach the encephalic structures contralateral to the retinas of origin (Easter et

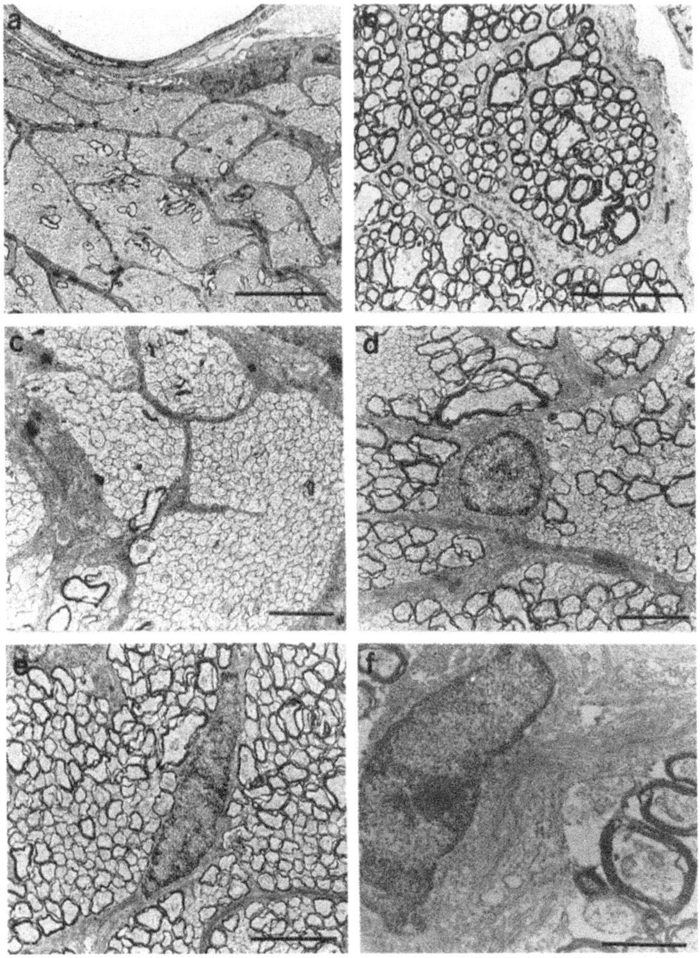

Figure 1-2. A, The germinal edge of the optic nerve. It has, principally, non-myelinated axons, although some myelinated axons appear among these. The astrocytic processes in this zone are very thin and are arranged in parallel to each other and are joined by desmosomes. Bar = 5,5 µm. B, Fascicles of axons separated by astrocytic processes. Cellular body of an astrocyte close to the glia limitans; its primary radial processes are directed to the glia limitans or separate fascicles of axons. The glia limitans is formed by astrocytic processes joined by desmosomes. Bar = 3,9 µm. C, Fascicles of non-myelinated axons of small diameter, with some disperse myelinated axons in the interior. The fascicles are separated by astrocyte processes joined by desmosomes forming a strong barrier in some sectors and in others only narrow layers of secondary processes. Bar = 2,4 µm. D, Small interfascicular oligodendrocyte, sheathing axons with small diameter. In each fascicle similar numbers of myelinated and non-myelinated axons are found. Bar = 1,3 µm. E, Astrocyte body elongated between several axon fascicles, with primary and secondary processes. Bar = 2,4 µm. F, Astrocyte cell body with a great primary process full of large packets of filaments and clear vesicles. Bar = 800 nm.

al., 1981; Bunt, 1982). Some individual axons vary their trajectory within the chiasm to reach the ipsilateral encephalic structures. Due to this change in the organisation, in a transversal section, nerve groups of fibres are present sectioned longitudinally or tangentially. On the periphery of this portion the presence of some non-myelinated axons is maintained, which also cross in their majority. The majority of the fascicles are, nonetheless, completely myelinated, with axons of diverse diameters and myelinated sheaths which are in general quite thick.

The glial cells have similar characteristics to those of the rest of the optic nerve, with no notable change in their morphology or distribution.

Optic Tract

In the post-chiasmic portion of the optic nerve the folds disappear and the tract acquires a rounded or oval shape (Bunt, 1982).

In goldfish (Lanners and Grafstein,1980; Bunt, 1982) and tench the fasciculation of the axons by the astrocytic processes relaxes in the tract, so that the fascicles are much more extense than in the rostral portions of the nerve. Both in *Carassius auratus* (Maggs and Scholes, 1986) and in tench the optic tract is crossed by a great number of endoneural capillaries, and others in contact with the projections of the conjunctiva which enter the nervous tissue and where the vascular prolongations of the astrocytes reach.

Electron microscopy shows that the radial processes of the astrocytes are longer and finer than in the rostral portions of the optic nerve. The great majority of the axons of the tract are myelinated and those of greater diameter are located in the centre of the tract (Rusoff, 1980; Easter et al., 1981; Stuermer and Easter, 1984), while those of lesser diameter are found in the peripheral zones (Lanners and Grafstein, 1980, in goldfish; Rusoff, 1984, in Perciforms). The non-myelinated axons are located in a small peripheral zone and in the proximities of the encephalon, they mix gradually with the myelinated ones.

In the post-chiasmic region the optic nerve is surrounded by an astrocytic lamina and a basal lamina with a deeply folded surface. The glia limitans in this portion of the nerve is formed by a great quantity of astrocytic processes arranged in laminas which make it particularly thick.

Both astrocytes and oligodendrocytes present similar characteristics to those described in more rostral portions of the optic nerve, but their distribution, conditioned by or conditioning the different fasciculation of this portion, is slightly different: cellular groupings are more frequent, especially in the peripheral zones. In transversal sections extense portions of glial cytoplasm are frequently identified, rich in organelles, which suggests a change in the usual orientation of the cellular bodies of these cells from that presented in other portions of the optic nerve.

FEATURES OF THE GLIAL CELLS IN THE TELEOST OPTIC NERVE

From the structural analysis of the optic nerve of teleosts it can be concluded that, as in the rest of the vertebrates, the neuroglia is comprised of astrocytes and oligodendrocytes (Maggs and Scholes, 1986, 1990; Levine, 1989; Lara et al., 1995), which, with the microglial cells, complete the glial population (Velasco et al., 1995).

In normal conditions, in the mature zones of the optic nerve of fish, the astrocytes provide the nerve with a structural reinforcement (Maggs and Scholes, 1986, 1990), while the oligodendrocytes myelinate the axons (Wolburg, 1981; Ransom et al., 1991; Bastmeyer et al., 1994).

The mature astrocytes, denominated reticular astrocytes, are associated through desmosomes in a network structure (Maggs and Scholes, 1986, 1990). Their abundant cytoskeleton is rich in cytokeratins (Giordano et al., 1989; Druger et al., 1992) and poor in GFAP (Blaugrund et al., 1990; Druger et al., 1992; Sivron et al., 1994). The presence in mature astrocytes of a type II keratin (Giordano et al., 1989), homologous to K-8 (Giordano et al., 1990), the first intermediate filament expressed in the embryo of mammals (Oshima et al., 1983), suggests that these glial cells maintain a relatively immature state (Sivron et al., 1994) and at the same time epithelioid characteristics that contribute to the elasticity and resistance to stretching essential for their structural functions (Abbot, 1995). Regarding the low immunoreactivity of the astrocytes of the optic nerve of teleosts to GFAP (Maggs and Scholes, 1986; Levine, 1989; Nona et al., 1989; Sivron et al., 1994), it has been demonstrated that these cells contain both mRNA and GFAP (Blaugrund et al., 1990), this, together with the different immunostaining of astrocytes of other parts of the CNS suggests the existence of diverse subtypes of GFAP in the CNS of teleosts (Bastmeyer et al., 1994).

In the germinal edge of the ribbon that constitutes the mediorbital portion of the optic nerve the only neuroglial elements identified are astrocytes, which is in accordance with the association of these cells with an environment adapted to the development of the axons, to which it can contribute by the secretion of growth promoter factors (Unsicker et al., 1987, Bixby and Harris, 1991), furthermore, the astrocytes are adapted to serve as guides to the growing axons (Drazba and Lemmon, 1990; Bixy and Harris, 1991; Brodkey et al., 1993). The capacity of the astrocytes to contribute (or permit) the axon growth changes as they mature and differentiate (Baehr and Bunge, 1990; Smith et al., 1990), so that as the astrocytes mature, or, which is the same, separate from the germinative zone their qualities as stroma for the axons development become less evident.

The available data allow the supposition that all the astrocytes of the optic nerve of teleosts, in normal conditions, originate from undifferentiated cells of the germinal edge and could be classified as type 1 astrocytes. In mammals the appearance of the first oligodendrocytes is late in relation to the identification of the first astrocytes (Barres et al., 1992). In the sequence of maturing from the young edge of the mediorbital ribbon, the first oligodendrocytes also begin to be identified in zones where the astrocytes already have a certain degree of maturity. It is considered that, as in the other vertebrates (Pfeiffer et al., 1994), the oligodendrocytes of the optic nerve of fish derive from O-2A progenitors (Jeserich and Stratmann, 1992), which "in vivo" only generate oligodendrocytes (Richardson et al., 1996), although in pathological conditions they could generate astrocytes (Goldman, 1996). As the oligodendrocytes mature they lose their proliferative and migratory capacities (Reynolds and Wilkin, 1991), although they can recover them in pathological conditions (Richardson et al., 1996). The presence of mature astrocytes in the zones of commencement of myelination could suppose that the differentiation of oligodendrocytes is mediated by the earlier differentiation of astrocytes and the interaction of these and the axons (Barres and Raff, 1996).

In the mature zones of the optic nerve of teleosts all the axons are myelinated, which supposes a high number of oligodendrocytes responsible for producing and maintaining great quantities of myelinated sheaths with the same structural characteristics as that of the myelin of the mammalian CNS (Jeserich and Waehneldt, 1986), nonetheless, they possess two proteins of myelin, IP1 and IP2 (Jeserich and Rauen, 1990), related to the protein P0 of the myelin of the Peripheral Nervous System (Schliess and Stoffel, 1991). Moreover, the oligodendrocytes of fish express a protein of myelin of 36K (Jeserich and Waehneldt, 1986) which is not present in undifferentiated cells (Bastmeyer et al., 1994).

With regard to the population of microglial cells, it is accepted that they derive from circulating monocytes (Streit et al., 1988; Gehrmann and Kreutzberg, 1991; Milligan et al., 1991) that later penetrate the nervous parenchyma and, in a process in which the astrocytes participate, they can transform to amoeboid or ramified microglia (Tanaka and Maeda, 1996). Furthermore, the estimations of the number of microglial cells in the optic nerve of fish carried out using tomato lectin labelling (Velasco et al., 1995) and with antimacrophage antibodies (Dowding et al., 1991) suggest a much greater population than that unequivocally identified ultrastructurally.

ACKNOWLEDGMENTS

We would like to thank Mr. G.H. Jenkins for revising the English version of the ms. This work was supported by Spanish DGICYT (PB94/1388) and Junta de Castilla y León (SA35/96) projects.

REFERENCES

Abbot, N.J. (1995) Morphology of nonmammalian glial cells: functional implications. In H. Kettermann and B.R. Ranson (eds): Neuroglia. Oxford: Oxford University Press, pp. 97-116.

Ariëns Kapers, C.U., G.C. Huber, and E.C. Crosby (1936) The comparative anatomy of the nervous system of the vertebrates including man. (reed. 1967, 3 vol.) New York: Hafner Publ. Comp.

Baehr, M., and R.P. Bunge (1990) Growth of adult retinal ganglion cell neurites on astrocytes. Glia *3*:293-300.
Barres, B.A., and M.C. Raff (1996) Axonal control of oligodendrocyte development. In K.R. Jessen and W.D. Richardson (eds): Glial cell development. Oxford: Bios Sci. Publ. pp. 71-83.
Barrres, B.A., I.K. Hart, H.S.R. Coles, J.F. Burne, J.T. Voyvodic, W.D. Richardson, and M.C. Raff (1992) Cell death and control of cell survival in the oligodendrocyte lineage. Cell *70*:31-46.
Bastmeyer, M., M. Bahr, and C.A.O. Stuermer (1993) Fish optic nerve oligodendrocytes support axonal regeneration of fish and mammalian retinal ganglion cells. Glia *8*:1-11.
Bastmeyer, M., M. Beckmann, M.E. Schwab and C.A.O. Stuermer (1994) Similarities and differences between fish optic nerve/tract oligodendrocytes and Schwann cells. Glia *11*:300-314.
Battisti, W.P., J. Wang, K. Bozek, and M. Murray (1995) Macrophages, microglia and astrocytes are rapidly activated after crush injury of the goldfish optic nerve: a light and electron microscopic analysis. J. Comp. Neurol. *354*:306-320.
Bixby, J.L., and W.A. Harris (1991) Molecular mechanisms of axon growth and guidance. Annu. Rev. Cell Biol. *7*:117-159.
Blaugrund, E., I. Cohen, Y. Shani, and M. Schwartz (1990) Glial fibrillary acidic protein in fish optic nerve. Glia *4*:393-399.
Braisted, J.E., T.F. Essman, and P.A. Raymond (1994) Selective regeneration of photoreceptors in goldfish retina. Development *12*:2409-2419.
Brodkey, J.A., M.A. Gates, E.D. Laywell, and D.A. Steindler (1993) The complex nature of interactive neuroregeneration-related molecules. Exp. Neurol. *123*:251-270.
Bunt, M. (1982) Retinotopic and temporal organization of the optic nerve and tracts in the adult goldfish. J. Comp. Neurol. *206*:209-226.
Büssow, H. (1980) The astrocytes in the retina and optic nerve head of mammals: a special glia for the ganglion cell axons. Cell Tissue Res. *206*:267-278.
Butt, A.M., A. Duncan, and M. Berry (1994) Astrocyte associations with nodes of Ranvier: ultrastructural analysis of HRP-filled astrocytes in mouse optic nerve. J. Neurocytol. *23*:486-499.
Butt, A.M., and B.R. Ransom (1993) Morphology of astrocytes and oligodendrocytes during development in the intact rat optic nerve. J. Comp. Neurol. *338*:141-158.
Dowding, A.J., A. Maggs, and J. Scholes (1991) Diversity amongst the microglia in growing and regenerating fish CNS: inmunohistochemical characterization using FL.1, an antimacrophage monoclonal antibody. Glia *4*:345-364.
Drazba, J., and V. Lemmon (1990) The role of the cell adhesion molecules in neurite outgrowth on Muller cells. Dev. Biol. *138*:82-93.
Druger, R.K., E.M. Levine, E. Glasgow, P.S. Jones, and N. Schechter (1992) Cloning a type I keratin from goldfish optic nerve: differential expression of keratins during regeneration. Differentiation *52*:33-43.
Duke-Elder, S. (1958) System of Ophthalmology. Vol.1: The eye in evolution. S. Duke-Elder (ed). London: Herry Kimpton Publs.
Easter, S.S.Jr., B. Bratton, and S.S. Scherer (1984) Growth-related order of the retinal fiber layer in goldfish. J. Neurosci. *4*:2173-2190.
Easter, S.S.Jr., A.C. Russoff, and P.E. Kish (1981) The growth and organization of the optic nerve and tract in juvenile and adult goldfish. J. Neurosci. *1*:793-811.
Gaze, R.M. (1970) The Formation of Nerve Connections. London:Academic Press.
Gehrmann, J., and G.W. Kreutzberg (1991) Characterization of two new monoclonal antibodies directed against rat microglia. J. Comp. Neurol. *313*:409-430.
Giordano, S., E. Glasgow, P. Tesser, and N. Schechter (1989) A type II Keratin is expressed in glial cells of the goldfish visual pathway. Neuron *2*:1507-1516.
Giordano, S., C. Hall, W. Quitschke, E. Glasgow, and N. Schechter (1990) Keratin 8 of simple epithelia is expressed in glia of the goldfish nervous system. Differentiation *44*:163-172.
Goldman, J.E. (1996) Developmental origins of astrocytes. In K.R. Jessen and W.D. Richardson (eds): Glial cell development. Oxford: Bios Sci. Publ. pp. 31-51
Hitchcock, P.F., and P.A. Raymond (1992) Retina regeneration. Trends Neurosci. *15*:103-108.
Jeserich G., and T. Rauen (1990) Cell cultures enriched in oligodendrocytes from the central nervous system of trout in terms of phenotypic expression exhibit parallels with cultured rat Schwann cells. Glia *3*:65-74

Jeserich, G., and A. Stratmann (1992) In vitro differentiation of trout oligodendrocytes: evidence for an A2B5-positive origin. Dev. Brain Res. *67*:27-35.

Jeserich G., and T.V. Waehneldt (1986) Characterization of antibodies against major fish CNS myelin proteins: Immunoblot analysis and immunohistochemical localization of 36K and IP2 proteins in trout nerve tissue. J. Neurosci. Res. *15*:147-158.

Johns, P.A.R.(1977) Growth of adult goldfish eye. III: Source of the new retinal cells. J. Comp. Neurol. *176*:343-358.

Johns, P.A.R. (1982) The formation of photoreceptors in the growing retinas of larval and adult goldfish. J. Neurosci. *2*:179-198.

Lanners, H.N., and B. Grafstein (1980) Effect of a conditioning lesion-regeneration of goldfish optic axons: ultrastructural evidence of enhanced outgrowth and pinocytosis. Brain Res. *196*:547-553.

Lara, J.M., A. Velasco, J.R. Alonso, and J. Aijón (1995) The neuroglia in the CNS of teleosts. In A. Vernadakis and B. Roots (eds): Neuron-glia interrelationships during phylogeny: I Phylogeny and ontogeny of glial cells. Totowa: Humana Press Inc. pp. 131-156.

Levine, R.L. (1989) Organization of astrocytes in the visual pathways of the goldfish: an immunohistochemical study. J. Comp. Neurol. *285*:231-245.

Lillo, C. (1996) Ultraestructura del nervio óptico de un teleósteo (*Tinca tinca* L., 1758). Minor Thesis. Univ. Salamanca.

Maggs, A., and J. Scholes (1986) Glial domains and nerve fiber patterns in the fish retinotectal pathways. J. Neurosci. *6*:424-438.

Maggs, A., and J. Scholes (1990) Reticular astrocytes in the fish optic nerve: macroglia with epithelial characteristics form an axially repeated lacework pattern, to which nodes of Ranvier are apposed. J. Neurosci. *10*:1600-1614.

Milligan, C.E., T.J. Cunninham, and P. Levitt (1991) Differential immunochemical markers reveal the normal distribution of brain macrophages and microglia in the developing brain. J. Comp. Neurol. *314*:125-135.

Monzón-Mayor, M., C. Yanes, J.L. James, and R.R. Sturrock (1990a) An ultrastructural study of the development of astrocytes in the midbrain of the lizard. J. Anat. *170*:33-41.

Monzón-Mayor, M., C. Yanes, J.L. James, and R.R. Sturrock (1990b) An ultrastructural study of the development of oligodendrocytes in the midbrain of the lizard. J. Anat. *170*:43-49.

Murray, M. (1982) A quantitative study of regenerative sprouting by optic axons in goldfish. J. Comp. Neurol. *209*:352-362.

Nona, S.N., S.A.S. Shehab, C.A. Stafford and J.R. Cronly-Dillon (1989) Glial fibrillary acidic protein (GFAP) from goldfish: its localisation in visual pathway. Glia *2*:189-200.

Oshima R.G., W.E. Howe, F.G. Klier, E.D. Adamson, and L.H. Shevinsky (1983) In vitro protein synthesis in the goldfish retinotectal pathway during regeneration: evidence specific axonal proteins of retinal origin in the optic nerve. J. Neurochem. *41*:1137-1142.

Peters, A., S.L. Palay, and H.F. Webster (1991) The fine structure of the nervous system. Neurons and their supporting cells. New York: Oxford University Press.

Peters, A. and J.E. Vaughn (1967) Microtubules and filaments in the axons and astrocytes of early postnatal rat optic nerves. J. Cell Biol. *32*:113-119.

Pfeiffer, S.E., A.E. Warrington, and R. Bansal (1994) The oligodendrocytes and its many cellular processes. Trends Cell Biol. *3*:191-197.

Ransom, B.R., A.M. Butt, and J.A. Black (1991) Ultrastructural identification of HRP-injected oligodendrocytes in the intact rat optic nerve. Glia *4*:37-45.

Raymond, P.A., M.J. Reifler, and P.K. Rivlin (1988) Regeneration of goldfish retina: rod precursors are a likely source of regenerated cells. J. Neurobiol. *19*:431-463.

Reynolds, R., and G.P. Wilkin (1991) Oligodendroglial progenitor cell but not oligodendroglia divide during normal development of the rat cerebellum. J. Neurocytol. *20*:216-224.

Richardson, W.D., N.P. Pringle, W.P. Yu, E.J. Collarini, and A. Hall (1996) Origins and early development of oligodendrocytes. In K.R. Jessen and W.D. Richardson (eds): Glial cell development. Oxford: Bios Sci. Publ. pp. 53-70.

Rusoff, A.C. (1980) Order in the optic nerve of goldfish. Science 208:311-312.

Rusoff, A.C. (1984) Paths of axons in the visual system of Perciform fish and implications of these paths for rules governing axonal growth. J. Neurosci. *4*:1414-1428.

Schliess, F., and W. Stoffel (1991) Evolution of the myelin integral membrane proteins of the central nervous system. Biol. Chem. H-S *372*:865-874.
Sivron, T., I. Cohen, and M. Schwartz (1994) Intermediate filaments reminiscent of immature cells expressed by goldfish (*Carassius auratus*) astrocytes and oligodendrocytes in vitro. Cell Tissue Res. *275*:327-337.
Skoff, R.P., D.L. Price, and A. Stocks (1976a) Electron microscopic autorradiographic studies of gliogenesis in rat optic nerve. I. Cell proliferation. J. Comp. Neurol. *169*:291-312.
Skoff, R.P., D.L. Price, and A. Stocks (1976b) Electron microscopic autorradiographic studies of gliogenesis in rat optic nerve. II. Time of origin. J. Comp. Neurol. *169*:313-334.
Skoff, R., P.E. Knapp, and W.P. Bartlett (1986) Astrocytic diversity in the optic nerve: a cytoarchitectural study. In S. Fedorof and A. Vernadakis (eds): Astrocytes Vol. 1: Development, morphology, and regional specialization of astrocytes. Orlando: Academic Press pp.269-291.
Smith, G.M., U. Rutishauser, J. Silver, and H. Miller (1990) Maturation of astrocytes in vitro alters the extent and molecular basis of neurites outgrowth. Dev. Biol. *138*:377-390.
Sperry, R.W. (1948) Patterning of central synapses in regeneration of the optic nerve in teleosts. Physiol. Zool. *23*:351-361.
Stensaas, S.S. (1977) The ultrastructure of astrocytes, oligodendrocytes, and microglia in the optic nerve of urodele amphibians (*A. punctatum, T. Pyrrhogaster, T. Viridescens*). J. Neurocytol. *6*:269-286.
Streit, W.J., M.B. Graeber, and G.W. Kreutzberg (1988) Functional plasticity of microglia: A review Glia *1*:301-307.
Stuermer, C.A.O., M. Bastmeyer, M. Bähr, G. Strobel, and K. Paschke (1992) Trying to understand axonal regeneration in the CNS of fish. J. Neurobiol. *23*:537-550.
Stuermer, C.A.O., and S.S.Jr. Easter (1984) Rules of order in the retinotectal fascicles of goldfish. J. Neurosci. *4*:1045-1051.
Tanaka, J., and N. Maeda (1996) Microglial ramification requires nondiffusibles factors derived from astrocytes. Exp. Neurol. *137*:367-375.
Tapp, R.L. (1973) The structure of the optic nerve of the teleost *Eugerres plumieri*. J. Comp. Neurol. *150*:239-252.
Uchiyama, H. (1989) Centrifugal pathways to the retina: Influence of the optic tectum. A review. Visual Neurosci. *3*:183-206.
Unsicker, K., H. Reichert-Preibsh, R. Schmidt, B. Pettmann, G. Labourdette, and M. Sensenbrenner (1987) Astroglial and fibroblast growth factors have neutrophic functions for cultured peripheral and central nervous system neurons. Proc. Natl. Acad. Sci. USA *84*:5459-5463.
Vanegas, H., and H. Ito (1983) Morphological aspects of the teleostean visual system: a review. Brain Res. Rev. *6*:117-137.
Velasco, A., E. Caminos, E. Vecino, J.M. Lara, and J. Aijón (1995) Microglia in normal and regenerating visual pathways of the tench (*Tinca tinca* L. 1758, Teleost): a study with tomato lectin. Brain Res. *705*:315-324.
Velasco, A., J.G. Briñón, E. Caminos, J.M. Lara, and J. Aijón (1997) S-100 positive glial cells are involved in the regeneration of visual pathways of teleosts. Brain Res. Bull. *43*:327-336.
Witowsky, P. (1971) Synapses made by myelinated fibers running to teleost and elasmobranch retinas. J. Comp. Neurol. *142*:205-222.
Wolburg, H. (1981) Myelination and remyelination in the regenerating visual system of the goldfish. Exp. Brain Res. *43*:199-206.

2

THE MACROGLIA OF TELEOSTS: CHARACTERIZATION, DISTRIBUTION AND DEVELOPMENT

Sofía M. Díaz-Regueira and Ramón Anadón

We present morphological, immunocytochemical and developmental studies on the macroglia of species of two groups of teleosts carried out in our laboratories. One of these species, the grey mullet (Chelon labrosus; Acanthopterygii), is an advanced euteleost. The other species are trouts (Salmo trutta fario, Oncorhynchus mykiss; Protacanthopterygii) which belong to the basal stock of euteleosts. We present our results on a specialized astroglial structure of the medulla oblongata of the adult grey mullet, the distribution of astroglial markers (vimentin, GFAP and S-100) in the brain of the adult mullet and their pattern of expression in juveniles of 5 to 70 mm in length, as well as results obtained with S-100 and GFAP immunocytochemistry in trout. In old adult mullet (circa 50 cm in length), we have studied with light and electron microscopy the diversity of oligodendrocytes of the lateral line area of the medulla oblongata. In the brain of old adult trout (circa of 36 cm in length), the NADPHd histochemical reaction was used to study the large population of oligodendrocytes. Finally, we studied the development of the oligodendrocytes in trout using an antibody to the 36K myelin protein. Our results in old adults of two species pertaining to different lines within teleosts showed that diversity of oligodendrocyte sizes in different regions of the brain is a shared characteristic of teleosts and mammals.

The nonneuronal elements of the central nervous system consist of different types of cell, which are generally classified as astroglia, oligodendroglia and microglia. In mammals, two main subtypes of astroglia are recognized, the fibrous and protoplasmic astrocytes (Andriezen, 1893; Cajal, 1909, 1913). In addition, the radial glia cells are considered as a third subtype of astroglia (Cajal, 1909; Achúcarro, 1915; Privat et al., 1995). Studies of Río Hortega (1919, 1921) have demonstrated that Cajal's "third element" (Cajal, 1913) consisted in fact of two different elements, the oligodendroglia and microglia. Río Hortega also proposed that while the astroglia and oligodendroglia originate from the embryonic neurectoderm, the microglia originates from mesoderm cells that invade the nervous tissue late in development. We will present morphological, immunocytochemical and developmental studies on the macroglia of species of two groups of teleosts carried out in our laboratories. One of these species, the grey mullet (*Chelon labrosus; Acanthopterygii*), is an advanced euteleost. The other species are trouts (*Salmo trutta fario, Oncorhynchus mykiss; Protacanthopterygii*) which belong to the basal stock of euteleosts.

ASTROGLIA

In adult mammals, the astroglia consists of protoplasmic and fibrous astrocytes, located in the grey and the white matter, respectively, and radial glia cells (cerebellar Bergmann's glia, Müller cells and tanycytes; Privat et al., 1995). Ultrastructural studies have shown the presence of gliofilaments (intermediate filaments) in these cells (Hendon, 1964; Mugnaini and Walberg, 1964). Eng et al. (1971) and Bignami et al. (1972) characterized biochemically the glial fibrillary acidic protein (GFAP), and immunocytochemical studies have demonstrated that this protein is found in intermediate filaments of astrocytes (Bignami and Dahl, 1974; Onteniente et al., 1983; Dahl et al., 1985). Another protein of intermediate filaments, vimentin, has been found in astroglial cells (Dahl et al., 1981; Schnitzer et al., 1981). This protein is abundantly expressed in immature astroglial cells. An abundant calcium-binding protein identified by Moore (1965) in the nervous tissue (S-100 protein), is preferentially expressed in astrocytes (Matus and Mughal, 1975). Antibodies against these proteins are currently used as immunocytochemical markers of astroglia. Radial glia cells extending through the thickness of the walls of the neural tube are abundant during development of mammalian brain, being considered as transitory elements in astrocyte maturation. Radial glia cells appear to have specific functions during development, forming a scaffold for neuronal migration (Rakic, 1971a, 1971b, 1972; Levitt and Rakic, 1980). In most species these cells express GFAP (Antanitus et al., 1976; Marcus and Easter, 1995).

Astrocytes have been found in the most primitive living vertebrates, the Agnatha (see Cajal, 1909) and in most other vertebrates. In teleosts, however, the most conspicuous astroglia is represented by ependymal tanycytes and radial glia cells (Achúcarro, 1915; Kruger and Maxwell, 1967; Stevenson and Yoon, 1982; Lara et al., 1989, 1995; Maggs and Scholes, 1990). More recently, astrocytes have been described in the teleost brain (Kruger and Maxwell, 1967; Lara et al., 1989, 1995), and reticular astrocytes with characteristics of epithelial cells are common in the optic nerves (Maggs and Scholes, 1990).

Most ependymal cells of adult teleosts possess long basal processes coursing to the external surface of the brain ending as subpial endfeet (tanycytes: Horstmann, 1954). These cells have been classically studied with metalic impregnations (see Lara et al., 1995), and the ultrastructure of these cells has been described in several teleosts (Kruger and Maxwell, 1966, 1967; Tapp, 1973; Laufer and Vanegas, 1974; Peter and Nagahama, 1976; Roots, 1978; Stevenson and Yoon, 1982; Díaz-Regueira et al., 1993). Modern immunocytochemical studies reported the presence of both GFAP (Onteniente et al. 1983; Anderson et al., 1984; Raamsdonk et al. 1984; Dahl et al. 1985; Levine, 1989; Nona et al. 1989; Cardone and Roots, 1990; Rubio et al. 1992; Díaz-Regueira et al., 1993; Marcus and Easter, 1995) and vimentin (Cardone and Roots 1990; Rubio et al. 1992; Díaz-Regueira et al., 1993) in ependymal and radial glia cells of teleosts. Using NADPH diaphorase histochemistry, Ma (1993) has described the regional distribution of these cells in the brain of the sunfish (*Lepomis* sp.). More recently, Marcus and Easter (1995) have studied the development of GFAP immunoreactive cells in the early zebrafish embryo.

In the following section, we first present our results on a specialized astroglial structure of the medulla oblongata of the adult grey mullet, then the distribution of astroglial markers (vimentin, GFAP and S-100) in the brain of the adult mullet and their pattern of expression in juveniles of 5 to 70 mm in length, and finally some results obtained with S-100 and GFAP immunocytochemistry in trout (Manso et al., 1997).

MARGINAL GLIA OF MULLET

In the grey mullet, we have studied with light and electron microscopy and immunocytochemistry a conspicuous glial region of the octavolateral area of the medulla oblongata which is associated with the lateral recess of the fourth ventricle (Díaz-Regueira et al., 1993). This glial region is located below the pial surface, ventrally to the cerebellar crest and medial, ventral and lateral to the lateral recess of the fourth ventricle that extends ventrally as a blind sac. The adjacent pial surface possesses coarse collagen fibers that are strongly PAS-positive.

Electron microscopy shows that the specialized glial region is mostly composed of an accumulation of ependymal and subependymal astroglia cells containing large amounts of gliofilaments (fig. 2-1A-B). No neuronal perikarya or processes are found among the glial cells. The ependymal cells are ciliated and show apical junctional complexes (zonula adhaerens and large gap junctions). The perinuclear cytoplasm and the basal process contain moderate to large amounts of gliofilaments. Subependymal cells generally are mostly filled with gliofilaments, and most organelles (rough or smooth endoplasmic reticulum, some mitochondria and occasional dense bodies) are found in perinuclear location. Both the cells and their processes are in frequent contact through gap junctions, and desmosomes can also be observed. Immunocytochemistry of this marginal glia has shown that it is strongly GFAP-immunoreactive (GFAP-ir) (fig. 2-1C-D). Thus, these cells share many characteristics of fibrous astrocytes.

As in astrocytes of mammals (Brightman and Reese, 1969; Braak 1975; Nabeshima et al. 1975), the processes of the mullet ependymal and subependymal

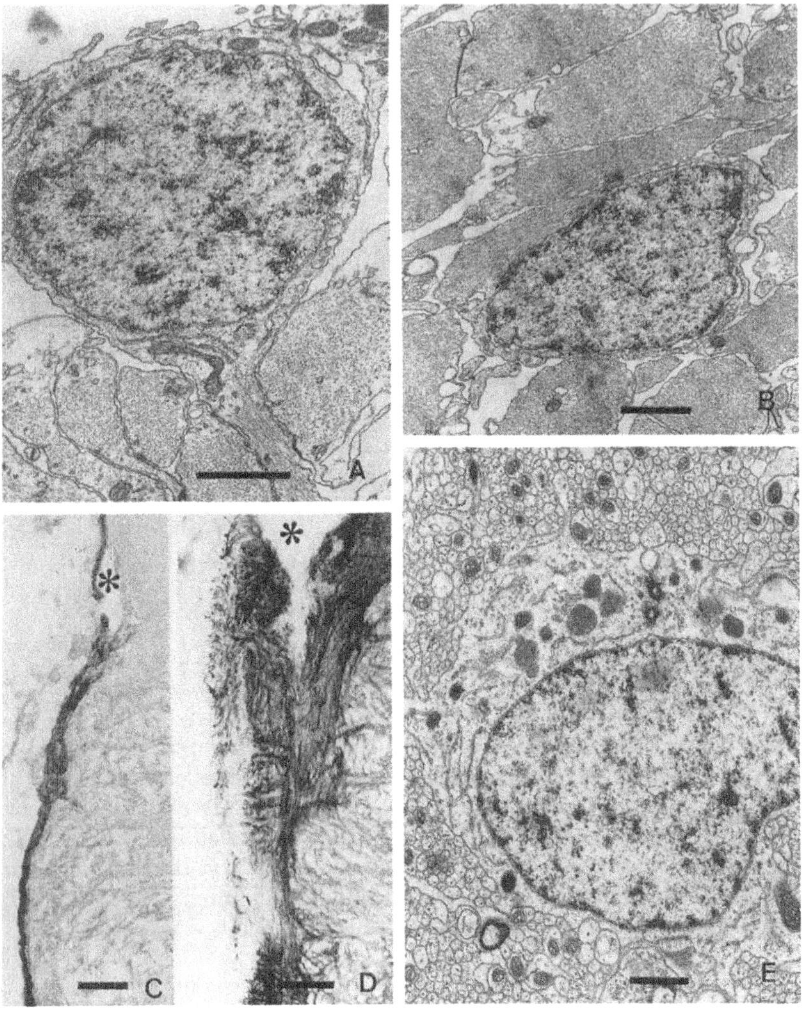

Figure 2-1. A-B. Ependymal and subependymal tanycytes of the marginal glial structure of the mullet. Note the abundance of gliofilaments. Electron microscopy. C-D. GFAP immunoreactive marginal glia of a 52 m. juvenile and old adult mullet, respectively. Asterisk, lateral recess. E. Astrocyte of the cerebellar crest of the lateral line area. Note emergence of thin lamellae and the scarcity of gliofilaments. (From Díaz-Regueira & Anadón, 1995). Scale bars = 1 µm. (A, B, E) and 50 µm. (C, D).

glial cells are connected by gap junctions (fig. 2-1B). In vitro studies have revealed a strong electrical coupling between mammalian astrocytes which may be interpreted as a metabolic syncytium designed to the transport of K+ as a buffer mechanism (Kettenmann and Ransom, 1988; Kimelberg and Kettenmann, 1990) and for the coordination of glial behavior (Ransom and Kettenmann, 1990). Gap junctions found between ependymal and subependymal cells might subserve similar functions.

A marginal glial region of the medulla oblongata similar but less developed than that of mullet has been described in the sunfish by Ma (1993), but is not present in trout (Manso et al., 1997). The processes of the gliofilament-rich cells of this region are mostly parallel to the surface forming a structure similar to but much thicker than the marginal glial layer in mammals (Bondareff and McLone 1973; Braak 1975; Nabeshima et al. 1975; Hajós and Kálmán 1989). While in mammals the marginal glia consists of astrocytes, in the mullet and sunfish a similar layer is formed by radial glia, suggesting that in mullet this type of cell replaces most mechanical functions of astrocytes, including those of the marginal glia of mammals. This is an example of the great evolutionary plasticity of radial glia.

In addition to the ependymal and subependymal marginal glia, there is ultrastructural evidence in mullets of the presence in the octavolateral area of scarce small cells exhibiting characteristics of protoplasmic astrocytes, i.e. cells with pale cytoplasm and a few dense bodies which practically lack intermediate filaments and that give rise to very thin laminar processes ensheathing dendrites and synaptic boutons (fig. 2-1E) (Díaz-Regueira and Anadón, 1995). These astrocytes of mullet are not revealed by vimentin or GFAP immunocytochemistry, but in late juvenile S-100 immunocytochemistry stains small glial cells in this region of the medulla that have the appearance of astrocytes and that probably correspond to the cells observed with electron microscopy. Similar small S-100 immunoreactive (S-100-ir) cells are widely distributed in the brain of mullet and here will be referred to as astrocytes.

Distribution and development of immunoreactivity to astroglial markers in the brain of mullet

Medulla oblongata and spinal cord

In the medulla oblongata of old adult mullet (circa 50 cm in length), strong GFAP immunoreactivity is observed in the ependymal and subependymal layer of the fourth ventricle (dorsolaterally to the medial longitudinal fascicle, and between the nucleus of the solitary tract and the octavolateral area). Intense vimentin immunoreactivity is found in raphe tanycytes and thick GFAP-ir processes course ventrally into the midline glial septum. Ependyma around the fourth ventricle are also vimentin-immunoreactive (Vim-ir) although much less intense. Other medullary areas, as well as the subpial endfeet, lack this immunoreactivity.

In the spinal cord, intense GFAP immunoreactivity is found in ependyma as well as in extensive ventrolateral and dorsal wings that are filled of GFAP-ir processes. Numerous subpial endfeet are strongly GFAP-ir. In the rostral cord, the dorsal midline shows thick processes ending "palisade-like" as finely indentated feet on the blood vessels of the area postrema. Vimentin is mostly restricted to the ventral ependyma and raphe processes and to subependymal cells, and Vim-ir subpial feets are scarce with the exception of the area postrema.

In the medulla of juvenile mullets, Vim-ir structures are more prominent than in adults. Numerous Vim-ir cells are found in the ventromedial wall of the fourth ventricle and their thick processes run in the midline glial raphe. In the spinal cord, strongly stained Vim-ir are also observed in the processes of the area postrema since 5-10 mm juveniles, and in ventral and lateral tanycytes and their processes and subpial endfeet (fig. 2-2A). In the spinal cord, GFAP immunoreactivity appears later than that vimentin. In 20-50 mm juveniles, however, GFAP has a distribution rather similar to that of vimentin, although there is an increasing tendency to accumulate a web of GFAP-ir processes in the area dorsal to the central canal. This accumulation is not observed for vimentin.

The marginal glia originates late in development from the lateral recess of the fourth ventricle. A thickened proliferative region (ventricular layer) is first observed in the corner between the cerebellum and the octavolateral region of juveniles of about 25-30 mm in length. Proliferation of glial cell and the caudal and lateroventral extension of the recess and associated glia occur later. This structure becomes Vim-ir in 25-50 mm mullets, but in juveniles of more than 50 mm in length loss this immunoreactivity and acquires the strong GFAP immunoreactivity characteristic of adults (fig. 2-1C).

In the medulla oblongata of 50-60 mm mullets, immunocytochemistry shows migrated S-100-ir radial glia in GFAP-ir areas extending away from the ventricle, because it reveals both the cell perikarya and processes (fig. 2-2E). Ependymocytes are also revealed throughout the ventricular surface. In addition, numerous small S-100-ir astrocytes are stained by this technique. These cells are scattered through many regions, without particular accumulations of this cell type in any region. In the rostral spinal cord, the thick glial processes abouting the area postrema exhibit strong S-100 immunoreactivity, as well as most ependymal cells and their radial processes. S-100-ir radial glia is found in the GFAP-rich region of the dorsal horn and in ventrolateral extensions of the central glial region, and small astrocyte-like cells are widely distributed through the grey and white matters.

Cerebellum

In the adult cerebellum, GFAP immunoreactivity is found in numerous thin irregular processes of the granule cell layer, both in the corpus and valvula cerebelli. In the ventral midline of the corpus, there are also intensely stained straight processes originated in the ependyma of the small dorsal recess of the fourth ventricle (cerebellar recess). Vim immunoreactivity, however, is very scarce, and some Vim-ir processes are found in the midline of the valvula and corpus cerebelli.

In juvenile mullets, Vim immunoreactivity is found in the end of the cerebellar ventricular recess (located between the auricular lobes and the corpus), and in a small raphe of the valvula and the corpus. This raphe is also GFAPir in the caudal corpus and rostral valvula of 20-30 mm mullets. In 50 mm mullets, subpial enfeets are GFAP-ir in the corpus but not in the valvula cerebelli. In this stage, thin GFAP-ir processes are also observed in the granule cell layer.

Mesencephalon

The mesencephalon exhibits strong vimentin immunoreactivity in tanycytes of the ventral midline and their processes. This immunoreactivity is also found in the endfeet of the region between the tectum and tegmentum. Prominent Vim-ir processes are associated with the large veins of dorsal tegmentum. Other regions practically lack vimentin. GFAP is very abundant in ependyma of the tegmental midline and torus semicircularis, and in ependymal cells and subependymal glial cells of the stratum album centrale and stratum periventriculare of the optic tectum and their proximal processes. In the medial region of the optic tectum, GFAP immunoreactivity is mainly found in ependymocytes, but in lateral regions it is found in subependymal glia cells while ependymal cells practically lack this immunoreactivity. No GFAP-ir subpial endfeet are observed along the outer surface of the optic tectum. The torus longitudinalis show thin radial GFAP-ir processes.

In 5-25 mm mullets, the optic tectum shows a Vim-ir raphe in the torus longitudinalis: this glial raphe continuates rostrally with the ependyma of the posterior commissure, and caudally with the raphe of the valvula cerebelli. The caudal and lateral end of the tectal ventricles also exhibit Vim-ir ependyma, and this immunoreactivity is also observed in ependyma of various tegmental regions (ventral raphe, nucleus lateralis valvulae). The raphe, subpial endfeet and radial processes of the tegmentum are intense Vim-ir. In 50 mm mullets, the mesencephalon shows vimentin immunoreactivity in these regions, but it is less intense than in early stages.

In the mesencephalon of 5-10 mullets, GFAP immunoreactivity is only observed in subpial endfeet of the tegmentum. In 20-30 mm mullets, the ependyma of the mesencephalic tegmentum is scarcely GFAP-ir, but this immunoreactivity is intense in subpial endfeet and distal radial processes. In the optic tectum, GFAP is observed only in some processes of the midline of the torus longitudinalis and in ependyma of most rostral region of tectal ventricles. GFAP immunoreactivity suffers a huge increase in 50 mm mullets. Intense GFAP immunoreactivity is found in ependyma and subependyma of the optic tectum (fig. 2-2D), except in its caudalmost portion, and GFAP-ir subpial enfeet are observed in the tectal suface. In the torus longitudinalis, immunoreactive processes are observed laterally but not in the raphe. In the tegmentum, the ependymal and subependymal regions are strongly GFAP-ir, with the exception of the nucleus lateralis valvulae.

Distribution of S-100 in the mesencephalon of 50-60 mm mullets is roughly similar to that of GFAP, but this marker reveals better the shape of cells. Large subependymal glial cells are observed close to the stratum album centrale in the lateral regions of the tectum, and in many tegmental regions as the lateral mesencephalic nucleus, in addition to S-100-ir ependymocytes (fig. 2-2F). Some S-100-ir astrocyte-like cells are observed in the tegmentum, but they are rare in the optic tectum. S-100 is practically not expressed in the mesencephalon of mullets below 40 mm in length.

Forebrain

The telencephalon shows Vim-ir ependymocytes in the walls of the preoptic recess and the ventral area of the telencephalon. Immunoreactive processes from these regions course laterally and rostrally, reaching the olfactory bulb and ending in the entrance of the olfactory nerve.

The diencephalon exhibits strongly Vim-ir ependyma in the thalamus just ventral to the magnocellular preoptic nucleus and its caudal extension. Coronet cells of the saccus vasculosus and bipolar ependymal cells of the organon vasculosum hypothalami and lateral and posterior recess nuclei are also Vim-immunoreactive. The ependyma of the dorsal and ventral thalami practically lack vimentin immunoreactivity. In addition, small groups (2-4 cells) of Vim-ir astrocytes with short irregular processes are found in a region between the lateral optic recess and the optic nerve. The optic nerve exhibits numerous Vim-ir endfeet on its external surface.

Distribution of GFAP immunoreactivity in the forebrain of old adult mullet is roughly similar to that of vimentin in some areas, but differs greatly in other regions. In the dorsal thalamus, the ependyma of some areas are strongly GFAP-ir but Vim-negative. In the organon vasculosum hypothalami, lateral and posterior recess nuclei, conversely, there is only faint GFAP immunoreactivity, and the saccus vasculosus lack GFAP immunoreactivity. Immunoreactivity to GFAP of the infundibular region and neurohypophysis is more intense than that to vimentin.

In 5-10 mm mullets the forebrain lacks GFAP immunoreactivity. However, the caudoventral telencephalon exhibits Vim-immunoreactive ependymal cells with thin processes ending in the ventral surface of the brain. The bottom of the preoptic recess is formed by a Vim-ir epithelium. In the epithalamus, cells in the ependymal roof of the ventricle exhibit strong immunoreactivity, originating long Vim-ir processes that course transversally to the habenula or longitudinally to the ventral surface of the optic tectum.

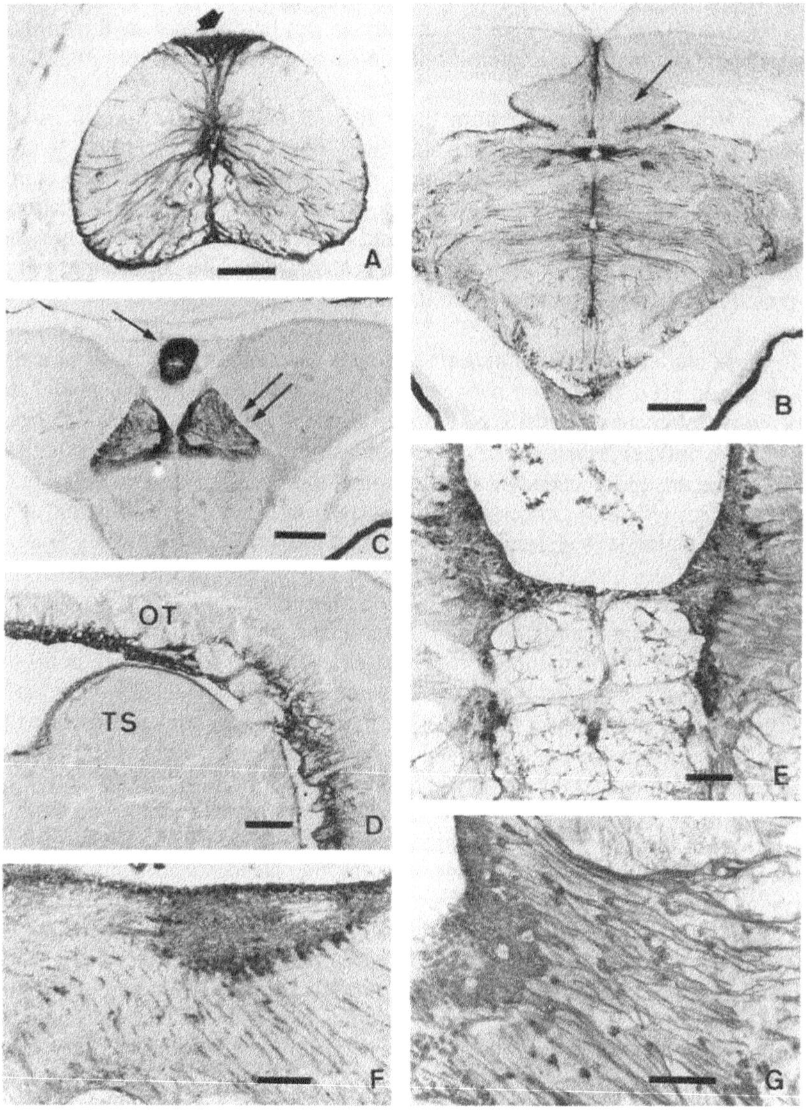

Figure 2-2. Examples of vimentin (A-C), GFAP (D) and S-100 (E-G) immunoreactivity. A, rostral spinal cord of a 15 mm. mullet. Arrow, area postrema; B, transverse section through the thalamus and pretectum of a 20 mm. mullet. Arrow, torus longitudinalis. C, section through the pineal (arrow) and habenula (double arrow) of a 20 mm. mullet. D, transverse section through the optic tectum (OT) and torus semicircularis (TS) of a 52 mm, mullet. E, transverse section through the caudal medulla of a 58 mm. mullet showing ependymal and subependymal tanycytes. F, section through the lateral mesencephalic nucleus showing the well-developed subependymal glia. 58 mm. mullet. G, Transverse section through the dorsal thalamus of a 60 mm. trout showing S-100-ir ependymal and radial glia. Scale bars = 50 µm. (A, E, F), 100 µm. (B, C, D) and 40 µm. (G).

In 10-20 mm mullets, vimentin immunoreactivity is more evident in the basal telencephalon. Intensely stained ependyma can be observed in the epithalamus-pretectal segment, and in areas of the dorsal and ventral thalamus (fig. 2-2B).

Long Vim-ir processes run from these regions to the optic tectum, medial habenulae, and lateral region of the thalamus. The walls of the third ventricle (ventral thalamus and hypothalamus) show more Vim-ir ependymal cells than in former stages, and in the optic region Vim-ir processes cross the first portion of the optic tracts and the pretectal nuclei to end in conspicuous endfeet in the basis of the optic chiasma and the lateral surface, respectively. The supporting cells of the pineal organ exhibit vimentin immunoreactivity (fig. 2-2C).

In these mullets, GFAP-immunoreactivity is observed in long basal processes ending as endfeet in the region between the dorsal and ventral telencephalic area, in perikarya of ependymal cells of the dorsal thalamus, and in basal processes crossing the optic chiasma. GFAP immunoreactivity is also observed in cells of the pineal organ, and in ependymal cells around the posterior commissure. In the hypothalamus, GFAP-ir processes are observed in the dorsal ependyma of the median hypothalamic lobes, and in pituicytes of the neural hypophysis.

In the telencephalon of 20-50 mm mullets, the immunoreactivity to vimentin extends to the medial ependyma of the dorsal telencephalic area, and to the caudal olfactory bulb. GFAP immunoreactivity increases in these regions and, in 50 mm mullets, the dorsal and lateral ependymal surfaces of the dorsal telencephalon are faintly GFAP-ir and many long processes from these cells converge towards the external sulcus between the dorsal and ventral telencephalic areas. The preoptic region, the thalamus and the pretectal region exhibit intense immunoreactivity to vimentin. In the OVH, there is a row of Vim-ir bipolar cells, with perikarya located away of the ventricle and an apical process that extend to the ventricular surface. GFAP immunoreactivity increases in the diencephalon and ependyma lateral and caudal to the subcommissural organ are GFAP-ir and show numerous processes that course through the posterior commissure to the external surface, while the epithalamus (habenula, subcommissural organ) practically lacks GFAP immunoreactivity. The dorsal thalamus exhibits a strong GFAP-immunoreactive thick ependyma in its ventral portion, but the region below the habenula and subcommissural organ and some other areas do not exhibit GFAP immunoreactivity.

S-100 immunocytochemistry of 50-60 mm mullets stains the GFAP-ir ependymal regions of the forebrain. As in the other regions this marker reveals the shapes of the cell better than GFAP.

In the telencephalon, most S-100-ir cells are ependymocytes, although laterally migrated subependymal cells are clearly demonstrated in the ventral telencephalic area. S-100-ir radial processes are abundant in the dorsal and ventral areas, but S-100-ir astrocyte-like cells are very scarce. In the diencephalon also the GFAP-rich ependymal areas show S-100 immunoreactivity, including areas of the thalamus, preoptic region and laterodorsal ependyma of the lateral hypothalamic recess. S-100-ir astrocytes are common in some diencephalic nuclei as the nucleus glomerulosus, and in the optic tracts.

S-100 immunoreactive astroglia of trout brain

In the brain of adult trout, immunocytochemistry with an antibody against S-100 protein stains both ependymocytes (tanycytes) and radial glia cells that have their perikarya located in subventricular regions or, in a few regions, far from the ventricle (fig. 2-2G) (Manso et al., 1997). The subpial endfeet and radial processes are also generally well stained. Ependymocytes were stained throughout the brain, except in a few areas of the ventral telencephalon, epithalamus, hypothalamus and brain stem. Unlike mullet, no S-100-ir astrocytes were observed. In the telencephalon, strongly stained ependyma is observed along two longitudinal grooves in the ventral and dorsal area ventralis. In the diencephalon, the epithalamus contained scarce S-100-ir structures, while the thalamus contained two regions with strongly S-100-ir cells in the dorsal and ventral thalamus, not only in ependymal and subependymal locations but also migrated laterally. The trout hypothalamus contains a highly complex system of ventricular diverticles and the distribution of S-100-ir ependyma is complex. The inner layers of the optic tectum contain tanycytes and sub-ependymal glial cells with long radial processes. In the valvula and the corpus cerebelli, numerous S-100-ir migrated glial cells are observed below the Purkinje cell layer. These cells appear to correspond to Bergmann's glia. In the medulla oblongata, S-100 immunnoreactive cells were observed in both ependymal and subependymal locations at the level of the sulcus limitans and of the intermedioventral and intermediodorsal grooves, and radial glia cells formed conspicuous bands parallel to the ventricle in the vagal sensory nucleus. All S-100-immunoreactive glial cells appear to be NADPHd-positive, although this enzymohistochemical reaction also reveals oligodendrocytes and some neurons (Pérez et al., 1996). This distribution of trout S-100-ir tanycytes is roughly similar to that of NADPHd-positive glia of sunfish (Ma, 1993). GFAP immunocytochemistry of the brain of adult trout reveals very scarce GFAP-ir ependymal cells, which mainly locate in the diencephalon, optic tectum and brainstem. In some areas (ventral spinal cord, ventral rhombencephalon, mesencephalic tegmentum, lateral optic tectum), moderate GFAP immunoreactivity was observed in the subpial endfeet, but this was scarcely appreciable in other regions. The radial processes lack generally GFAP immunoreactivity.

During trout development, the pattern of S-100 and GFAP immunoreactivity differs considerably from that of adults. In 9-12 mm embryos, S-100 immunoreactivity is only observed in a few ependymal regions of the spinal cord and rhombencephalon. The ventromedial glial septum (floor plate) was well stained in the rhombencephalon and caudal mesencephalon. In 18-22 mm larvae, S-100 immunocytochemistry stained the thick ependyma of the brainstem tegmentum and ependymal regions of the thalamus and posterior tubercle, but this immunoreactivity was undetectable in other regions. Only the ependymal cells were S-100-ir, and no stained radial processes or radial glia cells were observed at these stages with the exception of the ventromedial glial septum. In 60-65 mm juveniles, distribution of S-100 immunoreactivity is similar to that observed in adults. Unlike the adult trout, in 22 mm fry the ependymal cells were GFAP-ir in most brain regions. GFAP immunoreactivity was observed in thin fibers coursing from the ependymal perikarya to the pial surface. The distribution of GFAP-immunoreactive processes was rather similar to that of S-100-ir processes in adult trout, although no migrated radial glia was stained.

Results obtained in mullet and trout suggest that in teleosts expression of vimentin, GFAP and S-100 is developmental and regionally regulated. S-100 appear only in late stages and, at least in trout brain, develops in a clear caudorostral sequence. GFAP is practically not expressed in the smallest mullets investigated (larvae). This contrasts with results of Marcus and Easter (1995) in zebrafish, in which GFAP is early expressed in ependymal cells of embryos. In zebrafish, however, the expression of this protein descends in late embryos. Whether mullet embryos express GFAP in not known. In adult mullets, vimentin expression is conserved only in a few places, and GFAP immunoreactivity is generally conserved in glial perikarya, but tends to fade out in distal processes. The tendency to loss GFAP immunoreactivity in adult stages is much more marked in trout tanycytes. These differences are probably due to different contents of the various cytoskeletal structures. In teleosts radial processes of tanycytes may contain large amounts of microtubules, but scarce glio-filaments (Stevenson and Yoon, 1982), and this appears to be the case of most tanycytes of adult trout. As shown by electron microscopy in the marginal glia, tanycytes may contain large amounts of gliofilaments even in old adults.

OLIGODENDROGLIA

Most of our knowledge of the morphology and biological properties of oligodendrocytes has been obtained from studies carried out in mammals. This cell type, however, appears to occur in most vertebrates, with the noticeable exception of Agnatha (lampreys and hagfish). Río Hortega (1921) introduced the term oligodendroglia to describe a new type of central neuroglial cells with scarce number of processes. Using his silver carbonate and modified Golgi methods, Río Hortega found this glial type distributed in three locations: interfascicular, perineuronal or perivascular. Interfascicular oligodendrocytes are close related to myelinated fibers, and often form rows along the nerve fibers. Ultrastructural studies have confirmed that processes of these cells form the myelin sheaths in the central nerve system (Bunge, 1968). Other oligodendrocytes are found close to neuronal perikarya or in perivascular locations. Río Hortega (1928) distinguished four types of oligodendrocytes by their size and the appearance of their processes, without sharp boundaries between them. Type I oligodendrocytes (Robertson's oligodendrocytes) were small, with spherical or polygonal perikarya and numerous thin processes. This type is found in perineuronal, perivascular and, more often, in interfascicular locations associated with thin myelinated fibers. Type II oligodendrocytes (Cajal's oligodendrocytes) were found in the white matter closely associated with nerve fibers. These larger cells are polygonal or cuboid and have less and thicker processes than type I oligodendrocytes. The large type III oligodendrocytes (Paladino's oligodendrocytes) showed only three to four thick processes and are found associated with thick nerve fibers of the brain stem, cerebellar peduncles and the spinal cord. Type IV oligodendrocytes (schwannoid oligodendrocytes) are monopolar or bipolar cells closely associated with a single axon of medium or large size.

The use of electron microscopy circumvent the difficulty of staining oligodendrocytes for light microscopy, allowing an easy characterization of mammalian oligodendrocytes. These cells are characterized by the absence of intermediate filaments, scarcity of microtubules and the presence of a dense ground material in the cytoplasm. Ribosomes are abundant, both free and associated with endoplasmic reticulum, and the cytoplasm may contain several types of inclusions. These features, together with the round or oval dark nucleus, generally excentrical in the cell, distinguish oligodendrocytes from neurons, astrocytes and microglia cells (see Peters et al., 1976).

Mori and Leblond (1970) introduced a distinction of oligodendrocytes into light, medium and dark cells, indicating that light cells are abundant in young animals. According their observations, the light oligodendrocytes represent the more immature cells. Since that, several radioautographic experiments in the corpus callosum indicated that the three types probably represent successive stages in the maturation of oligodendrocytes from immature cells (Paterson et al., 1973; Imamoto et al., 1978; Paterson, 1981; Parnavelas et al., 1983). In the deep cortex and subcortical white matter of the rat, Parnavelas et al. (1983) have observed that oligodendrocytes suffer a sequential transformation from light to dark cells, and that most oligodendrocytes of mature animals are dark cells. Other in vitro and in vivo studies of the different types of oligodendrocytes suggest that the different morphology is associated with different functional specialization (Remahl and Hildebrand, 1990; Szuchet, 1995). Thus, light oligodendrocytes will be specialized in myelin production, while dark cells will be involved in myelin manteinance, which is also supported by biochemical studies (Szuchet et al., 1983, 1986).

Structural characterization of teleost oligodendrocytes

Despite the numerous studies on the morphology and function of oligodendrocytes existing in mammals, in only a few studies, mostly using electron microscopy, have been investigated in detail oligodendrocytes of other vertebrate classes (Kruger and Maxwell, 1967; Stensaas and Stensaas, 1968; Long et al., 1968; Lara et al., 1989; Monzón-Mayor et al., 1990; Díaz-Regueira et al., 1992). These studies have stressed the resemblances between vertebrate oligodendrocytes but, with the exception of those of Stensaas and Stensaas (1968) in toad, Monzón-Mayor et al. (1990) in lizard, and Díaz-Regueira et al. (1992) in mullet, the cell diversity of cells of this type has not been explored. The following results on these topics are based in our investigations on the trout and the mullet (Díaz-Regueira et al., 1992; Pérez et al., 1996; Anadón et al., in preparation).

In old adult mullet (circa 50 cm in length), we have studied with light and electron microscopy the oligodendrocytes of the lateral line area of the medulla oblongata (for a description of the organization of this area, see Díaz-Regueira and Anadón, 1995). Mullet oligodendrocytes are recognizable in semithin sections stained with toluidine blue by its roundish to oval shape, the eccentric round nucleus and basophilic cytoplasm. They were located in all regions of the lateral line area, although there were noticeable differences in size and appearance in the different parts. In the cerebellar crest there are only small oligodendrocytes (fig. 2-3A), while myelin-rich areas (nucleus medialis) show a great variety of sizes and densities of oligodendrocytes (fig. 2-3B-D). The size of the cells ranged from 4.5 to 18 µm. The appearance of nuclei varies between dark nuclei with rather condensed chromatin to light to very light euchromatic nuclei with a conspicuous nucleolus. Most oligodendrocytes are not associated with neuronal perikarya or with blood vessels.

Oligodendrocytes observed with electron microscopy display common features as the roundish shape of their cell bodies and nuclei, appearance of organules and intercellular junctions, although vary both in nuclear and cytoplasmic density and size, and in the amount of organules (fig. 2-3E-F). For a descriptive purpose, the continuous series of cells was subdivided into light, intermediate and dark oligodendrocytes. Dark oligodendrocytes are small cells with round nuclei with chromatin condensed in masses, mostly in peripheral regions, and an inconspicuous nucleolus. The cytoplasm is scarce and very dense, with few organules (fig. 2-3E). This subtype has very scarce mitochondria and dense bodies. Intermediate oligodendrocytes have nuclei with chromatin less condensed than in dark cells, and the cytoplasm contains more organules (fig. 2-3F). In addition to rough endoplasmic reticulum (RER) of tubular appearance and numerous mitochondria, this cell subtype possesses one or more Golgi dictyosomes and dense bodies of irregular shape and dense content. These bodies generally possess multilamellar inclusions (fig. 2-3F), and some of them exhibit a branched, reticular appearance. The light cells comprise the largest oligodendrocytes of the lateral line area (fig. 2-3D; see also Díaz-Regueira et al., 1992). The light round nucleus exhibit a very sparse chromatin and, occasionally, a conspicuous nucleolus not associated with clumped chromatin. The cytoplasm is extensive and contains a large amount of tubular RER, numerous small mitochondria of dense matrix similar to those of other oligodendrocyte subtypes, several dictyosomes and a few dense bodies.

In all subtypes, mitochondria are very thin (about 0.2 µm in diameter), with 2-3 tubular crests and a very dense matrix. Oligodendrocytes lack intermediate filaments, which can easily observed in neurons and in processes of glial cells of the astrocyte family (tanycytes, radial glia). All oligodendrocytes also form gap junctions with processes of cells of the astrocyte family (fig. 2-3E-F), although these junctions are much more abundant in intermediate and light cells. Contacts

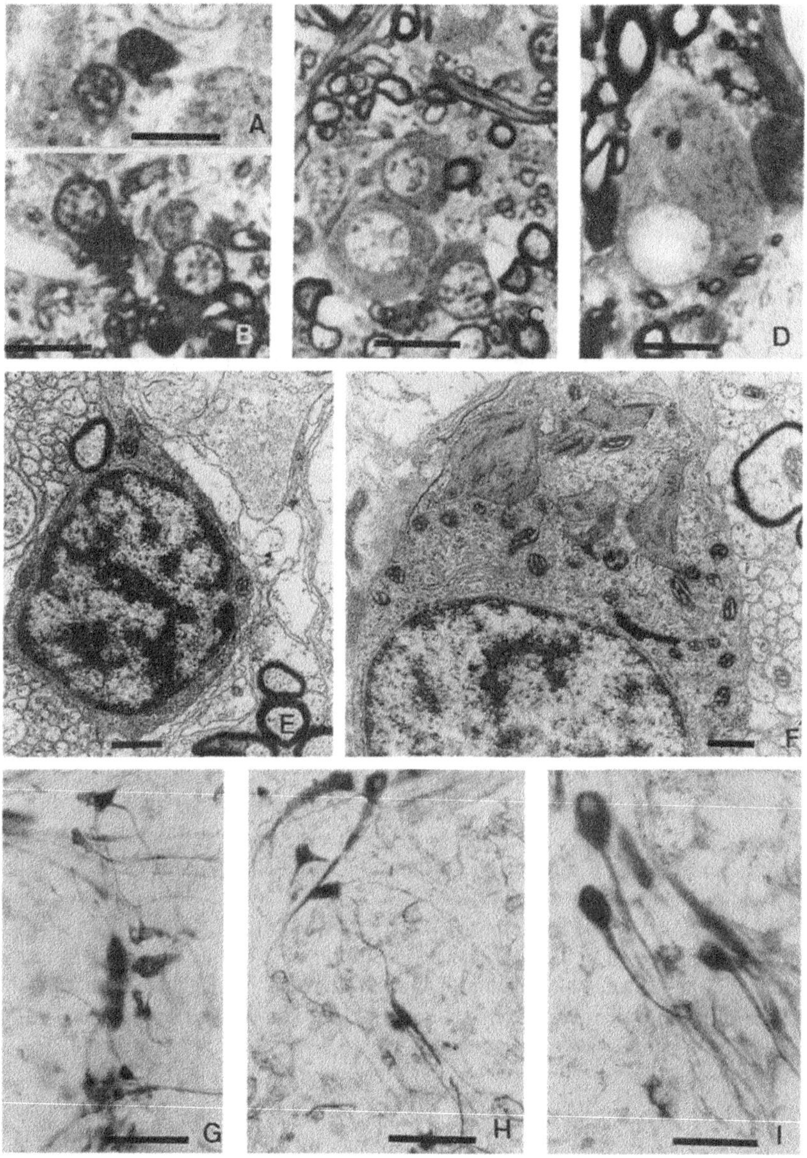

Figure 2-3. A-D semithin sections showing the spectrum of oligodendrocytes observed in the lateral line area of an old adult mullet. A, cerebellar crest, B-D, medial nucleus. Note the very thick myelinated axons in D. All figures at the same magnification. E-F electron microscope micrographs of dark and intermediate oligodendrocytes of the lateral line area (from Díaz-Regueira et al. 1992). G-I, large NADPH diaphorase positive oligodendrocytes associated to the medial longitudinal fascicle of the medulla of adult trout. Note in H processes forming rings around nerve fibers. Scale bars = 10 µm. (A-D), 1 µm. (E-F), 50 µm. (G-H) and 25 µm. (I).

between oligodendrocytes are unspecialized, although unconspicuous tight junctions and a few small puncta adherentia have been observed. No synapses were observed on oligodendrocytes.

Our results in the lateral line area of the mullet show that oligodendrocytes can be accurately identified by certain common ultrastructural characteristics, and in addition by their appearance in light microscopy. Moreover, unlike astroglia, these cells were S-100-, GFAP and vimentin-immunonegative throughout all stages investigated (5 mm juveniles to 50 cm adults). The sharp distinction found between oligodendrocytes and other classes of glial and nerve cell in the lateral line area of the mullet contrasts with reports suggesting a morphological continuum between oligodendrocytes and astrocytes and/or neuron in teleosts (Kruger and Maxwell, 1967; Lara et al., 1989), and with reports of vimentin immunoreactivity in cultured oligodendrocytes of goldfish (Sivron et al., 1994). Thus, our in vivo observations in the mullet appear to indicate that oligodendrocytes are so well-differentiated as a cell type as in the mammals (Vaughn and Peters, 1968; Mori and Leblond, 1970). A clear distinction between oligodendrocytes and astrocytes has been reported in elasmobranchs (Klatzo, 1967; Anadón, 1974), which appears to be a shared characteristic of vertebrates.

The use of brains of old adult mullets in our studies indicates that the wide range of sizes and densities of oligodendrocytes is a feature of the region studied. The presence of the largest oligodendrocytes was associated with that of bundles of thick myelinated fibers, while the smallest oligodendrocytes were observed in the cerebellar crest, a region that contains only very thin myelinated fibers (Díaz-Regueira and Anadón, 1995). The presence of complex dense bodies in light and intermediate oligodendrocytes which is probably age-related, indicates that they are mature cells. Complex dense bodies like those of the mullet have been found in oligodendrocytes of old rats and monkeys (Vaughan and Peters, 1974; Peters et al., 1991). These observations indicate that size and density differences noted in mullet oligodendrocytes are related to different functional specializations of mature cells (Díaz-Regueira et al., 1992). This is in line with suggestions of Rio Hortega (1928), Blakemore (1982), Remahl and Hildebrand (1990) and Szuchet (1995) that the morphological variety of subtypes of oligodendrocytes in adult mammals is related to functional specialization.

Distribution of NADPHd-positive oligodendrocytes in the brain of trout

Although electron microscopy provides valuable information on the characteristics of teleost oligodendrocytes (Díaz-Regueira et al., 1992), this technique is not well suited for studies of their distribution in the brain because the small volume of samples. Methods of methalic impregnation of oligodendrocytes, as Rio Hortega's silver carbonate stain, generally do not work well in nonmammals (Rio Hortega, 1928; Lara et al., 1989), although it has given good results in elasmobranchs (Klatzo, 1967). We have dealed with the distribution of oligodendrocytes in the trout using two different approaches. By one hand, studies in our laboratory in the trout indicated that the NADPHd histochemical reaction stained oligodendrocytes in a characteristic way, providing the fixation of the brain with 1% paraformaldehyde in phosphate buffer was short (1-4 hours) (Pérez et al., 1996). Using electron microscopy, similar evidence has been presented in the goldfish (Villani and Guarnieri, 1995). The very different location, appearance and staining of oligodendrocytes, tanycytes and neurons in the trout make this technique very useful for study of oligodendrocytes (Pérez et al., 1996). Note that in mammals the NADPHd reaction reveals only neurons (Scherer-Singler et al., 1983; Vincent et al., 1983) and that the NADPH diaphorase appears to be identical to the nitric oxide synthase (Bredt et al., 1991; Dawson et al., 1991). By the other hand, the biochemical characterization of trout myelin has allowed to obtain antisera against the myelin proteins (Jeserich and Waehneldt, 1986; Jeserich et al., 1990; Jeserich and Stratmann, 1992). These authors have shown that the antiserum directed against the 36K myelin protein of trout stains not only myelin but also developing trout oligodendrocytes. The 36K antiserum used in our lab was kindly supplied by Dr. G. Jeserich.

In the brain of old adult trout (circa of 36 cm in length), the NADPHd histochemical reaction revealed a large population of oligodendrocytes (Pérez et al., 1996). Characteristically, these cells appeared intensely and, often, very delicately stained (fig. 2-3G-I). They were small to medium sized cells (3-12 µm in transverse diameter), with round to oval-shaped body, that gives rise to one or several thin processes (less than 1-2 µm thick). These processes branch near the cell body, and often coursed among myelinated fibers. Processes of large oligodendrocytes are seen forming NADPHd-positive "rings" around axons (fig. 2-3H), as shown by Rio Hortega (1928) in mammals, and trout oligodendrocytes look very alike of those of dog and cat. The variation in oligodendrocyte size was similar to that found in the grey mullet, and the distribution of oligodendrocytes of different size was likewise heterogeneous: the largest cells were always associated with thick myelinated fibers, such as those of the medial and lateral longitudinal fascicles, or the cerebellar tracts. Large oligodendrocytes were pear-shaped, fusiform or stellate (fig. 2-3G-I), while smaller cells were round or oval. The morphology of large oligodendrocytes roughly corresponds to that Río Hortega's "Paladino" cells, many medium-sized oligodendrocytes to his "Cajal" cells, and small cells to his "Robertson" cells. The variety of sizes of trout oligodendrocytes is similar to those of the mullet (Díaz-Regueira et al., 1992). These results were obtained in old adults of two species pertaining to different evolutionary lines of teleosts. This indicates that the diversity of oligodendrocyte sizes in different regions of the brain is a shared characteristic of teleosts and mammals.

Development of oligodendroglia

Studies in mammals have indicated that oligodendrocytes and type 2-astrocytes originate from a common precursor, that is recognized by the A2B5 monoclonal antibody (Eisenbarth et al., 1979; Raff et al., 1983; Fok-Seang and Miller, 1994). These cells appear late in the embryonic development, before the onset of myelination, in caudorostral and ventrodorsal gradients (Verity and Campagnoni, 1988; Kagawa et al., 1996; Hajihosseini et al., 1996; Hall et al., 1996). Recent experimental studies have proposed that oligodendrocytes originate from a ventral region of the neuroepithelium, and then migrate to their definitive locations (Warf et al., 1991; Friedrich and Lazzarini, 1993; Noll and Miller, 1993; Orentas and Miller, 1996). Similar results have been obtained in chick (Ono et al., 1995). Immunocytochemistry with the 36K antibody stains myelin sheaths, and also demonstrates oligodendrocytes in developing trout (Jeserich and Waehneldt, 1986). Jeserich et al. (1990) have observed a caudorostral developmental gradient of myelination of fibers in the trout brain, as well as a temporal pattern of expression of the different myelin proteins (IP2, 36K, IP1) and galactocerebroside. Results obtained in our lab with the 36K antibody (Anadón et al., in preparation) confirm observations of these authors on gradients of fiber myelination in the brain, and extend their results to the development of 36K-ir oligodendrocytes in the different regions. In developing trout, PAP immunocytochemistry with this antiserum demonstrates delicate oligodendrocyte processes and their associated myelin segments. This technique also reveals a variety of shapes of trout oligodendrocytes and indicates that, by their sizes, first myelinating oligodendrocytes in each brain region form a rather homogeneous population of cells and that late in development some of them grow considerably.

First 36k-ir oligodendrocytes are observed in 11 mm embryos, where myelinating fibers and oligodendrocytes are only found in the caudal rhombencephalon and rostral spinal cord. Myelinogenesis was observed in two longitudinal systems: some fibers of the medial longitudinal fascicle (MLF) and, in the caudal medulla oblongata, a few fibers of a lateral fascicle. Oligodendrocytes, mostly perifascicular, were associated with these bundles (fig. 2-4A). These cells were either pearshaped, with a short and thick process that gives rise to thinner processes associated with myelinated fibers, or multipolar. In 15 mm prolarvae (1 day posthatching), the number of 36K-ir oligodendrocytes has increased considerably. They are associated with the FLM, the lateral longitudinal fascicles, and also with ventral fascicles and arcuate fibers. A few cells appear at the entrance of the oculomotor, trigeminal and octaval nerves. In the mesencephalon, a few oligodendrocytes were associated with tracts of the posterior commissure.

In 18 mm larvae (fry), myelination and oligodendrocytes have extended to many new areas. The MLF is the most conspicuous myelinated tract, but the number of its oligodendrocytes has not much changed. In the rhombencephalon, a large number of oligodendrocytes is now present in the octavolateral area, mostly in subpial location (fig. 2-4B). Myelinating fibers are also found in ventral, ventrolateral, laterolateral and dorsomedial medullary bundles, as well as in the cerebellar peduncles (granular eminences). The mesencephalic tegmentum exhibits myelinating fibers and some oligodendrocytes in both commissural and longitudinal systems. The optic tectum shows only a few 36K-ir oligodendrocytes and myelinating fibers. In the diencephalon, 36K-ir myelin and oligodendrocytes appear in the optic nerve and optic tracts, but the thalamus contains only sparse numbers of immunoreactive fibers and oligodendrocytes.

In the brainstem and spinal cord of 22-25 mm fry, the number of 36K-ir oligodendrocytes has increased very considerably. In the spinal cord and the rhombencephalon, specially in its octavolateral area, there are very numerous oligodendrocytes in subpial locations, just below the glial limiting membrane. In addition, the MLF and other longitudinal tracts appear heavily myelinated, with oligodendrocytes both in perifascicular and interfascicular locations. Most brainstem regions contain a number of myelinating fibers, but the facial-vagal viscerosensory column, the ascending secondary visceral tract, and the inferior olive are scarcely myelinated regions.

In the cerebellum of 22 mm fry there are a number of oligodendrocytes and myelinating fibers, mostly along spino-cerebellar tracts, in the middle of the cerebellar peduncles and in the granular eminences and granule cell layer of the rostral part of the corpus cerebelli, but only a few in the valvula cerebelli and the caudal corpus. The optic tectum show myelinating fibers and oligodendrocytes in the stratum album profundum, in the stratum opticum and between these regions. Distribution of these elements shows a conspicuous lateromedial and rostrocaudal decreasing gradient. The optic nerve and tract are now much more heavily myelinated than in the former stage, and the very numerous oligodendrocytes are often arranged in small rows among optic fibers. The pretectal region, the tract of the posterior commissure and thalamic fascicles have increased considerably the number of both fibers and oligodendrocytes (fig. 2-4E, F). Neither the telencephalon nor the hypothalamus caudal to the postoptic commissures show oligodendrocytes or myelinating fibers in this stage. While the shapes of oligodendrocytes are similar to those of previous stages (pearshaped and multipolar), a considerable increase of size is observed in some cells associated to the MLF and other fascicles of the medulla, which reach 16μm and 25-30μm in tranverse and longitudinal axes, respectively.

In 38 mm juveniles, myelination has progressed to new regions as the telencephalon caudal to the anterior commissure, that contains 36K-ir oligodendrocytes associated with fibers between the dorsal and ventral telencephalic areas and with the basal forebrain bundle. In the caudal diencephalon, the posterior tubercle and dorsolateral region of the lateral hypothalamic lobes shows a number of oligodendrocytes and myelinating fibers. The rostral hypothalamus practically lacks oligodendrocytes. The myelinating regions of former stages become more conspicuously myelinated and contain a higher number of oligodendrocytes (fig. 2-4C-D).

In the adult trout, the pattern of distribution of 36K-ir oligodendrocytes is similar to that of juveniles, but a few regions show a further development. The forebrain has numerous small 36K-ir oligodendrocytes in the ventrolateral and central regions of the dorsal telencephalic area and the dorsal region of the ventral telencephalic area, and they are also numerous in the granule cell layer of the olfactory bulb. In the hypothalamus, the nucleus diffusus of the lateral hypothalamic lobe shows numerous oligodendrocytes. In the mesencephalon, the torus longitudinalis shows some small 36K-ir oligodendrocytes and the optic tectum

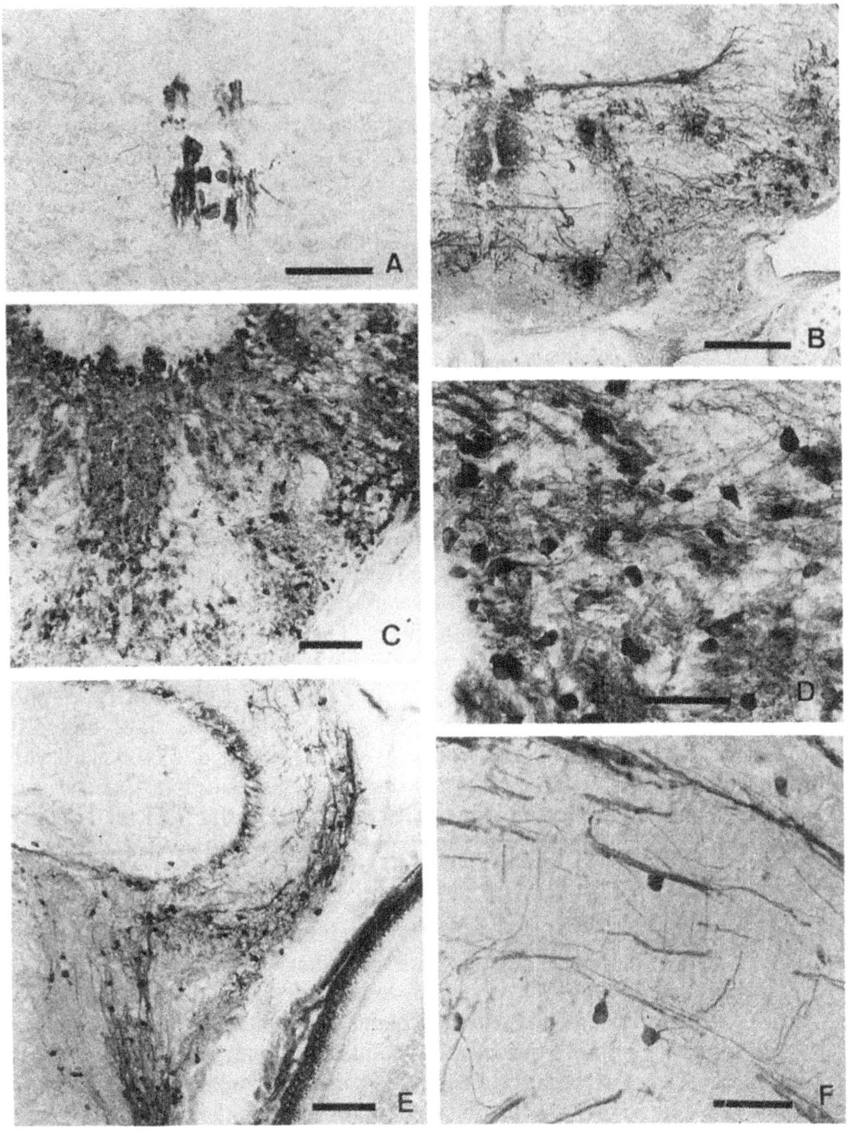

Figure 2-4. 36K immunoreactivity in the brain of developing trout. A, MLF and oligodendrocytes of a 11mm. embryo. B-C Transverse section of the medulla of a 22mm fry and a 38mm fry, respectively. D, detail of oligodendrocytes and myelinating fibers in the pretectum and optic tectum of a 25mm. fry. F, detail of oligodendrocytes in the optic tectum of a 25mm. fry showing their thin processes. Scale bars=50µm. (A, D, F) and 100µm. (B, C, E).

has these cells in all layers except the marginal layer. A few small oligodendrocytes appeared in the molecular layer of the cerebellum, in addition to the cells of other layers. In the medulla, 36K-ir oligodendrocytes appear in the facial-vagal sensory lobe and the cerebellar crest. The results obtained in adult trout with 36K immunocytochemistry agree well with those obtained with NADPHd histochemistry.

Immunocytochemistry with the 36K antibody in developing trout shows oligodendrocytes connected by thin processes to first myelinating fibers, generally external to the ventricular layer. Most of these cells have a tendency to migrate away the ventricular region, reaching often subpial locations. This behavior distinguishes oligodendrocytes from astroglial cells in trout. First oligodendrocytes may appear in disconnected areas, sometimes a single cell associated with short myelin segments of a few fibers. Like developing mammalian oligodendrocytes (Hardy and Friedrich, 1996), the trout cells appear distributed in a tract-specific manner.

The timing of appearance of oligodendrocytes in the different brain regions, thus, probably reflects functional maturation of their nerve fibers. The developmental gradients observed suggest that fibers related with brain stem circuits that control body movements (as the FLM, vestibular systems, etc.) are the first to mature, whereas most telencephalic systems have a late maturation. Our observations in trout show both caudorostral and ventrodorsal gradients of oligodendrocyte development, but whether these gradients are due to the origin of these cells from a ventral anlage associated with the notochord or with other factors is not known. Although, occasionally, faintly stained small bipolar cells have been observed in the ventral ventricular layer, indicating that 36K protein begins its expression slightly before than myelination, this does not demonstrates the region(s) of origin of oligodendrocytes or its routes of migration.

CONCLUSIONS

This study indicates that teleosts possess two well-differentiated types of macroglia, expressing different markers in vivo and having different origins and functions. These types show specializations to the different brain regions, which is probably regulated by local factors. The two well differentiated types of glia with morphology similar to those of mammals are also present in elasmobranchs (Klatzo, 1967; Long et al., 1968; Anadón, 1974), and this appears to be a shared characteristic of all gnathostome vertebrates.

ACKNOWLEDGMENTS

The immunocytochemical study of oli-godendrocytes was possible by the generous gift of the 36K antiserum by G. Jeserich. We thank M. Arochena, F. Adrio, M.A. Rodriguez, M.J. Manso, M. Becerra, I. Rodríguez-Moldes and S.E. Pérez for providing their results on mullet and trout, and helpful discussion. Work supported by grants of DGICYT (93-0527C0 201) and Xunta de Galicia (XUGA 20001B93).

REFERENCES

Achúcarro, N. (1915) De l'évolution de la névroglie, et spécialement de ses relations avec l'appareil vasculaire. Trab. Lab. Invest. Biol. Madrid, *13*:169-212.

Aloisi, F., C. Agresti, D. D'Urso and G. Levi (1988) Diferentiation of bipotential glial precursors into oligodendrocytes is promoted by interaction with type-1 astrocytes in cerebellar cultures. Proc. Natl. Acad. Sci. USA *85*:6167-6171.

Alvarez-Buylla, A., Buskrik and F. Nottebohm (1987) Monoclonal antibody reveals radial glia in the adult avian brain. J. Comp. Neurol., *264*:159-170.

Anadón, R. (1974) Núcleos electromotores de la tembladera (*Torpedo marmorata Risso*). I. Lóbulos electricos. Trab. Inst. Cajal Inv. Biol., *66*:103-131.

Anadón, R., F. Adrio, M.A. Rodríguez, G. Jeserich and I. Rodríguez-Moldes (in preparation) Development of oligodendrocytes and myelination in the brain of the brown trout: a 36K myelin protein immunocytochemical study.

Anderson, M.J., K.A. Swanson, S.G. Waxman, and L.F. Eng (1984) Glial fibrillary acidic protein in regenerating teleost spinal cord. J. Histochem. Cytochem., *32*:1099-1106.

Andriezen, W.L. (1893) The neuroglia elements of the brain. Br. Med. J., *2*:227-230.

Antanitus, D.S., B.H. Choi, and L.W. Lapham (1976) The demonstration of glial fibrillary acidic protein in the cerebrum of the human fetus by indirect immunofluorescence. Brain Res., *103*:613-616.

Arévalo, R., J. Alonso, E. García-Ojeda, J.G. Briñón, C. Crespo and J. Aijón (1995) NADPH-diaphorase in the central nervous system of the tench (Tinca tinca L, 1758). J. Comp. Neurol., *352*:398-420.

Bignami, A., and D. Dahl (1974) Astrocyte-specific protein and neuroglial differentiation. An immunofluorescent study with antibodies to the glial fibrillary acidic protein. J. Comp. Neurol., *153*:27-38.

Bignami, A., L.F. Eng, D. Dahl, and C.T. Uyeda (1972) Localization of the glial fibrillary acidic protein in astrocytes by immunofluorescence. Brain Res., *43*: 429-435.

Bitner, C., S. Benjelloun-Touimi, and P. Dupouey (1987) Palisading pattern of subpial astroglial processes in the adult rodent brain: relationship between the glial palisading pattern and the axonal and astroglial organization. Develop. Brain Res., *37*:167-178.

Blakemore, W.F. (1982) Myelination, demyelination and remyelination in the CNS. In N.Th. Smith and J.B. Cavanagh (eds.): Recent Advances in Neuropathology, Vol. 2. Edinburgh: Churchill-Livingstone, pp. 53-81.

Blakemore, W.F. and A.J. Crang (1989) The relationship between type-1 astrocytes, Schwann cells and oligodendrocytes following transplantation of glial cell cultures into demyelinating lesions in the adult rat spinal cord. J. Neurocytol., *18*:519-528

Bodega, G., B. Fernández, I. Suárez, and C. Gianonatti (1988) Glioarchitecture de la moelle épinière du crapaud (Bufo bufo L.): étude au microscope optique avec des techniques d'impregnation métallique. Can. J. Zool., *66*:2415-2420.

Bodega, G., I. Suárez, and B. Fernández (1990) Radial astrocytes and ependymocytes in the spinal cord of the adult toad (Bufo bufo L.). Cell Tiss. Res., *260*:307-314.

Bodega, G., I. Suárez, F. Oteruelo, B. Fernández, and C. Gianonatti (1990) Ultrastructure of the subpial glial limitans in the cerebellum of the lizard (Lacerta lepida). Anat. Anz., *171*:153-158.

Bodega, G., I. Suárez, M. Rubio, and B. Fernández (1990) Distribution and characteristics of the different astroglial cell types in the adult lizard (Lacerta lepida) spinal cord. Anat. Embryol., *181*:567-575.

Bondareff, W., and D.G. McLone (1973) The external glial limiting membrane in macaca: ultrastructure of a laminated glioepithelium. Am. J. Anat., *136*:277-296.

Bozhilova-Pastirova, A., K. Ichev, and V. Ovcharov (1989) Morphology of glial cells in the lower olivary complex of the pigeon (Columba livia). II. Oligodendrocytes. Eksp. Med. Morfol., *28*:26-31.

Braak, E. (1975) On the fine structure of the external glial layer in the isocortex of man. Cell Tiss. Res., *157*:367-390.

Bredt, D., and S.H. Snyder (1992) Nitric oxide, a novel neuronal messenger. Neuron, *8*:3-11.

Bredt, D.S., C.E. Glatt, P.M. Hwang, M. Fotuhi, T.M. Dawson, and S.H. Snyder (1991) Nitric oxide synthase protein and mRNA are discretely localized in neuronal populations of the mammalian CNS together with NADPH diaphorase. Neuron, *7*:615-624.

Bredt, D., P.M. Hwang, and S.H. Snyder (1990) Localization of nitric oxide synthase indicating a neural role for nitric oxide. Nature, *347*:768-770.

Brightman, M.W., and T.S. Reese (1969) Junctions between intimately apposed cell membranes in the vertebrate brain. J. Cell Biol., *40*:648-677.

Bunge, R.P. (1968) Glial cells and the central myelin sheath. Phys. Rev., *48*:197-251.

Cajal, S.R. (1909) Histologie du système nerveux de l'homme et des vertébrés. Paris: Maloine.

Cajal, S.R. (1913) Contribución al conocimiento de la neuroglia del cerebro humano. Trab. Lab. Invest. Biol. Univ. Madrid, *18*:225-315.

Cardone, B., and B.I. Roots (1990) Comparative immunohistochemical study of glial filament proteins (glial fibrillary acidic protein and vimentin) in goldfish, octopus, and snail. Glia, *3*:180-192.

Cohen, A., T. Sivron, R. Duvdevani and M. Schwartz (1990) Oligodendrocyte cytotoxic factor associated to fish optic nerve regeneration: implications for mam-malian CNS regeneration. Brain Res., *537*:24-32.

Dahl, D., D.C. Rueger, A. Bignami, K. Weber, and M. Osborn (1981) Vimentin, the 57,000 dalton protein of fibroblast filaments, is the major cytoskeletal component in immature glia. Eur. J. Cell Biol., *24*:191-196.

Dahl, D., C.J. Crosby, J. Sethi, and A. Bignami (1985) Glial fibrillary acidic (GFA) protein in vertebrates: immunofluorescence and immunoblotting study with monoclonal and polyclonal antibodies. J. Comp. Neurol., *239*:75-88.

Dawson, T.M., D.S. Bredt, M. Fotuhi, P.M. Hwang, and S.H. Snyder (1991) Nitric oxide synthase and neuronal NADPH diaphorase are identical in brain and peripheral tissues. Proc. Natl. Acad. Sci. USA, *88*:7797-7801.

Dermietzel, R. (1974) Junctions in the central nervous system of the cat. II. A contribution to the tertiary structure of the axonal-glial junctions in the paranodal region of the node de Ranvier. Cell Tiss. Res., *148*:577-586.

Dermietzel, R., D. Schünke, and A. Leibstein (1978) The oligodendrocytic junctional complex. Cell Tissue Res., *193*:61-72.

Díaz-Regueira, S.M., and R. Anadón (1995) Fine structure of the medullary lateral line area of Chelon labrosus (Order Perciformes), a nonelectroreceptive teleost. J. Comp. Neurol., *351*:429-440.

Díaz-Regueira, S.M., M. Becerra, and R. Anadón (1992) Light- and electron microscopic study of oligodendrocytes in the lateral line area of the medulla in Chelon labrosus (Teleostei). J. Hirnforsch., *33*:477-485.

Díaz-Regueira, S.M., R. Alvarez-Otero, and R. Anadón (1993) An immunocytochemical and ultrastructural study of a specialized glial region of the medulla oblongata in a teleost, Chelon labrosus. Tiss. Cell, *25*:657-668.

Eisenbarth, G.S., F.S. Walsh, and M. Nirenberg (1979) Monoclonal antibody to a plasma membrane antigen of neurons. Proc. Natl. Acad. Sci. U.S.A., *76*:4913-4917.

Eng, L.F., J.J. Vanderhaeghen, A. Bignami, and B. Gerstl (1971) An acidic protein isolated from fibrous astrocytes. Brain Res. *28*:351-354.

Fan, K., and L. McAlister (1985) Co-culture study of rat neuron-glial interaction: evidence of neuronal in-fluence on myelination. Neurosci. Lett., *59*:111-116.

Fok-Seang, J., and R.H. Miller (1994) Distribution and differentiation of A2B5+ glial precursors in the developing rat spinal cord. J. Neurosci. Res., *37*:219-235.

Franklin, R., A.J. Crang, and W.F. Blakemore (1991) Transplanted type-1 astrocytes facilitate repair of demyelinating lesions by host oligodendrocytes in adult rat spinal cord. J. Neurocytol., *20*:420-430.

Friede, R., K.H. Hu, and M. Johnstone (1969) Glial footplates in the bowfin. I. Fine structure and chemistry. J. Neuropathol. Exp. Neurol., *28*:513-539.
Friedrich, V.L. Jr., and R.A. Lazzarini (1993) Restricted migration of transplanted oligodendrocytes or their progenitors, revealed by transgenic marker M beta P. J. Neural Transplant Plast., *4*:139-146.
Giulian, D., V. Iwanij, and H. Stuckenbrok (1985) The response of optic tract glia during regeneration of the goldfish visual system. I. Biosynthetic activity within different glial population after transection of retinal ganglion cell axons. Brain Res., *339*:87-96.
Hajihosseini, M., T.N. Tham, and M. Dubois-Dalcq (1996) Origin of oligodendrocytes within the human spinal cord. J. Neurosci., *16*:7981-7994.
Hajós, F., and M. Kálmán (1989) Distribution of glial fibrillary acidic protein (GFAP) immunoreactive astrocytes in the rat brain. II. Mesencephalon, rhombencephalon and spinal cord. Exp. Brain Res., *78*:164-173.
Hall, A., N.A. Giese, and W.D. Richardson (1996) Spinal cord oligodendrocytes develop from ventrally derived progenitor cells that express PDGF alpha-receptors. Development, *122*:4085-4094.
Hardy, R.J., and V.L. Friedrich (1996) Oligodendrocyte progenitors are generated throughout the embryonic mouse brain, but differentiate in restricted foci. Development, *122*:2059-2069.
Hendon, R.M. (1964) The fine structure of the rat cerebellum. II. The stellate neurons, granule cells and glia. J. Cell. Biol., *23*:277-293.
Holmqvist, B., T. Östholm, P. Alm, and P. Ekström (1994) Nitric oxide synthase in the brain of a teleost. Neurosci. Lett., *171*:205-208
Hope, B., G.J. Michael, K.M. Knigge, and S.R. Vincent (1991) Neuronal NADPH diaphorase is a nitric oxide synthase. Proc. Natl. Acad. Sci. USA, *88*:2811-2814.Horstmann, E. (1954) Die Faserglia des Selachiergehirns. Z. Zellforsch., *39*:588-617.
Imamoto, K., J.A. Paterson, and C.P. Leblond (1978) Radioautographic investigation of gliogenesis in the corpus callosum of young rats. I. Sequential changes in oligodendrocytes. J. Comp. Neurol., *180*:115-138.
Jeserich, G., and A. Stratmann (1992) In vitro differentiation of trout oligodendrocytes: evidence for an A2B5-positive origin. Develop. Brain Res., *67*:27-35.
Jeserich, G., and T.V. Waehneldt (1986) Characterization of antibodies against major fish CNS myelin proteins: immunoblot analysis and immunohistochemical localization of 36K and IP2 proteins in trout nerve tissue. J. Neurosci. Res., *15*:147-158.
Jeserich, G., A. Müller, and C. Jacque (1990) Developmental expression of myelin proteins by oligodendrocytes in the CNS of trout. Dev. Brain Res., *51*:27-34.
Kagawa, T., A. Oba, S. Okumura, and K. Ikenaka (1996) Localization of mRNA for UDP-galactose: ceramide galactosyltransferase in brain during mouse development. Dev. Neurosci., *18*:309-318.
Kettenmann, H., and B.R. Ransom (1988) Electrical coupling between astrocytes and between oligodendrocytes studied in mammalian cell cultures. Glia, *1*:64-73.
Kimelberg H.K., and H. Kettenmann (1990) Swelling-induced changes in electrophysiological properties of cultured astrocytes and oligodendrocytes. I. Effects on membrane potentials, input impedance and cell-cell coupling. Brain Res., *529*:255-261.
Klatzo, I. (1967) Cellular morphology of the lemon shark brain. In P.W. Gilbert, R.F. Mathewson and Rall (eds.): Sharks, skates and Rays. Baltimore: John Hopkins Press, pp. 341-359.
Kruger, L., and D.S. Maxwell (1966) The fine structure of ependymal processes in the teleost optic tectum. Am. J. Anat., *119*:479-498.
Kruger, L., and D.S. Maxwell (1967) Comparative fine structure of vertebrate neuroglia: Teleosts and reptiles. J. Comp. Neurol., *129*:115-142.
Lara, J.M., J.R. Alonso, E. Vecino, R. Coveñas, and J. Aijon (1989) Neuroglia in the optic tectum of teleosts. J. Hirnforsch., *30*:465-462.
Lara, J.M., A. Velasco, J.R. Alonso, and J. Aijon (1995) Neuroglia in the CNS of teleosts. In A. Vernadakis and B. Roots (eds.): Neuron-glia interrelations during phylogeny:. I. Phylogeny and ontogeny of glilal cells. Totowa (NJ): Humana Press, pp. 131-156.
Laufer, M. and H. Vanegas (1974) The optic tectum of a perciform teleost. II. Fine structure. J. Comp. Neurol. *154*:61-69.

Lázaro, A. (1991) Desarrollo postnatal de la astroglia medular en ratón blanco Apodemus apodemus. Doctoral Thesis. Madrid: Universidad Complutense.
Leonhardt, H. (1980) Ependym und circumventriculäre Organe. In A. Oksche, and L. Vollrath (eds): Handbuch der mikroskopischen Anatomie des Menschen, vol 4/10, Neuroglia I. Berlin: Springer, pp 177-666
Levine, R.L. (1989) Organization of astrocytes in the visual pathways of the goldfish: an immunocytochemical study. J. Comp. Neurol., *285*:231-245.
Levitt, P. and P. Rakic (1980) Immunoperoxidase localization of glial fibrillary acidic protein in radial glial cells and astrocytes of the developing rhesus monkey brain. J. Comp. Neurol. *193*:815-840.
Ling, E.A. and C. Leblond (1973) Investigation of glial cells in semithin sections. II. Variations with age in the numbers of the various glial cell types in rat cortex and corpus callosum. J. Comp. Neurol. *149*:73-82.
Ling, E., J.A. Paterson, A. Privat, S. Mori, and C.P. Leblond (1973) Investigation of glial cells in semithin sections. I. Identification of glial cells in the brain of young rats. J. Comp. Neurol., *149*:43-72.Long, D.M., T.S. Bodenheimer, J.F. Hartmann, and I. Klatzo (1968) Ultrastructural features of the shark brain. Am. J. Anat., *122*:209-236.
Lotan, M., and M. Schwartz (1992) Postinjury changes in platelet-derived growth factor-like activity in fish and rat optic nerves. J. Neurochem., *58*:1637-1642.
Ludwin, S. (1984) The function of perineuronal satellite oligodendrocytes: an immunohistochemical study. Neuropathol. Appl. Neurobiol., *10*:143-149.
Ma, P.M. (1993) Tanycytes in the sunfish brain: NADPH-diaphorase histochemistry and regional distribution. J.Comp. Neurol., *336*:77-95.
Maggs, A., and J. Scholes (1990) Reticular astrocytes in the fish optic nerve: macroglia with epithelial characteristics form an axially repeated lacework pattern, to which nodes of Ranvier are apposed. J. Neurosci., *10*:1600-1614.
Manso, M.J., M. Becerra, M. Becerra, and R. Anadón (1997) Expression of a low-molecular-weight calcium binding protein (9-10k) in glial cells of the brain of the trout (Teleosts). Anat. Embryol (in press)
Marcus, R.C., and S.S. Easter (1995) Expression of glial fibrillary acidic protein and its relation to tract formation in embryonic zebrafish (Danio rerio). J. Comp. Neurol., *359*:365-381.
Matus, A. and S. Mughal (1975) Immunohistochemical localization of S-100 protein in brain. Nature, *258*:746-748.
Massa, P. and E. Mugnaini (1982) Cell junctions and intramembrane particles of astrocytes and oligodendrocytes: a freeze-fracture study. Neuroscience, *7*:523-538.
Monteiro, R. (1983) Do the Purkinje cells have a special type of oligodendrocyte as satellites?. J. Anat., *137*:71-83.
Monzon-Mayor, M., C. Yanes, J.L. James, and R.R. Sturrock (1990) An ultrastructural study of the development of oligodendrocytes in the midbrain of the lizard. J. Anat., *170*:43-49.
Moore, B.W. (1965) A soluble protein characteristic of the nervous system. Biochem. Biophys. Res. Commun., *6*:739-744.
Morales, R., and D. Duncan (1975) Specialized contacts of astrocytes with astrocytes and with other cell types in the spinal cord of the cat. Anat. Rec., *182*:255-266.
Mori, S., and C.P. Leblond (1970) Electron microscopic identification of three classes of oligodendrocytes and a preliminary study of their proliferative activity in the corpus callosum of young rats. J. Comp. Neurol., *139*:1-30.
Mugnaini, E. and F. Walberg (1964) Ultrastructure of neuroglia., Ergeb. Anat. Entwicklungsgesch *37*:194-236.
Nabeshima, S., T.S. Reese, D.M.D. Landis, and M.W. Brightman (1975) Junctions in the meninges and marginal glia. J. Comp. Neurol., *164*:127-170.
Noll, E., and R.H. Miller (1993) Oligodendrocyte precursors originate at the ventral ventricular zone dorsal to the ventral midline region in the embryonic rat spinal cord. Development, *118*:563-573.
Nona, S.N., S.A.S. Shehab, C.A. Stafford, and J.R. Cronly-Dillon (1989) Glial fibrillary acidic protein (GFAP) from goldfish: its location in visual pathway. Glia, *2*:189-200.

Ogawa,Y., S.Eins and J.Wolff (1985) Oligodendrocytes in the pons and middle cerebellar peduncle of the cat. Topographical relations to neurons and transverse axon bundles. Cell Tiss. Res. *240*:541-552.

Ono, K., R. Bansal, J. Payne, U. Rutishauser, and R.H. Miller (1995) Early development and dispersal of oligodendrocyte precursors in the embryonic chick spinal cord. Development, *121*:1743-1754.

Onteniente, B., H. Kimura, and T. Maeda (1983) Comparative study of the glial fibrillary acidic protein in vertebrates by PAP immunohistochemistry. J. Comp. Neurol., *215*:427-436.

Orentas, D.M. and R.H. Miller (1996) The origin of spinal cord oligodendrocytes is dependent on local influences from the notochord. Dev. Biol., *177*:43-53.

Parnavelas, J.G., R. Luder, S.G. Pollard, K. Sullivan, and A.R. Lieberman (1983) A qualitative and quantitative ultrastructural study of glial cells in the developing visual cortex of the rat. Phil. Trans. R. Soc. London B, *301*:55-84.

Paterson, J.A. (1981) Postnatal development of oligodendrocytes. In S. Fedoroff (ed.): XI International Congress of Anatomy, Part A. Glial and Neuronal Cell Biology. New York: Alan R. Liss, pp. 83-92.

Paterson, J.A., A. Privat, E.A. Ling, and C.P. Leblond (1973) Investigation of glial cells in semithin sections. III. Transformation of subependymal cells into glial cells, as shown by radioautography after 3H-thymidine injection into the lateral ventricle of the brain of young rats. J. Comp. Neurol., *149*:83-102.

Pérez, S.E., F. Adrio, M.A. Rodríguez, I. Rodríguez-Moldes, and R. Anadón (1996) NADPH-diaphorase histochemistry reveals oligodendrocytes in the rainbow trout (Teleost). Neurosci. Lett., *205*:83-86.

Peter, R.E. and Y. Nagahama (1976) Light and electron microscopic study of the structure of the nucleus preopticus and nucleus lateral tuberis of the goldfish Carassius auratus. Can. J. Zool., *54*:1423-37.

Peters, A., K. Josephson, and S.L. Vincent (1991) Effects of aging on neuroglial cells and pericytes within area 17 of the rhesus monkey cerebral cortex. Anat. Rec., *229*:384-398.

Peters, A., S.L. Palay, and H. de F. Webster (1976) Fine Structure of the Nervous System: The Neurons and Supporting Cells. Philadelphia: W. Saunders.

Pixley, S. R., and J. De Vellis (1984) Transition between immature radial glia and mature astrocytes studied with a monoclonal antibody to vimentin. Develop. Brain Res., *15*:201-209.

Privat, A., M. Giménez-Ribotta, and J.-L. Ridet (1995) Morphology of astrocytes. In H. Kettenmann and B.R. Ramsom (eds.): Neuroglia. New York: Oxford University Press, pp. 3-22.

Raamsdonk, W. Van, C.Heyting, C.W.Pool, M.J.Smit-Onel, and J.L. Groen (1984) Differentiation of neurons and radial glia in the spinal cord of the teleost Brachydanio rerio (the zebrafish): An immunocytochemical study. Int. J. Dev. Neurosci., *2*:471-481.

Raff, M.C. R.H. Miller, and M. Noble (1983) A glial progenitor cell that develops in vitro into an astrocyte or an oligodendrocyte depending on culture medium. Nature, *303*:390-396.

Rakic,P. (1971a) Guidance of neurons migrating to the fetal monkey neocortex. Brain Res., *33*:471-476.

Rakic, P. (1971b) Neuron-glia relatioship during granule cell migration in developing cerebellar cortex: a Golgi and electron-microscopic study in Macacus rhesus. J. Comp. Neurol., *141*:283-312.

Rakic, P. (1972) Mode of cell migration to the superficial layers of fetal monkey neocortex. J. Comp. Neurol., *145*:61-84.

Ransom, B.R., and H. Kettenmann (1990) Electrical coupling, without dye coupling, between mammalian astrocytes and oligodendrocytes in cell culture. Glia, *3*:258-266.

Remahl,S., and C.Hildebrand (1990) Relation between axons and oligodendroglial cells during initial myelination. I. The glial unit. J. Neurocytol., *19*:313-328.

Río Hortega, P. del (1919) El tercer elemento de los centros nerviosos. Bol. Soc. Esp. Biol., *9*:68-83.

Río Hortega, P. del (1921) Estudios sobre la neuroglia. La glía de escasas radiaciones (oligodendroglia). Bol. Real Soc. Esp. Hist. Nat., *21*:63-92.

Río Hortega, P. del (1928) Tercera aportación al conocimiento morfológico e interpretación funcional de la oligodendroglia. Mem. Real Soc. Esp. Hist. Nat.,*14*:5-122.

Roots, B.I. (1978) A phylogenetic approach to the anatomy of glia. In "Dynamic Properties of Glial Cells" (E. Schoffeniels, B. Franck, L. Hertz and D.B. Towers, eds.) Pergamon. New York. pp. 45-54

Rubio, M., I. Suárez, G. Bodega, and B. Fernández (1992) Glial fibrillary acidic protein and vimentin immunohistochemistry in the posterior rhombencephalon of the Iberian barb (Barbus comiza). Neurosci. Lett., *134*:203-206.

Scherer-Singler, U., S.R. Vincent, H. Kimura, and E.G. McGeer (1983) Demonstration of a unique population of neurons with NADPH-diaphorase histochemistry. J. Neurosci. Meth., *9*:229-234.

Schnitzer,J., W.Franke and M.Schachner (1981) Immunocytochemical demonstration of vimentin in astrocytes and ependymal cells of developing and adult mouse nervous system. J. Cell Biol., *90*:435-47.

Sivron, T., I. Cohen, and M. Schwartz (1994) Intermediate filaments reminiscent of immature cells expressed by goldfish (Carassius auratus) astrocytes and oligodendrocytes in vitro. Cell Tiss. Res., *275*:327-337.

Stensaas, L.J., and S.S. Stensaas (1968) Astrocytic neuroglial cells, oligodendrocytes and microgliacytes in the spinal cord of the toad. II. Electron microscopy. Z. Zellforsch., *86*:184-213.

Stensaas, L (1977) The ultrastructure of astrocytes, oligodendrocytes, and microglia in the optic nerve of urodele amphibians (A. punctatum, T. pyrrhogaster, T. viridescens). J. Neurocytol., *6*:269-286.

Stevenson, J.A., and M.G. Yoon (1982) Morphology of radial glia, ependymal cells, and periventricular neurons in the optic tectum of goldfish (Carassius auratus). J. Comp. Neurol., *205*:128-138.

Suárez Nájera, I., B. Fernández Ruiz and L. M. García Segura (1980) Specialized contacts of astrocytes with astrocytes and with other cell types in the hypothalamus of the hamster. J. Anat., *130*:55-61.

Szuchet, S. (1995) The morphology and ultrastructure of oligodendrocytes and their functional implications. In H. Kettenmann and B.R. Ramsom (eds.): Neuroglia. New York: Oxford University Press, pp. 23-43.

Szuchet, S., S.H. Yim, and S. Monsma (1983) Lipid metabolism of isolated oligodendrocytes maintained in long-term culture mimics events associated with myelinogenesis. Proc. Natl. Acad. Sci. USA., *80*:7019-23.

Szuchet, S., P.E. Polak, and S.H. Yim (1986) Mature oligodendrocytes cultured in the absence of neurons recapitulate the ontogenic development of myelin. Dev. Neurosci., *8*:208-221.

Tabira, T., M. Cullen, P.J. Reier, and H. de F. Webster (1978) An experimental analysis of interlamellar tight junctions in amphibian and mammalian C.N.S. myelin. J. Neurocytol., *7*:489-503.

Tani, E., T. Itagaki, and M. Nakano (1977) Tight junctions of oligodendrocytes. Cell Tiss. Res. *184*:139-42.

Tapp, R.L. (1973) The structure of the optic nerve of the teleost Eugerres plumieri. J. Comp. Neurol., *150*:239-252.

Thomas, E., and A. E. Pearse (1961) The fine localization of dehydrogenases in the central nervous system. Histochemistry, *2*:266-282.

Vaughan, D.W. (1984) The structure of neuroglial cells. In E.G. Jones and A. Peters, (Eds.): Cerebral Cortex, Vol. 2. New York: Plenum, pp. 285-329.

Vaughan, D.W., and A. Peters (1974) Neuroglial cells in the cerebral cortex of rats from young adulthood to old age: an electron microscopic study. J. Neurocytol., *3*:405-429.

Vaughn. J. (1969) An electron microscopic analysis of gliogenesis in rat optic nerve. Z. Zellforsch., *94*:293-327.

Vaughn, J.E., and A. Peters (1968) A third neuroglial cell type: an electron microscopic study. J. Comp. Neurol., *133*:269-288.

Vaysse, P-J., and J.E. Goldman (1990) A clonal analysis of glial lineages in neonatal forebrain development in vitro. Neuron, *5*:227-235.

Verity, A.N., and A.T. Campagnoni (1988) Regional expression of myelin protein genes in the developing mouse brain: in situ hybridization studies. J. Neurosci. Res., *21*:238-248.

Villani, L., and T. Guarnieri (1995) Ultrastructural localization of NADPH-diaphorase in the goldfish brain. Brain Res., *705*:332-336.

Vincent,S., O.Johansson, T.Hökfelt, L.Skirboll, RElde, L. Terenius, J. Kimmel, and M. Goldstein (1983) NADPH-diaphorase: selective histochemical marker for striatal neurons containing both somatostatin and avian pancreatic polypeptide (APP)-like immunoreactivities. J. Comp. Neurol., *217*:252-263.

Warf, B.C., J. Fok-Seang, and R.H. Miller (1991) Evidence for the ventral origin of oligodendrocyte precursors in the rat spinal cord. J. Neurosci., *11*:2477-2488.

Wolburg, H., R. Kästner, and G. Kurz-Isler (1983) Lack of orthogonal particle assemblies and presence of tight junctions in astrocytes of the goldfish (*Carassius carassius*). Cell Tissue Res., *234*:389-402.

Zamora, A. and M. Mutin (1988) Vimentin and glial fibrillary acidic protein filaments in radial glia of the adult urodele spinal cord. Neuroscience, *27*:279-288.

3

ORGANIZATION AND FUNCTION OF ASTROCYTES IN HUMAN RETINA

José M. Ramírez, Alberto Triviño,
Ana I. Ramírez and Juan J. Salazar

The functioning of the central nervous system (CNS) has been attributed to the neurons, but it is known that the astrocytes perform numerous activities which ensure that the neurons function and survive. The retina is a portion of the CNS that is composed of a large number of neurons and two types of macroglial cells: astrocytes and Müller cells. Both types perform similar functions, but the Müller glia are located troughout the retina while the astrocytes only appear in the vascular zones. The morphology and distribution of astrocytes varies according to the specie, the greatest morphological variety occurring in the rabbit retina. Because of the peculiar arrangement of the (epiretinal) blood vessels, the rabbit requires a specific type of astroglia to anchor these to the retinal parenchyma.

Human retinal astrocytes acquire two morphologies: elongated and star-shaped. Star-shaped astrocytes are located between the ganglion cell layer and the inner nuclear layer; they lie close to or over the blood vessels, their processes joining to form a honeycomb structure. Elongated astrocytes lie over the nerve fiber layer and follow the course of the axons, although they also send processes to the capillaries in this zone. Both astrocytes are variants or subclasses of a single type of astrocyte, specializing in different ways. Star-shaped astrocytes specialize more in relations with the blood vessels and elongated astrocytes with the axons.

The functioning of the Nervous System has generally been attributed to the activity of the neurones, which have been seen as "noble cells" having the role of processing information by exchanging electrical signals. However, a major portion of the cells of the Nervous System are not neuronal but glial cells, and of these astrocytes are the most numerous group (Kimelberg and Norenberg, 1989).

It is now known that the astrocytes perform a variety of essential roles for the normal phy-siology of the nervous system and that they maintain a close and permanent relationship with the neurones (Kuffler and Nicholls, 1976). These roles include the following:

Role of the astrocytes in neuronal metabolism: When the metabolic balance is disturbed by neuronal activity, the astrocytes restore equilibrium in all its aspects (energy metabolism, metabolism of neurotransmitters and regulation of ions and metabolites) enabling neuronal activity to be continued or repeated.

The astrocytes store glycogen in their cytoplasm and release glucose to the neurones as required (Ripps and Witkovsky, 1985).

They also capture the carbon dioxide (CO_2) generated by neuronal metabolism, converting it to bicarbonate (HCO_3^-) by means of the enzyme carbonic anhydrase (Muser and Rosen, 1973; Kumpulainen, 1980).

At the same time they play an important role in the uptake and recycling of various neurotransmitters released into the synaptic cleft. These include glutamate, aspartate, GABA, taurine and others (Lam and Hollyfield, 1980; Sato et al., 1991; Kalloniatis, 1995).

These also possess the capacity to control the level of substrate availabe for neuronal protein synthesis, since the level of accumulated free aminoacids is higher in the glial than the neuronal cells (Kimelberg and Kaltz, 1985; Hamberger, 1971).

Finally, it is known that the astroglial cells have a role in the ionic composition of the region around the neurones, removing the potassium (K^+) that accumulates as a result of neuronal activity (Trachtenberg, 1970; Ripps and Witkovsky, 1985; Karwoski et al., 1989). These cells further present a large variety of ionic channels, incluiding channels for Na^+, Ca^{2+}, Cl^- (Newman, 1986a; Berwald-Netter et al., 1986; MacVicar et al., 1991; Ritchie, 1992).

Another function of the astrocytes is to guide the neurones during development. Astrocytes or their cellular precursors further posses the capacity to secrete growth factors (e.g. neuronal growth factor, laminine and fibronectine) (Liesi and Silver, 1988) which aid neuronal survival and enable neurones to grow and differentiate (Thano and Thiel, 1990; Chang-Ling, 1994).

These cells also secrete fibroblastic growth factor (FGF) and transforming growth factor-ß (TGF-ß) which modulate angiogenesis (Jiang et al., 1994; Jiang et al., 1995).

The astrocytes also have a role in pathological processes in the Nervous System. Whenever damage occurs and there is neuronal destruction, the astrocytes proliferate to form a glial scar (Lindsay, 1986) and they are capable of phagocytosis (Mano and Puro, 1990). These reactive cells present more processes and intermediate filaments than normal and a higher level of metabolic activity (Cook and Wisniewski, 1973; Dusart et al., 1991). Cell proliferation may be caused by the release of growth factors which trigger multiplication of astrocytes (Nieto-Sampedro et al., 1985).

Finally, the astrocytes induce the blood-brain barrier and blood-retinal barrier (Janzer and Raff, 1987; Risau and Wolburg, 1990). The characteristics of these barriers are determined by the junctions between the endothelial cells of the cerebral or retinal capillaries across these cells. These barrier properties are induced in the endothelial cells of cerebral capillaries by the astrocytes that form their glia limitants. However, the nature of this induction signal, whether it requires cell contact or is a soluble factor is still unresolved (Stewart and Coomber, 1986; Lobrinus et al., 1992).

All these astrocyte functions enable us to view the response of the Central Nervous System as a functional whole comprising the following types (Sole et al., 1992): (i) neuronal response: immediate and effect-producing, based on electrical phenomena; and (ii) astroglial response: delayed, serving to modulate and restore, based on metabolic phenomena.

The two main types of macroglial cell in the retina are the Müller cell and the astrocyte. Although both are neuroectodermal, their origins are different. The Müller glia are generated by retinal neuroepithelial cells and may share a common progenitor with the neurones (Turner and Cepko, 1987), whereas the astrocytes are generated outside the retina and invade it via the optic nerve, at around the 13th week of gestation (Vaca and Wend, 1992).

The astrocytes enter the retina along the route of the spindle cells, which are the mesenchymatous precursors of retinal vascularization (Gariano et al., 1996a) (fig. 3-1). It has been established that these precursors do not occur in animal species possesing an avascular retina, and again astrocytes do not occur in retinas lacking these precursors. This situation is readily verified in animal species having a partially vascular retina, such as the rabbit, where the astrocytes are restricted to the medullary nerve fiber region, which is the only vascularized part of the retina (Schnitzer, 1985; Schnitzer and Karschin, 1986; Schintzer, 1987, 1988a, Triviño et al., 1992).

In all retinas generally, the astrocytes do not invade regions where there are no vascular cell precursors, such as the fovea and the outermost layers of the retina (Gariano et al, 1996a, 1996b).

The vascular precursors attract the astrocytes to the retina, where the latter cells induce the expression of fibronectine in the endothelial cells and vasculogenesis is initiated (Jiang et al., 1994, 1995) (fig. 3-1). For this to happen, there must first be metabolic demand, giving rise to hypoxia and hence to the secretion of angiogenic factors such as: fibroblastic growth factors (FGF), the inhibiting transforming growth factor beta (TGF-ß) or vascular endothelial growth factor (VEGF), which is synthesized by the astrocytes and the Müller glia (Chang-Ling, 1994) (fig. 3-1).

The ocurrence of metabolic demand and hypoxia is not sufficient to produce astrocyte degeneration, but it is sufficient for new vessels to form; these will be surrounded by astrocyte processes which induce the formation of blood-retinal barrier glia (Chang-Ling, 1994) (fig. 3-1).

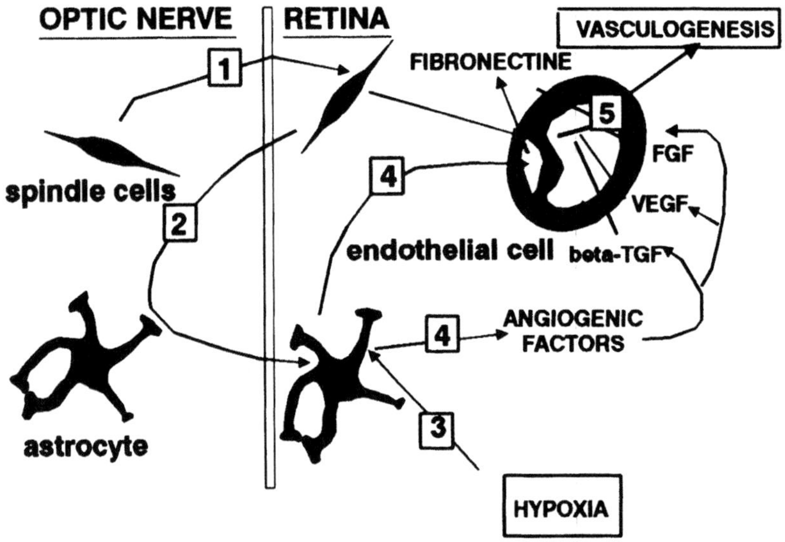

Figure 3-1. Origin of retinal astrocytes.

MORPHOLOGY OF HUMAN RETINAL ASTROCYTES

In terms of ultrastructure, human retinal astrocytes resemble other astrocytes in the Central Nervous System. They possess a clear cytoplasm in which there are abundant glycogen particles and intermediate glial filaments of 10 nm (fig. 3-2A).

These cells differ from the Müller glia in, that their electron density is lower. They have a relatively dark nucleus with abundant chromatin and heterochromatin accumulations (figs. 3-2A, B). Their cytoplasms contain a variety of cytoplasmic organelles, including: rough endoplasmic reticulum, free ribosomes, mitochondria, Golgi apparatus, glycogen particles and dense packs of intermediate glial filaments. In older people, astrocytes also contain dense bodies and lipofuscin. In their processes, the cytoplasmic organelles are reduced to a few intermediate filaments and a few mitochondria (Ramírez et al., 1996) (fig. 3-2A). Glial filaments are composed of a protein of 50 kD (glial fibrillary acidic protein, GFAP) (Liem et al., 1978, 1981). The GFAP antibody selectively binds to the glial filaments of the astrocytes and constitutes a good marker for these (Bignami et al., 1972; Ghandour et al., 1979).

In morphological terms, there are two types of astrocyte in the human retina: star-shaped astrocytes and elongated astrocytes (Ramírez et al., 1994).

Star-shaped astrocytes are cells possessing high GFAP immunoreactivity (fig. 3-2C). They have a small cell body from which an ovoid nucleus protrudes. From 4 to 8 primary processes issue from the cell body and subdivide into secondary processes of smaller calibre. In most astrocytes the processes issue from the cell body, lending the astrocyte a star-like appearance. These processes, which can be very long, make contact with other astrocytes or with the blood vessels to form an astroglial plexus with the appearance of a honeycomb. These cells are essentially located in the ganglion cell layer (GCL), although processes can be found entering the retinal parenchyma vertically and running with the blood vessels up to the internal nuclear layer (INL). The authors have observed star-shaped astrocytes in the nerve fiber layer (NFL) in the region of the horizontal raphe temporal to fovea along the horizontal median (Ramírez et al., 1994).

Elongated astrocytes likewise possess high GFAP immunoreactivity (fig. 3-2D). The cell body is small and oval with a small protruding ovoid nucleus. From 2 to 4 long, straight primary processes issue from the cell body and follow the course of the ganglion cell axons. There can also be other short processes issuing from the cell body (fig. 3-2D); these either penetrate the retinal parenchyma in the direction of the GCL or else extend perpendicularly from the cell body and run towards other nearby astrocytes. The cell body of these astrocytes is located in the NFL between the axon bundles and the longitudinal radial peripapillary capillaries described by Michaelson and Campbell (1940) and Henkind (1967, 1969). The processes from these cells are not normally arranged individually but tend to join up with others to form bundles of astroglial processes which separate the axon bundles running to the optic disk. These processes also make contact with the capillaries in this layer (Ramírez et al., 1994).

The morphological types of astrocyte found in the human retina have also been observed in other animal species such as monkeys and cats (Karschin et al., 1986a, 1986b; Schnitzer, 1988b; Chang-Ling and Stone, 1991a, 1991b). On the other hand, in the rat retina (Stone and Dreher, 1987) the astrocytes are mainly star-shaped, although they may exhibit a tendency to follow the course of the axons in the vicinity of the optic disk, where the axon bundles are thicker.

The greatest variety of retinal astroglial cell morphologies is found in the rabbit. This animal possesses two large astrocyte groups: one associated with the ganglion cell axons, and the other associated with the vascular system (Schnitzer,

1985; Schnitzer and Karschin, 1986; Schnitzer, 1987; Triviño et al., 1990, 1992, 1997) (fig. 3-2D). The astrocytes associated with the axons fall into three subgroups: astrocytes with thick processes forming astroglial bundles, astrocytes with thin processes, and intermediate astrocytes. The last two do not align with the axon bundles (Triviño et al, 1997).

In the group of astrocytes associated with the vascular system we find three morphologically differentiated types (Triviño et al., 1992, 1997): Type I astrocytes, equivalent to the perivascular astrocytes described by Wolter (1955-1961) in the human retina (fig. 3-2E); these cells are located over the medium-sized blood vessels and capillaries, surrounding them with processes without making contact with other structures. Type II astrocytes, which are star-shaped with a basal cone from which processes issue towards the medium and large blood vessels. And finally, type III astrocytes; these are star-shaped astrocytes whose processes emerge directly from the cell body and arrange themselves among the capillaries of the GCL to form a network.

The morphology of retinal astrocytes in different animal species is determined by the way their processes adapt to the surrounding structures (Chan-Ling and Stone, 1991a, b; Holländer et al., 1991; Ramírez et al., 1994). Thus, the astrocytes located in the NFL, which associate essentially with the longitudinal ganglion cells, are of the elongated type, whereas the astrocytes located in the GCL, which form interconnections with the various blood vessels, tend to be star-shaped.

To account for the elongated morphology of the NFL astrocytes, Büssow (1980) suggested that the monkey and cat astroglia might be a cell type peculiar to the ganglion cell axons. Karschin, Wässle and Schnitzer (1986a, b) supported this theory, observing that the elongated astrocytes in the cat retina became star-shaped after provoking axon degeneration; they therefore suggested that the characteristic star shape of the astrocyte is in fact a "lack of shape" which the cells adopt in the absence of other structures (Chan-Ling and Stone, 1991a, b). The force which determines elongated morphology in NFL astrocytes could be the affinity between astrocyte processes and axon specializations equivalent to nodes (Holländer et al., 1991). Each axon bundle is surrounded by a continuous set of astrocytes over its entire course (Chan-Ling, 1994). This agrees with findings in the rat retina, whose NFL astrocytes are star-shaped (Stone and Dreher, 1987), in that the connections with structures equivalent to axon nodes are realized by the Müller glia (Holländer et al., 1991).

Astrocytes also have a pronounced affinity for the blood vessels and can divert their processes towards these, adopting a less star-like shape in so doing. It has been demonstrated that following degeneration of all the retinal blood vessels in kittens exposed to an atmosphere of 70-80% oxygen for 40 days and degeneration of all the ganglion cell axons induced by a nerve crush at birth, the astrocytes adopt the classic star-shaped morphology (Chan-Ling and Stone, 1991a, b; Chan-Ling, 1994). We therefore believe that all retinal astrocytes may be variations or subclasses of the same atroglial type but specializing in different degrees, so that the elongated astrocytes are more specialized in associating with the axons, while the more star-shaped astrocytes (located in the GCL) tend to associate more with the blood vessels (Ramírez et al., 1994).

Wolter (1955-1961) described a special kind of astroglial cell in the human retina which associated only with the blood vessels; he called these perivascular glia. These cells are situated over the blood vessels, surrounding them with processes without contacting any other structures. Most subsequent authors assumed that these cells existed in the human retina (Erskine, 1963) however, since no structural differences were observed in humans between astrocytes located over the blood vessels and astrocytes distant from them, researchers began to doubt the existence of such cells (Hogan et al., 1971; Ikui et al., 1976).

Subsequent research on the rabbit retina using immunohistochemical techniques detected the presence of perivascular glia (Schnitzer, 1985, Schnitzer et al., 1986, 1988a, b; Robinson and Dreher, 1989, Triviño et al., 1992, 1997) (fig. 3-2E), thus implicitly supporting their existence in the human retina. Research carried out at this laboratory using both immunohistochemical techniques and electron microscopy (Ramírez et al., 1994, 1996; Triviño et al, 1997) detected no "Wolter perivascular astrocytes". We believe that the silver technique used by Wolter (1955-1961), which would stain the cell surface, could have marked the web of glial processes over the blood vessels, lending them a cell-like appearance which could have influenced his interpretation of the images. Moreover, no author has observed Wolter's perivascular glia in retinas whose astroglial distribution resembles that of humans (cat, monkey, rabbit) (Schnitzer, 1988b, 1988d; Thano and Thiel, 1990; Chan-Ling and Stone, 1991a, b; Holländer et al., 1991). The fact that these cells have been found only in the rabbit retina may be simply explained by the arrangement of the blood vessels. The blood vessels in the rabbit retina are located epiretinally (Hyvärinen, 1967; Triviño et al., 1988, 1991), whereas in man, monkey and cat their location is intraretinal (Michaelson, 1954; Toussaint et al., 1961). This epiretinal vascular arrangement favours the appearance of glial types specializing in the anchoring of blood vessels to the retinal parenchyma (Robinson and Dreher, 1990; Triviño et al., 1992, 1997).

Thus, studies of the rabbit retina have revealed three types of astroglial cell associated with the vascular system (Triviño et al., 1992, 1997). Type I (fig. 3-2E) and type II astrocytes, anchoring the blood vessels in the vitreous humor to the retina (hence the true perivascular astrocytes), and type II astrocytes, located in the retinal parenchyma, which make contact with the intraparenchymatous retinal capillaries. This last cell type would be the equivalent of the star-shaped astrocytes in the ganglion cell layer of the human retina. The existence of "perivascular glia" as vascular anchoring cells would therefore make no sense in retinas with intraretinal vascularization. These cells occur in pathologies like retinopathy of prematurity or in the epiretinal membranes (Hiscott et al., 1984; Hui et al., 1988; Sramek et al., 1989) where there are intravitreous vessels.

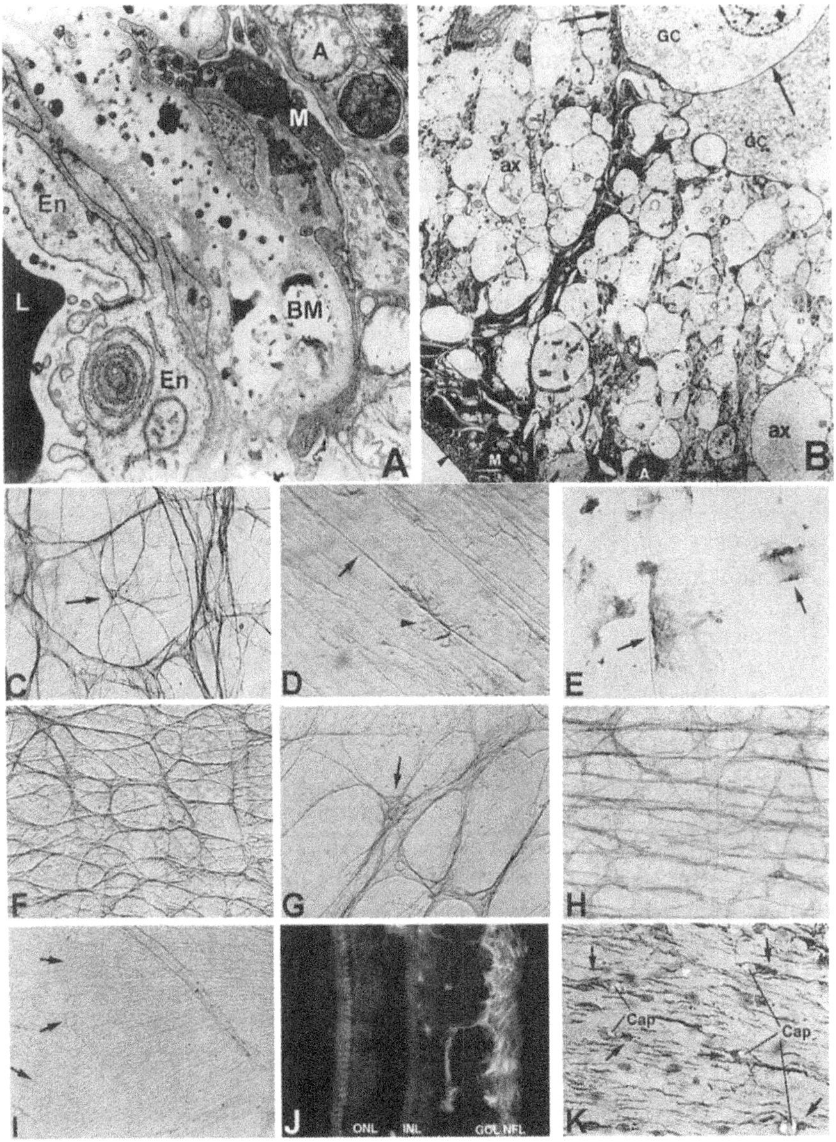

RELATIONSHIP BETWEEN ASTROCYTES AND THE VARIOUS RETINAL STRUCTURES

The processes of human retinal astrocytes tend to join together to form astroglial process bundles. The distribution of these bundles varies according to the retinal layer in which they are located.

In the GCL, astroglial bundles interconnect with the various blood vessels, forming a horizontal plexus with a honeycomb morphology (fig. 3-2F). The ganglion cell somas are located inside the honeycomb cells (Triviño et al, 1997; Ramírez et al., 1994). The ganglion cells may lie adjacent to one another, but they are surrounded by glial processes, belonging largely to the Müller glia, which form a narrow dividing line between the neurones (fig. 3-2B). However, there are also astrocyte processes, located between the neurones but without forming a continuous envelope (Ramírez et al., 1996). Höllander (1991) suggested that the Müller glia was the cell that specialized in isolating the ganglion cell somas in a myelin-free environment. The astroglial processes forming the bundles join up by means of desmosomes (Ikui et al., 1976; Ramírez et al., 1996) or gap-junctions (Ramírez et al., 1996). Moreover, the astrocytes also maintain contact with the Müller glia; desmosomes have been observed between the two cell types (Ramírez et al., 1996). Hence, the astroglial processes form structures which are interconnected both with one another and with the Müller glia and could serve to strengthen the capillary mesh and support the neurones in the mesh (Ikui et al., 1976).

Astrocytes located in this layer maintain a close association with the blood vessels. The processes from these cells surround the blood vessels in a dense mesh, forming crook-shaped structures parallel to them (Wolter, 1955b, 1961; Ramírez et al., 1994) (fig. 3-2F). Electron microscopy has shown that the glial processes surrounding the blood vessels belong to both astrocytes and Müller glia (Hogan et al., 1971; Ikui et al., 1976; Penfold et al., 1990; Provis et al., 1991; Ramírez et al., 1996) (fig. 3-2A). The processes from these cells lie alternately over the blood vessels (Ramírez et al., 1996). In the capillaries the glial processes lie over the basal membrane (fig. 3-2A), but in the case of larger calibre vessels the alternating glial cell layers lie over the vessel's adventitia (Hogan et al., 1971; Pollock et al. 1986; Penfold et al., 1990; Ramírez et al., 1996).

Figure 3-2. Retinal astroglia. Human:A, B, C, D, F, G, H, I, J, K; Rabbit: E. (C, D, E, F, G, H, I: anti-GFAP-PAP; J: anti-GFAP-inmunofluorescence FITC); K: anti-GFAP-PAP counterstained with hematoxylin). A, Retinal capillary in the GCL. Astrocytes (A) and Müller glia (M) processes alternate and lie over the basal membrane (BM). (En: endothelial cell, L: lumen; 21,600x). B, Müller glia reaches the GCL and completely envelopes a ganglion cell separating if from its neighbour (arrow). (A: astrocytes, GC: ganglion cell; ax: axons; arrowhead: ILM; 1980x). C, Star-shaped astrocyte (arrow) located within a honeycomb in the GCL (250x). D, Elongated astrocyte with some processes reaching into the GCL (arrowhead) and others running parallel to NFL axons (arrow) (400x). E, Type I perivascular astrocyte in rabbit retina (310x). F, GCL astroglial bundles with a honeycomb morphology (125x). G, Astrocyte clusters (arrow) forming part of GCL astroglial bundle (400x). H, Longitudinal astroglial bundles within the NFL (250x). I, Retinal whole-mount showing the progressive dissapearance of the NFL astroglial bundles (arrows) (32x). J, Vertical retina section showing GFAP(+) fundamentally within the NFL and the GCL; some capillaries reach the INL surrounded by astroglial processes (250x). K, Horizontal section at NFL level. The elongated astrocytes (arrows) lie parallel to one another, enveloping the capillaries (Cap) with their processes, 250x). [A,B,H,K: Reproduced from Ramírez et al. (1996); C,D,F,I,J: Reproduced from Ramírez et al. (1994); E: Reproduced from Triviño et al. (1992)].

The astrocyte cell bodies may either lie inside the honeycomb cells (fig. 3-2C), form part of the astroglial bundles or lie over the surface of the blood vessels (fig. 3-2G). In some cases clustered astrocytes have been observed in the vicinity of the blood vessels (fig. 3-2G).

In the NFL the processes from elongated astrocytes tend to combine in bundles which fasciculate the axons (fig. 3-2H); they also arrange themselves among the blood vessels in this layer. Electron microscopy has shown that the formation of the retina's internal limiting membrane does not only involve Müller glia processes; ramifications from astrocytes and even some cell bodies of this glial type have also been found (Cohen, 1961; Hogan et al., 1971; Büssow, 1980; Höllander et al., 1991; Ramírez et al., 1996) (fig. 3-2B). The astroglial and Müller cell processes surround the ganglion cell axon bundles which make up the NFL (Hogan et al., 1971; Büssow, 1980; Höllander et al., 1991) (fig. 3-2B).

As in the GCL, desmosomes and gap-junctions are apreciable between two astroglial processes. Desmosomes have also been found between the Müller glia and the astrocyte processes in retinas of primates (Büssow, 1980) and cats (Büssow, 1980; Höllander et al., 1991).

The astroglial bundles in the NFL are not uniformly arranged throughout the retina, but only in an area of 175mm^2 centered around the optic disk and taking in the peripapillar region and part of the macular areas (fig. 3-2I). The disappearance of these bundles in the rest of the retinal tissue is due to the progressively diminishing thickness of the NFL as a consequence of the disappearance of the nerve fiber capillary plexus and the decreasing number of axons running to the disk (fig. 3-2I): there is therefore a point at which the NFL astrocytes merge with the GCL astrocytes and become difficult to observe (Ramírez et al., 1994) (fig. 3-2I).

At the macular level the astroglial bundles are arranged concentrically around the fovea, as reported in the monkey (Diestler et al., 1993) and the cat retina (Karschin et al., 1986a, b; Schintzer, 1988b). In the region of the papillomacular bundle, on the other hand, the astroglial bundles are straight and densely packed (Ramírez et al., 1994) owing to the high density of the axons at this level (Pollock and Miller, 1986). The NFL astroglial bundles disappear in the vicinity of the fovea, and hence the astrocytes observed there belong to the GCL (Ramírez et al., 1994). The astrocytes in the horizontal raphe are star-shaped (Ramírez et al., 1994) owing to their location in an axon crossing zone (Pollock and Miller, 1986).

DISTRIBUTION OF THE RETINAL ASTROCYTES

The astrocytes in the human retina are located mainly in the NFL and the GCL, although their processes and some cell bodies may sometimes reach the INL following the capillaries (fig. 3-2J). These cells are present throughout the retina and disappear 0.4 mm from the centre of the fovea and in the vicinity of the ora serrata, where there are no blood vessels either (Ramírez et al., 1994). These astrocytes are found in the same regions as the blood vessels. They are situated in the GCL and form a plexus which surrounds the vessels (as noted earlier) (fig. 3-2F); in the NFL the astrocytes run parallel to the course of the axons and make contact with the radial peripapillary capillaries (fig.3-2K), which terminate abruptly in the vicinity of the main arterial trunks 4 mm nasal (Toussaint et al., 1961; Henkind, 1967, 1969; Hogan et al., 1971). In the NFL the astrocytes and 7 mm temporal to the optic disk decrease gradually in number, disappearing approximately 6 mm nasal, 10 mm temporal, 5 mm superior and 6 mm inferior the optic disk (fig. 3-2I).

Astrocyte/blood vessel coexistence is supported by observations in other animal species which possess a different vascularization pattern. In anangiotic or avascular retinas (hippopotamus, rhinoceros and some marsupials) (Kalt, 1905) the astrocytes are located only in the optic disk (Schnitzer, 1987, 1988b); in parangiotic retinas, or retinas having very restricted vascularization with vessels only around the optic disk (horse and guinea-pig) (Kalt, 1905), astrocytes are only found in a very small area adjacent to the optic disk (Schnitzer, 1988b); in merangiotic or partially-vascularized retinas, typical of lagomorphs (rabbit) (Kalt, 1905), blood vessels are found only in two wing-shaped zones known as medullary nerve fibers (Hyvärinen, 1967; Triviño et al., 1990), like the astrocytes, which are found only on these medullary nerve fibers and are absent in the rest of the retina (Schnitzer, 1985; Schnitzer and Karschin, 1986; Reichenbach, 1987; Schnitzer, 1988a, 1988c; Robinson and Dreher, 1989; Triviño et al., 1990; Triviño et al., 1988, 1992, 1997). And finally, in holoangiotic or totally vascularized retinas (Kalt, 1905), the vessels are distributed throughout the retina except for the fovea and the zone close to the ora serrata, in the same way as the astroglial cells (Ogden, 1976; Büssow, 1980; Dickson and Eng, 1981; Karschin et al., 1986a, 1986b; Schnitzer, 1987, 1988b; Chan-Ling and Stone 1991a, 1991b; Höllander et al., 1991; Ramirez et al., 1994). This group embraces most mammals (rats, mice, cats and primates).

In avascular retinas, neurone-glia connections are made only through the Müller glia, whereas in partially and fully vascularized retinas connections are made through both the Müller glia and the astrocytes (Höllander et al., 1991, Ramírez et al., 1994, 1996). Moreover, it has been demonstrated physiologically that both cell types perform equivalent functions in the retina: both store glucogen and release glucose to the neurones (Ripps and Witkovsky, 1985; Newman, 1986b), regulate the levels of extracellular K^+ (Orkand et al., 1973; Newman, 1985a, b, 1986a, b; Karwoski, 1989), intervene in neurotransmitter metabolism (Lam and Hollyfield, 1980; Newman, 1986b; López-Colome and Romo-de-Vivar, 1991), help remove retinal CO_2 by conversion to bicarbonate (Kumpulainen, 1980), and lastly both types of cell induce barrier properties in the vascular endothelial cells (Trachtenberg and Pollen, 1970).

If both types of cell perform the same functions, why are astrocytes needed in the retinas of species with partial or full vascularization? Where retinal thickness is less than 130 μm, the oxygen and nutrients from the choroids can spread throughout the thickness of the retina (Michaelson, 1954) and hence blood vessels are unnecessary; in such retinas the Müller glia could act as a barrier, blocking the entrance of astrocytes from the optic disk (Höllander et al., 1991). This type of glia would then serve to maintain neuronal physiology. Where retinal thickness exceeds 130 mm, blood vessels are necessary; the appearance of vascular precursors in the retina could open up gaps in the Müller glia arrangement on the inner retinal surface, thus admitting the astrocytes (Chan-Ling et al., 1990; Chan-Ling and Stone, 1991a, 1991b). These would follow the precursors and would not penetrate the zones where these precursors were absent, such as the fovea or the outer layers of the retina (Jiang et al., 1994, 1995; Gariano et al., 1996a, b).

All the evidence therefore suggests that the reason for the presence of astrocytes in vascular retinas (like the human retina) is that this type of retina requires additional macroglial cell input, since the Müller glia existing from the initial growth stages would not suffice to perform all the physiological functions in which these cells are involved (Ramírez et al., 1994, 1996; Triviño et al., 1997).

REFERENCES

Berwald-Netter, Y., Koulakoff, A., Nowak, L., Ascher, P. (1986). Ionic channels in glial cells. In S. Fedoroff, A. Vernadakis (Eds): Astrocytes. Vol. 2: Biochemistry, physiology, and pharmacology of astrocytes. Orlando: Academic Press INC, pp: 51-75.

Bignami, A., Eng, L.F., Dahl, D., Uyeda, C.T. (1972). Localization of the glial fibrillary acidic protein in astrocytes by immunofluorescence. Brain Res. *43*:429-435.

Büssow, H. (1980). The astrocytes in the retina and optic nerve head of mammals: a special glia for the ganglion cell axons. Cell Tissue Res. *206*: 367-378.

Cohen, A.I. (1961). Electron microscopic observations of the internal limiting membrane and optic fiber layer of the retina of the Rhesus monkey (M. mulatta). Am. J. Anat. *108*: 179-197.

Cook, R.D., Wisniewski, H.M. (1973). The role of oligodendroglia and astroglia in Wallerian degeneration of the optic nerve. Brain Res. *61*: 191-206.

Chan-Ling, T., Halasz, P., Stone, J. (1990). Development of retinal vasculature in the cat: Processes and mechanisms. Curr. Eye Res. *9*: 459-478.

Chan-Ling, T., Stone, J. (1991a). Factors determining the migration of astrocytes into the developing retina: migration does not depend on intact axons or patent vessels. J. Comp. Neurol. *303*: 375-386.

Chan-Ling, T., Stone, J. (1991b). Factors determining the morphology and distribution of astrocytes in the cat retina: a 'contactspacing' model of astrocyte interaction. J. Comp. Neurol. *303*: 387-399.

Chan-Ling, T. (1994). Glial, neuronal and vascular interactions in the mammalian retina. Prog. Retin. Eye Res. *13*: 357- 389.

Diestler, C., Weigel, H., Hoffmann, K.P. (1993). Glia cells of the monkey retina. I.-Astrocytes. J. Comp. Neurol. *333*: 134-147.

Dixon, R.G., Eng, L.F. (1981). Glial fibrillary acidic protein in the retina of the developing albino rat: an immunoperoxidase study of paraffin-embedded tissue. J. Comp. Neurol. *195*: 305-321.

Dusart, I., Marty, S., Peschanski, M. (1991). Glial changes following and excitotoxic lesion in the CNS. II.- Astrocytes. Neuroscience *45*: 541-549.

Erskine, C.A. (1963). Neuroglial relationships in the human retinal capillary network. Br. J. Ophthalmol. *47*: 601-610.

Gariano, R.F., Sage, E.H., Kaplan, H.J., Hendrickson, A.E. (1996a). Development of astrocytes and their relation to blood vessels in fetal monkey retina. Invest. Ophthalmol. Vis. Sci. *37*: 2367-2375.

Gariano, R.F., Kalina, R.E., Hendrickson, A.E. (1996b). Normal and pathological mechanism in retinal vascular development. Surv. Ophthalmol. *40*: 481-490.

Ghandour, M.S., Langley, O.K., Vincendon, G., Gombos, G. (1979). Double labeling immunohistochemical technique provides evidence of the specificity of glial cell markers. J. Histochem. Cytochem. *27*: 1634-1637.

Hamberger, A. (1971). Amino acid uptake in neuronal and glial cell fractions from rabbit cerebral cortex. Brain Res. *31*: 169-178.

Henkind, P. (1967). Radial peripapillary capillaries of the retina. I. Anatomy: human and comparative. Br. J. Ophthalmol. *51*: 115-123.

Henkind, P. (1969). Microcirculation of the peripapillary retina. Trans. Am. Acad. Ophthalmol. Otolaryngol. *73*: 890-897.

Hiscott, P.S., Grierson, I., Trombetta, C.J., Rahl, H.S., Marshall, J., McLeod, D. (1984). Retinal and epiretinal glia: an immunohistochemical study. Br. J. Ophthalmol. *68*: 698-707.

Hogan, M.J., Alvarado, J.A., Wendell, J.E. (1971). Histology of the human eye: an atlas and textbook. Toronto: W.B. Saunders Company Ed., pp: 523-606.

Holländer, H., Makarov, F., Dreher, Z., van Driel, D., Chan-ling, T.L., Stone, J. (1991). Structure of the macroglia of the retina: sharing and division of labour between astrocytes and Müller cells. J. Comp. Neurol. *313*: 587-603.

Hui, Y.N., Goodnight, R., Zhang, X.J., Sorgente, N., Ryan, S.J. (1988). Glial epiretinal membranes and contraction. Immunohistochemical and morphological studies. Arch. Ophthalmol. *106*: 1280-1285.

Hyvärinen, L. (1967). Vascular structures of the rabbit retina. Acta Ophthalmol. *45*: 852-861.

Ikui, H., Uga, S., Kohno, T. (1976). Electron microscope study on astrocytes in the human retina using ruthenium red. Ophthalmic Res. *8*: 100-110.

Janzer, R.C., Raff, M.C. (1987). Astrocytes induce blood-brain barrier properties in endothelial cells. Nature *325*: 253-257.

Jiang, B., Liou, G.L., Bezhadian, M.A., Caldwell, R.B. (1994). Astrocytes modulate retinal vasculogenesis: effects on fibronectin expression. J. Cell Sci., *107*: 2499-2508.

Jiang, B, Bezhadian, M.A., Caldwell, R.B. (1995). Astrocytes modulate retinal vasculogenesis: effects on endothelial cell differentiation. Glia *15*: 1-10.

Kalloniatis, M. (1995). Amino acids in neurotransmission and disease. J. Am. Optom. Assoc. *66*:750-757.

Kalt, M.E. (1905). Anatomie et physiologie comparées de l'appareil oculaire. In F. Lagrange, E. Valude (Eds.): Encyclopédie française d'ophtalmologie, Tome II. Paris: Octave Doin, pp: 685-933.

Karschin, A., Wässle, H., Schnitzer, J. (1986a). Shape and distribution of astrocytes in the cat retina. Invest. Ophthalmol. Vis. Sci. *27*: 828-831.

Karschin, A., Wässle, H., Schnitzer, J. (1986b). Immunocytochemical studies on astroglia of the cat retina under normal and pathological conditions. J. Comp. Neurol. *249*: 564-576.

Karwoski, C.J., Lu, H-K., Newman, E.A. (1989). Spatial buffering of light-evoked potassium increases by retinal Müller (glial) cells. Science, *244*: 578-580.

Kimelberg, H.K., Kaltz, D.M. (1985). High-affinity uptake of serotonin into immunocytochemically identified astrocytes. Science, *228*: 889-891.

Kimelberg, H.K., Norenberg, M. (1989). Astrocytes. Sci. Amer. *260*: 44-52.

Kuffler, S.W., Nicholls, J.G. (1976). Physiology of neuroglial cells. In: From neuron to brain: a cellular approach of the nervous system.; Sunderland: Sinauer Associates, INC, pp: 255-263.

Kumpulainen, T. (1980). Carbonic anhydrase isoenzime C in the human retina. An immunohistochemical study. Arch. Ophthalmol. *58*: 397-405.

Lam, D.M.K., Hollyfield, J.G. (1980). Localization of putative amino acid neurotransmitters in the human retina. Exp. Eye Res. *31*: 729-732.

Liem, R.K.H., Yen, S.H., Solomon, G.D., Shelanski, M.L. (1978).: Intermediate filaments in nervous tissues. J. Cell Biol. *79*: 637-645.

Liem, R.K.H., Keith, C.H., Leterrier, J.F., Trenkner, E., Shelanski, M.L. (1981). Chemistry and biology of neuronal and glial intermediate filaments. Cold Spring Harb. Symp. Quant. Biol. *46*:341-350.

Liesi, P., Silver, J. (1988). Is astrocyte laminin involved in axon guidance in the mammalian CNS?. Dev. Biol. 130: 774-785.

Lindsay, R.M. (1986). Reactive gliosis. In S. Fedoroff, A. Vernadakis, (Eds.): Astrocytes . Vol 3: Cell biology and pathology of astrocytes. Orlando: Academic Press INC, pp: 231-262.

Lobrinus, J.A., Juillerat-Jeanneret, L., darekar, P., Schlosshaver, B., Janzer, R.C. (1992). Induction of the blood-brain barrier specific HT7 and neurothelin epitopes in endothelial cells on the chick chorioallantoic vessels by a soluble factor derived from astrocytes. Dev. Brain Res. *70*: 207-211.

López-Colome, A.M., Romo-de-Vivar, M. (1991). Excitatory amino acid receptors in primary cultures of glial cells from the retina. Glia *4*: 431-439.

MacVicar, B.A., Hochman, D., Delay, M.J., Weiss, S. (1991). Modulation of intracellular Ca++ in cultured astrocytes by influx through voltage-activated Ca++ channels. Glia, *4*: 448-455.

Mano, T., Puro, D.G. (1990). Phagocytosis by human retinal glial cells in culture. Invest. Ophthalmol. Vis. Sci. *31*: 1047-1055.

Michaelson, I.C., Campbell, A.C. (1940). The anatomy of the retinal vessels, and some observations on their significance in certain retinal diseases. Trans. Ophthalmol. Soc. U.K. *60*: 71-112.

Michaelson, I.C. (1954). Retinal circulation in man and animals. Springfield: Charles C. Thomas.

Musser, G.L., Rosen, S. (1973). Localization of carbonic anhydrase activity in the vertebrate retina. Exp. Eye Res. 15: *105*-119.

Newman, E.A. (1985a). Regulation of potassium levels by glial cells in the retina. TINS *8*: 156-159.

Newman, E.A. (1985b). Voltage-dependent calcium and potassium channels in retinal glial cells. Nature *317*: 809-811.

Newman, E.A. (1986a). High potassium conductance in astrocyte endfeet. Science, *233*: 453-454.

Newman, E.A. (1986b). The Müller cell. In S. Fedoroff, A. Vernadakis (Eds.): Astrocytes. Vol I: Development, morphology and regional specialization of astrocytes, Orlando: Academic Press INC, pp: 149-171.

Nieto-Sampedro, M., Saneto, P.R., De Vellis, J., Cotman, C.W. (1985). The control of glial populations in brain: changes in astrocytes mitogenic and morphogenic factors in response to injury. Brain Res. *343*: 320-328.

Ogden, T.E. (1976). Nerve fiber layer astrocytes of the primate retina: morphology, distribution, and density. Invest. Ophthalmol. Vis. Sci. 17: 499-510.

Orkand, P.M., Bracho, H., Orkand, R.K. (1973). Glial metabolism: alteration by potassium levels comparable to those during neuronal activity. Brain Res. *55*: 467-471.

Penfold, P.L., Provis, J.M., Madigan. M.C., van Driel, D., Billson, F.A. (1990). Angiogenesis in normal human retinal development: the involvement of astrocytes and macrophages. Graef. Arch. Clin. Exp. Ophthalmol. *228*: 255-263.

Pollock, S.C., Miller, N.R. (1986). The retinal nerve fiber layer. Int. Ophthalmol. Clin. *26*: 201-221.

Provis, J.M., Penfold, P.L., Madigan, M.C., van Driel, D., Billson F.A. (1991). Angiogenesis in human retinal development: astrocytes, macrophages and early lumen formation. In C.Y. Khoo, B.C. Ang, W.M. Cheah, P.T.V. Chew, A.S.M. Lim (Eds.): New frontiers in Ophthalmology. Amsterdam: Excerpta Medica, pp: 608-612.

Ramírez, J.M., Triviño, A., Ramírez, A.I., Salazar, J.J., García-Sánchez, J. (1994). Immunohistochemical study of human retinal astroglia. Vision Res. *34*: 1935-1946.

Ramírez, J.M., Triviño, A., Ramírez, A.I., Salazar, J.J., García-Sánchez, J. (1996). Structural specializations of human retinal glial cells. Vision Res. *36*: 2029-2036.

Reichenbach, A. (1987). Quantitative and qualitative morphology of rabbit retinal glia. A light microscopical study on cells both in situ and isolated by papaine. J. Hirnforsch. *28*: 213-220.

Ripps,H., Witkovsky, P. (1985). Neuron-glia interaction in the brain and retina. Prog. Retin. Res. *4*:181-219.

Risau, W., Wolburg, H. (1990). Development of the blood-brain barrier. TINS *13*: 174-178.

Ritchie, J.M. (1992). Voltage-gatedion channels in Schwann cells and glia. TINS *15*: 345-351.

Robinson, S.R., Dreher, Z. (1989). Evidence for three morphological classes of astrocyte in the adult rabbit retina: functional and developmental implications. Neurosci. Letters *106*: 261-268.

Robinson, S.R., Dreher, Z. (1990). Müller cells in adult rabbit retinae: morphology, distribution and implications for function and development. J. Comp. Neurol. *292*: 178-192.

Sato, K., Yoshida, S., Fujiwara, K., Tada, K., Tohyama, M. (1991). Glycine cleavage system in astrocytes. Brain Res. *567*: 64-70.

Schnitzer, J. (1985). Distribution and immunoreactivity of glia in the retina of the rabbit. J. Comp. Neurol. *240*: 128-142.

Schnitzer. J., Karschin, A. (1986). The shape and distribution of astrocytes in the retina of the adult rabbit. Cell Tissue. Res. *246*: 91-102.

Schnitzer, J. (1987). Retinal astrocytes: their restricition to vascularized parts of the mammalian retina. Neurosci. Letter *78*: 29-34.
Schnitzer, J. (1988a). The development of astrocytes and blood vessels in the postnatal rabbit retina. J. Neurocytol. *17*: 433-449.
Schnitzer, J. (1988b): Atrocytes in mammalian retina. Prog. Retin. Res. *7*: 209-231.
Schnitzer, J. (1988c). Immunocytochemical studies on the development of astrocytes, Müller (glial) cells, and oligodendrocytes in the rabbit retina. Dev. Brain Res. *44*: 59-72.
Schnitzer, J.(1988d): Astrocytes in the guinea pig, horse, and monkey retina: their occurrrence coincides with the presence of blood vessels. Glia, *1*: 74-89.
Sole, P, Dalens, H., Gentou, C. (1992). Biophtalmologie. Libro VI: Le Système Nerveux. Chapitre I: Generalites sur le système nerveux; Paris: Masson, pp: 3-72.
Sramek. S.J., Wallow, I.H., Stevens, T.S., Nork, T.M. (1989). Immunostaining of preretinal membranes for actin, fibronectin, and glial fibrillary acidic protein. Ophthalmol. *96*: 835-841.
Stewart, P.A., Wiley, M.J. (1981). Developing nervous tissue formation of blood-brain barrier characteristics in invading endothelial cells: a study using quail-chick transplatation chimeras. Dev. Biol. *84*: 183-192.
Stewart, P.A., Coomber, B.L. (1986). Astrocytes and the blood-brain barrier. In S. Fedoroff, A. Vernadakis (Eds.): Astrocytes; Vol. I: Development, morphology and regional specialization of astrocytes. London.Academic Press INC, pp: 311-328.
Stone, J., Dreher, Z. (1987). Relationship between astrocytes, ganglion cells and vasculature of the retina. J. Comp. Neurol. *255*: 35-49.
Thano, S., Thiel, H-J. (1990). Regenerative and proliferative capacity of adult human retinal cells in vitro. Graef. Arch. Clin. Exp. Ophthalmol. *228*: 369-376.
Toussaint, D., Kuwabara, T., Cogan, D.G. (1961). Retinal vascular patterns. Part II. Human retinal vessels studied in three dimensions. Arch. Ophthalmol. *65*: 575-581.
Tout, S., Dreher, Z., Chang-Ling, T.L., Stone, J. (1993). Contact-spacing among astrocytes is independent of neighbouring structures: in vivo an in vitro evidence. J. Comp. Neurol. *332*: 433-443.
Trachtenberg, M.C., Pollen, D.A. (1970). Neuroglia: biophysical properties and physiologic function. Science *167*: 1248-1252.
Triviño, A., Ramírez, A.I., Ramírez, J.M., Salazar, J.J., Rivera, M., García-Sánchez, J. (1988). Estudio de las relaciones vaso-gliales en la retina del conejo albino (Oryctolagus cuniculus). Arch. Soc. Esp. Oftalmol. Invest *1*: 103-110.
Triviño, A., Ramírez, A.I., Salazar, J.J., Ramírez, J.M., García-Sánchez, J. (1990). Astrocitos retinianos relacionados con los axones de las células ganglionares. Estudio inmunohistoquímico. Arch. Soc. Esp. Oftalmol. Invest *3*: 87-92.
Triviño, A., Ramírez, A.I., Ramírez, J.M., Salazar, J.J., García-Sánchez, J. (1991). Estudio mediante el método de diafanización de la vascularización retiniana del conejo. St. Ophthalmol. *10*: 13-18.
Triviño, A., Ramírez, J.M., Ramírez, A.I., Salazar, J.J., García-Sánchez, J. (1992). Retinal perivascular astroglia: an immunoperoxidase study. Vision Res. *32*: 1601-1607.
Triviño, A., Ramírez, J.M., Ramírez, A.I., Salazar, J.J., García-Sánchez, J. (1997). Comparative study of astrocytes in human and rabbit retinae. Vision Res. *37*: 1607-1611.
Turner, D.L., Cepko, C.L. (1987). A common progenitor for neurons and glia persist in rat retina late in development. Nature *328*: 131-136.
Vaca, K., Wend, T.E. (1992). Divergent effects of astroglial and microglial secretions on neuron growth and survival. Exp. Neurol. *118*: 62-72.
Wolter, J.R. (1955a). The astroglia of the human retina: and other glial elements of the retina under normal and pathologic conditions. Am. J. Ophthalmol. 40: *88*-100.
Wolter, J.R. (1955b). The cells of Remark and the astroglia of the normal human retina. Arch. Ophthalmol. *53*: 832-838.
Wolter, J.R. (1956a). Perivascular glia of the blood vessels of the human retina. Am. J. Ophthalmol. *44*: 766-773.
Wolter, J.R. (1956b). Reactions of the elements of retina and optic nerve. In common morbid entities of the human eye. Am. J. Ophthalmol. *42* (pt. II): 10-27.
Wolter, J.R. (1956c). Über Besondere Astroglia an der Innenfläche der Retina. Klin. Monstsbl. Augenheilkd. *129*: 224-230.
Wolter, J.R. (1959). Glia of the human retina. Am. J. Ophthalmol. *48* (pt. II): 370-393.

Wolter, J.R. (1961). Silver carbonate techniques for the demonstration of ocular histology. In G.K. Smelser (Ed): The structure of the eye. New York: Academic Press, pp: 117-138.

4

ASTROGLIAL ARCHITECTURE OF THE HUMAN OPTIC NERVE: FUNCTIONAL ROLE OF ASTROCYTES

Alberto Triviño, José M. Ramírez,
Juan J. Salazar, Ana I. Ramírez

The optic nerve is the structure that transmits information from the retina to the brain. It is composed of ganglion cell axons, astroglial cells, oligodendrocytes, microglia, blood vessels and connective tissue. The astroglia cells perform important functions, including formation of astroglial limiting membranes and organisation of the ganglion cell axons of the retina into bundles throughout their course along the nerve. There are two morphological types of astrocytes, thick and thin, whose morphologies change according to the region of the nerve that they occupy. On the surface nerve fibre layer (SFNL) the thin-bodied astrocytes issue long processes which follow the course of the axons and make contact with the capillaries derived from the central retinal artery: in the anterior prelaminar region (PR) the thin-bodied astrocytes assume a star shape and their processes are arranged in a basket formation. In the posterior PR the astrocyte are thick bodied and make up glial tubes inside which the nerve fibres travel towards the laminar region (LR). In the LR, the thick bodied astrocytes cover the inner surface of the pores of the lamina cribosa; and finally in the retrolaminar region (RR) they separate the nerve tissue from the connective tissue. The glial limiting membranes are composed of thick-bodied astrocytes.

The optic nerve (ON) is a projection of white matter from the brain, It may be considered part of the Central Nervous System rather than a peripheral nerve (Raynaud and Milin-Cioni, 1992). Two characteristics of the ON support these statements. First, the ON fibres are myelinized by oligodendrocytes and not by Schwann cells (Reeh, 1981). Second, the ON is enveloped by the cerebral meninges: duramater, arachnoides and piamater (Wolf-Heidegger, 1954). The length of the human ON varies from 35 to 55 mm, while its diameter increases from 1.5 mm (in the area closest to the retina) to 3-4 mm (beyond the lamina cribosa) (Duke-Elder and Wybar, 1961; Kestenbaum, 1963; Raynaud and Milin-Cioni,

1992). The ON is composed of: ganglion cell axons; glial cells (astrocytes, oligodendrocytes and microglia); blood vessels; and connective tissue.

Four classic regions of the humans ON have been described (Duke-Elder and Wybar, 1961; Wollf, 1961; Kestenbaum, 1963; Hogan et al., 1971; Reeh, 1981; Wirtschafter, 1981; Jonas and Naumann, 1993): intraocular or intrascleral region (0.5-1 mm), intraorbital or orbital region (25 mm), intracanular region (6-7.5 mm) and an intracranial region (10-12 mm.). This chapter deals with the "optic nerve head" (ONH), which is composed by the intraocular region and the anterior part of the intraorbital region of the ON (Hayreh and Vrabec, 1966; Hayreh, 1974, 1978; Ramírez et al., 1984; Hayreh, 1987, 1989; Jonas and Naumann, 1993; Hayreh, 1994). The ONH may be divided into four regions: Superficial Nerve fibre Layer, Prelaminar Region, Laminar Region, and Retrolaminar Region.

Superficial nerve fibre layer (SNFL)

The SNFL is composed by the ganglion cells axons and two glial limiting membranes of astrocytes. The axons from the retinal ganglion cells converge on the optic disk thus forming the nerve fibre layer. In this region of the ONH the axons and the connective tissue are separated by two glial limiting membranes: Elschnig's limiting membrane and Kuhnt's central meniscus (Anderson and Hoyt, 1969; Anderson, 1973; Büssow, 1980; Ramírez et al., 1989; Ramírez and Triviño, 1989; Triviño et al., 1996; Ramírez et al., 1997).

Prelaminar region

This term is used to define the portion of the ONH known by many authors as the choroid or anterior part of the lamina cribosa (Wolter, 1957; Blunt et al., 1965; Levitzky and Henkind, 1969; Ffrench-Constant and Raff, 1986b). Commencing at the margin of the optic disk, the axons from the nerve fibre layer bend 90°, grouping into bundles, which are surrounded by astroglial processes (Wolter, 1957, 1961; Anderson, 1969a; Anderson and Hoyt, 1969). The astrocyte cell bodies display a columnar arrangement at this point (Liss, 1956; Ortiz-Miterrer, 1974; Skoff et al., 1986). As in the SNFL, the axons, which are still unmyelinated, are isolated from the adjacent retina and choroid by glial limiting membranes: the Kuhnt's intermediate tissue (Anderson, 1970; Okinami et al., 1976; Senelick, 1982; Triviño et al., 1996) and Jacoby's tissue respectively (Anderson and Hoyt, 1969; Anderson, 1973; Triviño et al., 1996).

Laminar region (LR)

This region of the ONH encompasses what many authors refer to as the posterior or scleral part of the lamina cribosa (Wolter, 1957; Anderson, 1969a, 1973; Elkington et al., 1990). The lamina cribosa is formed by an extension of collagen bundles (Anderson, 1969a) and elastic fibres from the inner two-thirds of the sclera across the optic nerve canal. The bundles of scleral collagen and elastic tissue extend from one side of the canal to the other intersecting at various angles to form a sieve with many openings traversed by the axons bundles (Wilczek, 1947; Wollf, 1961; Morrison et al., 1989a, 1989b).

The astrocytes in this region line the pores of the lamina cribosa, enveloping and protecting the axons, some of which start to become myelinated at the end part of the lamina (Anderson, 1969a; Hogan et al., 1971; Anderson, 1973; Elkington et al., 1990).

Retrolaminar region (RR)

The RR is a transition zone from the end of the lamina cribosa up to the entry-point of the central retinal artery in the ON (Ramírez et al., 1984).

At this level the ONH is divided into axonal bundles that are separated by connective septa. These extend either from the piamater connective tissue that penetrates the ON accompanying the ciliary system pial arteries or from the septa of the connective tissue surrounding the central vessels and sheath the arterioles exiting the central retinal artery.

Oligodendrocytes are also present in a columnar fashion enveloping the axons in a sheath of myelin (Kestenbaum, 1963; Hogan et al., 1971; Ortiz-Miterrer, 1974; Butt and Ramson, 1989; Butt et al., 1994).

The ONH in this region is surrounded by three meningeal sheaths: duramater, arachnoides and piamater (Anderson, 1969a; Hayreh, 1984). Separating the piamater from the axons there is a limiting membrane composed by astrocytes. This is called Fuchs's or Greef's peripheral glial mantle piamater (Anderson and Hoytt, 1969; Anderson, 1970, 1973; Hayreh, 1984).

MORPHOLOGY AND ULTRASTRUCTURE OF THE HUMAN OPTIC NERVE ASTROCYTES

As in other parts of the Central Nervous System (Roots, 1986), the ultrastructure of the ON astrocytes in humans (Anderson et al., 1967; Cohen, 1967; Anderson 1969a, 1969b, 1969c, 1970, 1973; Elkington et al., 1990) and non-hominid primates (Anderson, 1969b; Anderson and Hoyt, 1969; Cook et al., 1973; Quigley, 1977; Büssow, 1980) is characterized by a cytoplasm consisting of mitochondria, free ribosomes, rough endoplasmic reticulum, Golgi apparatus, glycogen granules and many intermediate glial filaments' 8-10 nm in diameter composed of protein subunits (gliofibrillary acidic protein [GFAP]) of 50 kD molecular weight (Lazarides, 1980).

Astrocytes of the human optic nerve can be classified according to the thickness of the cell body (Triviño et al., 1996). Thin-bodied astrocytes (fig. 4-1 A2, A3, B2, B3) are cells with high GFAP immunoreactivity. They have an oval cell body with an ovoid nucleus. Two to four very long primary processes extend from the cell body and divide into thinner secondary processes. Occasionally short processes extend directly from the cell body. Electron microscopy has revealed scant perinuclear cytoplasm and cytoplasmic organelles concentrated at the cell poles. These astrocytes are located in the SNFL and the anterior PR (Triviño et al., 1996).

The morphology of these cells varies from elongated to star-shaped. Within the latter group there is a morphological variety with 7 or 8 short radial processes extending from the cell body as well as a single long straight process that follows the course of the axons; this last type of cell is located in the PR and LR (Triviño et al., 1996).

Thick-bodied astrocytes (fig. 4-1 C2, C3, D2, D3) are cells with greater GFAP immunoreactivity than thin-bodied astrocytes. They possess an oval nucleus with a large perinuclear cytoplasm, in which the characteristic astrocyte organelles are visible under electron microscopy. These cells possess many intermediate glial filaments that are grouped in dense packs in both cell body and processes. Normally 6 to 8 primary processes extend almost radially from the cell body depending on its location within the optic nerve . The primary processes divide into thinner secondary ones. However, thin processes may also leave the cell body. These cells are distributed throughout the length of the optic nerve from the SNFL to the RR (Triviño et al., 1996).

The only structural difference we have found between thick-bodied and thin-bodied astrocytes is the amount of cytoplasm surrounding the nucleus and the density of the filament bundles in the cytoplasm (Triviño et al., 1996).

The perikarion of the astrocytes located in the limiting membranes in the posterior PR and RR (all considered thick-bodied) is larger than the cytoplasm found in astrocytes in the SNFL and the anterior PR (considered thin-bodied). The difference between both cytoplasmic sizes could explain why thin-bodied astrocytes are not found with a larger GFAP(+) cell body. An additional reason for the strong immunoreactivity displayed by the processes of the posterior PR and RR astrocytes (both thick-bodied) is the high density of filament bundles arranged in parallel formation inside them. The processes from these astrocytes look denser and darker when observed by electron microscope.

Other than these features, there is no significant difference regarding the number and type of cytoplasmic organelles between both types of astrocytes; all both exhibit inclusions of lipofuchsin and dense bodies (Anderson et al., 1967; Cohen, 1967; Hogan et al., 1971).

These small differences between the astrocytes of the ON are thus determined by location. The aim of our classification into groups is just to reflect the ability of these cells to adjust to the specific characteristics of the environment, which could decide the functions they may perform. Thick-bodied astrocytes possess a soma rich in glial filaments, which are supposed to confer tensile strength (Elkington et al., 1990) and thereby an important structural role in isolating and protecting the axons from the other ON structures. This supposition is based on the location of these cells in the ON, since they are the constituents of the glial limiting membranes that insulate the optic nerve of the retina, choroids and meninges. They also form glial tubes that conduct the axons to the pores of the lamina cribosa in the LR, and again they serve to separate the axons from the connective tissue in the RR (Triviño et al., 1996). Thin-bodied astrocytes, on the other hand, possess fewer glial filaments, suggesting less tensile strength. These cells may be responsible for the organization of the axons as they enter the nerve in the SNFL and anterior PR (Triviño et al., 1996). They could be considered continuations of the retinal astrocytes since they exhibit similar morphological features (Ramírez et al., 1994).

The ON astrocytes of primates have been found ultrastructurally similar to those to the rat (Malmfors, 1963; Skoff et al., 1980, 1986; Suárez and Raff, 1989) and the cat (Wendell-Smith et al., 1966; Büssow, 1980). In rabbit and rat have all reported the existence of astrocytes whose cell bodies send astroglial processes which traverse the nerve both horizontally and vertically (Blunt et al., 1965; Vaughn and Peters, 1967; Büssow, 1980; Dickson and Eng, 1981; Skoff et al., 1986). Until now none has considered that there may be different classes of astrocyte.

However, the application of metal impregnations (Miller and Raff, 1984), electron microscopy (Ffrench-Constant et al., 1986; Miller et al., 1989b), immunohistochemical techniques using different antibodies (Miller et al., 1984, 1986; Ramírez et al., 1997), and injection of intracellular colorants (Miller et al., 1989b, Butt and Ransom, 1989; Butt et al., 1994), have all shown that the astrocytes in rabbit and rat conform two different morphological patterns depending on the orientation of the processes. Thus, on the one hand there are astrocytes with numerous ramifications extending radially in all directions and giving the cell a star-like appearance. On the other hand, there are also astrocytes with processes running parallel to one another following the direction of the nerve axis (Butt and Ransom, 1989; Miller et al., 1989b; Ramírez et al., 1989; Butt et al., 1994, Ramírez et al., 1997).

The main point of discussion, however, is to establish the correlation between the two astroglial strains of astrocytes (type 1 astrocytes and type 2 astrocytes) defined in cultures of rat ON (Raff et al., 1983a, 1983b, 1984; Raff, 1989; Miller et al., 1989a) with the astrocytes founded in intact rat Ons. Using the antibody A2B5, Miller and Raff (1984) first differentiated between GFAP(+) astrocytes in the rat nerve: A2B5(+) and A2B5(-) astrocytes. These authors reported that GFAP(+)/ A2B5(-) cells that form the glial limiting membranes and make contact

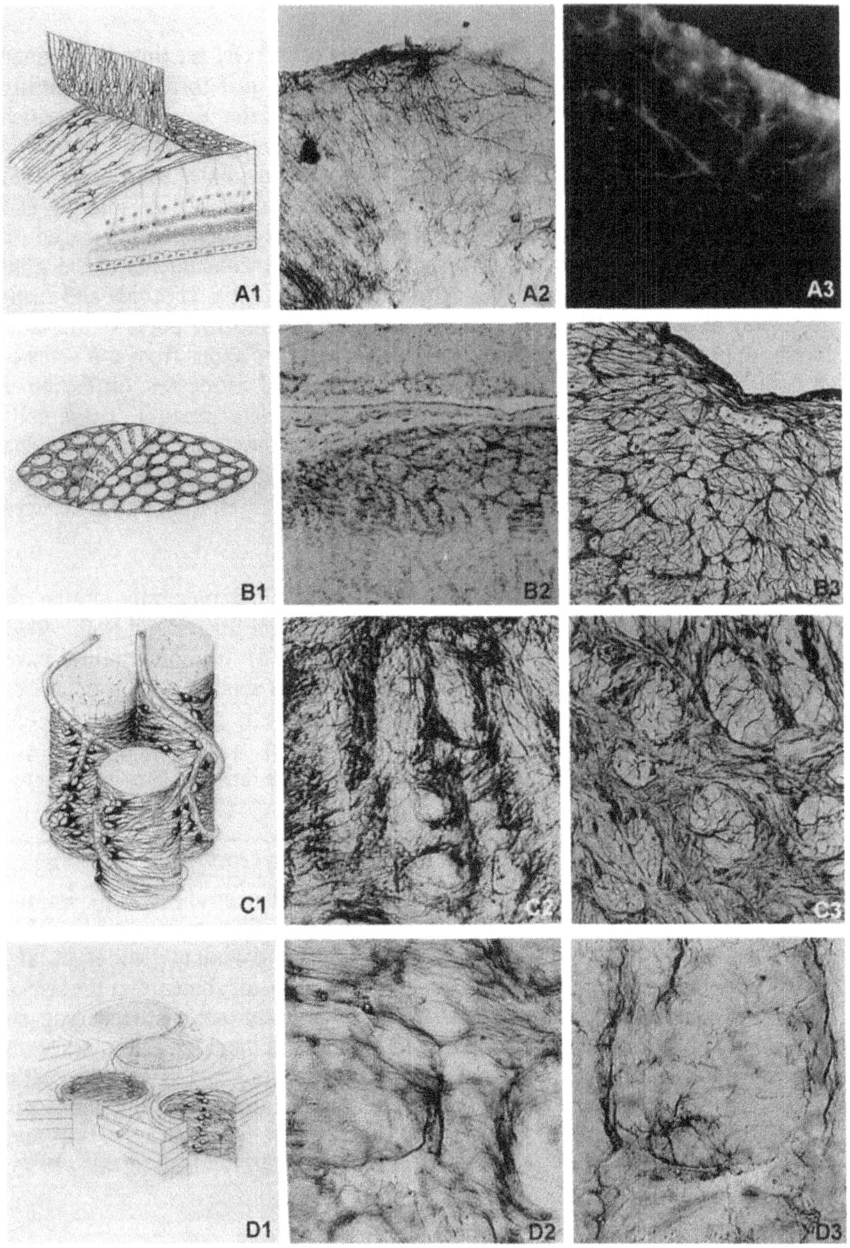
A1
A2
A3
B1
B2
B3
C1
C2
C3
D1
D2
D3

with the blood vessels would be type 1 and could be considered protoplasmic astrocytes; the GFAP(+)/A2B5(+) cells, which are associated with the nerve fibres, would be type 2 and could be considered fibrous astrocytes. With the use of new antibodies (Ffrench-Constant and Raff, 1986a, 1986b), this initial hypothesis has been abandoned. All these authors now consider type 1 astrocytes to be fibrous and type 2 astrocytes to be a new kind of glial cell (Ffrench-Constant et al., 1986; Miller et al., 1989a, 1989b; Raff, 1989).

Type 1 astrocytes were thought to correspond to the astrocytes described earlier as having multiple radial processes, and to be responsible for the formation of the sub-pial limiting membrane and the perivascular envelope; type 2 astrocytes would be the cells described as having processes running parallel to the nerve fibres (Miller et al., 1989b) and from which extend the paranodal processes observed in electron microscopy (Hidelbrand and Waxman, 1984; Waxman and Black, 1984; Black and Waxman, 1985; Ffrench-Constant and Raff, 1986b) and immunohistochemistry fibres (Ffrench-Constant et al., 1986; Miller et al., 1989b; Black and Waxman, 1988; Black et al., 1989a, 1989b).

However, it has been found that the division of functions between type 1 and type 2 astrocytes is so strict as initially though. The sub-pial and perivascular astrocytes (supposedly type 1) can also exhibit paranodal processes occasionally (Suárez and Raff, 1989). This agrees with our own observations in human ON. We have found thick-bodied astrocytes with a more or less star-shaped morphology that fit with the type 1 category. They were connecting equally with the connective septum, the peripheral limiting membrane, the blood vessels, the nerve fibres or the oligodendrocytes (Triviño et al., 1996).

Figure 4-1. Morphology and distribution of human optic nerve astrocytes.
(A2, B2, C2, D2, D3: anti.GFAP-PAP; A3: anti-GFAP inmunofluorescence-FITC; B3, C3: anti-GFAP-PAP counterstained with hematoxylin).
A, Superficial Nerve Fiber Layer. A1, Three-dimensional drawing of the SNFL. Elschning's limiting membrane is folded back. The elongated astrocytes lying below this membrane become stellate towards the PR. A2-A3, Thin bodied astrocyte with elongated morphology. (A2: 400x, A3: 500x). B, Anterior Prelaminar Region. B1, Three-dimensional drawing of the anterior PR glioarchitecture. B2-B3, Stellate thin bodied astrocytes from compartments that surround and bundle the axons making up a basket-like structure (B2:63x; B3: 400x). C, Posterior Prelaminar Region. C1, Three-dimensional drawing of the three astroglial tubes in the posterior PR. C2-C3, The thick bodies astrocytes take the form of crescent moon and make up the wall of the tubes through which the axons run. (C2, Longitudinal section, 400x; Transverse section, 500x). D, Laminar Region. D1, Three-dimensional drawing ot the position of the astrocytes that cover the inner wall of the lamina cribosa pores. D2-D3, Thick bodied astrocytes only cover the interior perimeter of the laminar canal (D2: 500x; D3: 400x).

[A1, A3, B1, C1, D1: Reproduced from Triviño et al. (1996)]

GLIAL ARCHITECTURE OF THE OPTIC NERVE: AXON FASCICULATION

Surface nerve fibre layer (SFNL)

The astrocytes in this region (fig. 4-1, section A) are thin-bodied and have long, straight processes that give them a star-like appearance (Triviño et al., 1996).

The cause of this morphology is that these astrocytes follow the course of the ganglion cell axons, which run from the retina to the optic disk and lie parallel to the surface of the adjacent vitreous humour.

Prelaminar region (PR)

In this region the axons change course, bending 90° in the direction of the optic chiasm and the astrocytes group axons into bundles or fascicles. The astrocytes gradually undergo a series of changes in both morphology and distribution, from the anterior to the posterior PR.

In the anterior PR we find star-shaped thin bodied astrocytes (fig. 4-1 B1, B2, B3). Their arrangement is closely related to the distribution pattern of the vascular system. The astrocyte nuclei mostly lie over the blood vessel walls, sending out processes in all directions to form an astroglial network that begins to regroup and fasciculate the axons. This network form structures that resemble honeycomb compartments in cross sections (Triviño et al., 1996).

Inside these compartments are the ganglion cell axons, which are gathered into bundles by the action of the astrocytes. These axon bundles subdivide in turn into smaller groups under the influence of large numbers of astroglial processes crossing over the compartments in all directions. The spatial arrangement of these honeycomb-like compartments is such that a three-dimensional view of the glial framework in this zone resembles a wicker basket (Wolter, 1957) (fig. 4-1 B1, B2). The basket arrangement of the astrocytes in the anterior PR reflects their function of supporting and protecting the unmyelinated fibres at the point where they turn 90° (Anderson et al., 1967, Anderson, 1970). It can also play an important mechanical role in avoiding crushing or friction among the axons, as it is a more elastic structure than the rigid lamina cribosa (Wolter, 1957). This elasticity can lessen or prevent irreparable damage to nerve fibres in case of swelling of the disk caused by papilloedema or neuritis.

This basket structure is replaced in the posterior PR by glial tubes (fig. 4-1 C1, C2, C3). The astrocytes in this zone become more robust, presenting thick cell bodies. These cells are arranged very precisely on the ON to form actual cylindrical glial tubes inside which the nerve fibres run; the blood vessels are distributed over the surface of the glial tubes walls (fig. 4-1 C1).The glial tubes are in contact with one another andrun parallel to the ON axis in an anterior-posterior direction (Triviño et al., 1996).

In the transitional zone between this and the LR we can see how the openings in the prelaminar glial tubes are situated exactly opposite to the laminar pores. In that way the nerve fibre bundles are conducted by the glial tubes into the pores of the lamina cribosa (Triviño et al., 1996).

These glial tubes may perform a mechanical function of resisting the pressures that originate at this level when the eye move. This idea may be supported by the finding of important amount of GFA protein in the astroglial processes, which would provide them with tensile strength (Elkington et al., 1990).

Additionally, the presence of desmosomes and gap junctions (Cook et al., 1973; Quigley, 1977) also support this possibility. Both types of junctions may have an important role in maintaining the astroglial network through which the axons, since even under high osmotic pressures the gap junctions remain intact (Quigley, 1977). Moreover, the glial architecture of this zone organizes the axonal bundles in preparation for their entry to the laminar region; this is clearly appreciable in the transition zone between the two regions, where the glial tubes are perfectly aligned opposite the pores of the lamina cribosa (Triviño et al., 1996).

Laminar Region (LR)

A striking feature of the glial architecture of this region (fig. 4-1 section D) is the pronounced diminishment of the glial tissue, which here covers only the inner surface of the cribose pores (fig. 4-1 D2). This glial tissue is composed of thick-bodied astrocytes like those in the posterior prelaminar region, but here it forms a single layer covering the inner surface of the cribose pore (Triviño et al., 1996) (fig. 4-1 D1).

The reason for the smaller number of astrocytes may be that the connective tissue of the lamina cribosa form a framework that separates the axons into bundles, a function that is performed by the astrocytes in the posterior PR. In this region only one layer of astrocytes is required to isolate the nerve fibres and separate them from the collagen in the walls of the cribose pores.

Retrolaminar Region (RR)

In this region, the astrocytes form a glial tissue that lines the connective septum and separate it from the axons. These cells can be located either between or within axonal bundles. In the first case the astrocyte's morphology is slightly less stellate (Ortiz-Miterrer, 1974).

The latter astrocytes sent out primary processes in a more radial formation, which gives them a star-like appearance; the processes emerge from the soma and spread out in all directions, subdividing into thinner secondary branches that interpose themselves between the axons in the bundle. These processes form a dense network that completely subdivides the interior of the bundles. The processes from these cells may run (i) to the glial walls, contributing to their formation; (ii) to the capillaries, forming part of their astroglial sheath or, (iii) to the oligodendrocytes (Triviño et al., 1996).

The astrocytes in this region serve to isolate the axons in a bundle and regroup them into packs. They do not play the predominant role that they had in the PR, since the connective septa in the RR already serve to fasciculate the axons.

However, in the last few years considerable importance has been attached to the role of the astrocytes in this myelinated region of the ON. They have been associated with maintenance of the functionality and structure of the Ranvier nodes. The perinodal astroglial processes in the ON have been clearly marked (Ffrench Constant et al., 1986). The close association between these perinodal astroglial processes and the axon membrane suggest that the astrocytes play an important part in the physiology of the nodes (Ffrench-Constant et al., 1986, Black and Waxman, 1988, Black et al., 1989a).

GLIAL LIMITING MEMBRANES IN THE HUMAN OPTIC NERVE

All regions of the human ON are GFAP immunoreactive. The stronger immunoreactivity is located at the level of the membranes that isolate the ON from the surrounding tissues. Elschnig's limiting membrane appeared as a narrow band strongly GFAP(+) lying parallel to the disk surface. This band represent a direct continuation of the inner limiting membrane of the retina. It separates the more superficial tissue of the optic disk from the surface of the adjacent vitreous body (Anderson et al., 1967; Anderson and Hoyt, 1969; Triviño et al., 1996). Toward the centre of the optic disk, this limiting membrane thickens as it penetrates the intraneuronal tissue forming the Kuhnt's central meniscus (Anderson et al., 1967; Anderson and Hoyt, 1969; Triviño et al., 1996; Ramírez et al., 1997). This new limiting membrane skirt the connective tissue that covers the central blood vessels (retinal artery and central vein) from the point where they penetrate the nerve (in the RR) to the point where they reach the optic disk (Triviño et al., 1996). Both, Elschnig and Kuhnt limiting membranes, are composed of thick-bodied astrocytes densely packed in two or three layers, giving the appearance of a continuous band.

The other blood vessels in the ON are also surrounded by perivascular astroglial processes. This relationship between the astrocytes and the blood vessels is detected at the level of the optic nerve from the time when the blood vessels enter the optic nerve so on after birth. At this time the astrocytes send out processes in the direction of the blood vessels, isolating them from the ganglion cells (Dickson and Eng, 1981). A main function of the astrocytes surrounding the vessels is to induce blood-brain and blood-optic barrier properties in the endothelial cells (Janzer and Raff, 1987; Lobrinus et al., 1992). The astrocytes secrete soluble factors (Lobrinus et al., 1992; Janzer et al., 1993) which induce barrier properties like tight junctions and molecular transport polarity in the endothelial cells (Tout et al, 1993).

On the edges of the ON, separating the nerve tissue from the retinal tissue, there is another intense GFAP(+) region. This GFAP(+) region continues in the posterior PR, but here it separates the ON tissue from the surrounding choroidal tissue. These dense immunoreactive zones correspond to Kuhnt's intermediate tissue and Jacoby's limiting membrane respectively (Anderson, 1970; Hogan et al., 1971; Triviño et al., 1996; Ramírez et al., 1997).

In the RR there is a new astroglial limiting membrane. This membrane separates the nerve tissue from the meningeal sheaths surrounding it at this level; it corresponds to Greef's glial peripheral mantle (Anderson, 1969b; 1970; Triviño et al., 1996).

All these limiting membranes are composed of 3 to 4 layers of thick-bodied astrocytes that sending out straight branches join up with the processes from the neighbouring astrocytes. These processes, which are linked by desmosomes and gap junctions, run parallel to one another, leaving practically no intervening space. The cells that make up the innermost layers (close to the nerve fibres) tend to be more star-shaped, whereas the outermost ones (close to the surrounding tissues) appear flatter and more closely packed together (Triviño et al., 1996).

Electron microscopy has shown many dense bodies in the cytoplasm of these cells. Moreover, abundant phagocytosed melanin granules have been found in the astrocytes that form Kuhnt's intermediate tissue and Jacoby's limiting membrane. The principal role that has been attributed to these membranes in the PR is the formation of a barrier to the passage of molecules between the ON and the adjacent tissues. This function has been studied mainly in Kuhnt's intermediate tissue (Tso et al., 1975; Tsukahara and Yamashita, 1975; Okinami et al., 1976; Flage, 1980; Flage and Ringvold., 1980). The proposed barrier function is supported by the existence of close junctions between the astrocytes of Kuhnt's intermediate tissue and the cells of the pigmentary epithelium in the retina as well as the adherence areas between the astrocytes and the outer limiting membrane (Tso et al., 1975; Okinami et al., 1976). This barrier function could account for the numerous phagocytosed fragments of myelin and dense bodies found in the interior of the astrocytes that form these limiting membranes, which could be in the process of degradation. Although it has been shown that the astrocytes making up Kuhnt's tissue perform an important function by acting as a barrier between the ON and the retina, recent research has noted the existence of minor defects at the junctions between the retinal pigmentary epithelium and the astrocytes of Kuhnt's tissue (Cohen, 1973; Flage, 1980; Flage and Ringvold., 1980).

Another basic function accomplished by Kuhnt's tissue and Jacoby's membrane is to act as bearings, cushioning the friction that occurs in minor displacements of the ON upon movement of the eye. This prevents damage to the nerve fibres that penetrate the peripheral zones of the nerve. This function is borne out by the parallel arrangement of the astroglial branches, which are linked together by numerous desmosomes, gap junctions and tight junctions (Tso et al., 1975; Tsukahara and Yamashita, 1975; Okinami et al., 1976). Many intermediate filaments are detected in the astroglial branches, giving them a certain amount of tensile strength.

A membrane specialization has been detected in the limiting membranes of the ONH. Invaginations in the plasmatic membrane of the subpial astrocytes, described as "vesicles caveolae" (Massa, 1982), which contain contractile microfilaments. The function of the vesicles is to initiate contraction of the peripheral limiting membrane in response to the strain occurring in the meningeal sheaths of the ON.

REFERENCES

Anderson, D.R., Hoyt, W.F., Hogan, M.J. (1967). The fine structure of the astroglia in the human optic nerve and optic nerve head. Trans. Am. Ophthalmol. Soc. *65*: 274-305.

Anderson, D.R., Hoyt, W.F. (1969). Ultrastructure of intraorbital portion of human and monkey optic nerve. Arch. Ophthalmol. *82*: 506-530.

Anderson, D.R. (1969a). Ultrastructure of meningeal sheaths: normal human and monkey optic nerves. Arch. Ophthalmol. *82*: 659-674.

Anderson, D.R. (1969b). Ultrastucture of human and monkey lamina cribosa and optic nerve head. Arch. Ophthalmol. *82*: 800-814.

Anderson, D.R. (1970): Ultrastructure of the optic nerve head. Arch.Ophthalmol. *83*: 63-73.

Anderson D.R. (1973): Fine structure and function of ocular tissues: The optic nerve. In K.M. Zinn (Ed.): Ocular fine structure for the clinician. Boston, Little, Brown and Company, International Ophthalmology Clinics, *13*: 229-242.

Black, J.A., Waxman, S.G. (1985). Specialization of astrocytic membrane at glia limitans in rat optic nerve: freeze-fracture observations. Neurosci. Letters *55*: 371-378.

Black, J.A., Waxman, S.G. (1988). The perinodal astrocyte. Glia *1*: 169-183.

Black, J.A., Friedman, B., Waxman, S.G., Elmer, L.W., Angelides, K.J. (1989a). Immuno-ultrastructural localization of sodium channels at nodes of Ranvier and perinodal astrocytes in rat optic nerve. Proc. Roy. Soc. Med. London (B) *238*: 39-51.

Black, J.A., Waxman, S.G., Friedman, B., Elmer, L.W., Angelides, K.J. (1989b). Sodium channels in astrocytes of rat optic nerve in situ: immuno-electron microscopic studies. Glia *2*: 353-369.

Blunt, M.J., Wendell-Smith, C.P., Baldwin, F. (1965). Glia-nerve fibre relationships in mammalian optic nerve. J. Anat. *99*: 1-11.

Büssow, H: (1980). The astrocytes in the retina and optic nerve head of mammals: A spacial glia for the ganglion cell axons. Cell Tissue Res. *206*: 367-378.

Butt, A.M., Ransom B.R. (1989). Visualization of oligodendrocytes and astrocytes in the intact rat optic nerve by intracellular injection of Lucifer Yellow and horseradish peroxidase. Glia *2*: 470-475.

Butt, A.M., Colquhoun K., Berry M. (1994). Confocal imaging of glial cells in the intact rat optic nerve. Glia *10*: 315-322.

Cohen, A.I. (1967). Ulstrastructural aspects of human optic nerve. Invest. Ophthalmol. *6*: 294-308.

Cohen, A.I. (1973). Is there a potential defect in the blood-retinal barrier at the choroidal level of the optic nerve canal?. Invest. Ophthalmol. *12*: 513-519.

Cook, R.D., Raine, C.S., Wisnieswski, H.M. (1973): On perivascular astrocytic membrane specializations in monkey optic nerve. Brain Res. *57*: 491-497.

Dixon, R.G., Eng, L.F. (1981): Glial fibrillary acidic protein in the optic nerve of the developing albino rat: an immunoperoxidase study of paraffin-embedded tissue. J. Comp. Neurol. *201*: 15-24.

Duke-Elder S., Wybar, K.C. (1961): The optic nerve. In S. Duke-Elder (Ed.): System of Ophthalmology. Vol. II: The anatomy of the visual system; London , Henry Kimpton; pp: 273-294.

Elkington, A.R., Inman, C.B., Steart, P.V., Weller, R.O. (1990): The structure of the lamina cribosa of the human eye: an immunocytochemical and electron microscopical study. Eye *4*: 42-57.

Ffrench-Constant, C., Raff, M.C. (1986a). Proliferating bipotential glial progenitor cells in adult rat optic nerve. Nature *319*: 499-502.

Ffrench-Constant, C., Raff, M.C. (1986b). The oligodendrocyte-type-2 astrocyte cell lineage is specialized for myelination. Nature, *323*: 335-338.

Ffrench-Constant, C., Miller, R.H., Kruse, J., Schachner, M., (1986). Molecular specialization of astrocyte processes at nodes of Ranvier in rat optic nerve. J. Cell Biol. *102*: 844-852.

Flage, T. (1980): A defect in the blood-retina barrier in the optic nerve head region in the rabbit and the monkey. Acta Ophthalmol. *58*: 645-651.

Flage, T., Ringvold, A. (1980): Demonstration of a diffusional pathway between the subretinal space and the juxtapapillary connective tissue. An in vitro experiment using horseradish peroxidase as a tracer. Acta Ophthalmol. *58*: 899-907.

Hayreh, S.S., Vrabec, Fr. (1966). The structure of the head of the optic nerve in Rhesus monkey. Am. J. Ophthalmol. *61*: 136-150.

Hayreh, S.S. (1974). Anatomy and physiology of the optic nerve head. Trans. Am. Acad. Ophthalmol. Otolaryngol. *78*: 240-274.

Hayreh, S.S. (1978). Structure and blood supply of the optic nerve. In K. Heilmann, K.T. Richardson (Eds.) : Glaucoma: Conceptions of a disease. Pathogenesis, diagnosis, therapy. Stuttgart, Georg Thieme Publishers; pp: 78-96.

Hayreh, S.S. (1984). The sheath of the optic nerve. Ophthalmologica *189*: 54-63.

Hayreh, S.S. (1987). Factors determining the glaucomatous optic nerve head damage. In G.K. Krieglstein (Ed): Glaucoma Update III. Berlin, Springer-Verlag; pp: 40-46. .

Hayreh, S.S. (1989). Blood supply of the optic nerve head in health and disease. In G.N. Lambrou; E.L. Greve (Eds): Ocular bood flow in glaucoma: means, methods and measurements. ; Amstelveen, Kugler & Ghedini; pp: 3-54 .

Hayreh, S.S. (1994). Progress in the understanding of the vascular etiology of glaucoma. Curr. Op. Ophthalmol. *5(II)*: 26-35.

Hidelbrand, C., Waxman, S.G. (1984). Postnatal differentiation of rat optic nerve fibers: electron microscopic observations on the development of nodes of Ranvier and axoglial relations. J. Comp. Neurol. *224*: 25-37.

Hogan, M.J., Alvarado, J.A., Weddell, J.E. (1971). Histology of the human eye. An atlas and textbook. Chapter 10: Optic nerve. Philadelpia, W.B. Saunders Company.; pp: 523-606.

Janzer, R.C., Raff, M.C. (1987). Atrocytes induce blood-brain barrier properties in endothelial cells. Nature, *325*: 253-257.

Janzer, R.C., Lobrinus, J.A., Darekar, P., Juillerat, L. (1993). Astrocytes secrete a factor inducing the expression of HT7-protein and neurothelin in endothelial cells of chorioallantoic vessels. Adv. Exp. Med. Biol. *331*: 217-221.

Jonas, J.B., Naumann, G.O.H. (1993). The optic nerve: Its embryology, histology, and morphology. In R. Varma, G.L. Spaeth, K.W. Parker (Eds.): The Optic Nerve in Glaucoma. Philadelphia, J.B. Lippincott Company(; pp: 3-26.

Kestenbaum, A. (1963).: Applied anatomy of the eye. Chapter 5: The optic nerve. New York, Grune & Stratton; pp: 129-147.

Lazarides, E. (1980). Intermediate filaments as mechanical integrators of cellular space. Nature, *283*: 249-256.

Levitzky, M., Henkind, P. (1969). Angioarchitecture of the optic nerve II: Lamina cribosa. Am. J. Ophthalmol. *68*: 986-996.

Liss, L. (1956). The astroglia of the human optic nerve, chiasma and tract. A study with silver-carbonate. J. Neurol. *105*: 151-160.

Lobrinus, J.A., Juillerat, J.L., Darekar, P., Schlosshauer, B., Janzer, R.C. (1992). Induction of the blood-brain barrier specific HT7 and neurothelin epitopes in endothelial cells of the chick chorioallantoic vessels by a soluble factor derived from astrocytes. Dev. Brain Res. *70*: 207-211.

Malmfors, L. (1963). Electron microscopic description of the glial cells in the nervus opticus in mice. J. Ultrastruct. Res. *8*: 193.

Massa, P.T. (1982). Plasmalemmal vesicles (caveolae) of fibrous astrocytes of the cat optic nerve. Am. J. Anat. *165*: 69-81.

Miller, R.H., Raff, M.C. (1984). Fibrous and protoplasmic astrocytes are biochemically and developmental distinct. J. Neurosci. *4*: 585-592.

Miller, R.H., Abney ,E.R., David, S., Ffrench-Constant, C., Lindsay, R., Patel, R., Stone, J., Raff, M.C. (1986). Is reactive gliosis a property of a distinct subpopulation of astrocytes. J. Neurosci. *6*: 22-29.

Miller, R.H., Ffrench-Constant, C., Raff, M.C. (1989a). The macroglial cells of the rat optic nerve. Ann. Rev. Neurosci. *12*: 517-534.

Miller, R.H., Fulton, B.P., Raff, M.C. (1989b). A novel type of glial cell associated with nodes of Ranvier in rat optic nerve. Eur. J. Neurosci. *1*: 172-180.

Morrison, J.C., L'Hernault, N.L., Jerdan, J.A., Quigley, H.A. (1989a). Ultrastructural location of extracellular matrix components in the optic nerve head. Arch. Ophthalmol. *107*: 123-129.

Morrison, J.C., Jerdan, J.A., Dorman-Pease, M.E., Quigley, H.A. (1989b). Structural proteins of the neonatal and adult lamina cribosa. Arch. Ophthalmol. *107*: 1220-1224.

Okinami, S., Ohkuma, M., Tsukahara, T. (1976). Kuhnt intermediary tissue as a barrier between the optic nerve and retina. Graef. Arch. Clin. Exp. Ophthalmol. *201*: 57-67.

Ortiz-Miterrer, G. (1974). Contribución al estudio de la neuroglia del nervio óptico humano. Arch. Soc. Esp. Oftalmol. *34*: 105-128.

Quigley, H.A. (1977). Gap junctions between optic nerve head astrocytes. Invest. Ophthalmol. Vis. Sci. *16*: 582-585.

Raff, M.C., Abney, E.R., Cohen, J., Lindsay, R., Noble, M. (1983a). Two types of astrocytes in cultures of developing rat white matter: differences in morphology, surface gangliosides, and growth characteristics. J. Neurosci. *3*: 1289-1300.

Raff, M.C., Miller, R.H., Noble, M. (1983b). Glial cell lineages in the rat optic nerve. Colds Spring Harb. Symp. Quant. Biol. *48*: 569-572.

Raff, M.C., Williams, B.P., Miller, R.H. (1984). The in vitro differentiation of a bipotential glial progenitor cell. EMBO J. *3*: 1857-1864.

Raff, M.C. (1989). Glial cell diversification in the rat optic nerve. Science, *243*: 1450-1455.

Ramírez, J.M., Triviño, A., García-Sánchez, J. (1984). Vascularización de la cabeza del nervio óptico en el hombre.Arch. Soc. Esp. Oftalmol. *46*: 413-426.

Ramírez, J.M., Salazar, J.J., Triviño, A., Ramírez, A.I., García-Sánchez, J. (1989). Membranas limitantes gliales y fasciculación axonal en la cabeza del nervio óptico en el conejo. Glaucoma *3*: 111-119.

Ramírez, J.M., Triviño, A. (1989). Anatomofisiología del nervio óptico. In F. Gómez-Ulla, F. Marín, J.M. Ramírez, A. Triviño (Eds): La circulación coroidea. Barcelona, Edika-Med; pp: 31-40.

Ramírez, J.M., Triviño, A., Ramírez, A.I., Salazar, J.J., García-Sánchez, J. (1994). Immunohistochemical study of human retinal astroglia. Vision Res. *34*: 1935-1946.

Ramírez, J.M., Salazar, J.J., Ramírez, A.I., Andrés M.V., Triviño, A. (1997). Morfología y distribución de los astrocitos del nervio óptico en el conejo albino. Arch. Soc. Esp. Oftalmol. *72*: 153-160.

Raynaud, C., Milin-Cioni, A.V. (1992). La cellule retino-thalamique. In P. Sole, H. Dalens, C. Gentou (Eds): Biophtalmologie. Livre VI, Le Systeme Nerveux, Chapitre IV. Paris, Masson; pp: 135-172.

Reeh, M.J. (1981). The optic nerve: definition and description. In M.J. Reeh, J.L. Wobig, J.D. Wirtschafter (Eds): Ophthalmic anatomy. A manual with some clinical applications, Part 2: The globe. ; California, American Academy of Ophthalmology; pp: 146-151.

Roots, B.I. (1986). Phylogenetic development of astrocytes. In: S. Fedoroff, A. Vernadakis (Eds): Astrocytes vol. 1: Development, morphology, and regional specialization of astrocytes. Orlando, Academic Press; pp: 1-34.

Senelick, R.C. (1982). Optic nerve. In T.D. Duane, E.A. Jaege (Eds): Biomedical Foundations of Ophthalmology; Vol. II. Philadelphia, Harper & Row Publishers; pp: 1-34.

Skoff, R.P., Toland, D., Nast, E. (1980). Pattern of myelination and distribution of neuroglial cells along the developing optic system of the rat and rabbit. J. Comp. Neurol. *191*: 237-253.

Skoff, R.P., Knapp, P.E., Bartlett, W.P. (1986). Astrocytic diversity in the optic nerve: a cytoarchitectural study". In S. Fedoroff, A. Vernadakis (Eds): Astrocytes. Vol. 1: Development, morphology, and regional specialization of astrocytes. Orlando, Academic Press, INC; pp: 269-291.

Suárez, I., Raff, M.C. (1989). Subpial and perivascular astrocytes associated with nodes of Ranvier in the rat optic nerve. J. Neurocytol. *18*: 577-582.

Triviño A., Ramírez J.M., Salazar J.J., Ramírez A.I., García-Sánchez, J. (1996). Immunohistochemical study of human optic nerve head astroglia. Vision Res. *36*: 2015-2028.

Tso, M.O.M., Shih, C.Y., McLean, M.I.W. (1975). Is there a blood-brain barrier at the optic nerve head?. Arch. Ophthalmol. *93*: 815-825.

Tsukahara, I., Yamashita, H. (1975). An electron microscopic study of the blood-optic nerve and fluid-optic nerve barrier. Graef. Arch. Clin. Exp. Opthalmol. *196*: 239-246.

Tout, S., Dreher, Z., Chan-Ling, T.L., Stone, J.(1993). Contact-spacing among astrocytes is independent of neighbouring structures: in vivo an in vitro evidence. J. Comp. Neurol. *332*: 433-443.

Vaughn, J.E., Peters, A. (1967). Electron microscopy of the early postnatal development of fibrous astrocytes. Am. J. Anat. *121*: 131-152.

Waxman, S.G., Black, J.A. (1984). Freeze-fracture ultrastructure of the perinodal astrocyte and associated glial junctions. Brain Res. *308*: 77-87.

Wendell-Smith, C.P., Blunt M.J., Baldwin, F. (1966). The ultrastructural characterization of macroglial cell types. J. Comp. Neurol. *127*: 219-240.

Wilczek, M. (1947). The lamina cribosa and its nature. Br. J. Ophthalmol. *31*: 551-565.

Wirtschafter, J.D. (1981). The visual pathway. In M.J. Reeh, J.L. Wobic, J.D. Wirtschafter (Eds): Ophthalmic anatomy. A manual with some clinical applications, Part 3: Ophthalmic neuroanatomy.California, American Academy of Ophthalmology; pp: 155-186.

Wollf, E. (1961). Anatomy of the eye and orbit. Including the central connections, development, and comparative anatomy of the visual apparatus (Fifth Ed., Revised by R.J. Last); Chater VII: The visual pathway. Philadelphia, W.B. Saunders Company; pp: 295-315.

Wolf-Heidegger, G. (1954). Anatomía de las formaciones extrabulbares del órgano de la visión". In M. Amsler, A. Brückner, A. Franceschetti, H. Goldmann, E.B. Streiff, (Eds): Oftalmología.Barcelona, Salvat Editores S.A; pp: 197-218.

Wolter, J.R. (1957). The human optic papilla. A demonstration of new anatomic and pathologic findings. Am. J. Ophthalmol. *44* (Pt. II): 48-64.

Wolter, J.R. (1961). Silver carbonate techniques for the demonstration of ocular histology. In G.K. Smelser (Ed): The structure of the eye. New York, Academic Press; pp: 117-138.

5

GLIAL CELLS ONTOGENY IN THE TELENCEPHALON AND MESENCEPHALON OF THE LIZARD *Gallotia galloti*

Maximina Monzón-Mayor, Carmen M. Yanes, María M. Romero-Alemán, Jean de Barry, Robert R. Sturrock and Giorgio Gombos

In contrast to the extensive literature on mammalian gliogenesis, relatively little attention has been paid to lower vertebrates and particularly to reptiles. Along this chapter we have documented the distribution of macroglial cells and the modifications of their precursors during ontogeny in the central nervous system (CNS) of the lizard Gallotia galloti by using histological, immunohistological methods, and electron microscopy. Our observations showed that most glial cell phenotypes in Gallotia galloti and the sequence of their morphological differentiation and maturation do not greatly differ from those of mammals. However there are some differences that should be pointed. The first striking feature is the different glial distribution between telencephalon and mesencephalon as demonstrated by double-labelling immunocytochemistry. Another remarkable feature is the large proportion of radial glial cells present in the adult brain. Finally, another particularity of special interest, is the presence of immature cells in some ventricular sulci throughout lizard lifetime. Probably, the abundance of immature cells in mesencephalic and telencephalic sulci of adult nuclei could be related to the regenerative capacity of reptilian CNS.

Most studies on glial cells and gliogenesis in the central nervous system (CNS) carried out by using ultrastructural (Blunt et al., 1972; Caley and Maxwell, 1968; Fedoroff, 1986; Immamoto et al., 1978; Lord and Duncan, 1987; Parnavelas et al., 1983; Phillips, 1973; Privat and Leblond, 1972; Skoff et al., 1976a,b; Sturrock, 1974, 1975, 1976, 1978, 1980, 1982a, 1986; Vaughn and Peters, 1967; Vaughn, 1969) and/or immunohistochemical methods (Antanitus et al., 1976; Bignami et al., 1982; Cameron and Rakic, 1991; Choi, 1988; Dixon and Eng,

1981; Ghandour et al., 1983; Luskin et al., 1993; Noble et al., 1989; Skoff and Knapp, 1991) were performed in mammals. Also the older studies with the Golgi impregnation method were essentially in mammals (Kölliker, 1893; Lenhosséck, 1893; Ramón y Cajal ,1894). Such studies have shown that the major ultrastructural feature of mature mammalian astrocytes is the presence of specific intermediate filaments or gliofilaments (Schultz, 1964; Vaughn and Peters, 1967; Mori and Leblond, 1969). Fibrous astrocytes contain a large number of gliofilaments, protoplasmic astrocytes contain few gliofilaments and cells with an intermediate gliofilament equipment also can be found (Sturrock, 1986; Wilkin et al., 1990). Gliofilaments are made of Glial Fibrillary Acidic Protein (GFAP) (Eng et al., 1971; Bignami et al., 1972, 1980; Dahl and Bignami, 1973; Schachner et al., 1977; Ghandour et al., 1979, 1983; Eng, 1980; Levitt and Rakic, 1980; Shaw et al., 1981; Choi et al., 1983; Onteniente et al., 1983; Dahl et al., 1985; Didier et al., 1986). Astrocyte precursors in most animal species contain intermediate filaments made of vimentin that are gradually substituted by gliofilaments made of GFAP as the cells mature. Dahl et al. (1981, 1982) observed that in rat encephalon and spinal cord, vimentin decreases and GFAP increases during the second and third postnatal week, while Pixley and De Vellis (1984) have shown that in developing rat brain vimentin disappears with radial glia (see below) and GFAP appears with astrocyte maturation. The quantity of intermediate filaments varies in different animal species and in different regions of the same species in adult animals. This is one aspect of the morphological heterogeneity of astrocytes which might represents one facet of their response to different microenviroments (Privat and Rotaboul, 1986). Glial Fibrillary Acidic Protein is present not only in "classical" astrocytes but also in tanycytes and radial glia (RG). Phylogenetically, RG is the most primitive form of glia (Kruger and Maxwell, 1966; Friede and Stone, 1969; Schachner et al., 1977; Polak et al., 1982; Onteniente et al., 1983; Miller and Liuzzi, 1986) and the first to appear during ontogeny (Ramón y Cajal, 1911; Levitt and Rakic, 1980) and contains also vimentin. In mammals (Rakic, 1971; Schmechel and Rakic, 1979; Eckenhoff and Rakic, 1984) and in chick spinal cord (Ramón y Cajal, 1894) RG disappears during ontogeny. Rakic (1972, 1982) and Choi et al. (1983) have described this phenomenon in the monkey and in the human, respectively. These authors suggested that radial glia is transformed into other glial cell types since the disappearance of the radial glia is accompanied by the appearance of cells with an intermediate aspect between RG and astrocytes (Rakic, 1982; Choi et al., 1983) or RG and oligodendrocytes (Choi et al., 1983).

Other mammalian glial cells, the ependymal cells, which coat the ventricles and display microvilli and cilia projecting in the ventricular lumen (Wood and Cohen, 1980), often contain intermediate filaments but these are not composed of glial fibrillary acidic protein (GFAP).

In addition to GFAP, another protein, preferentially localized, with few exceptions, in mammalian astrocytes "in situ" (Martinez-Hernández et al., 1977; Norenberg and Martinez-Hernández, 1979; Didier et al., 1986), in Müller cells in the retina (Riepe and Norenberg, 1977; Sarthy and Lam, 1978) and in astrocytes in primary mouse brain cell cultures (Hallermayer et al., 1981), is the enzyme Glutamine synthetase (GS). Astrocytes preferentially or exclusively expressing either GS or GFAP, however, were described in rat brain (Didier et al., 1986) and Hallermayer and Hamprecht (1984) described GS-negative rat astrocytes in primary cultures, whereas Patel et al. (1985) reported that the ratio between GFAP and GS varied in the different regions of the adult rat CNS. In mammals, oligodendrocytes, the other macroglial cells, do not contain GFAP while the absence of GS is not certain in all cases. Galactocerebrosides and carbonic anhydrase isoenzyme II are good markers of mammalian oligodendrocytes. Oligodendrocytes are responsible for myelination in the CNS and it is therefore not surprising that the rapid increase in glial cell number which precedes the onset of myelination and continues during the phase of rapid myelination (Roback and Scherer, 1935) is mainly due to proliferation and differentiation of oligodendrocytes (Vaughn, 1969; Sturrock, 1974, 1982a,b; Skoff et al., 1976 a,b). The myelinated nerve tracts can be detected, and myelination can be followed, by using an antibody directed against myelin basic protein (MBP) present in CNS and peripheral nervous system (PNS) myelin, (Matthieu et al., 1973; Sternberger et al., 1978a,b; Eng and Bigbee, 1978). In many species, MBP has also been transiently detected, during myelination, in immature oligodendrocyte cell bodies (Mattieu et al., 1973; Sternberger et al., 1978a,b; Eng and Bigbee 1978) and, in this way, oligodendrocyte formation was followed (Campagnoni et al., 1989). Several authors have published studies on the identification and/or localization of MBP in the CNS, particularly in mammals, amphibians and birds (Mendel and Withaker, 1978; Hartman et al., 1979; Norton, 1981; Nussbaum and Roussel, 1983) but not in reptiles.

GLIA AND GLIOGENESIS IN LIZARD

In contrast to the extensive literature on mammalian gliogenesis relatively little attention has been paid to lower vertebrates (i. e. non mammals), and particularly in reptiles. The few data available on glia and gliogenesis in lower vertebrates however, show that all the different types of macroglial cells are present in lower vertebrate though they exhibit some differences from mammalian macroglia together with some similarities. In addition to its intrinsic interest, a comparative studies on glial cells of different animal species at different evolutive level might help to clarify a few questions, including the suggested origin of astrocytes from radial glia. We have documented the distribution of macroglial cells and the modifications of their precursors during ontogeny in the CNS of *Gallotia galloti* (Monzón-Mayor et al., 1990a-d, 1991, 1997; Yanes et al., 1988 a,b, 1990, 1992, 1997) by using immunohistological methods and electron microscopy.

GLIA AND GLIOGENESIS IN LIZARD: ASTROCYTES

The development of astroglial cells and their origin from spongioblasts were first studied by Kölliker (1893), Lenhosséck (1893) and Ramón y Cajal (1894) by using the Golgi impregnation method but only few studies were carried out to examine, with the same method, the ontogeny of neural cells in lower vertebrates, particularly in the lizard (Díaz et al., 1990a,b; Medina et al., 1990, 1991; Báez, 1992).

Consequently, we studied, with Golgi techniques, the differentiation sequence of the cells of the astroglial line in lizard from the embryonic stage 35 (E.35) to adult age (Romero-Alemán et al., 1995).

Radial glia and astroglial cell development studied with Golgi techniques

In adult *Gallotia galloti* the RG is present and appears to be similar to that in other lower vertebrates (Ramón y Cajal, 1911) and mammals (Schmechel and Rakic, 1979). Astrocytes are also present in the midbrain while the telencephalon is characterized by the absence of impregnated mature astroglial cells. Impregnated immature astroglia, however, is present in the telencephalon throughout adult life. This observation is consistent with ultrastructural (Yanes et al., 1997) and immunohistochemical observations (Yanes et al., 1990) (see below).

During ontogeny, radial glial cells, all of them in mammals but only few in lizard midbrain (Romero-Alemán et al., 1995), are transformed into protoplasmic astrocytes. The course of this transformation in lizards is similar to that observed in mammals (Rakic, 1971; Schmechel and Rakic, 1979; Eckenhoff and Rakic, 1984) and in chick spinal cord (Ramón y Cajal, 1894). Some of the *Gallotia galloti* protoplasmic astrocytes like those in monkey telencephalon (Schmechel and Rakic, 1979) retain some very long processes that could be considered as vestigial RG processes (fig. 5-3).

Golgi impregnated cells, appearing as transitional stages between RG and fibrous astrocytes as those in mammals (Rakic, 1971; Schmechel and Rakic, 1979), do not occur in *Gallotia galloti*. Cells with intermediate characteristics between "displaced radial glia" and immature protoplasmic astrocytes do occur, however, suggesting that the latter originate from the former and ultimately from radial glia.

Glioblasts, the early precursors of astrocytes and oligodendrocytes, are also present in *Gallotia galloti* CNS. Their distribution and time of formation, determined in Golgi-impregnated tissue, agree with the results obtained in ultrastructural studies (see below) (Monzón-Mayor et al., 1990c). Since glioblasts are considered as stem-cells of astrocytes and oligodendrocytes, we proposed that they are precursors of a line of astroglial cells different from that originating from RG. The presence of Golgi impregnated cells exhibiting irregular cytoplasmic protrusions and processes similar to those described for "small glioblasts" in rats (Vaughn, 1969) and the presence of star-shaped astroblasts not differing from the astroblasts described in rats (Skoff et al., 1976a) corroborate our hypothesis.

Golgi impregnated mature and immature astroglial cells were frequently observed in the same area of *Gallotia galloti* brain suggesting a certain asynchrony in the maturation of astroglial cells in a given zone; this agrees with Vaughn's data (1969) in rats. In *Gallotia galloti*, mature protoplasmic astrocytes already appear at early embryonic stages whereas the first mature fibrous astrocytes appear at late embryonic stages. This sequence agrees with that observed in monkeys (Schmechel and Rakic, 1979).

Electron microscopy

Astrocytes, characterized by the presence of abundant gliofilaments in the cell body were found in the midbrain (fig. 5-1F) of *Gallotia galloti* but never in telencephalon. Their ultrastructure is similar to that of mammalian astrocytes. In general, astrocytes and ependymal cells of non-mammalian vertebrates contain greater amounts of glycogen than mammalian astrocytes (Kruger and Maxwell, 1967; Roots, 1986) but in *Gallotia galloti*, glycogen is only present in small quantities in some developing astrocytes and in processes of fibrous astrocytes in contact with blood vessels or at the pia-glial interface (Monzón-Mayor et al., 1990c).

In the rat optic nerve (Vaughn, 1969; Skoff et al., 1976a) and in the mouse commissura anterior (Sturrock, 1974) the whole sequence of mammalian glial cell maturation and the ultrastructural differences between glial cells at each major stage of maturation have been described. Such differences were based on the electron density of the nucleus, of the cytoplasm, and on the characteristics of organelles. Vaughn (1969) described the presence of short strand of dilated rough endoplasmic reticulum filled with floculent material as one of the earliest signs of astrocyte differentiation, preceding the appearance of gliofilaments or glycogen. In cultured astroglial cells, microtubules appeared before gliofilaments and both were assembled in the centriolar region (Fedoroff, 1986). The characteristic endoplasmic reticulum described by Vaughn (1969) in rat optic nerve was also found in astrocyte precursors by Skoff et al. (1976a) in the rat optic nerve, by Parnavelas et al. (1983) in the rat visual cortex and by Sturrock (1975) in the human optic nerve. In the mouse, astrocytes and astrocyte precursors do not have dilated rough endoplasmic reticulum but contain an electron dense substance (Sturrock, 1986).

Indeed the different precursors appear in lizard telencephalon and midbrain according to the same sequences as in mammals: from "early glioblasts" originate "light" and "dark glioblast"; the "light glioblasts" mature in "oligodendroblasts" and the "dark glioblast" in "astroblast" (Sturrock, 1974).

Astroblast can be identified in the lizard brain (Monzón-Mayor et al., 1990c) by the presence of dilated endoplasmic reticulum filled with a material similar to that seen in mammalian astrocyte precursors. In lizards, astroblasts are first seen at E.35 in the midbrain and at E.37 in the telencephalon as very immature astrocytes which contain small quantities of gliofilaments (fig. 5-1E) and, more rarely, glycogen granules.

Immature astrocytes exhibing few gliofilaments were detected at E.35 in the midbrain and E.37 in telencephalon of *Gallotia galloti* (Monzón-Mayor et al., 1990c; Yanes et al., 1997). Astroglia seems to be unable to attain full maturity in adult lizard forebrain.

Glial development studied with immunohistochemistry

We have also characterized the astroglial cell development in the telencephalon and mesencephalon of *Gallotia galloti* (Yanes et al., 1990; Monzón-Mayor et al., 1990b) by using immunohistochemistry with antibodies directed against vimentin, GFAP or GS. As early as 1985 Dahl et al. had shown that astrocytes and Bergmann glia in the cerebellum of the lizard *Anolis carolinensis* were GFAP-positive.

Glial Fibrilary Acidic Protein

We have shown (Yanes et al., 1990; Monzón-Mayor et al., 1990b) that the GFAP-positive radial glia is present in the adult lizard CNS. Scattered GFAP positive cell bodies are also present in the adult mesencephalon (fig. 5-1B), but not in the telencephalon. In the adult GFAP-immunoreactive cell fibers (radially or tangentially disposed or criss-crossing some areas), perivascular and subpial end feet are, however, present not only in mesencephalon but also in telencephalon, including cortex and striatum ventralis. The origin of these fibers , i.e. from RG or scattered astrocytes, was not accertained. Interspecific differences in GFAP-immunoreactive cell distribution in reptiles are striking: in a snake (Onteniente et al., 1983) and in a turtle (Kriegstein et al., 1986) GFAP-positive astrocytes are absent in the tegmentum and present in hippocampus (Onteniente et al., 1983) while GFAP-positive cells are present in the optic tracts of these species.

In mesencephalon, GFAP immunoreactivity (GFAPir) appears first in ventricular walls, radial fibers, subpial and perivascular end feet, increased between E34 and E37, then decreased in radial fibers and increased in astrocytes with age (Monzón-Mayor et al., 1990). In telencephalon, GFAP appears first at E35 in end feet in the marginal zone and then appears in radial and sinuous fibers and in fibers originating from the sulci and invading ventral striatum (fig. 5-1A), and septum (Yanes et al., 1990). The intensity of GFAPir increases with age. Immunoreactivity in adult telencephalon remains weak in cortex, amygdala and in the Anterior Dorsal Ventricular Ridge.

Vimentin

In *Gallotia galloti*, as in other species (Bignami et al., 1982) vimentin is the major component of the intermediate filament of immature glia before GFAP becomes detectable. Its cellular localization has been described in mammals by

Lazarides (1980, 1982), Schnitzer et al. (1981), Bignami et al. (1982), and Pixley and De Vellis (1984). Vimentin-containing RG in the lizard is very similar to that in mammals. However vimentin-positive scattered cells are absent in *Gallotia galloti* telencephalon.

In birds and mammals, the expression of vimentin coincides with neuronal proliferation and migration (Dahl et al., 1981; Schnitzer et al., 1981; Pixley and De Vellis, 1984; Quitschke et al., 1985; Schiffer et al., 1986). Simi-larly in *Gallotia galloti* midbrain the maximal expression of vimentin (*e.g.* at E.34-E.35) apparently coincides with neuronal migration as suggested by the presence of stacks of immunonegative cells piled up in between strongly vimentin positive fibers. In the telencephalon the first wave of neuroblast migration and the maximal expression of vimentin occur between embryonic stages E.35 and E.36 (Yanes, 1985; Yanes et al., 1987, 1989). Axonal and dendritic growth also occurs between embryonic stages E.35 and E.37 (Díaz et al., 1990a).

It is not know if the simultaneous occurrence of the two events (maximal expression of vimentin and neuronal migration) has any meaning *i.e.* expresses a yet unknown aspect of glial/neuronal cooperation.

Another role for vimentin-containing fibers was proposed by Pixley and De Vellis (1984) who suggested their involvement in transport processes between "the extracellular fluid compartment" (*i.e.* extracellular space, cerebrospinal fluid, blood) and the adjacent tissue parenchyma. This hypothesis was based on the following considerations: all cultured cells express vimentin; all the cells which express vimentin in the adult in vivo are cells in contact with extracelullar fluids; and above all, in the CNS vimentin-positive radial fibers disappear at the time of the dramatic reduction of extracellular space. This hypothesis cannot be extended to *Gallotia galloti* and other lower vertebrates, however, since in these species the developmental change of extracellular space is not associated with the disappearance of radial glial fibers and vimentin-positive end-feet are present in the adult.

In *Gallotia galloti* midbrain the critical phase when vimentin begins to decrease and GFAP starts to accumulate occur at E35. However, some fibers and particularly end-feet are still vimentin-positive in the adult and we do not know if these processes belong to radial glia or to astrocytes. In *Gallotia galloti* telencephalon, vimentin immunoreactivity decreases starting from E.37 until it almost completely disappears in the adult except in few end-feet. The great decrease of vimentin in RG is associated with the accumulation of GFAP in these fibers. Thus in *Gallotia galloti* midbrain (fig. 5-1B) and telencephalon (fig. 5-1A), although RG does not disappear, the composition of its intermediate filaments greatly changes. Between E.37 and E.40 both vimentin and GFAP are present either in the same or in contiguous fibers. This is not an unique phenomenon, in fact vimentin and GFAP coexist in the same cell in mammalian cultured astrocytes and, in some cases, in astrocytes in situ (Schnitzer et al., 1981; Shaw et al., 1981; Dahl et al., 1982; Lazarides, 1982; Schiffer et al., 1986; Chiu et al., 1988).

In the adult mesencephalon, GFAP-positive star-shaped astrocytes (fig. 5-1B) are present, thus it is possible that the scattered vimentin-positive cells seen at early developmental stages are immature astrocytes, which become GFAP-positive cells at a more mature stage of development. If this is the case, the vimentin to GFAP shift is common to radial glia and astrocytes and occurs almost simultaneously in the two cell types. The coincidence of these two phenomena is surprising if it involves cells derived from RG but, possibly, it essentially involves glioblasts-derived astrocytes.

Glutamine synthetase

Cellular localization of Glutamine synthetase in adults:

Glutamine synthetase (GS) localization during ontogeny in reptile CNS complements our immunohistological studies on astroglial development (Monzón-Mayor et al., 1990a).

In the adult *Gallotia galloti* telencephalon and mesencephalon, GS is present in radial glial processes and in ependymal glial cell bodies. This cellular distribution is similar to that of GFAP (Yanes et al., 1990; Monzón-Mayor et al., 1990b) although in subpial end-feet GS immunoreactivity is less intense than that of GFAP. GS-positive star-shaped cells with the morphology of typical astrocytes are abundant in the striatum ventralis, amygdaloid complex and commissura anterior, all areas without GFAP immunoreactive perikarya. They are also present in tectum and mesencephalic tegmentum (fig. 5-1D) areas where GFAP immunoreactive perikarya are present.

In addition to these star-shaped cells, GS-positive ovoid or pear shaped cells (as indicated by immunofluorescence or HRP stain) are scattered throughout mesencephalon and telencephalon. Since GFAP-positive perikarya are absent in telencephalon the GS-positive cell bodies in the same area do not contain GFAP. In all midbrain areas, GS-positive ovoid cells are much more abundant than the GFAP-positive cells in the same area (Monzón-Mayor et al., 1990a, b; Yanes et al., 1990). The ultrastructural features of these cells and the presence of processes originating from their cell bodies and terminating as end-feet on blood vessels suggest that they are astrocytes (fig. 5-1C), thus the absence of star-like shape alone is not a sufficient criterion for doubting the astrocytic nature of these cells.

Glutamine synthetase ontogeny:

During embryonic development, GS immunoreactivity appears in the ependymal glial cell bodies and in radial glial processes at about the same time as, or slightly after, that of GFAP (thus later than that of vimentin), and it increases in parallel with the formation and maturation of ependymal and radial glia. This is a

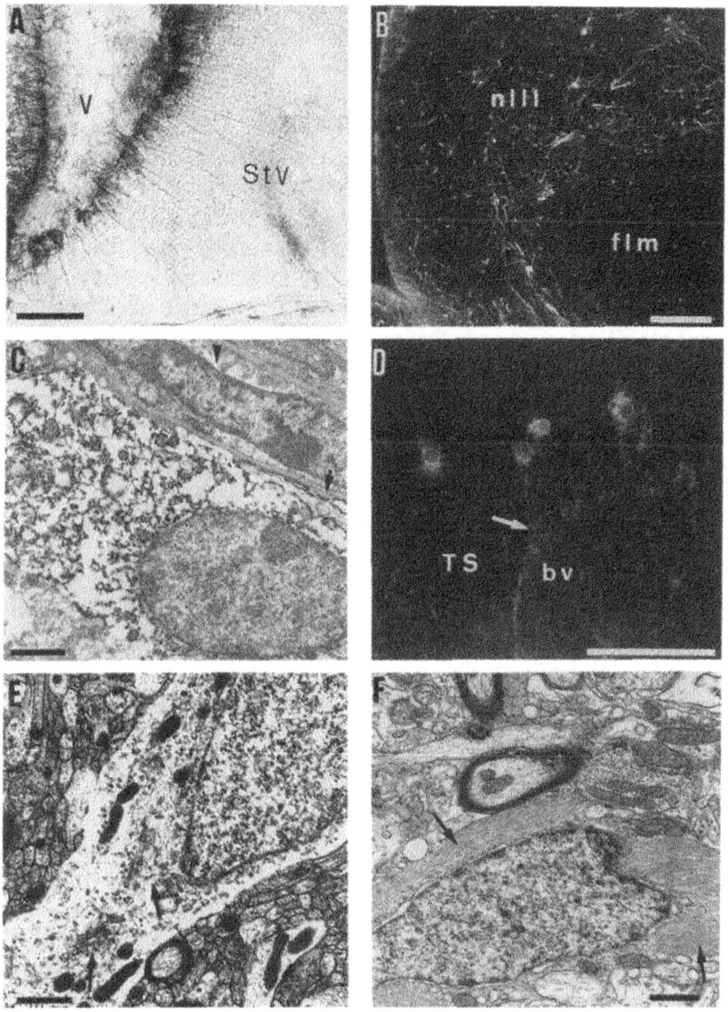

Figure 5-1. Indirect immunofluorescence with anti-GFAP immunoserum in adult *Gallotia galloti* telencephalon and midbrain transverse sections at the level of the Sulcus ventralis (A) and the oculomotor nuclei (B). Notice the radial orientation of stained fibers in the striatum ventralis (vs) and the strongly immunofluorescent astrocytes in tegmentum (B) which contrasts with the virtual absence of immunofluorescence in fasciculus longitudinalis medialis (flm). Scale bars: A=50μm B=100μm. Transverse sections of *Gallotia galloti* telencephalon showing glutamine synthetase immunoreactivity as revealed by immunoperoxidase reaction in electron microscopy (C). The immunoprecipitate is localized in the cytoplasm of an astrocyte. Note the positive perivascular end-foot (arrow) and the negative vascular endotelial cells (arrowhead). In (D) notice with immunofluorescence method the positive cells apparently "migrating" out of the ventricular zone at E.40 and the perivascular glial processes (arrow) in the Torus semicircularis (TS) of the midbrain. Scale bars: C=100μm D=100μm. Electronmicrograph of an immature astrocyte in telencephalon (E). Notice a few gliofilament (arrow) and fibrous astrocyte with a cytoplasm packed with gliofilaments (arrow) (F) in the Adult lizard mesencephalon. Scale bars: E = 100μm, F = 100μm.

general feature of glial maturation, since it was also observed in mammalian astrocytes in situ and in cultures in vitro (Patel, 1986). Topographically, GS, like GFAP, reveals a caudorostral progression of astrocytes maturation during development (Table 5-1).

Also the immature glial cells (Vim-, GFAP- and GS- positive) are present in "proliferative zones" at the level of the mesencephalic sulcus. These zones could be a reservoir from which immature glial cells migrate before they mature throughout the lizard life (Monzón-Mayor et al., 1990a,b). After hatching, GS immunoreactivity clearly decreased, particularly in radial fibers, but it is unclear if the regression of GS immunoreactivity is due to a reduction of the cellular GS concentration or to a decrease of the number of GS-positive ependymal glial cell bodies or both. We can suppose that during cell multiplication, when the need for amino groups for the synthesis of nuclei acid bases is high, the need for glutamine is also high; thus, sustained GS activity is necessary. In this case radial glial fibers, besides having a role in cell migration, may also contribute to cell formation by supplying the needed glutamine to mitotic cells. When cell multiplication is finished, the need for glutamine decreases; hence high GS activity is no longer required, and GS concentration decreases in radial glia.

The earliest star-shaped GS-positive cells in telencephalon appeared in the striatum and in the amygdala with no indication of the presence or the migration of astrocyte precursors (i.e., vimentin-positive perikarya) and in the absence of GFAP-positive cell bodies in the same areas (Yanes et al., 1990). Possibly these GS-immunopositive cells migrated out of the ependymal zone and reached their adult position before expressing GS. In spite of the absence of GFAP immunoreactivity we think that these star-shaped cells represent a specialized type of astrocytes because their ultrastructure is that of astrocytes (fig. 5-1C).

Simultaneous presence of GS and GFAP

The immunohistochemical results indicated that the distribution of GS and GFAP immunoreactivities in *Gallotia galloti* CNS do not always overlap. This raised some doubts on an exclusively astroglial localization of GS. Indeed Linser (1985) has shown that the distribution of GFAP- and GS-positive cells in the various layers of avian optic tectum is different from that of GFAP-negative/GS-positive cells.

The GFAP-negative/GS-positive cells are protoplasmic astrocytes, although they apparently contain also carbonic anhydrase II, which in mammals is an oligodendrocyte marker. Norenberg (1983), using an anti-sheep GS antiserum, has shown the presence of GFAP-negative/GS-positive ependymoglial cells in toad and goldfish. Kriesgstein et al. (1986), using a monoclonal antibody raised against turtle glial cells and an anti-GFAP immunoserum, described the presence of GFAP-negative astrocyte cell bodies in turtle striatum.

Table 5-1. *Immunoreactivity in ependymal glia, radial fibers, and glia limitans.*

		Mesencephalon			Telencephalon		
		VIM	GFAP	GS	VIM	GFAP	GS
A							
	E32	+			+		
	E34	+++	±		+++		
	E35	+++	+	±	+++		
	E36	++	++	±	++	±	±
	E40	+	++	++	+	++	+
	Hatching	#	+++	+++		++	++
	PND 5	#	+++	++		++	+++
	Adult	#	+++	+		++	+
B							
	E32						
	E34						
	E35	*		*			
	E36	*		*			
	E40	*		*			*
	Hatching			*			*
	PND5		*	*			*
	Adult		*	*			*

Schematic representation of the average immunoreactivity in ependymal glia, radial fibers, and glia limitans when antibodies directed against vimentin, glial fibrillary acidic protein or glutamine synthetase were used (A) and the presence during development of scattered astrocytes exhibiting immunolabelling with the different antibodies (B).
GFAP = glial fibrillary acidic protein; GS = glutamine synthetase; VIM = vimentin. The arbitrary scale is from very weak (±) to very strong (+++). # = persistente of vimentin reactivity in some perivascular and subpial end-feet. * = immunolabelling by different antibodies. Note the constant absence of VIM- and GFAP-positive scattered astrocytes in telencephalon.

This indicates that the diversity of astrocyte marker expression is much greater in birds and reptiles than in mammals in order to further examine the differential distribution. To clarify some of these questions described before we have tested if the presence of GS is restricted to astrocytes in these animal species. On this matter, we have studied the segregation of GS and GFAP in *Gallotia galloti* CNS by using a double labelling immunohistochemical technique with antibodies recognizing GFAP and GS (Monzón-Mayor et al., 1997).

In telencephalon ependymal glial cell bodies and many radial processes are GFAP-negative and GS-positive, and in the medial cortex GS-positive fiber bundles are differently oriented than GFAP-positive fiber bundles (Yanes et al., 1990). Some GS-negative/GFAP-positive fibers are also present. Numerous GS-positive/GFAP-negative cells, either ovoid or star-shaped, were observed in the dorsomedial cortex, amygdaloid complex, and striatum ventralis. In these area GFAP positive cells were practically absent but GFAP-positive processes, probably originating from the GFAP-positive/GS-positive ependymal glial cell bodies were detected (Yanes et al., 1990).

In the midbrain some of the cells bodies scattered are GFAP-positive hence they are astrocytes. They are not very abundant and while some of them exhibit a classical astrocytic morphology and are strongly GFAP-positive and weakly GS-positive, other strongly GS-positive and weakly GFAP-positive cell bodies are ovoid or pyriform but the presence of GFAP designates them as astrocyte cell bodies in spite of the unusual shape (Monzón-Mayor et al., 1997).

It appears that not all astrocytes in the adult *Gallotia galloti* are GFAP positive. In addition, the distribution of the cell bodies and fibers suggests that the astrocytes in this lizard express different combination of GS and GFAP.

This diversity seems greater in mesencephalon than in telencephalon were almost no GFAP-positive cell bodies were observed. In a previous work (Yanes et al., 1988a) we observed immature astrocytes in the adult telencephalon and suggested that such immature cells were migrating from the proliferative zones in the sulci. A possible explanation for the absence of GFAP-positive cells in *Gallotia galloti* telencephalon (Yanes et al., 1990) could be the permanent immaturity of astrocytes that express very little GFAP. Astrocyte immaturity together with permanent presence of glioblasts, could be an indication of the regenerative capacity of the glial cell population in these animals, which persists through life.

In the midbrain, other pyriform cell bodies either scattered or congregated in small clusters, aligned in short rows or juxtaposed to vascular walls are GS-positive/GFAP negative. Our previous data (Monzón-Mayor et al., 1990a,b) indicated that they are also vimentin-negative. These GFAP-negative ovoid and pyriform cell bodies look so much like the GFAP-positive ones, that can be an immunological (GFAP-negative) variant of the astrocytes exhibiting a particular, ovoid phenotype. The present ultrastructural immunocytochemistry results, however indicate that these GS-positive/GFAP-negative cells display characteristics of light and medium electron-dense oligodendrocytes. Some authors have described GS-positive ovoid galactocerebroside containing cells, in situ and in cultures (Cammer, 1990; Tansey et al., 1991) that were considered to be oligodendrocytes.

Thus in conclusion some of the dispersed and/or clustered ovoid or round cells are astrocytes displaying a particular phenotype (the GS-positive/GFAP-positive cells) while others are oligodendrocytes (GS-positive/GFAP-negative cells).

Our present results indicate a certain heterogeneity in the immunological and morphological phenotypes of glial cells in the lizard. Such an heterogeneity has also been described previously in the chicken CNS (Linser, 1985), where, however, GS- and GFAP- immunorectivity distribution was different, and in both cases different from what has been observed in mammalian CNS.

GLIA AND GLIOGENESIS IN LIZARD: OLIGODENDROCYTES

Few immunochemical studies deal with the oligodendrocyte development in the CNS of amphibians (Stensaas and Stensaas, 1968a,b; Schonbach, 1969; Stensaas, 1977) and teleosts (Kruger and Maxwell, 1967) and none in reptiles such as lizards.

We have used an antibody directed against carbonic anhydrase II, an oligodendrocyte marker in mammals, but could not detect any immunoreaction, probably because of interspecific differences of the enzyme molecule. We used, also unsuccesfully, an antibody against galactocerebroside. We have performed however an immunohistochemical study with an antibody recognizing MBP, a marker of myelin, the very important oligodendrocyte product (Yanes et al., 1992). Thus our studies on oligodendrocytes are essentially ultrastructural. However recently we have stained the mature oligodendrocytes with the antibody raised against S100 protein (Romero-Alemán et al., 1996).

Oligodendrocyte ultrastructure in the adult lizard

On the basis of their cytoplasm electron density, mature mammalian oligodendrocytes were classified as "light", "medium" and "dark" oligodendrocytes (Kruger and Maxwell, 1966; Mori and Leblond, 1970; Peters et al., 1970; Sturrock, 1974, 1976; Ling et al., 1973; Vaughan, 1984). Similar differences are found in oligodendrocytes in adult *Gallotia galloti* mesencephalon and telencephalon (Monzón-Mayor et al., 1990d; Yanes et al., 1997). In mammals the number of "dark oligodendrocytes" increases with age while that of "light oligodendrocytes" decreases, suggesting that the former derive from the latter (Imamoto et al., 1978; Parvanelas et al., 1983). We observed the same cell types when we investigated oligodendrocyte development in the midbrain and telencephalon of the lizard *Gallotia galloti* by electron microscopy.

Blakemore (1982) stated that the electron density of oligodendrocyte cytoplasm is directly related to size and metabolic activity of the oligodendrocyte; thus he proposed that the oligodendrocytes which provide myelin to large axons exhibit a paler cytoplasm that those which provide myelin for small axons. This seems to be the case also in the adult lizard midbrain (Monzón-Mayor et al., 1990d). Satellite (perineuronal) oligodendrocytes were present in *Gallotia galloti* (Yanes et al., 1997) while Stensaas and Stensaas (1968a) report that these cells are absent in amphibians. This might suggest that these cells appear in the phylogenetic scale in reptiles for the first time.

Oligodendrocyte ultrastructure during Gallotia galloti development

In *Gallotia galloti* telencephalon and midbrain at different stages of development, we have found cells displaying the same features as in mammals (Vaughn, 1969;

Sturrock, 1974; Skoff et al., 1976a); thus apparently the gliogenesis sequence in lizard is the same as in mammals. In *Gallotia galloti* oligodendroblasts were present from E.35 in midbrain and from E.39 in telencephalon. At E.36 in midbrain and at E.39 in telencephalon "light oligodendroblasts" were numerous and rapidly became "immature oligodendrocytes". At E.37 in midbrain and at E.40 in telencephalon myelinization begins and oligodendrocytes displaying long parallel stacks of rough endoplasmic reticulum (RER) and a nucleus with dispersed chromatin and a prominent nucleolus appear. These cells are similar to the"'medium oligodendrocytes" described by Mori and Leblond (1970) and Parnavelas et al., (1983) on one hand and to the "active oligodendrocytes", on the other hand. The term "active oligodendrocytes" (fig. 5-2H) designates cells appearing in the phase of intense protein synthesis that occurs during myelination; these cells were described in the rat optic nerve (Vaughn and Peters, 1971) and in the mouse commissura anterior (Sturrock, 1974).

Notwithstanding some similarities, the cells observed in developing lizard CNS are structurally different from the medium oligodendrocytes found in the adult lizard midbrain. In medium oligodendrocytes the nuclear chromatin is more clumped and the nucleus is spherical or elliptical, while it is irregularly shaped in "active oligodendrocytes"; also the RER is not very abundant in medium oligodendrocytes. In the lizard, therefore, active oligodendrocytes present during ontogeny and medium oligodendrocytes in the adult are quite different cells. A confirmation of the validity of this interpretation is the fact that "active oligodendrocytes" appear in lizard midbrain at E37, practically simultaneously with the myelin marker MBP (Monzón-Mayor et al., 1990d; Yanes et al., 1992) and were present until hatching (Monzón-Mayor et al., 1990d; Yanes et al., 1997).

Mature oligodendrocytes can be found already at E.37 in the midbrain but only after hatching in the telencephalic nuclei, (Monzón-Mayor et al., 1990d; Yanes et al., 1997); this indicates that, also in lizard, oligodendrocytes mature in a caudo-rostral sequence. In the midbrain, light oligodendrocytes were present at all ages from E.37 to adult while medium oligodendrocytes first appeared at E.40. In telencephalon medium oligodendrocytes appeared in the immediate postnatal period. Dark oligodendrocytes (fig. 5-2G) were present at all postnatal ages in both regions.

Myelin

MBP immunoreactivity in the adult Gallotia galloti

Hartman et al. (1979) determined that MBP in chick CNS is present only in areas corresponding to myelinated nerve fibers, and was absent in neuronal somata or dendrites, in astrocyte cell bodies or processes and in microglia. In *Gallotia galloti* MBP-like immunoreactivity was present, with different degrees of intensity, in many nerve fibers (isolated, in tracts and in commissurae) in telencephalon and, more abundantly, in the mesencephalon (fig. 5-2 I, J). Thus MBP-

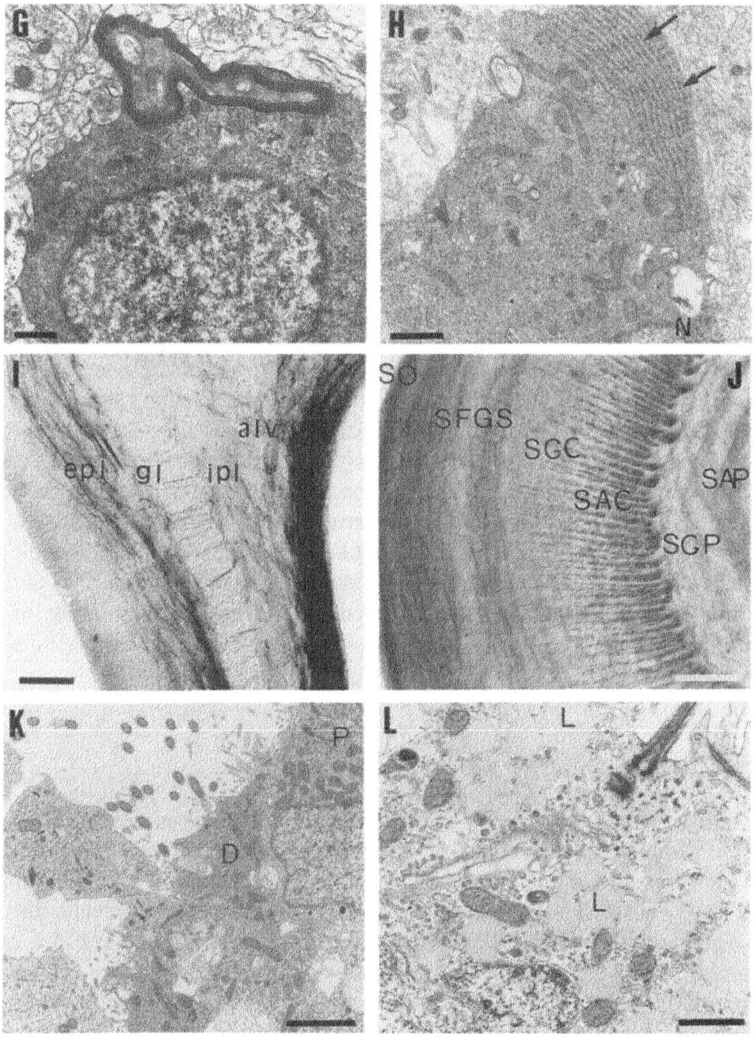

Figure 5-2. Electronmicrograph of a dark oligodendrocyte with a thick myelin ring in the lateral forebrain bundle (lfb) of the adult telencephalon (G). Active oligodendrocyte (H) with an electron-dense cytoplasm at E.40 in mesencephalon. Note the long parallel stacks of rough endoplasmisc reticulum (arrow). Scale bars: G=100μm H=100μm. Details at higher magnification of the cross sections of adult MBP telencephalon (I) shown the alveus (alv), in the internal (ipl) and external (epl) plexifom layers and in the granular layer (gl). MBP-immunopositive fibers of tectum mesencephalic (J) in adult lizard demonstrated by PAP method with the six strata defined by Huber and Crosby (1933a,b). Scale bars: I=50μm J=100μm. Electronmicroghaph shows in the midbrain of the lizard at E.39 (K) a large cytoplasmic protusion, a dark cell (D) and a paler cell (P). At E.37 (L) a pale cell has a large number of lipids droplets (l) in the cytoplasm on the ventricular side of the nucleus. Scale bars: J= 100μm, K= 100μm.

immunoreactivity designed the path of nerve fibers isolated or forming commissures and bundles. As far as we can judge all myelinated tracts previously described in the literature are MBP-immunopositive (Yanes et al., 1992).

Onset of myelination

In the rat, glial cell proliferation is very intense before the onset of myelination, and decreases during myelination (Bignami and Dahl, 1973), while myelination of nerve fibers (Hartman et al., 1979) begins after the establishment of nerve connections.

During ontogeny, the earliest MBP-like immunoreactivity was detected at E.36 in few tracts in mesencephalon and appeared at E.40 in telencephalon (Yanes et al., 1992). This confirms our former morphological data (Yanes, 1985) and coincides with the previously reported first appearance of active oligodendrocytes (Monzón Mayor et al., 1990d). Thus E.36 and E.40 possibly correspond to the onset of myelinization in each of the two regions. Accumulation of MBP continued after hatching. In contrast to many reports in mammals (Sternberger et al., 1978a,b) chick or amphibians but in agreement with Bjelke and Seiger (1989) who did not find MBP-positive oligodendrocytes cell bodies in developing rat CNS, we never detected immunoreactive oligodendrocyte cell bodies in *Gallotia galloti* not even at the youngest ages studied. Possibly the concentration of antigen in the lizard oligodendrocyte cell body is much lower than in other myelinating animal species.

Regional chronology of myelination

Myelination in the chick begins at incubation day 14, and is completed 2 or 3 days before hatching, while in mammals brain myelinogenesis is mainly postnatal, and is completed in the rat at the end of the third postnatal week (Hartman et al., 1979; Bjelke and Seiger, 1989). During chick or rat development, the first fibers reacting with an anti-MBP immune serum are those located in the spinal cord, and finally in the cerebral cortex (Hartman et al., 1979). In *Gallotia galloti* myelination begins at late prenatal stages and progresses after hatching at least until the 15th postnatal day (Yanes et al., 1992). As in the chick and rat, the lizard myelination proceeds from the most caudal regions to the most rostral and, at each level, from the ventral (or basal) to the dorsal (or alar) regions.

Myelination and astrocytes

As in the rat, the glial cell population in *Gallotia galloti* (Yanes et al., 1990; Monzón-Mayor et al., 1990a,b,c,d) increases noticeably before the beginning of myelination, and decreases in the post-hatching period when myelination is well advanced (Monzón-Mayor et al., 1990d., Yanes et al., 1997). Immunohistochemical experiments (Monzón-Mayor et al., 1990a,b,c; Yanes et al., 1990) and electron microscopy has shown the presence not only of oligodendrocytes, but also of astroblasts and astrocytes in myelinating tracts and commissures in the lizard midbrain (Monzón-Mayor et al., 1990c,d) and telencephalon (Yanes et al., 1997). At later ages abundant astrocytes are seen along the nerve fiber tracts. This was noticed by early authors who suggested a role for astrocytes in myelinogenesis (Nagashima, 1979; Okado, 1982).

Bignami and Dahl (1973) suggested that in rat cerebral cortex the GFAP protein appears simultaneously with or just before the beginning of myelin formation. In our case, we have also observed that during the development of *Gallotia galloti* brain the apparition of GFAP and GS in glial cells and the formation of myelin in the same region are concomitant. Whether this is a simple coincidence or is the indication of a partnership between oligodendrocytes and astrocytes (or a class of astrocytes) in the process of myelination remains to be ascertained.

GLIA AND GLIOGENESIS IN LIZARD: EPENDYMAL CELLS

The telencephalic vesicles in *Gallotia galloti* begin to form at E.30; thus the ontogeny of ependyma was studied in the midbrain and forebrain from E.31 to adult age (Yanes et al., 1988a,b; Monzón-Mayor et al., 1991). In this lizard, at the earliest stages of ontogeny (E.31 to E.34) the ventricular wall is covered by a thick homogenous cellular layer consisting of an undifferentiated pseudostratified columnar neuroepithelium similar to that described in mammalian embryos (Stensaas and Gilson, 1972; Shoukimas and Hinds, 1978; Sturrock, 1981). The differentiation of the telencephalic vesicle, from the first ontogenic stages until the adult ependyma is formed, occurs according to a caudo-rostral and ventro-dorsal progression; localized particular aspects are also present.

During early developmental stages (until E.35 in telencephalon), cells in the ventricular zones and the intercellular space between them present homogeneous cytological and ultrastructural characteristics and the same vimentin immunoreactivity in the whole ventricular system (Monzón-Mayor et al., 1990b; Yanes et al., 1990). Vimentin forms a characteristic perinuclear ring. Numerous mitosis, indicating cell proliferation, are present; and when cell multiplication attains the highest rate, cells migrate out of the ventricular layer. At the ultrastructural level the cytological characteristics of the *Gallotia galloti* cells in the ventricular proliferative zones (E.32, E.35 in telencephalon), are similar to those described by Tennyson and Pappas (1962) in humans, Allenspach and Roth (1967) in chick medulla, Stensaas and Gilson (1972) in rabbit, and Marrero (1986) in lizard.

Between E.35 and E.37 three types of cell were differentiated in the ependymal zones in midbrain (Monzón-Mayor et al., 1990b). The first cell type consisted of columnar cells containing a rough endoplasmic reticulum filled by an amorphous substance similar to that of astrocytes. The second type was represented by broad cells containing numerous lipid droplets. The third cell type exhibited a large

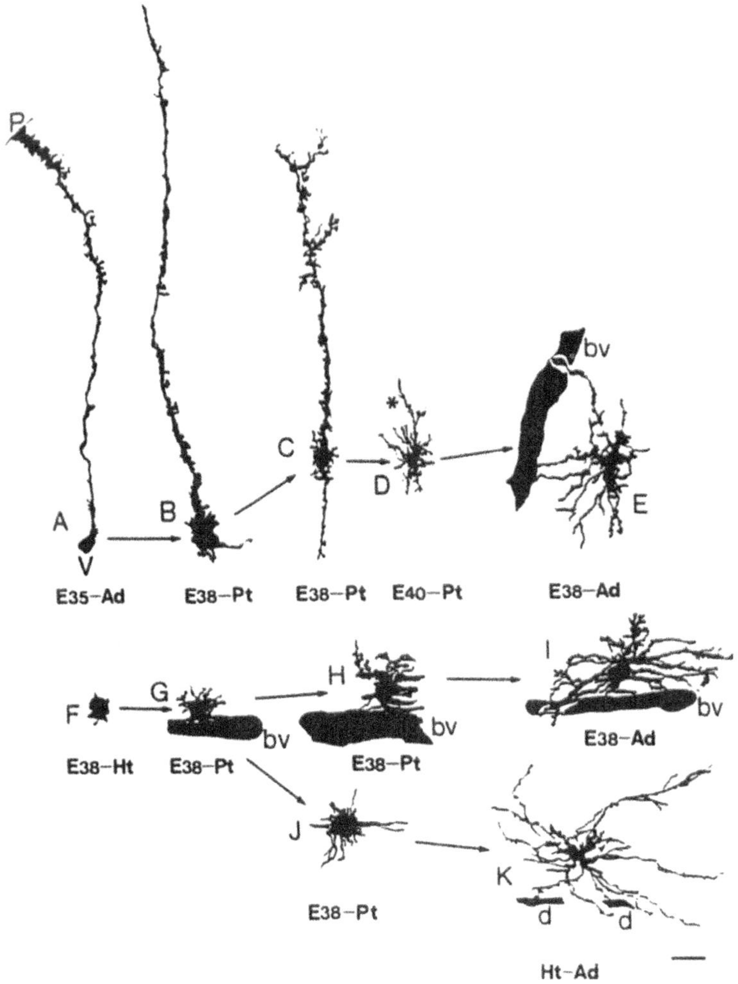

Figure 5-3. *Gallotia galloti* schematic drawings illustrating astroglial cell developmental sequence. The cellular types were impregnated between the ages: embryonic stage 35 (E.35)-adult (Ad); embryonic stage 38 (E38) early postnatal (Pt); embryonic stage 38 (E.38)-adult (Ad); hatching (Ht)-adult (Ad). A, B, C, D, and E show astrocytes derived from radial glia. A. Radial glia cell extended from ventricular layer (V) to the pia matter (P). B: "Displaced radial glia" with a single process. C: Bipolar "displaced radial glia". D: Mature protoplasmic astrocyte displaying a long process (asterisk) reaching the pia matter. E: Mature protoplasmic astrocyte in the adult. Notice the endfeet on a blood vessel (bv). F, G, H, I, J and K show astrocytes derived from glioblasts. F: a glioblast with small superficial protrusions. G: Astroblast adjacent to a blood vessel (bv). H: Immature protoplasmic astrocyte adjacent to a blood vessel (bv). I: Mature protoplasmie astrocyte. The processes are thicker than those of the immature protoplasmic astrocyte (see fig. 5-3H). Some processes contact a blood vessel (bv). J: Immature fibrous astrocyte. Note its smooth and slender processes, compared to those of an immature protoplasmic astrocyte (see fig. 5-3H). K: Mature fibrous astrocyte. Note the processes in the neighborhood of a dendrite (d). Bar = 8μm.

process, broad at the basis and extending radially to the pia mater, microtubules, numerous glycogen granules and few intermediate filaments are present. At E.35 in the midbrain and E.37 in the telencephalon a GS like immunoreactivity appears in the ventricular cells (Yanes et al., 1988a,b; Monzón-Mayor et al., 1991).The onset of a faint GFAP (Monzón-Mayor et al., 1990b) and GS immunoreactivity in ventricular cells (Monzón-Mayor et al., 1990a) also occurs between E.35 and E.37.

The ependyma from E.37 in the midbrain ventricle and aqueduct and from E.38 in telencephalon shows marked regional variation: much of it is formed by a single layer of moderately dark cuboidal cells similar to the cuboidal ependymal cells seen in the ventricular system of mammals (Westergaard, 1970). Another cell type in lizards midbrain are the columnar cells, which contain varying amounts of lipid that often have cytoplasmic protrusions on the ventricular surface which are characteristic of tanycytes (fig. 5-2K) (Flament-Durand and Brion, 1985; Rafols, 1986). Despite the presence of both vimentin and GFAP immunoreactivity, intermediate filaments are only rarely found in ventricular cells or their processes. Radial glial processes at this stage contain microtubules, scattered glycogen granules and very few filaments.

From E.35 to E.38 in telencephalon, the ependymal characteristics appear clearly defined: cubic, prismatic and secretory cells are identified and the intercellular spaces disappear. The existence of prismatic, cubic and secretory cells in lizard CNS has also been observed by Hetzel (1977). In telencephalon the matrix gradually differentiates in ependymal zones and in four sulci (s. lateralis, s. terminalis, s. ventralis and s. septomedialis) and the mesencephalon differentiates in ependymal zones and in the sulcus limitans. Mitotic activity decrease in the future ependymal zones, while, at all the levels of the four telencephalic sulci, the proliferative activity begins at E.32, becomes quite intense at E.34 and always are present in adult specimens. It should be noted that cells in the sulci are still highly homogeneous and intercellular spaces are numerous. A decreases in the number of GS-positive ependymal cell bodies apparently occurs in the midbrain possibly because they migrate out of the periventricular zone. The migration of GS-positive cells out of the sulci is indicated the fact that ovoid GS-positive cells appear at E.40 along the path of radial fibers, at first around the sulcus limitans in the Silvius aqueduct; at hatching these cells have spread, always along radial fibers, in the optic tectum and pretectum to more than halfway toward the pial surface. These cells in migration are Vim-, GFAP- and GS-positive. The migration starts at E.40 and finishes in adult specimens (Monzón-Mayor et al., 1990a,b).

Cell differentiation in telencephalic sulci becomes evident at E.38, but mitotic activity continues until postnatal and adult ages, specially in the s. terminalis. The different ependymal zones, thus, begin to assume the adult structure at E.38 ; this corresponds to the prediction by Das (1979), i.e. that in an animal species ependymal differentiation begins half way through embryonic CNS development.

In the adult telencephalon the ependymal lining varied from cuboidal to low columnar or prismatic. The different cytological characteristics showing the high degree of regionalization in *Gallotia galloti* ependyma are similar to those reported by Krautgartmer et al. (1984), in *Tinca tinca.*

Cilia and lipid droplets are characteristic of ependymal cell differentiation. Similarly to what described by Fleischhauer (1972) in the mouse, cilia density increases during *Gallotia galloti* embryonic development from early developmental phases to perinatal ones, above all in the s. ventralis and the s. terminalis. Schultz (1969), in *Lacerta agilis*, pointed out the presence of cilia in the ventral germinative zone. Hetzel (1977), also declared that "the great ciliar diversity of these proliferative zones, seem to be related to the nutritive properties of the cerebrospinal fluid". Ependymal cells in the region of the sulcus limitans and those of the adjacent subventricular layer were packed with lipid droplets.

Lipid droplets

One feature of the developing ependymal cells in lizards is the presence of abundant lipid droplets (fig. 5-2L). This abundance suggests that lipids constitute an energy source; this is supported by the fact that lipid droplets are most numerous in cells like those in the sulcus limitans, which can proliferate not only during development but also in the adult. Lipid droplets are presents in adult mouse ependymal cells but not during development and cell differentiation (Sturrock, 1981). In mammals glycogen rather than lipids seems to be stored for energetic needs. Indeed at midgestation ependymoglial cells in the dorsal and ventral parts of the mouse spinal cord close to the midline are packed with glycogen (Sturrock, 1981). These cells appear to have a secretory or transport function (Yoshioka and Tanaka, 1989) and probably are the mammalian equivalent of the lipid-rich cuboidal cells observed in the present study from E.37 to hatching. The subventricular zone around the sulcus limitans may correspond to the subependymal layer of mammals (Allen, 1912; Smart, 1961) and birds (Goldman and Nottebohm, 1983).

The presence of lipid droplets in embryonic lizard ependymal cells, apparently equivalent of the glycogen grain found in embryonic mammalian ependyma (Sturrock, 1981; Gadisseux et al., 1990), probably reflects differences in metabolism between oviparous and placentate embryos. In fact, oviparous relies on yolk as the sole source of nutrition. The continuing presence of lipid droplets in cells of the proliferative sulcal zone in adult lizards may indicate that, even in the adult reptile, lipid remains as an energy store in the ependyma and subependymal layer.

Meaning of the sulci

The five sulci exhibited a great cellular homogeneity between E.32 and E.35. Sulci and ventricular zones begin to assume the adult ependymary structure from

E.38. From this stage on cellular hetereogeneity becomes a persisting sulcus feature. One factor contributing to this hetereogeneity is the coexistence of cells at different stages of differentiation and maturation (dark and early glioblasts, ependymal cells, etc, were observed.).

The proportion of these cells however is different in the different sulci. Together with the morphological hetereogeneity another aspect makes the reptilian sulci unique (Fleischhauer, 1957): The mitotic capacity of all the sulci is not lost during ontogeny: the s. lateralis show matricial characteristics from E.40 to adult, while the s. terminalis and ventralis exhibits proliferative activity during postnatal life and even in the adult. The sulcus terminalis and sulcus limitans presents the highest number of mitosis from mid-development stages (E.37, E.38) to the postnatal ones and in the adult. The proportion of mitosis found by us in *Gallotia galloti* is similar to that described by Hetzel (1974) in *Lacerta sicula.* Thus cells at different stages of maturation mitotic activity is a persisting feature of reptilian ventricular sulci while in mammals this is a transient phenomenon.

Fleischhauer (1957) and Kirsche (1967), supposed that all sulci in adult reptiles are made of matricial rests. Kahle (1951) considers the sulcus terminalis as an advanced matricial waste. Our observations indicate that the five studied sulci in *Gallotia galloti* indeed represent embryonic residues that survive as such for a long period of the lizards life, but the capacity for cell proliferation and differentiation is different for each sulcus. The survival of sulci in the adult may account for the regenerative capacity of reptilian CNS.

Interspecific and regional differences in ependymal glia

In *Gallotia galloti*, ependymal cells and radial processes are GFAP- and GS-positive; as in other reptiles (Kirsche, 1972; Cruce and Nieuwenhuys, 1974) the ependymal layer is multicellular at the level of the "proliferative zones" of the sulci. Also the cell layer in the ventricular wall from which radial fibers originate shows regional and interspecies differences, this layer can be monocellular or multicellular with different intermingled or stratified cell types.

GFAP cellular localization can differ between the animal species or can change with age as the following examples indicate. In adult mammals both astrocytes and radial glia are GFAP-positive while ependymocytes are GFAP-positive during early development (Levitt and Rakic, 1980) and become GFAP-negative in the adult (Dahl and Bignami, 1973). Mammalian ependymocytes, however, do express other astrocyte markers such as S100 protein (De Vitry et al., 1980; Didier et al., 1986). In adult birds the ependymal layer is GFAP-positive (Onteniente et al., 1983), but no radial fibers are detectable, and in the finch the GFAP-positive cells in the ependymal layer have the morphology of typical astrocytes (Onteniente et al., 1983).

In the optic tectum of Testudo GFAP-positive radial fibers spread out from a thick GFAP-positive multicellular layer around the ventricle (Dahl and Bignami, 1973) but several GFAP negative cell bodies are also present in this layer. In teleosts the GFAP-positive layer is the subependymal layer. In goldfish (Stevenson and Yoon, 1982) the periventricular wall at the level of the optic tectum is multilayered : ependymocytes are located at the ventricular surface (ependymal layer) and their processes (in contrast to those of ependymocytes in adult mammals) are GFAP-positive and form a local plexus which does not extend into the tectum further than the ependymal layer. Long radial processes called "periependymal radial glia" originate from these "periventricular cells". These processes do not contain intermediate filaments, are GFAP-negative, but form perivascular and subpial end-feet. These characteristics, however, might be specific to goldfish, which shows several particularities concerning the glial population: although typical astrocytes were also observed, radial glia contains neither GFAP nor gliofibrils, Bergman glia is absent in cerebellum (Dahl and Bignami, 1973) and optic nerves do not contain GFAP-positive cells. In the mouse spinal cord ependymoglial cells remain in small numbers in the midline and have processes packed with microfilaments similar to those seen in the adult lizard radial glia (Sturrock, 1981).

Tanycytes

Among the different cubic and prismatic cell types of the ependyma, cells with another morphology are present. These cells are similar to the tanycytes described in fetal hamster third ventricle and in the circumventricular organs in the adult rat (Mitro and Kiss, 1977). They display long periventricular somas and long, radially disposed, processes. These cells were first described by Schultz (1969) and then by Hetzel (1977), and were called tanycytes. Schmechel and Rakic (1979), Stevenson and Yoon (1982), called them ependymal astrocytes and Kriegstein et al. (1986), named them ependymal glial cells.

Cell processes emanating from the ependyma

In reptiles, as in mammals, immature ependymoglial cells, which retain contact with the pial surface, apparently mature in GFAP rich RG, from which filament-packed cell processes emerge, and in GS-rich tanycytes with GS-rich cell processes. Indeed our double labelling studies have shown that tanycyte-like ependymocytes in *Gallotia galloti* are GS positive (Monzón-Mayor et al., 1997). The difference in distribution of GFAP positive and GS positive fibers observed in *Gallotia galloti* (Monzón-Mayor 1990a,b) may reflect the differences between RG processes and tanycyte processes, particularly in the telencephalon where mature astrocytes are absent. This may reflect a particular feature of reptiles since rat tanycytes have been described as GS-negative (Noremberg, 1979).

At the ultrastructural level two types of ependymal process extending to the pia mater could be identified. One type was packed with intermediate filaments probably gliofilaments GFAP-positive in radial glial cells, while the other type contained a core of microtubules and scattered glycogen granules probably ependymoglial cells.

Ependymal cells and radial processes in *Gallotia galloti* are GFAP- and GS-positive; thus, the scattered GS-positive/GFAP-negative cells in *Gallotia galloti* midbrain apparently are not equivalent to the "transitional" cells in mammalian brain on two counts: first, they are immature since they do not further differentiate even at advanced ages; second, they are GFAP-negative while the cells they supposedly derive from and the cells in which they should mature are GFAP-positive. It is possible that they attain adult phenotype only in special circumstances (stand-by immature astrocytes).

CONCLUSIONS

Most of glial cell phenotypes and the sequence of their morphological differentiation and maturation in *Gallotia galloti* do not greatly differ from those of mammals: myelin and myelinization are the same, dark, light and medium oligodendrocytes are present, and all intermediate cell types between glioblasts and oligodendrocytes or astrocytes were found. Differences between the lizard and mammals, however, were also present and here we summarize these differences.

The first striking feature is the different astrocyte distribution between telencephalon and mesencephalon: mature GFAP-immunoreactive astrocytes being absent in telencephalon and present in midbrain. In both regions, GFAP positive cell bodies are present: these are the perikarya of radial glia which persist in the adult lizard CNS and from which GFAP-immunoreactive radial fibers originate. These fibers terminate with subpial end feet; perivascular end feet also originate from branches of these fibers. This appears to be the only origine of GFAP immunoreactive fibers in lizard telencephalon indeed ependymocyte and tanycytes in *Gallotia galloti* are GPAP-negative but GS-positive. Some ovoid GFAP-negative, GS-positive cells in *Gallotia galloti* telencephalon exhibited ultrastructure of immature astrocytes, others in mesencephalon were clearly oligodendrocytes. Thus oligodendrocytes are GS positive in this lizard.

The situation is similar in mesencephalon with however a major difference: classical GFAP-positive astrocytes are present together with GFAP-positive ovoid perikarya, presumably odd shaped astrocytes.

Another particularity of *Gallotia galloti* is the presence of immature cells in some ventricular sulci throughout lizard lifetime. During ontogeny, cells are formed in the ventricular sulci and migrate out of them. Some of these cells then become GFAP-positive or GS-positive or both. Also a large proportion of radial glia is present throughout lizard lifetime. Whether astrocytes are formed by a transformation of some radial glia element, transformation that in mammals seems to involve the whole radial glia, is difficult to say. In fact those cells, which in *Gallotia galloti* are morphologically similar to the mammalian cells intermediate between radial glia and astrocytes, remain immature in the adult. In any case, if astrocytes are formed from radial glia they account for one minor lineage of astroglia, most of which apparently comes from the sulci. Probably the abundance of immature cells in sulci, in mesencephalic and telencephalic nuclei is related to the regenerative capacity of reptilian CNS (Lang et al., 1998).

ACKNOWLEDGMENTS

This work was carried out thanks to the support of Convenio de Colaboración del Gobierno y Caja General de Ahorros de Canarias (Canary Islands), Ministerio de Educación y Ciencia (Spain), F. Univ. de Las Palmas de G.C., Spanish and French Ministries of Education (Acciones integradas 245 A and HF-015), DGICYT, PB-94-0432, Gobierno de Canarias nº978. We are grateful to Mrs Yolande Schladenhaufen, J. James for their technical assistance and Lourdes Ramirez Quintana for her typewriting assistance.

REFERENCES

Allen, E. (1912) Cessation of mitosis in the central nervous system of the albino rat. J. Comp. Neurol. *22*:547-568.

Allenspach, A.L., and L.E. Roth (1967) Structural variations during mitosis in the chick embryo. J. Cell Biol. *33*:179-196.

Antanitus, D.S., B.H. Choi, and L.W. Lapham (1976) The demonstration of glial fibrillary acidic protein in the cerebrum of the human fetus by indirect immunofluorescence. Brain Res. *103*:613-616.

Báez, J. (1992) Formación y citodiferenciación del techo óptico del lagarto *Gallotia galloti*. Tesis doctoral, Facultad de Ciencias Médicas y de la Salud, Universidad de Las Palmas de Gran Canaria, Canary Islands, Spain.

Bignami, A., L.F. Eng, D. Dahl, and C.T. Uyeda (1972) Localization of the glial fibrillary acidic protein in astrocytes by immunofluorescence. Brain Res. *43*:429-435.

Bignami, A., and D. Dahl (1973) Differentiation of astrocytes in the cerebellar cortex and the pyramidal tracts of the newborn rat: an immunofluorescence study with antibodies to a protein specific to astrocytes. Brain Res. *49*:393-402.

Bignami, A., and D. Dahl (1980) Glial fibrillary acidic protein (GFA) in normal neural cells and pathological conditions. In S. Fedoroff and L. Hertz (eds): Advances in Cellular Neurobiology. New York: Academic Press, pp. 286-313.

Bignami, A., T. Rasu, and Dahl D. (1982) Localization of vimentin, the non-specific intermediate filament protein, in embryonal glia and early differentiating neurones. Dev. Biol. *91*:286-295.

Bjelke, B., and A. Seiger (1989) Morphological distribution of MBP-like immunoreactivity in the brain during development. Int. J. Dev. Neurosci. *7*:145-164.

Blakemore, W. (1982) Myelination, demyelination and remyelination in the CNS. In N. Thomas Smith and J.B. Cavanagh (eds): Recent Advances in Neuropathology vol 2. Edinburgh, London, Melbourne and New York: Churchill Livingstone, pp. 53-81.

Blunt, M.J., F. Baldwin, and C.P. Wendell-Smith (1972) Gliogenesis and myelination in kitten optic nerve. Z. Zell. mikrosk. Anat. *124*:293-310.

Caley, D.N. and D.S. Maxwell (1968) An electron microscopic study of the neuroglia during postnatal development of the rat cerebrum. J. Comp. Neurol. *133*:45-70.

Cammer, M. (1990) Glutamine synthetase in central nervous system is not confined to astrocytes. J. Neuroimmunol. 26:173-78.

Cameron, R.S., and P. Rakic (1991) Glial cell lineage in the cerebral cortex: A review and synthesis. Glia. *4*:124-137.

Campagnoni A.T., J.M. Roth, M. Hunkeler, P.J. Pretorius, and C.W. Campagnoni (1989) Expression of myelin basic protein genes in the developing mouse brain. In: P. Evrard, A. Minkowski (eds): Develop. Neurobiology. Nestlé nutrition workshop series. New York: Raven Press, pp. 95-109.

Chiu, F.C., R.S. Sacchi, L. Claudio, S. Kobayashi, and K. Suzuki (1988) Coexpression of glial fibrillary acidic protein and vimentin in the central and peripheral nervous system of the Twitcher mutant. Glia. *1*:105-112.

Choi, B.H., R.C. Kim, and L.W. Lapham (1983) Do radial glia give rise to both astroglial and oligodendroglial cells?. Dev. Brain Res. 8:119-130.

Choi, B.H.(1988) Prenatal gliogenesis in the developing cerebrum of the mouse. Glia. *1*:308-316.

Cruce, W.L.R. and R. Nieuwenhuys (1974) The cell masses in the brain system of the turtle Testudo hermanni; a topographical and topological analysis. J. Comp. Neurol. *156*:277-306.

Dahl, D., and A. Bignami (1973) Immunochemical and immunofluorescence studies of the glial fibrillary acidic protein in vertebrates. Brain Res. *61*:279-293.

Dahl, D., D.C. Rueger, A. Bignami, K. Weber, and M. Osborn (1981) Vimentin, the 57,000 dalton protein of fibroblast filaments, is the major cytoskeletal component in immature glia. Eur. J. Cell Biol. *24*:191-196.

Dahl, D., P. Strocchi, and A. Bignami (1982) Vimentin in the central nervous system. Differentiation *22*:189-190.

Dahl, D., C.J. Crosby, A. Sethi, and A. Bignami (1985) Glial fibrillary acidic (GFA) protein in vertebrates: Immunofluorescence and immunoblotting study with monoclonal and polyclonal antibodies. J. Comp. Neurol. *239*:75-88.

Das, G.D. (1979) Gliogenesis and ependymogenesis during embryonic development of the rat. J. Micrological Sci. *43*:193-204.

De Vitry, F., R. Picart, C. Jacquet, L. Legault, P. Dupouey, and A. Tixier-Vidal (1980) Presumptive common precursor for neuronal and glial cell lineages in mouse hypothalamus. Proc. Natl. Acad. Sci. U.S.A. *77*:4165-4169.

Díaz, C., C. Yanes, L. Medina, M. Monzón-Mayor, C.M. Trujillo, and L. Puelles (1990a) Golgi study of the anterior dorsal ventricular ridge in a lizard. I. Neuronal typology in the adult. J. Morphol. *203*:301-310.

Díaz, C., C. Yanes, L. Medina, M. Monzón-Mayor, C.M. Trujillo, and L. Puelles (1990b) Golgi study of the anterior dorsal ventricular ridge in a lizard. II Neuronal cytodifferentiation. J. Morphol. *203*:293-300.

Didier, M., M. Harandi, M. Aguera, B. Bancel, M. Tardy, C. Fages, A. Calas, M. Stagaard, K. Mollgard, and M.F. Belin (1986) Differential immunocytochemical staining for glial fibrillary acidic (GFA) protein, S100 protein and glutamine synthetase in the rat subcomissural organ, non-specialized ventricular ependyma and adjacent neuropil. Cell. Tissue Res. *245*:343-351.

Dixon, R.G., and L.F. Eng (198l) Glial fibrillary acidic protein in the retina of the developing albino rat: an immunoperoxidase study of paraffin-embedded tissue. J.Comp.Neurol.*195*:305-332

Eckenhoff, M.F., and P. Rakic (1984) Radial organization of the hippocampal dentate gyrus: A Golgi, ultrastructural and immunocytochemical analysis in the developing Rhesus monkey. J. Comp. Neurol. *223*:1-21.

Eng, L.F., J.J. Vanderhaeghen, A. Bignami, and B. Gerstl (1971) An acidic protein isolated from fibrous astrocytes. Brain Res. *28*:351-354.

Eng, L.F., and J.W. Bigbee (1978) Immunochemistry of nervous system specific proteins, In B.W. Agranoff and M.H. Aprison (eds): Advances in Neurochemistry. New York: Plenum Publishing Corp., pp. 43-98.

Eng, L.F. (1980) The glial fibrillary acidic (GFA) protein, In R.A. Bradshaw and D.M. Schneider (eds): Proteins of the Nervous System. New York: Raven Press, pp. 85-117.

Fedoroff, S. (1986) Prenatal ontogenesis of astrocytes. In S. Fedoroff and A. Vernadakis (eds): Astrocytes vol 1. New York: Academic Press, pp. 35-44.

Flament-Durand, J., and J.P. Brion (1985) Tanycytes: morphology and functions: a review. In. Rev. Citol. *96*:121-155.

Fleischhauer, K. (1957) Untersuchungen am Ependym des Zwischen und Mittelhirns der Landschildkröte (Testudo graeca). Z. Zellforsch. *46*:729-765.

Fleischhauer, K. (1972) Ependyma and subependymal layer. In: The structure and function of nervous system VI, pp. 1-46.

Friede, R.L., and M.T. Stone (1969) Glial foot plates in the bowfine. I Fine structure and Chemistry. J. Neuropathol. Exp. Neurol. *28*:513-539.

Gadisseux, J.F., H.J. Kadhim., P. Van Den Bosch De Aguilar., V.S. Caviness, and P. Evrard (1990) Neuron migration within the radial glial fiber system of the developing murine cerebrum: an electron microscopic autoradiographic analysis. Dev. Brain Res. *52*:39-56.

Ghandour, M.S., O.K. Langley, G. Vincendon, and G. Gombos (1979) Double labeling immunohistochemical technique provides evidence of the specificity of glial cell markers. J. Histochem. Cytochem. *27*:1634-1637.

Ghandour, M.S., O.K. Langley, and J. Clos (1983) Immunohistochemical and biochemical approaches to the development of neuroglia in the CNS, with special reference to cerebellum. Int. J. Develop. Neurosci. *1*:411-425.

Goldman, S.A., and F. Nottebohm (1983) Neuronal production, migration and differentiation in a vocal control nucleus of the adult female canary brain. Proc. Natl. Acad. Sci. USA *80*:2390-2394.

Hallermayer, K., G. Harmening, and B. Hamprecht (1981) Cellular localization and regulation of glutamine synthetase in primary cultures of brain cells from newborn mice. J. Neurochem. *37*:43-52

Hallermayer, K. and B. Hamprecht, B. (1984) Cellular heterogeneity in primary cultures of brain cells revealed by immunocytochemical localization of glutamine synthetase. Brain Res. *295*:1-11.

Hartman, B.K., H.C. Agrawal, S. Kalmbach, and W.T. Shearer (1979) A comparative study of the immunohistochemical localization of basic protein to myelin and oligodendrocytes in rat and chicken brain. J. Comp. Neurol. *188*:273-290.

Hetzel, W. (1974) Die Ontogenese des Telencephalons bei Lacerta sicula (Rafinisque) mit besonderer Berücksichtigung der pallianlen Entwicklung. Zool. Beitr. N. S. *20*:361-458.

Hetzel, W. (1977) Das Ependym der Seitenventrikel von Acanthodactylus pardalis (Reptilia, Lacertidae). Acta. Anat. *97*:68-80.

Huber, G.C. and E.C. Crosby (1933a) The reptilian optic tectum. J. Comp. Neurol. *57*:57-163.

Huber, G.C. and E.C. Crosby (1933b) A phylogenic consideration of the reptile-tectum. Proc. Natl. Acad. Sci. USA *19*:15-22.

Imamoto, K., J.A. Paterson, and C.P. Leblond (1978) Radioautographic investigation of gliogenesis in the corpus callosum of young rats. I. Sequential changes in oligodendrocytes. J. Comp. Neurol. *180*:115-138.

Kahle, W. (1951) Studien über die Matrixphasen und die örtlichen Reifungsunterschiede im embryonalen menschlichen Gehirn. Dtsch. Z. Nervenheilk. *166*:273.

Kirsche, W. (1967) Über postembryonale Matrixzonen im Gehirn veschiedener Vertebraten und deren Bezichung zur Hirn-Bauplaniehre. Z. mikrosk.-anat. Forsch. *77*:313-406.

Kirsche, W. (1972) Die Entwicklung des Telencephalons der Reptilien und deren Beziehung zur Hirn-Bauplanlehre. In: Johann Ambrosious Barth (ed.). Leipzig. Nova Acta Leopoldina. *1*:37-78. Humboldt-Universität Berlin.

Kölliker, R.A. (1893) Handbuch der Gewebelehre des Menschen. Vol. 2.2nd ed. Wilhem Engelmann. Leipzig.

Krautgartmer, K., K. Kotrschal, and A. Lamtschwa-ndtner (1984) The ependyma of Acpenser ruthenus and Tinca tinca. 8th European Congress, Elec. Micros., Budapest.

Kriegstein, A.R., J.M. Shen, and N. Eshhar (1986) Monoclonal antibodies to the turtle cortex reveal neuronal subjets antigenic cross reactivity with the mammalian neocortex and forebrain structures sharing a pallial derivation. J. Comp. Neurol. *254*:330-340.

Kruger, L. and D.S. Maxwell (1966) The fine structure of ependymal processes in the teleost optic tectum. Am. J. Anat. *11*:479-498.

Kruger, L. and D.S. Maxwell (1967). Comparative fine structure of vertebrate neuroglia: teleosts and reptiles. J. Comp. Neurol. *129*:115-142.

Lang, D. M., M. Monzón-Mayor, C. E. Bandtlow and C .A. O. Stuermer (1998). Retinal axon regeneration in the lizard *Gallotia galloti* in the presence of CNS myelin and oligodendrocytes. Glia, (in press).

Lazarides, E. (1980) Intermediate filaments as mechanical integrators of cellular space. Nature (Lond.) *283*:249-256.

Lazarides, E. (1982) Intermediate filaments: A chemically heterogeneous, developmentally regulate class of proteins. Annu. Rev. Biochem. *51*:219-250.

Lenhossék, M. (1893) Der feinere Bau des Nervensystems im Lichte neuester Forschungen. In: H. Kornfeld (ed.). Fischer's Medicinische Buchhandlung. Berlin.

Levitt, P., and P. Rakic (1980) Immunoperoxidase localization of glial fibrillary acidic protein in radial glial cells and astrocytes of the developing Rhesus monkey brain. J. Comp. Neurol. *193*:815-840.

Linser, P.J. (1985) Multiple marker analysis in the avian optic tectum reveals three classes of neuroglia and carbonic anhydrase-containing neurons. J. Neurosci. *5*:2388-2396.

Ling, E.A., J.A. Paterson, A. Privat, S. Mori, and C.P. Leblond (1973) Investigation of glial cells in the brain of young rats. J. Comp. Neurol. *149*:43-72.

Lord, K.E., and I.D. Duncan (1987) Early postnatal development of glial cells in the canine cervical spinal cord. J. Comp. Neurol. *265*:34-46

Luskin, M.B., J.G. Parnavelas, and S.A. Barbifield (1993) Neurons, astrocytes and oligodendrocytes of the rat cerebral cortex originate from separate progenitor cells: An ultrastructural analysis of clonally related cells. J. Neurosci. *13*:1730-1750

Marrero, A. (1986) Ontogenia de la corteza cerebelar de *Gallotia galloti* (Reptil Lacertidae). Estudio estructural y ultraestructural. Tesis Doctoral. Universidad de La Laguna. Tenerife. Canary Islands. Spain.

Martinez-Hernández, A., K.P. Bell, and M.D. Norenberg (1977) Glutamine synthetase: Glial localization in brain. Science. *195*:1131-1358.

Matthieu J.M., S. Widmer, N. Herschkowitz (1973) Biochemical changes in mouse brain composition during myelination. Brain Res. *55*:391-402.

Medina, L., C.M. Trujillo, C. Díaz, A. Martín, and L. Puelles (1990) Neuronal differentiation in the thalami area triangularis of a lizard. J. Morphol. *205*:123-134.

Medina, L., C.M. Trujillo, C. Díaz, A. Martín, and L. Puelles (1991) Neuronal typology of the thalami area triangularis of *Gallotia galloti* (Reptilia Sauria). J. Morphol. *205*:113-121.

Mendel, J.R., and J.N. Whitaker (1978) Immunocytochemical studies of myelin basic protein. J. Cell Biol. *76*:502-511.

Miller, R.H., and F.J. Liuzzi (1986) Regional specialization of the radial glial cells of the adult frog spinal cord. J. Neurobiol. *15*:181-96.

Mitro, A., and A. Kiss (1977) The ependyma of the ventriculus mesencephali in the rat. Acta Morphol. Hungarica. *25*:1-8.

Monzón-Mayor, M., C. Yanes, G. Tholey, J. de Barry, and G. Gombos (1990a) Immunohistochemical Localization of Glutamine Synthetase in Mesencephalon and Telencephalon of the Lizard *Gallotia galloti* During Ontogeny. Glia *3*:81-97.

Monzón-Mayor, M., C. Yanes, M.S. Ghandour, J. de Barry, and G. Gombos (1990b) Glial Fibrillary Acidic Protein and Vimentin Immunohistochemistry in the Developing and Adult Midbrain of the Lizard *Gallotia galloti*. J. Comp. Neurol. *295*:569-579.

Monzón-Mayor, M., C. Yanes, J.L. James, and R.R. Sturrock (1990c) An ultrastructural study of the development of astrocytes in the midbrain of the lizard. J. Anat. *170*:33-41.

Monzón-Mayor, M., C. Yanes, J.L. James, and R.R. Sturrock (1990d) An ultrastructural study of the development of oligodendrocytes in the midbrain of the lizard. J. Anat. *170*:43-49.

Monzón-Mayor, M., C. Yanes, J.L. James, and R.R. Sturrock (1991) An ultrastructural study of ependymal cell differentiation during lizard (*Gallotia galloti*) mesencephalon development. J. Anat. *174*:251-261.

Monzón-Mayor, M., C. Yanes, J. de Barry, C. Capdevilla-Carbonel, J. Renau-Piqueras, G. Tholey, and G. Gombos (1997) Heterogeneous immunoreactivity of glial cell in the mesencephalon of a lizard: a double labeling immunohistochemical study. J. Morphol. *235*: 2-26.

Mori, S., and C.P. Leblond (1969) Identification of microglia in light and electron microscopy. J. Comp. Neurol. *135*:57-80.

Mori, S., and C.P. Leblond (1970) Electron microscopic identification of three classes of oligodendrocytes and a preliminary study of their proliferative activity in the corpus callosum of young rats. J. Comp. Neurol. *139*:1-30.

Nagashima, R. (1979) Ultrastuctural study of myelinating cells and subpial astrocytes in developing rat spinal cord. J. Neurol. Sci. *44*:1-12.

Noble, M., G. Wolswijk, and D. Wren (1989) The complex relationship between cell division and the control of differentiation in oligodendrocytes type-2 astrocyte progenitor cells isolated from perinatal and adult rat optic nerves. Prog. G. F. Res. *1*:179-194.

Norenberg, M.D. and Martínez-Hernández, A. (1979) Fine structural localization of glutamine synthetase in astrocytes of rat brain. Brain Res. *161*:303-310.

Norenberg, M.D (1979) The distribution of glutamine synthetase in the rat central nervous system. J. Histochem. Citochem. *27*:756-762.

Norenberg, M.D. (1983) Immunohistochemistry of glutamine synthetase. In: L. Hertz, E. Kvamme, E.G. McGeer, and A. Schousboe (eds): Glutamine, Glutamate, and GABA. New York: Alan R Liss, pp. 95-111.

Norton, W.T. (1981) Formation structure and biochemistry of myelin. In: G.J. Siegel, R.W. Albers, R. Katzman, B.W. Agranoff (eds): Basic Neurochemistry. Little. Boston: Brown and Co, pp. 63-92.

Nussbaum J.L. and G. Roussel (1983) Immunocytochemical demonstration of the transport of myelin proteolipids throught the Golgi apparatus. Cell Tissue Res. *234*:547-559.

Okado, N. (1982) Early myelin formation and glial cell development in the human spinal cord. Anat. Rec. *202*:483-490.

Onteniente, B., H. Kimura, and T. Maeda (1983) Comparative study of the glial fibrillary acidic protein in vertebrates by PAP immunohistochemistry. J. Comp. Neurol. *215*:427-436.

Parnavelas, J. G., R. Luder, S.G. Pollard, K. Sullivan and A.R. Lieberman (1983) A qualitative and quantitative ultrastructural study of glial cells in the developing visual cortex of the rat. Phil. Trans. R. Soc. Lond. *301*:55-84.

Patel, A.J., M.D. Weir, A. Hunt, C.S.M. Tahourdin, and D.G.T. Thomas (1985) Distribution of glutamine synthetase and glial fibrillary acidic protein and correlation of glutamine synthetase with glutamate decarboxylase in different regions of the rat central nervous system. Brain Res. *331*:1-9.

Patel, A.J. (1986) Development of astrocytes: In vivo studies. In T. Grisar, G. Frank, L. Hertz, M. Sensenbrenner, and D.M. Woodbury (eds): Advances in the Biosciences, vol 61, Dynamic Properties of Glial Cells: Cellular and Molecular Aspects. Oxford: Pergamon Press, pp. 87-96.

Peters, A., S.L. Palay, and H de F Webster (1970) The fine structure of the nervous system. New York: Harper and Row.

Phillips, D.E. (1973) An electron microscopic study of macroglia and microglia in the lateral funiculus of the developing spinal cord in the foetal monkey. Z. Zell. Mikrosk. Anat. *140*:145-167.

Pixley, S.K.R., and J. De Vellis (1984) Transition between immature radial glia and mature astrocytes studies with a monoclonal antibody to Vimentin. Dev. Brain. Res. *15*:201-209.

Polak, M., W. Haymaker, J.E. Johnson, and F. D'Amelio (1982) Neuroglia and their reactions. In W. Haymaker and R. Adams (eds): Histology and Histopathology of the Nervous System. Illinois: Thomas CC. Pub., pp. 363-445.

Privat, A., and C.P. Leblond (1972) The subependymal layer and neighbouring region in the brain of the young rat. J. Comp. Neurol. *146*:277-302.

Privat, A., and P. Rotaboul (1986) Fibrous and protoplasmic astrocytes. In S. Fedoroff and A. Vernadakis (eds): Astrocytes, vol I. New York: Academic Press, pp. 105-129.

Quitschke, W., P.S. Jones, and N. Schechter (1985) Survey of intermediate filament protein in optic nerve and spinal cord: Evidence for differential expression. J. Neurochem. *44*:1465-1476.

Rafols, J.A. (1986) Ependymal tanycytes of the ventricular system in invertebrates. In S. Fedoroff and A. Vernadakis (eds): Astrocytes, vol I. New York: Academic Press, pp. 131-148.

Rakic, P. (1971) Neuron-glia relationship during granule cell migration in developing cerebellar cortex. A Golgi and electron microscopic study in Macacus rhesus. J. Comp. Neurol. *141*:283-312.

Rakic, P. (1972) Mode of cell migration to the superficial layers of fetal monkey neocortex. J. Comp. Neurol. *145*:283-312.

Rakic, P. (1982) The role of neuronal-glial cell interaction during brain development. In T.A. Sears (ed): Neuronal-glial cell interaction. New York: Springer Verlag, pp. 25-38.

Ramón y Cajal, S. (1894) Les nouvelles idées sur la structure du système nerveux chez l'homme et chez les vertébrés. La Névroglie. pg. 151. C. Reinwald, Libraires-Editèur. Paris.

Ramón y Cajal, S. (1911) Histologie du systéme nerveux de l'homme et des vertébrés. Tome 2. Paris: Maloine. Reprinted by Consejo Superior de Investigaciones Científicas, Madrid, 1955.
Riepe, R.E. and M.D. Norenberg (1977) Müller cell localization of glutamine synthetase in rat retina. Nature. *268*:654-655.
Roback, H.N., and H.J. Scherer (1935) Über die feiner Morphologie des frühkindlichen Hirnes unter besonderer Berücksichtigung der Gliaentwicklung. Virchows Archiv. für pathologishe Anatomie und Physiologie. *294*:365-431.
Romero-Alemán, M.M., M. Monzón-Mayor, M.L. Plaza-Pérez, and C. Yanes (1995) Development of astroglial cells in the encephalon of *Gallotia galloti*: A Golgi technique study. J. Morphol. *223*:13-20.
Romero-Alemán, M.M., M. Monzón-Mayor, M.L. Plaza-Pérez, A. Velasco, J.F. Arbelo-Galván, and C. Yanes (1996) S100 immunoreactive glial cells in the mesencephalon and telencephalon of a lizard. Second European Meeting on Glial Cell Function in Health and Disease. B56, pp. 79 (abstract).
Roots, B.I. (1986) Phylogenetic development of astrocytes. In Astrocytes, vol 1. (ed. S. Fedoroff and A. Vernadakis), pp. 1-34. New York: Academic Press.
Sarthy, P.V. and D.M.K. Lam (1978) Biochemical studies of isolated glial (Müller) cells from turtle retina. J. Cell. Biol. *78*:675-684.
Schachner, M., E.T. Hedley-White, D.W. Hsu, R.G. Shoonmaker, A. Bignami (1977) Ultrastructural localization of glial fibrillary acidic protein in mouse cerebellum by immunoperoxidase labeling. J. Cell Biol. *75*:67-73.
Schiffer, D., M.T. Giordana, A. Migheli, G. Giaccone, S. Pezzota, and A. Mauro (1986) Glial fibrillary acid protein and vimentin in the experimental glial reaction of the rat brain. Brain Res. 374:110-118.
Schmechel, D.E., and P. Rakic (1979) A Golgi study of radial glial cells in developing monkey telencephalon: Morphogenesis and transformation into astrocytes. Anat. Embryol. *150*:115-152.
Schnitzer, J., W.W. Franke, and M. Schachner (1981) Immunocytochemical demonstration of Vimentin in astrocytes and ependymal cells of developing and adult mouse nervous system. J. Cell. Biol. *90*:435-447.
Schonbach, C. (1969) The neuroglie in the spinal cord of the newt Triturus viridescens. J. Comp. Neurol. *135*:93-120.
Schultz, R. L. (1969) Zur postnatalen Biomorphose des Ependyms im Telencephalon von Lacerta agilis agilis. Z. mikrosk. -anat. Forsch. *81*:111-152.
Shaw, G., M. Osborn, and K. Weber (1981) An immunofluorescence microscopical study of the neurofilament triplet proteins, vimentin and glial fibrillary acid protein within the adult rat brain. Eur. J. Cell. Biol. 26:68-82.
Shoukimas, G.M., and J.W. Hinds (1978) The development of the cerebral cortex in the embryonic mouse: an electron microscopic serial section analysis. J. Comp. Anat. *179*:795-830.
Skoff, F.P., Price D.L., and A. Stocks (1976a). Electron microscopic autoradiographic studies of gliogenesis in rat optic nerve. I. Cell proliferation. J. Comp. Neurol. *169*:291-312.
Skoff, R.P., D.L. Price, and A. Stocks. (1976b). Electron microscopic autoradiographic studies of gliogenesis in rat optic nerve. II. Time of origin. J. Comp. Neurol. *169*:313-334.
Skoff, R.P., and P.E. Knapp (1991) Division of astroblast and oligodendroblast in postnatal rodent brain: Evidence of separate astrocyte and oligodendrocyte lineages. Glia. *4*:165-174.
Smart, 1. (1961) The subependymal layer of the mouse brain and its cell production as shown by autoradiography after thymidine-H3 injection. J. Comp. Neurol. 116:325-347.
Stensaas, L.J., and S. Stensaas, (1968a) Astrocytic neuroglial cells, oligodendrocytes and microgliocytes in the spinal cord of the toad. II. Electron microscopy. Z. Zell. mikrosk. Anat. *86*:184-213.
Stensaas, L.J., and S.S. Stensaas (1968b) Light microscopy of glial cells in turtles and birds. Z. Zell. mikrosk. Anat. *91*:315-340.
Stensaas, L.J., and B.C. Gilson (1972) Ependymal and subependymal cells of the caudato-pallial junction in the lateral ventricle of the neonatal rabbit. Zeits. Zell. mikrosk. Anat. *132*:297-322.
Stensaas, L.J. (1977) The ultrastructure of astrocytes, oligodendrocytes and microglia in the optic nerve of urodele amphibians (A. punctatum, T. pyrrhogaster, T. viridescens). J. Neurocytol. *6*:269-286.

Sternberger, N.H., Y. Itoyama, M.W. Kies, and H.F. Webster (1978a) Immunocytochemical method to identify basic protein in the myelin-forming oligodendrocytes of newborn rat CNS. J. Neurocytol. 7:251-263.
Sternberger, N.H., Y. Itoyama, M.W. Kies, and H.F. Webster (1978b) Myelin basic protein demonstrated immunocytochemically in oligodendroglia prior to myelin sheath formation. Proc. Natl. Acad. Sci. USA *75*:2521-2524.
Stevenson, J.A., and M.G. Yoon (1982) Morphology of radial glia, ependymal cells and periventricular neurons in the optic tectum of goldfish (Carassius auratus). J. Comp. Neurol. *205*:128-139.
Sturrock, R.R. (1974) Histogenesis of the anterior limb of the anterior commissure of the mouse brain. III. An electron microscopic study of gliogenesis. J. Anat. *117*:37-53.
Sturrock, R. R. (1975) A light and electron microscopic study of proliferation and maturation of fibrous astrocytes in the optic nerve of the human embryo. J. Anat. *119*:223-234.
Sturrock, R.R. (1976) Light microscopic identification of immature glial cells in semithin sections of the developing mouse corpus callosum. J. Anat. *122*:521-537.
Sturrock, R.R. (1978) Development of the indusium griseum II. A semithin light microscopic and electron microscopic study. J. Anat. *125*:433-445.
Sturrock, R.R. (1980) A developmental study of the mouse neostriatum. J. Anat. *130*:243-261
Sturrock, R.R. (1981) An electron microscopic study of the development of the ependyma of the central canal of the mouse spinal cord. J. Anat. *132*:119-136.
Sturrock, R.R. (1982a) Gliogenesis in the prenatal rabbit spinal cord. J. Anat. *134*:771-793.
Sturrock, R.R. (1982b) Cell division in the normal central nervous system. Adv. Cell. Neurobiol. *3*:3-33.
Sturrock, R. R. (1986). Postnatal ontogenesis of astrocytes. In S. Fedoroff and A. Vernadakis (eds): Astrocytes. New York: Academic Press, pp. 765-103.
Tansey, F.A., M. Farooq, and W. Cammer (1991) Glutamine synthetase in oligodendrocytes and astrocytes: new biochemical and immunochemical evidence. J. Neurochem. *56*:266-272
Tennyson, V.M., and G.D. Pappas (1962) An electron microscope study of ependymal cells of the fetal, early postnatal and adult rabbit. Z. Zellforsch. *56*:595-618.
Vaughan, D.W. (1984) The structure of neuroglial cells. In cerebral Cortex, vol.2. Functional Properties of Cortical Cells (ed. E.G. Jones & A. Peters), pp. 285-329. New York and London: Plenum Press.
Vaughn, J.E., and A. Peters (1967). Electron microscopy of early postnatal development of fibrous astrocytes. Am. J. Anat. *121*:131-152.
Vaughn, J.E. (1969) An electron microscopic analysis of gliogenesis in rat optic nerve. Z. Zell. mikrosk. Anat. *94*:293-327.
Vaughn, J.E., and A. Peters (1971) The morphology and development of neuroglial cells. In D. Pease (ed): Cellular Aspects of Neural Growth and Differentiation. Los Angeles: University of California Press, pp. 103-140.
Westergaard, E. (1970) The lateral cerebral ventricles and the ventricular walls. MD thesis, Aarhus University.
Wilkin, G.P., D.R. Marriott, and A.J. Cholewinski (1990) Astrocyte heterogeneity. Tins *13*:43-46.
Wood, M.R., and Cohen, M.J. (1980) Synaptic regeneration and glial reactions in the transected spinal cord of the lamprey. J. Neurocytol. *10*:57-79.
Yanes, C. (1985) Ontogenia de las Areas basales telencefálicas de *Gallotia galloti* (reptil Lacertidae). An. Univ. La Laguna *2*:11-66.
Yanes, C.M., M.A. Pérez-Batista, J.M. Martín-Trujillo, M. Monzón-Mayor, and A. Marrero (1987) Anterior Dorsal Ventricular Ridge in Lizard: Embryonic Development. J. Morphol. *194*:55-64.
Yanes, C., J.M. Martin, M.A. Pérez, M. Monzón-Mayor, M., and A. Marrero (1988a) Ependymogenesis of the lizard basal areas. I. Ependymal zones. Z. mikrosk. -anat. Forsch. Leipzig *102*:555-572.
Yanes, C., J.M. Martín, M.A. Pérez, M. Monzón-Mayor, and Marrero, A. (1988b) Ependymogenesis of the lizard basal areas. II. Sulcus. Z. mikrosk. -anat. Forsch. Leipzig *102*:573-589.
Yanes, C.M., M.A. Pérez-Batista, J.M. Martín-Trujillo, M. Monzón-Mayor, and A. Rodriguez (1989) Development of the ventral striatum in the lizard *Gallotia galloti*. J.Anat.*164*:92-100.
Yanes, C., M. Monzón-Mayor, M.S. Ghandour, J. de Barry, and G. Gombos (1990) Radial glia and astrocytes in developing and adult telencephalon of the lizard *Gallotia galloti* as revealed by immunohistochemistry with anti-GFAP and anti-vimentin antibodies. J. Comp. Neurol. *295*:559-568.

Yanes, C., M. Monzón-Mayor, J. de Barry, and G. Gombos (1992) Myelin and myelination in the telencephalon and mesencephalon of the lizard *Gallotia galloti* as revealed by the immunohistochemical localization of myelin basic protein. Anat. Embryol. *185*:475-487.
Yanes, C, M. Monzón-Mayor, G. Gombos, J. Renau-Piqueras, and R.R. Sturrock (1997) Glial cells in the lizard subpallial nuclei during ontogeny: an ultrastructural study. J. Morphol. *233*:1-13.
Yoshioka, T., and O. Tanaka (1989) Ultrastructural and cytochemical characterization of the floor plate ependyma of the developing rat spinal cord. J. Anat. *165*:87-100.

6

ROLE OF THYROID HORMONE ON THE OLIGODENDROCYTE TYPE 2-ASTROCYTE LINEAGE

Angeles Rodríguez-Peña,
Nieves Ibarrola and Sonia Vega

It is well known the need of systemic thyroid hormone for the normal central nervous system (CNS) development. One of the most striking features of the hypothyroid brain is a reduction in the total myelin content. This effect is correlated with a decrease in the amount of the myelin protein transcripts. Since the expression of the myelin protein genes is confined to the differentiated oligodendrocytes generated from their progenitors, the O-2A cells, we proposed that the major action of thyroid hormone on myelination is the promotion of the oligodendrocyte differentiation, leading to a correlation between the relative levels of myelin transcripts with the number of oligodendrocytes formed. In order to find out if this is the case, we have studied the cell composition in the optic nerve of eu- and hypothyroid rats. We have found that the oligodendrocyte number was reduced 50% in hypothyroid animals, but the number of progenitors and astrocytes were unchanged. Next, to obtain insights into how the timely generation of these cells is regulated by thyroid hormone (T3) we have used an "in vitro" model of oligodendrocyte differentiation from pure cultures of O2A progenitor cells. Our results show that T3 decreases the division of the O-2A progenitors and increases their differentiation to oligodendrocyte. In contrast with the existing model, the oligodendrocyte generation is associated with asymmetric differentiation controlled by cell-extrinsic signals, and their capability to generate at least one oligodendrocyte "in vitro" at a time equal to the rat's day of birth is independent of T3 and seems to be a cell-intrinsic property.

Origin of the oligodendrocyte lineage

Development is driven by a very precise schedule of spatio-temporal events involving proliferation, migration, differentiation, and survival of different cell lineages. In the vertebrate central nervous system (CNS), multipotential stem cells could be isolated from many regions of the developing neuroepithelium. These stem cells proliferate symmetrically to produce uncommitted daughter cells that form a major component of neuroepithelium. From lineage analysis *in vivo* and *in vitro*, there is evidence for the presence of lineage restrictions, including bipotential neuronal and oligodendrocyte precursors and committed neuronal progenitors (Luskin, et al., 1993; Davis and Temple, 1994; Williams, 1995). This lineage restriction has been clearly demonstrated with the identification of the bipotential oligodendrocyte precursor cell, O-2A, from the rat postnatal optic nerve (fig. 6-1). The general consensus is that oligodendrocyte progenitors arise from the subventricular zone and migrate out from there into the developing white matter, differentiating and proliferating "en route" (Altman, 1966; Goldman and Vaysse, 1991). Until very recently, there was no information available on the instructive signal(s) responsible for the induction of the lineage restriction. Interestingly, some of such signals have been identified *in vitro*. Thus, the transient exposure of stem cells to ciliary neurotrophic factor (CNTF) or thyroid hormone (T3) is sufficient to initiate a stable switch to a glial fate, generating clones of astrocytes or oligodendrocytes, respectively (Johe, et al., 1996). Once the lineage has been restricted, the O-2A progenitors proliferate in response to growth factors, and are committed to differentiate into oligodendrocytes since growth factors deprivation irreversibly induce their differentiation. Although thyroid hormone might play a role in the very early determination of the stem cells to the oligodendrocyte fate, we are going to described the role of this hormone, once this event has been accomplished, i.e., on the O-2A lineage.

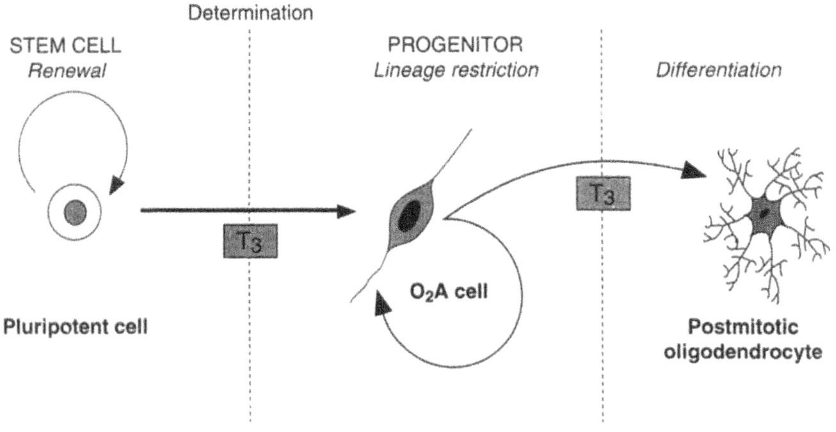

Figure 6-1. Cell-type transitions of CNS stem cells, using the O-2A lineage as model. The drawing illustrates the distinct effects of extracellular factors. Mitogenic factors for stem and progenitor cells, and differentiation factors that can be subdivided in two kinds: factors that promote differentiation and factors that cause lineage restriction. T3, as differentiation factor fulfill both types of actions.

Cell types and antigenic markers of the O2A lineage

Oligodendrocytes were first defined precisely by Del Rio-Hortega in 1920 (Somjen, 1988). Since then, the ontogeny of this special type of macroglial cell has been profusely studied, and it was not before 1983 that the studies of Raff and coworkers on the development of the glial cells from the rat optic nerve, unambiguously established that there were a parental relationship between a type of astrocytes, named type 2 and the oligodendrocytes (Raff, et al., 1983; Raff, et al., 1984; Temple and Raff, 1985). They showed that, *in vitro*, a progenitor cell (the O-2A cell) present in the rat optic nerve was able to generate, in response to either platelet derived growth factor (PDGF) or serum, oligodendrocytes and type-2 astrocytes, respectively. In contrast, type-1 astrocytes, that are flat fibroblast-like cells *in vitro*, are present very early in the developing rat optic nerve and have a distinct developmental origin.

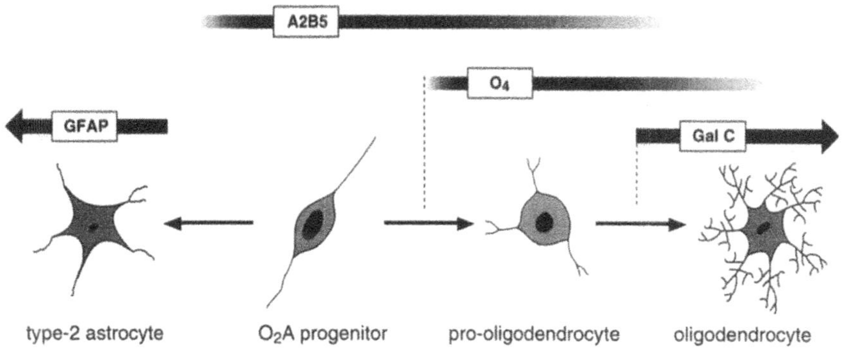

Figure 6-2. Squematic representation of the O-2A lineage. The different stages of the oligodendrocyte differentiation are recognized with specific antibodies to surface membrane components. Astrocytes are recognized by the anti-glial fibrillary acid protein (GFAP).

The characterization of this lineage has been possible because the different stages of these nerve cells can be distinguished by their morphology and by their expression of different antigenic markers (fig. 6-2; for review see Barres and Raff, 1994a). Thus, glial progenitor cells (O-2A) can be identified in cultures of rat optic nerve and brain by their bipolar morphology and by its expression of a tetrasialogangliosides recognized by the A2B5 monoclonal antibody. This antigen is also present on neurons (Eisenbarth, et al., 1979), and therefore, only in the optic nerve that lacks intrinsic neurons, A2B5 is a useful marker specifying only the O-2A lineage. Newly differentiated oligodendrocytes express the myelin glycolipid, galactocerobroside (GalC), in their plasma membrane and lose of immunoreactivity for A2B5. In the progression towards the differentiated phenotype, appears an additional antigen on the cell surface named for its corresponding monoclonal antibody O4 (Gard and Pfeiffer, 1990), and has been used to trace this lineage in other CNS areas. Finally, both type-1 and type-2 astrocytes are labeled by antibodies to glial fibrillary acidic protein (GFAP). Thus, by staining freshly dissociated rat optic nerve cells in suspension with A2B5, GalC and anti-GFAP antibodies, the development of the various glial cell types in the optic nerve *in vivo* has been charted: Type-1 astrocytes first develop around embryonic day 16 (E16), with small numbers of proliferating O-2A progenitors at that time. They begin to give rise to postmitotic oligodendrocytes around the day of birth (E21). A week or so later, cells with the antigenic phenotype of type-2 astrocytes start to appear. Some O-2A progenitors seem to undergo yet another developmental transition giving rise to a different type of O-2A progenitor cell that persists in the adult optic nerve (Wolswijk and Noble, 1989).

Factors that control proliferation, differentiation and survival in the O-2A lineage

Characterization of the factors that regulate oligodendrocyte proliferation, differentiation and survival has been facilitated by the ability to generate and analyze cells at different stages of the oligodendroglial lineage *in vitro*, using chemically defined medium. O-2A progenitors or mixed populations of rat optic nerve cells are cultured on poly-L-lysine (PLL) coated surfaces. The basal medium used is Dulbecco's modified Eagles's medium (DMEM) supplemented with 25μg/ml gentamicine, 2mM glutamine, 1μg/ml bovine pancreas insulin, 100 μg/ml human transferrin, 0.2μM progesterone, 0.10nM putrescine, 0.224μM selenium (Bottenstein and sato, 1979; DMEM-BS). Interestingly all the experiments carried out previously also contained a mixture of thyroid hormones (T3 and T4) in this basal medium.

It has been shown that oligodendrocytes are generated with a timing which reproduces that observed *in vivo* when embryonic optic nerve or spinal cord cultures are grown in DMEM-BS that contains thyroid hormones and is further supplemented with PDGF (Raff, et al., 1988; Bögler, et al., 1990; Noll and Miller, 1994). Since then, it has been widely accepted that PDGF is a potent mitogen for purified O-2A progenitors cells. This factor is produced by type-1 astrocytes and is present in the astrocytes-conditioned medium. The mitogenic response of the O-2A cells to PDGF is lost during the time in culture, in parallel with their differentiation to oligodendrocyte. The combined application of PDGF and basic fibroblast growth factor (bFGF) induces O-2A progenitor cells to undergo repeated division, in a type of self-renewal cycle that will not end until removal of the growth factors (Bögler, et al., 1990; Bögler and Noble, 1994; McKinnon, et al., 1990). Other reported mitogenic signal is neurotrophin-3 (NT-3) (Barres, et al., 1994b). Although, in our hands, NT-3 is not required for the PDGF mitogenic action, but when present it partially suppresses oligodendrocytes generation in O-2A progenitor populations induced to divide by PDGF (Ibarrola, et al., 1996). Finally, conditioned medium of different neuroblastoma cell lines contain mitogens for oligodendrocytes and their progenitors, such activity corresponds to the member of the neuregulins family, the glial growth factor (GGF) (Canoll, et al., 1996). The extent of oligodendrocytes generation is modulated by signals such as ciliary neurotrophic factor (CNTF; Mayer, et al., 1994), insulin growth factor-I (IGF-I; McMorris and Dubois, 1988), transforming growth factor ß (TGF-ß; McKinnon, et al., 1993) or thyroid hormone (fig. 6-3). These factors can interact *in vitro*, and presumably *in vivo*, in complex ways that are determined by the combination of the factors presented (Ibarrola, et al., 1996). Both O-2A and oligodendrocytes require external signals to survive. Approximately 50% of the daily generated oligodendrocytes in the rat optic nerve die by apoptosis within the 2-3 next days (Barres and Raff, 1993). Many of the factors with activity on division and differentiation function as well as survival factors (for a review, see Barres and Raff, 1994).

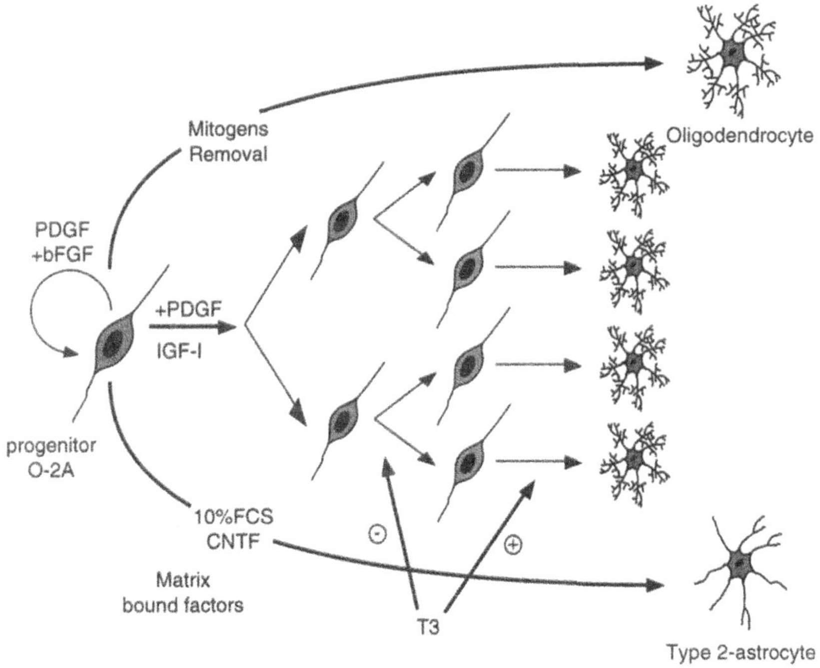

Figure 6-3. Factors that regulate the development of the O-2A lineage *in vitro*. Proliferation of O-2A progenitors is promoted by PDGF-AA, possibly with IGF-I as mitogenic cofactor and/or survival factor. IGF-I, TGF-ß and/or hydrophobic hormone/factors as T3 provide permissive signals that allow O-2A progenitors to carry out their default differentiation program into oligodendrocytes. CNTF from type-1 astrocytes seems to initiate the type-2 astrocyte differentiation pathway, while unidentified, matrix-bound factors from mesenchymal cell are required to stabilized the type-2 astrocyte phenotype. Combined addition of PDGF and bFGF blocks oligodendrocyte differentiation and stimulates the endless O-2A progenitors division.

Mechanism of thyroid hormone action

The effects of T3 is mediated by the intracellular thyroid hormone receptors (TR), members of the type II steroid superfamily of receptors that act as ligand-modulated transcription factors. This superfamily is defined by a highly conserved 66-68 amino acid domain predicted to form zinc finger structures that are necessary for sequence-specific interactions with cis-acting regulatory elements of targets genes. In addition, a less-well-conserved C-terminal region of approximately 220 amino acids functions as ligand-binding domain and facilitates receptor dimerization. This superfamily of receptors appears to be an ancient family that is represented in all species (Mangelsdorf, et al., 1995).

The TRs are encoded by two distinct genes, c-erbA α and c-erbA ß (fig. 6-4). The c-erbA α gene is alternatively spliced into α1 and α2 (hormone and non-hormone binding isoforms, respectively). The c-erbA ß gene is alternatively spliced into ß1 and ß2. The α1, α1 and ß2 isoforms are functional receptors, whereas α2 is an isoform that is non-functional and exerts a dominant negative effect on the function of the other receptors. Intriguingly, the c-erbA α locus has also been demonstrated to contain an overlapping transcription unit utilizing coding information on the opposite strands (rev-erbAα). *In situ* studies of c-erbA transcripts have shown that TRα mRNAs are widely expressed in the embryonal CNS whereas TRß are expressed predominantly in the postnatal period and with a more restricted spatial distribution (Bradley et al., 1992). TRs bind to response elements containing tandem repeats of the AGGTCA motif with spacing of three or four nucleotides, and regulate gene transcription. Although TRs can function as homodimers, they normally require "accessory/ auxiliary" factors for high-affinity binding to their cognate sequences. The retinoid X receptor (RXR) family is one of these accessory proteins that is activated by 9-cis retinoic acid. The RXRs heterodimerized with TRs and function to selectively target the high-affinity binding of these receptors to their cognate elements (Brent, et al., 1991).

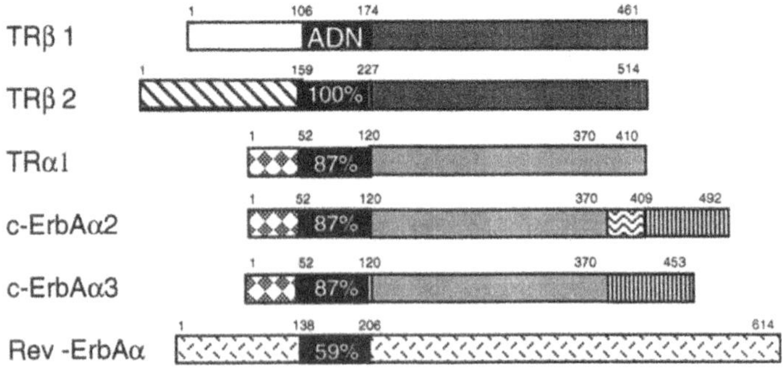

Figure 6-4. Thyroid hormone receptor isoforms. The percentage of homology of each isoform with the DNA binding domain of the TRß1 is shown.

Thyroidal trophic influence on myelination

It has been known for some time that the presence or absence of systemic thyroid hormone affects the normal myelination process (Balazs, et al., 1969; Dussault and Ruel, 1987). The initial studies described that hypothyroidism decreased the amounts of cholesterol, cerebrosides, sulfatides, glycolipids, sulfolipid and gangliosides in the myelin sheets (Bhat, et al., 1979; Flynn, et al., 1977; Shanker, et al., 1987) as a consequence of the reduction of the enzymatic activity involved in myelin lipid synthesis, such as cerebroside sulfotransferase and galactosyl transferases (Bhat, et al., 1979). Hypothyroid rat brains showed a reduction in the total myelin content and a developmental retardation of 1-2 days in the myelin protein composition, although a mature myelin protein composition was attained (Walters and Morell, 1981).

Moreover, it was also shown that thyroid hormone regulates myelin synthesis in cultures of mouse oligodendrocytes (Bhat, et al., 1981a,b). T3 transcriptionally regulates the myelin basic protein (MBP) gene, and a functional thyroid hormone responsive element has been localized in its promoter (Farsetti, et al., 1991). However, there are experimental evidences suggesting the existence of other mechanisms of T3 action. On one hand, we have shown that thyroid hormone has no effect on the transcriptional rate of the myelin associated glycoporotein (MAG) gene (Rodriguez-Peña, et al., 1993), and we and others have failed to localize a thyroid responsive element in the proteolipid protein (PLP) promoter. On the other hand, lack of thyroid hormone does not impair the increase in the myelin gene transcription during brain development and affects the expression of each myelin gene following the same pattern: a transient decrease in the accumulation of their mRNAs in most brain regions. The more affected regions are those that myelinate late, as striatum and cortex, whereas the caudal regions are only affected in early postnatal life. Moreover, gene expression specific for the oligodendrocyte progenitor cells, such as the expression of platelet-derived growth factor receptor α is not affected by hypothyroidism (Ibarrola and Rodríguez-Peña, 1997). Thus, our *in vivo* results showed that the lack of thyroid hormone only affects gene expression of the differentiated oligodendrocytes. We proposed that, *in vivo*, the major action of thyroid hormone on myelination be the promotion of the oligodendrocyte differentiation, leading to a correlation between the relative levels of myelin transcripts with the number of oligodendrocytes formed.

GENERATION OF OLIGODENDROCYTES IN VIVO AND IN OPTIC NERVES OF HYPOTHYROID ANIMALS

If the hypothesis proposed above would be true, in hypothyroid animals there should be a reduction in the number of oligodendrocytes formed during early developmental periods, but this reduction might not be very severe in some brain regions. To examine this possibility, pregnant rats were treated with methylmercatoimidazol (MMI) to ensure fetal hypothyroidism, and pups born to these rats were treated with MMI and thyroidectomized shortly after birth; this complete set of manipulations lead to nearly undetectable levels of thyroid hormones in the neonatal brain (Rodriguez-Peña, et al., 1993). In single cell suspensions prepared from optic nerves of these animals using collagenase, trypsin, EDTA as described (Williams, et al., 1991), we compared the glial cell composition of the optic nerves of control and hypothyroid animals at 2, 6 and 8 days after birth. As shown in fig. 6-5, pups born on MMI-treated mothers consitently had a reduced representation of oligodendrocytes in their optic nerves compared with control animals. In contrast, no differences between hypothyroid and control animals were observed in the proportional representation of (A2B5+ GalC-) O-2A progenitors cells, nor in (GFAP+) astrocytes. Thus, it appears that O-2A progenitors, but not oligodendrocytes, were being normally produced *in vivo* in optic nerves of hypothyroid rats.

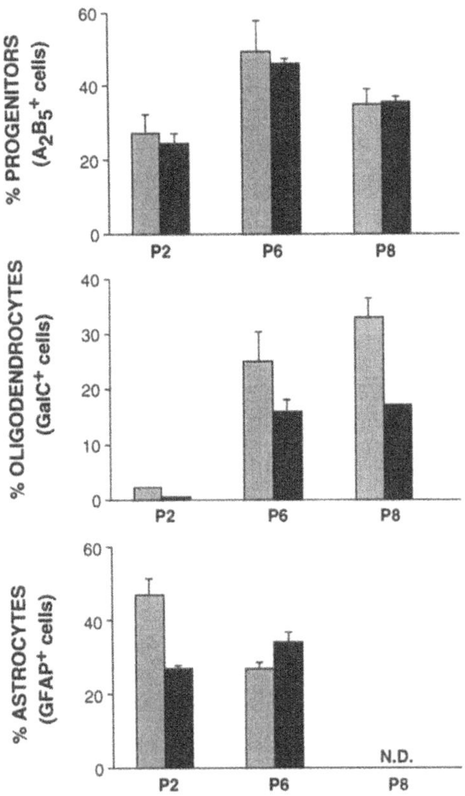

Figure 6-5. Hypothyroid animals show a reduced generation of oligodendrocytes. Single cell suspension were stained with one of the antibodies shown, and the proportion of the total population labeled with the antibody was determined. The values represent the means ± SD of three independent experiments. For each time point a minimum of 600 randomly chosen cell were counted on four separate coverslips. The proportion of oligodendrocytes (GalC+cell) was consistently reduced in the optic nerves of hypothyroid animals (dark bars) when compared with their controls (gray bars).N.D. not determined.

The partial impairment of the oligondedrocyte generation in the hypothyroid animal described led us to study with more detail the effects of T3 on the biology of the O-2A lineage. With this aim we initiated the following studies to ask two simple questions.We first wishes to know if thyroid hormone was required as a survival factor for the newly formed oligodendrocytes and second if the presence of thyroid hormone promoted oligodendrocyte generation. Further, we wished to determine if T3 was required for the appropriately timed generation of oligodendrocytes *in vitro*.

IS T3 A SURVIVAL FACTOR FOR OLIGODENDROCYTES?

One possible explanation of the reduction in the oligodendrocyte number found in the hypothyroid optic nerve would be that thyroid hormone is necessary as survival signal for the newly formed oligodendrocytes. Thus, the lack of T3 would increase the programmed cell death, with the final result of lower oligodendrocyte numbers in the hypothyroid brain. One method to experimentally study the activity, as a survival factor, of any defined molecule is to perform the survival MTT assay. The addition to the cultures of 0.5 mg/ml of MTT (3-(4,5-dimethylthiazol-2yl)-2,5-diphenyl tetrazolium bromide) allow to count viable cells, since their active mitochondria cleave the tetrazolium ring to form a visible dark blue formazan reaction product. The percentage of viable cells could be counted by bright-field phase microscopy.

Table 6-1. *Thyroid hormone does not enhance survival of the mature oligodendrocytes*

Conditions	MTT+	% survival
DMEM-BS	177±40	100
DMEM	64±20	37±11
+500nM T3	55±18	32±5
+50nM T3	80±15	48±9
+5nM T3	28±16	19±15
+5nM T3+CNTF 0.5ng/ml	65±20	37±23
+5nM T3+IGFI 0.1ng/ml	39±10	23±6

O-2A progenitors purified from P7 optic nerve were differentiated in defined medium. They were cultured additionally 3 days in the indicated conditions and received MTT for the last 2 hours. Cells were fixed and stained with anti-GalC antibody. The MTT+ column shows the number of cells that survived in each condition, counted in at least 3 different coverslips. The survival value for the DMEM condition is considered the lowest possibly obtained.

For our oligodendrocyte survival assay, 1500 oligodendrocytes /coverslip were obtained after spontaneous differentiation of pure O-2A progenitor cells from a P8 optic nerves. Oligodendrocytes were maintained in DMEM (the lowest rate of survival) with increasing concentrations of T3. The results were compared with the effect of the IGF-I and CNTF that increase the % of surviving cells up to 65 and 50 % respectively. As it is showed in Table 6-1, T3 was completely ineffective as survival factor for oligodendrocytes, either alone or in combination with suboptimal concentrations of IGF-I or CNTF. Thus, the reduction in the oligodendrocyte number in the hypothyroid brain should be due to an impairment in their differentiation.

EFFECT OF T3 ON DIVISION AND DIFFERENTATION OF PURIFIED O-2A PROGENITOR CELLS

We have studied the division and differentiation of purified O-2A progenitor cells isolated from P1 and P7 rat optic nerves. The purification of the O-2A progenitors was performed by immunopanning as described (Mayer, et al., 1994). This method is based in immunoselection of the mixed population. Cell suspensions are allowed sequentially to bind to culture dishes covered with specific antibodies, Ran2 for astrocytes and GalC for oligodendrocytes. The non-adherent cells are finally selected in A2B5 or O4 covered dishes, depending of the origin of the culture, optic nerve or other brain regions, respectively. The method is illustrated in fig. 6-6. In the final culture, the contaminating number of A2B5- cells (type-1 astrocyte and oligodendrocytes) represented <0.5% of the total cells while O-2Aprogenitors were at >99.5%. In order to follow clones of cells on daily basis, progenitors cells were plated at a density of 500-1000 cells in 25 cm^2 flasks. These plating densities yield a range of 100-150 clones per 25 cm^2 flasks (i.e., an average density of < 1 clone/ 15 mm^2). After 24 hr, cells in all conditions were analyzed visually to identify single cells. Clones were marked and numbered on the plastic to allow identification and were examined each day for the duration of the experiment. The cultures were grown in DMEM-BS /T3- supplemented with PDGF ± T3. These experiments, by using purified cells isolated from optic nerves of postnatal rats, allow us to categorize cells as oligodendrocytes or progenitors cells on the basis of morphology. The validity of distinguishing oligodendrocytes from progenitors on the basis of morphology was confirmed at the end of the experiment by staining cultures with a combination of A2B5 and anti-GalC antibodies.

O-2A progenitors isolated from P1 rats and grown in the presence of PDGF+T3 divided effectively and generate 3 different types of clones (fig. 6-6). There were homogeneous clones consisting wholly of bipolar progenitors, heterogeneous clones consisting both multipolar oligodendrocytes, and bipolar progenitors and homogenous clones consisting wholly of multipolar oligodendrocytes. In these experiments, by day 4 only 3.5% of colonies were composed purely of O-2A progenitors, and 45% were composed wholly of multipolar oligodendrocytes. The remaining 57.5% contained a mixture of bipolar progenitor

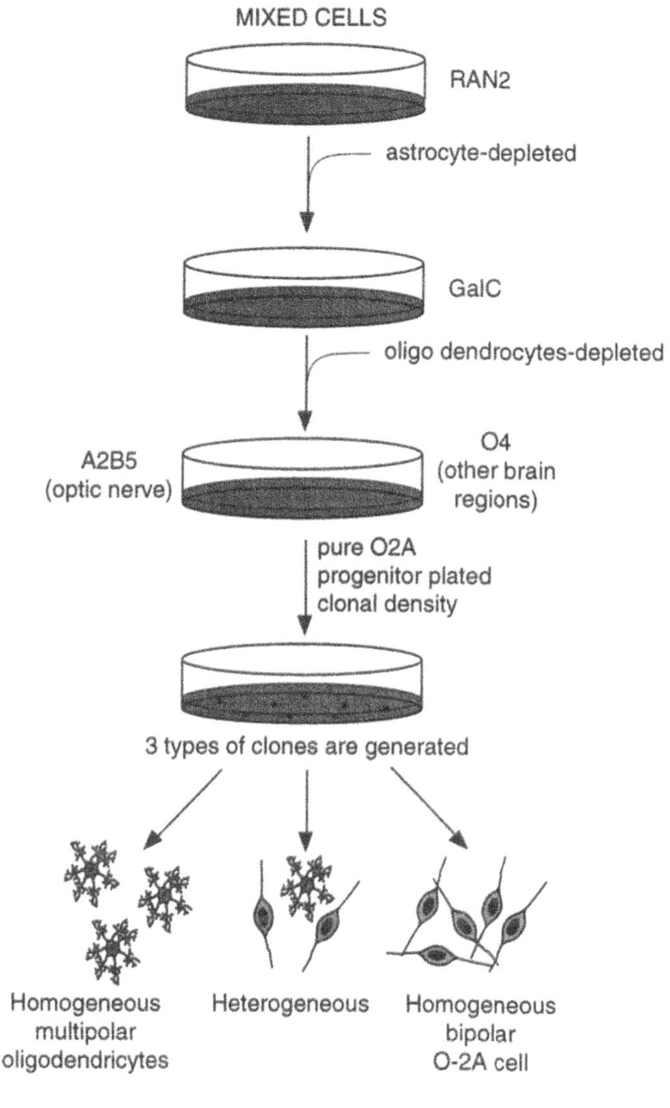

Figure 6-6. Immunopanning method for purification of the O-2A progenitors.
Dissociated cells from optic nerve or other brain region are allowed to bind, sequentially, to different culture dishes coated with specific antibodies. First, they are allowed to bind to Ran2-coated dish, to obtain astrocytes-depleted cells, then they are transferred into the GalC-coated dish(to deplete from oligodendrocytes). Finally, the remaining unbound cells are positively selected after binding to A2B5 for optic nerve cultures or O4 for other brain regions. Pure O2A progenitors are seeded at very low density to allow clonal division and differentiation. Three types of clones are observed: homogeneous clones of progenitors O-2A or oligodendrocytes, or heterogenous clones composed by the two types of cells.

cells and multipolar oligodendrocyte. Division continued within at least some of the progenitor-containing colonies for at least 2 more days, as maximal colony sizes of the heterogeneous clones increased from 15 to 22 cells. At this final time point, 17% of the clones still were heterogeneous in their composition, with several of these clones consisting predominantly of bipolar progenitor-like cells. No homogenous progenitor clones were observed on day 5 (not shown) or day 6 (fig. 6-7).

When P1-derived O-2A progenitors cells were exposed to PDGF but not to T3, we did not observe any alteration in the probability of an individual clone of cells generating at least one oligodendrocyte within 4 days in vitro growth.

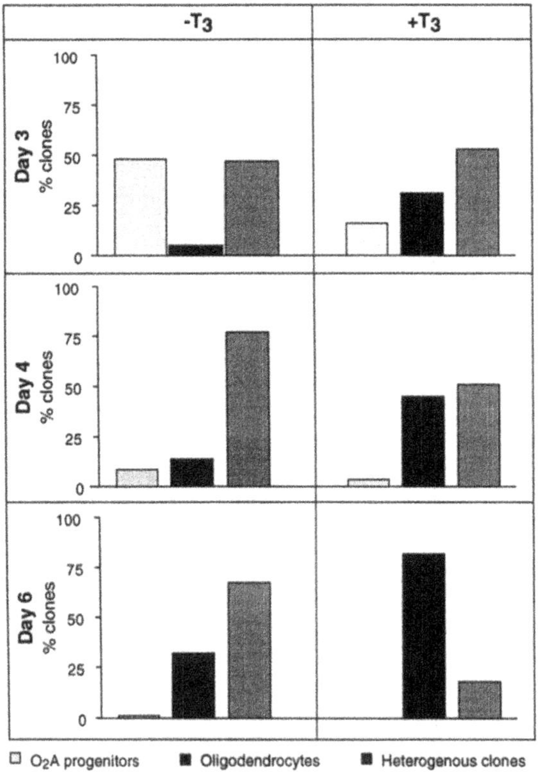

Figure 6-7. Thyroid hormone promotes oligodendrocyte differentiation of purified O-2A progenitors from P1 rat optic nerve. The proportion of homogeneous colonies of progenitors O-2A, oligodendrocytes and heterogeneous colonies in the absence (-T3) or in the presence of T3 (+T3) are shown after different times of *in vitro* growth (day 3, 4 and 6). The addition of T3 enhanced the extent of oligodendrocyte generation within individual clones, such that cultures grown in the presence of T3 had higher probability of containing oligodendrocytes.

We did, however, observe an enhancement of progenitor self-renewal within heterogeneous colonies and a reduction in the formation of clonal families of oligodendrocytes. Minor differences between $T3^-$ and $T3^+$ conditions were seen after 2 days of in vitro growth (data not shown), with differences in clonal composition becoming more clearly observable from Day 3 onward (fig. 6-7). The percentage of clones that were composed wholly of oligodendrocyte was reduced in cultures grown in the absence of T3 and was, for example, 15% on Day 4 and 38% on day 6 (compared with numbers of 45 and 83%, respectively, if T3 was present). In addition, the heterogenous colonies contained greater representation of O-2A progenitors in the absence of this hormone than in its presence (average values of 58% vs. 27%, respectively, for day 4 and 38% vs. 9% on day 6).

As would be expected if differentiation of dividing progenitors into non dividing oligodendrocytes occurred less frequently, there was also an increase in average colony sizes in cultures grown in the absence of T3; for example on day 6 the average colony size was 8.3 cells, compared with a value of 6.5 for comparable cultures grown in the presence of T3. It is critical to note, however, that almost every single clone of cells examined from day 4 onward contained clearly recognizable oligodendrocyte; although some clones contained as many as only 1 oligodendrocyte, other clones contained as many as 9 such cells. Thus, the absence of T3 did not preclude the generation of oligodendrocytes from dividing O-2A progenitors cells. Next, we performed the same type of analysis on P7-derived O-2A progenitors cells, (the time most frequently used in the preparation of glial cells cultures from rat optic nerve). These purified O-2A progenitor cells grown in DMEM-BS supplemented with PDGF±T3 behaved similarly to cells derived from optic nerves of P1 rats. Virtually all clones of cells contained oligodendrocytes by the fourth day of *in vitro* growth. There were generally fewer progenitors per clone if T3 was present, but heterogeneous clones containing bipolar progenitors were still present on Day 6 even in the presence of T3 (result not shown). Thus, T3 modulates the extent of oligodendrocyte generation within a clone of dividing progenitors and reduces the division of O2A cells. It is noteworthy that the absence of T3, however, did not alter the probability of a clone generating at least one oligodendrocyte, which suggested that T3 was not controlling the time of the oligodendrocyte differentiation. This is an important question because T3 has been proposed to be a "timer" for correct CNS development. The observation that the morphology of the hypothyroid brain resemble the one of a control animal of 2-3 days younger, led to think that the lack of thyroid hormone retarded events required for a proper CNS development. Thus, we decided to directly address this question.

THYROID HORMONE IS NOT REQUIRED FOR THE TIMED GENERATION OF OLIGODENDROCYTES

To determine whether the presence of thyroid hormone is required for the appropriately timed generation of oligodendrocytes *in vitro*, we examined the differentiation of O-2A progenitor cells in cultures prepared from brains of 15-day-old rat embryos (E15) and grown in DMEM-BS/T3-, the basal medium used in all experiments in this studies, supplemented with 10 ng/ ml PDGF ± 45 nM T3. It has been previously shown that oligodendrocytes are generated with a timing which reproduces that observed in vivo when embryonic optic nerve or spinal cord cultures are grown in DMEM-BS that contains T3 and is further supplemented with PDGF (Raff, et al., 1988; Bögler, et al., 1990; Noll and Miller, 1994).

Cultures were grown at a density that yielded individual colonies of O-2A lineage cells well separated from each other. On fifth, sixth and seventh days of in vitro growth, cultures were stained with O4 and anti-GalC antibodies to identify colonies of O-2A lineage cells. The O4 antibody is the only pre-oligodendrocyte marker that specifically labels O-2A lineage cells in the CNS (except for in the olfactory bulb; Barnett, et al., 1993) and has been previously shown to label O-2A progenitors cells in embryonic cultures (Noll and Miller, 1994). As the O4 antibody appears to label a subset of the A2B5+GalC- O2A progenitor population, separate cultures also were labeled with a combination of A2B5 and anti-GalC antibodies. The results obtained with O4 and A2B5 antibodies were indistinguishable, except for the fact that cultures stained with A2B5 antibody exhibited still higher numbers due to the ability of this antibody to label pre-O4+ O-2A progenitor cells.

Oligodendrocytes first appeared at a time corresponding to the time of their first appearance in vivo (i.e., the day of birth of the rat), regardless of whether T3 was added to the cultures. Moreover, the presence or absence of T3 was not associated with any differences in the percentage of total colonies that contained oligodendrocytes. When examined 5 days after plating, a time corresponding to 1 day before birth, cultures exposed to T3 contained a total of five oligodendrocytes, of several thousand O-2A progenitors growing in well-isolated colonies. Just 1 day later, however, all cultures contained substantial numbers of oligodendrocytes. Colonies were randomly chosen for detailed analysis on the basis of whether they contained cells with the morphology of O-2A progenitors. In cultures exposed to T3, 40/60 randomly chosen clones contained at least one oligodendrocyte. Similarly, 35/60 of the randomly chosen colonies that developed in the absence of T3 also contained at least one oligodendrocyte (Table 6-2).

Almost all colonies of cells in the cultures of embryonic rat brain that contained oligodendrocytes also contained many unipolar and bipolar O4+ GalC- O2A progenitor cells. Of the 75 oligodendrocyte- containing colonies examined, only 2 contained no detectable progenitor cells. Both of these homogenous oligodendrocyte colonies developed in cultures exposed to T3.

The issue of whether the observation of the appearance of heterogeneous clones containing O-2A progenitors and oligodendrocytes, that we have observed during these studies, indicated asymmetric generation of oligodendrocytes has been taken and studied by itself. We have confirmed that oligodendrocytes are being generated within single clonal - families of cells in association with asym-

Table 6-2. *Growth of embryonic O-2A progenitor cells in the presence or absence of thyroid hormone does not alter the appropriately timed generation of oligodendrocytes*

Conditions	+T3	-T3
Total number of cells	9386	11718
Number of O-2A lineage cells	6268	6850
Number of oligodendrocytes	1013	257
Proportion of O-2A cells (%)	67	59
Total number of counted clones	60	60
O-2A cell containing clones	58	57
Pure oligodendrocyte-clones	2	0
Pure progenitor-clones	16	15
Mixed clones	38	35
Cell Types in mixed clones (%)		
bipolar progenitors (O4+/GalC-)	65	80
mulitpolar progenitors (O4+/GalC-)	19	16
oligodendrocytes (O4+/GalC+)	16	4

metric division and differentiation using analysis of embryonic brain cultures labeled at limiting dilution with ß-galactosidase expressing (BAG) retrovirus (Price, et al., 1987; Williams, et al., 1991). In contrast with the previous model of the timely generation of oligodendrocytes, in which dividing O-2A progenitors cell clones undergo symmetric division and synchronous differentiation within a limited number of cell division (fig. 6-3).

We proposed a revised model: O-2A progenitor cells from the embryonic rat brain divide and generate more progenitors *in vitro* until a time equivalent to the rat's day of birth. At this point oligodendrocytes are generated within single clones of O-2A progenitors in association with asymmetric division and differentiation (fig. 6-8; Ibarrola, et al., 1996). It is important to note that clones of O-2A progenitors derived from optic nerve of postnatal rats have the ability to undergo symmetric and synchronous differentiation (Temple and Raff, 1985; Raff, et al., 1988; Wren, et al., 1992; Zhang and Miller, 1995). But such experiments have been carried out in heterogeneous cultures containing astrocytes (a potential source of CNTF and TGF-ß) and also supplemented with T3, conditions which could induce complete clonal differentiation due to the action of these signaling molecules. In other words, the apparently symmetric differentiation observed in these conditions may have been promoted by the action of "cell-extrinsic" signals.

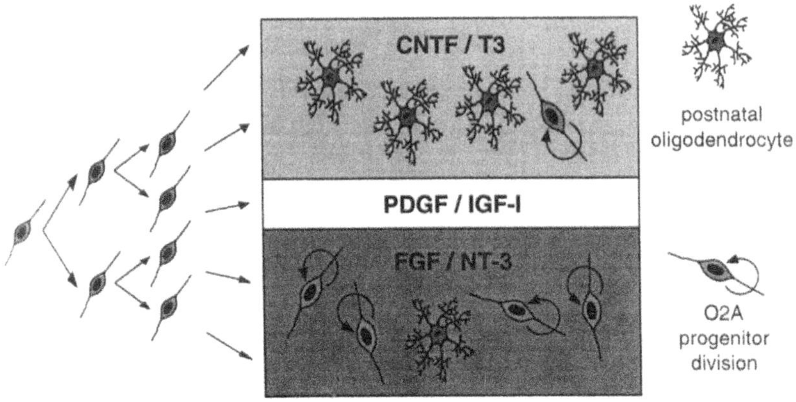

Figure 6-8. Revised model of the control of the timely generation of oligodendrocytes. O-2A progenitors from embryonic rat brain divide and generate more progenitors *in vitro* in association with asymmetric division and differentiation. After the day of birth, progenitors can either undergo self-renewing divisions or can generate oligodendrocytes. The probability of either event is modulated by the factors present in the extracellular environment. CNTF and T3 increase the probability of oligodendrocyte generation, and NT-3 and FGF increase the probability of self-renewing division.

CONCLUSIONS AND FUTURE DIRECTIONS

We believe that our data highlights the major role of thyroid hormone in the determination of the oligodendrocyte phenotype. The actions of this hormone on the O-2A lineage could be summarized as follows: T3 reduces the division capability of the O-2A progenitors and promotes their differentiation to oligodendrocytes. Whether these actions are irreversibly linked or can be dissociated from each other is one of the questions open for the future.

During the course of the presented experiments, we faced the unexpected finding that two types of clones were generated: homogeneous clones wholly composed of oligodendrocytes or O-2A progenitors, and heterogeneous clones with both types of cells. These results are most consistent with the hypothesis that the generation of oligodendrocytes from the O-2A progenitors is associated to an asymmetrical differentiation. Accordingly, we found that in cultures of embryonic rat cortex grown in the presence of PDGF, oligodendrocytes appeared in a timely manner in the absence of clonal differentiation. Furthermore, in contrast with previous suggestions, the presence or the absence of T3 did not alter the probability of individual clones of O-2A progenitors generating at least one oligodendrocyte; instead, T3 increased the proportion of oligodendrocytes generated within clones. Therefore, we propose a new model in which extracelullar factors, as T3, modulate the number of oligodendrocytes asymmetrically generated within a clone. Finally thyroid hormone does not enhance the survival of postmitotic oligodendrocytes *in vitro*.

The molecular bases for the described effects of T3 are still completely unknown. One point of interest, that still has not been addressed properly, is the different actions that the TR isoforms could exert in the O-2A lineage. It has been described that the expression de TRß is restricted to the differentiated oligodendrocytes, meanwhile TRα is expressed by both O-2A progenitors and oligodendrocytes. Thus, it is possible that the expression of TRß is necessary for the T3 actions either in the reduction of the O-2A cell division or oligodendrocytes differentiation or both. Besides, we should nor forget that the mechanism of the T3 action is the regulation of gene transcription by activation of its nuclear receptors. Thus, gene expression should be regulated by thyroid hormone in the O-2A cell in its progression towards oligodendrocyte. In this context, the application of the new technology of PCR-based detection of low abundant genes, hopefully will provide us with the " key " genes responsible for the induction of the oligodendrocyte phenotype.

REFERENCES

Altman, J. (1966) Proliferation and migration of undifferentiated precursor cell in the rat postnatal gliogenesis. Exp. Neurol. *16*: 263-278.

Balazs, R., B. W. L. Brooksbank, A. N. Davison, J. T. Eayrs and D. A. Wilson (1969) The effect of neonatal thyroidectomy on myelination in the rat brain. Brain Res. *15*: 219-232.

Barnett, S. C., A. M. Hutchins and M. Noble (1993) Purification of olfactory nerve ensheathing cells from the olfactory bulb. Dev Biol *155*: 337-50.

Barres, B. A. and M. C. Raff (1993) Proliferation of oligodendrocyte precursor cells depends on electrical activity in axons. Nature *361*: 258-60.

Barres, B. A. and M. C. Raff (1994a) Control of oligodendrocyte number in the developing rat optic nerve. Neuron *12*: 935-42.

Barres, B. A., M. C. Raff, F. Gaese, I. Bartke, G. Dechant and Y. A. Barde (1994b) A crucial role for neurotrophin-3 in oligodendrocyte development. Nature *367*: 371-5.

Bhat, N. R., G. S. Rao and R. A. Pieringer (1981a) Investigations on myelination in vitro. Regulation of sulfolipid synthesis by thyroid hormone in cultures of dissociated brain cells from embryonic mice. J. Biol. Chem. *256*: 1167-71.

Bhat, N. R., L. L. Sarlieve, G. Subba Rao and R. A. Pieringer (1979) Investigations on myelination in vitro: Regulation by thyroid hormone in cultures of dissociated cells from embryonic mice. J. Biol. Chem. *254*: 9342-9344.

Bhat, N. R., G. Shanker and R. A. Pieringer (1981b) Investigations on myelination in vitro: regulation of 2,3'-cyclic nucleotide 3'-phosphohydrolase by thyroid hormone in cultures of dissociated brain cells from embryonic mice. J. Neurochem. *37*: 695-701.

Bögler, O. and M. Noble (1994) Measurement of time in oligodendrocyte-type-2 astrocyte (O-2A) progenitors is a cellular process distinct from differentiation or division. Dev. Biol. *162*: 525-38.

Bögler, O., D. Wren, S. C. Barnett, H. Land and M. Noble (1990) Cooperation between two growth factors promotes extended self-renewal and inhibits differentiation of oligodendrocyte-type-2 astrocyte (O-2A) progenitor cells. Proc. Natl. Acad. Sci. USA *87*: 6368-72.

Bottenstein, J. E. and G. H. sato (1979) Growth of a rat neuroblastoma cell line in serum-free supplemented medium. Proc. Natl. Acad. Sci. USA *76*: 514-517.

Bradley, D.J., H.C. Towle, W.S. Young III (1992) Spatial and temporal expresion of a- and b-thyroid hormone mRNAs, including the b2-subtype, in the developing mammalian nervous system.J. Neurosci. *12*: 2288-2302.

Brent, G. A., D. D. Moore and P. R. Larsen (1991) Thyroid Hormone Regulation of Gene Expression. Ann. Rev. Physiol. *53*: 17-35.

Canoll, P. D., J. M. Musacchio, R. Hardy, R. Reynolds, M. A. Marchionni and J. L. Salzer (1996) GGF/Neuregulin is a neuronal signal that promotes the proliferation and survival and inhibits the differentiation of oligodendrocyte progenitors. Neuron *17*: 229-243.

Davis, A. A. and S. Temple (1994) A self-reneweing multipotential stem cell in embryonic rat cerebral cortex. Nature *372*: 263-266.

Dussault, J. H. and J. Ruel (1987) Thyroid hormones and brain development. Ann. Rev. Physiol. *49*: 321-334.

Eisenbarth, G. S., F. S. Walsh and M. Nirenberg (1979) Monoclonal antibody to a plasma membrane antigen of neurons. Proc Natl Acad Sci U S A *76*: 4913-7.

Farsetti, A., T. Mitsuashi, B. Desvergne, J. Robbins and V. Nikodem (1991) Molecular basis of thyroid hormone regulation of myelin basic protein gene expression in rodent brain. J. Biol. Chem. *266*: 23226-23232.

Flynn, T. J., D. S. Deshmukh and R. A. Pieringer (1977) Effect of altered thyroid function on galactosyl diacylglycerol metabolism in myelinating rat brain. J. Biol. Chem. *252*: 5864-5870.

Gard, A. L. and S. E. Pfeiffer (1990) Two proliferative stages of the oligodendrocyte lineage (A2B5+O4- and O4+GalC-) under different mitogenic control. Neuron *5*: 615-25.

Goldman, J. E. and P. J. Vaysse (1991) Tracing glial cell lineages in the mammalian forebrain. Glia *4*: 149-56.

Ibarrola, N., M. Mayer-Pröschel, A. Rodríguez-Peña and M. Noble (1996) Evidence for the existence of at least two timing mechanisms that contribute to oligodendrocyte generation in vitro. Dev. Biol.*180*: 1-21.

Ibarrola, N. and A. Rodríguez-Peña (1997) Hypothyroid coordinately and transiently affects myelin protein gene expression in most brain regions during postnatal development. Mol. Brain Res. *752*: 285-293.

Johe, K. L., T. G. Hazel, M. M. Dugich-Djorjevic and R. D. McKay (1996) Single factors direct the differentiation of stemm cells from the fetal and adult central nervous system. Genes & Development *10*: 3129-3140.

Luskin, M. B., J. G. Parnavelas and J. A. Barfield (1993) Neurons, astrocytes, and oligodendrocytes of the rat cerebral cortex originate from separate progenitor cells: an ultrastructural analysis of clonally related cells. J Neurosci *13*: 1730-50.

Mangelsdorf, D. J., C. Thummel, M. Beato, P. Herrlich, G. Schütz, K. Umesono, B. Blumberg, P. Kastner, M. Mark, P. Chambon and R. M. Evans (1995) The nuclear receptor superfamily: The second decade. Cell *83*: 835-839.

Mayer, M., K. Bhakoo and M. Noble (1994) Ciliary neurotrophic factor and leukemia inhibitory factor promote the generation, maturation and survival of oligodendrocytes in vitro. Development *120*: 143-53.

McKinnon, R. D., T. Matsui, D. M. Dubois and S. A. Aaronson (1990) FGF modulates the PDGF-driven pathway of oligodendrocyte development. Neuron *5*: 603-14.

McKinnon, R. D., G. Piras, J. J. Ida and D. M. Dubois (1993) A role for TGF-beta in oligodendrocyte differentiation. J Cell Biol *121*: 1397-407.

McMorris, F. A. and D. M. Dubois (1988) Insulin-like growth factor I promotes cell proliferation and oligodendroglial commitment in rat glial progenitor cells developing in vitro. J Neurosci Res *21*: 199-209.

Noll, E. and R. H. Miller (1994) Regulation of oligodendrocyte differentiation: a role for retinoic acid in the spinal cord. Development *120*: 649-60.

Price, J., D. Turner and C. Cepko (1987) Lineage analysis in the vertebrate nervous system by retrovirus-mediated gene transfer. Proc. Natl. Acad. Sci. USA *84*: 156-160.

Raff, M. C., L. E. Lillien, W. D. Richardson, J. F. Burne and M. D. Noble (1988) Platelet-derived growth factor from astrocytes drives the clock that times oligodendrocyte development in culture. Nature *333*: 562-565.

Raff, M. C., R. H. Miller and M. Noble (1983) Aglial progenitor cell that develops into astrocytes or an oligodendrocyte depending on the culture medium. Nature *303*: 390-396.

Raff, M. C., B. P. Williams and R. H. Miller (1984) The in vitro differentiation of a bipotential progenitor cell. EMBO J *3*: 1857-1864.

Rodriguez-Peña, A., N. Ibarrola, M. A. Iñiguez, A. Muñoz and J. Bernal (1993) Neonathal hypothyroidism affects the timely expression of myelin-associated glycoprotein in the rat brain. J. Clin. Invest. *91*: 812-818.

Shanker, G., A. T. Campagnoni and R. A. Pieringer (1987) Investigations on myelinogenesis in vitro: Developmental expression of myelin basic protein mRNA and its regulation by thyroid hormone in primary cerebral cell cultures from embryonic mice. J. Neurosci. Res. *17*: 220-224.

Somjen, G. G. (1988) Nervenkitt: Notes on the history of the concept of neuroglia. Glia *1*: 2-9.

Temple, S. and M. C. Raff (1985) Differentiation of a bipotential glial progenitor cell in single cell microculture. Nature *313*: 223-225.

Walters, S. N. and P. Morell (1981) Effects of altered thyroid states on myelinogenesis. J Neurochem. *36*: 1792-801.

Williams, B. P. (1995) Precursor cell types in the germinal zone of the cerebral cortex. Bioessays *17*: 391-3.

Williams, B. P., J. Read and J. Price (1991) The generation of neurons and oligodendrocytes from a common precursor cell. Neuron *7*: 685-93.

Wolswijk, G. and M. Noble (1989) Identification of an adult-specific glial progenitor cell. Development *105*: 387-400.

Wren, D., G. Wolswijk and M. Noble (1992) In vitro analysis of the origin and maintenance of O-2A adult progenitor cells. J. Cell Biol. *116*: 167-76.

Zhang, H. and R. H. Miller (1995) Asynchronous differentiation of clonally realted spinal cord oligodendrocytes. Mol. Cell. Neurosci. *6*: 16-31.

II. MOLECULAR AND BIOCHEMICAL PROPERTIES

7

CHARACTERIZATION OF RAT NEURAL STEM CELLS FROM EMBRYONIC STRIATUM AND MESENCEPHALON DURING IN VITRO DIFFERENTIATION

Eulalia Bazán, Miguel A. López-Toledano,
Carolina Redondo, Alberto Alcázar, Maria A. Mena,
Carlos L. Paíno and Antonio S. Herranz

Recent studies in vivo and in vitro suggest that a population of neural stem cells, with self-renewal and multilineage potential, are present in the embryonic through the adult mammalian central nervous system. These cells can be expanded in culture using a serum free medium and a mitogenic factor such as epidermal growth factor. We have cultured epidermal growth factor-responsive cells from embryonic rat striatum and mesencephalon. These cells differentiate in vitro into neurons, astrocytes and oligodendrocytes. This study shows the ability of some neurotransmitters to activate functional receptors coupled to second messenger systems. Noradrenaline, glutamate, and carbachol, increase the levels of inositol monophosphate, thus indicating the existence of functional receptors for these neurotransmitters linked to the phosphoinositide lipid hydrolysis. Dopamine and noradrenaline increase the levels of cyclic AMP, during in vitro neural stem cells differentiation. We have also studied the endogenous content of amino acids which are well known as trophic factors. Significant levels of glutamate, taurine, and glycine are found in neural stem cells at any time studied (3, 10, and 17 days post plating). However, γ-aminobutyric acid is only present between 10 and 17 days post plating, when the maximum number of neurons is observed. We conclude that the in vitro differentiation of neural stem cells could be, in part, regulated by neurotransmitters and amino acids released into the medium which activate functional receptors linked to second messenger systems. Furthermore, these results open the possibility to manipulate the differentiation of neural stem cells by acting on functional neurotransmitter receptors, and suggest that neural stem cells may be used as a tool to study the molecular mechanisms involved during brain development.

The adult mammalian central nervous system (CNS) contains a variety of cell types that arise over a short period of time from a small number of cells in the neural tube, but the specific molecular signals that initiate and terminate this process are not well known. The generation of neurons and glia during different developmental periods (Frederiksen and McKay, 1988; Skoff, 1990) suggests that these two cell types are derived from separate precursors. However, recent *in vivo* and *in vitro* studies (Reynolds, et al., 1992a, b; Weiss, et al., 1996; Suhonen, et al., 1996) have shown that neural stem cells (NSC), with self-renewal and multilineage potential, are present from the embryonic through the adult mammalian CNS. In culture, these cells grow as spheric aggregates (neurospheres) in defined medium supplemented with mitogens such as epidermal growth factor (EGF), transforming growth factor-α, or basic fibroblast growth factor (Gritti, et al., 1995; Gritti, et al., 1996). Following removal of the mitogen and exposure to a suitable substrata, NSC can differentiate into the three major cell phenotypes; neurons, astrocytes, and oligodendrocytes. These cells are potentially useful for neural transplantation (Gage, et al., 1995).

In the adult nervous system (NS), neurotransmitters (NTs) operate as chemical mediators of intercellular communication by acting on specific receptors which can activate second messenger systems in postsynaptic cells. Various observations suggest that a growing number of NTs can play a role as trophic or differentiating agents during CNS development (Lauder, 1993; Leslie, 1993; Legido, 1995). These transmitters are catecholamines noradrenaline (NA) and dopamine (DA), serotonin, acetylcholine (Ach) and amino acids (AAs) (γ-aminobutyric acid (GABA) and glutamate (Glu). The effects of NTs on neural development are mediated by specific receptors linked to concrete second messengers (Mattson, 1990; Lauder, 1993; Legido, 1995). One type of transduction mechanism uses cyclic adenosine monophosphate (cAMP) as a second messenger (Strada, et al., 1990). Another type of transduction mechanism employs an inositol lipid as a precursor to generate the second messengers diacylglycerol (DAG), which activates protein kinase C (PKC), and inositol 1,4,5-trisphosphate (IP_3), which mobilizes Ca^{2+} (Berridge, 1993; Furuichi and Mikoshiba, 1995).

The present report outlines some phenotypical and pharmacological characteristics of striatal and mesencephalic neural stem cells during the process of *in vitro* differentiation by studying: i) the presence of different cell types in cultures at several times post plating; ii) the functionality of some neurotransmitter receptors; and iii) the content in endogenous amino acids that have a role as neurotransmitters.

CULTURING NEURAL STEM CELLS

Striata and mesencephalons from E15 Sprague-Dawley rat embryos were dissected and mechanically dissociated. Cell suspensions were grown in a defined medium (DF12), composed of Dulbecco's modified Eagle's medium and Ham's F-12 (1:1), L-glutamine (2 mM), sodium piruvate (1 mM) (all from GIBCO), glucose

(0.6%), insulin 25 μg/ml, progesterone 20 nM, putrescine 60 μM, and sodium selenite 30 nM (all from Sigma), human transferrin 100 μg/ml and EGF 20 ng/ml (both from Boehringer-Mannheim). After a minimum of five passages, cells were plated at a density of 40000 cells/cm^2 on poly-l-ornithine coated glass coverslips (ø 12 mm) or plastic dishes (ø 35 mm). Cultures were maintained in DF12 for 72 hours and then switched to DF12 without EGF for longer culture periods. Immunocytochemical and biochemical studies were carried at 3, 10 and 17 days post plating (dpp).

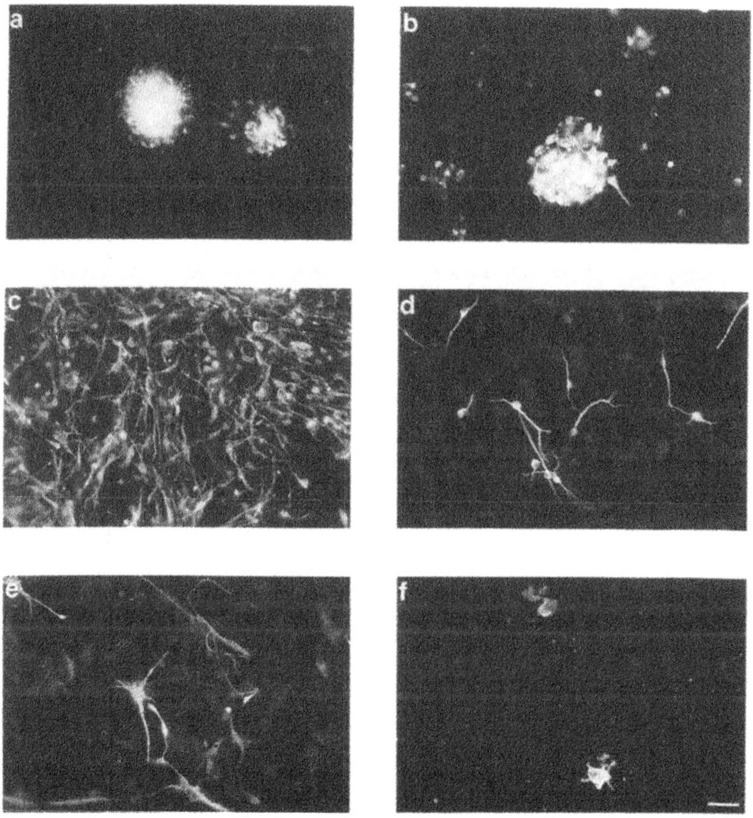

Figure 7-1. Immunocytochemical characterization of neural stem cells and their progeny. At 2 hours post plating, most of the cells in striatal (A) and mesencephalic (B) neurospheres are positive for the intermediate filament nestin. At 3 dpp, NSC have migrated from the neurosphere, as can be seen in C with immunostaining for nestin. Neurons (D) and astrocytes (E) in their initial stages of differentiation appear in striatal-derived cultures at 3 dpp, as assessed by their immunostaining for ß-tubulin isotype III or GFAP, respectively. The O1 antibody (F) shows the presence of some oligodendrocytes before the third day post plating. Scale bar: 25 μm.

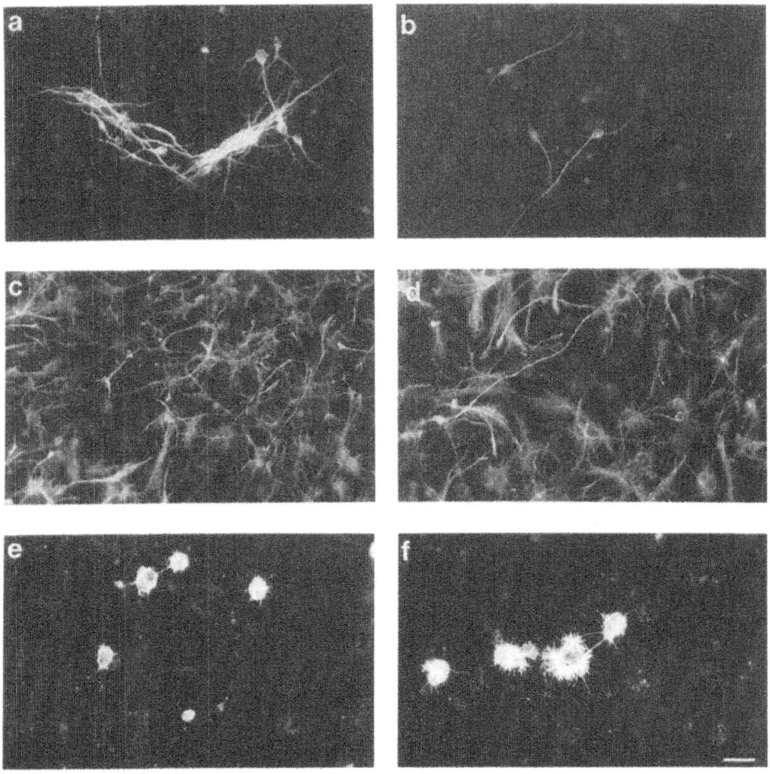

Figure 7-2. Immunocytochemical characterization at 10 dpp.
Striatal (A, C and E) and mesencephalic (B, D and F) cells immunostained for ß-tubulin isotype III (A, B), GFAP (C, D) and the O1 antibody (E, F). At this time period NSC progeny presented qualitatively similar morphologies, although mesencephalic NSC produced smaller number of neurons. Scale bar: 25 μm.

IMMUNOCYTOCHEMICAL CHARACTERIZATION

Plating neurospheres from E15 rat striatum and mesencephalon on an adhesive substrata induced a phenotypic differentiation (Figs. 7-1, 7-2, and 7-3). Cells on coverslips were fixed in 4% paraformaldehyde for 10 min, and post-fixed and permeabilized in ethanol-acetic (19:1) for 10 min at -20°C. Dual immunolabelling was performed by overnight incubation at 4°C in a mixture of mouse anti ß-tubulin isotype III (1:400, Sigma), anti nestin (1:100, DSHB University of Iowa), and anti glial fibrillary acid protein (GFAP) (1:500, DAKO). Immunostaining with O1 monoclonal antibody (1:10, Sommer and Schachner, 1981) was performed for 1 hour at room temperature, previous to postfixation and permeabilization. Secondary antibodies (Anti-mouse IgG-FITC 1:40, Boehringer-Mannheim or Anti-rabbit IgG-LRSC 1:200, Jackson) were incubated for 30 min at 24°C. Coverslips were washed three times in PBS and mounted in a medium containing p-phenylenediamine and bisBenzimide (Hoechst 33342, Sigma).

At two hours post-plating most of the cells were nestin$^+$ (fig. 7-1A, B), an intermediate filament abundantly expressed in neuroephitelial stem cells and extinguished in almost all mature CNS cells (Tohyama, et al., 1992). At 3 dpp nestin$^+$ cells were still abundant. (fig. 7-1C).

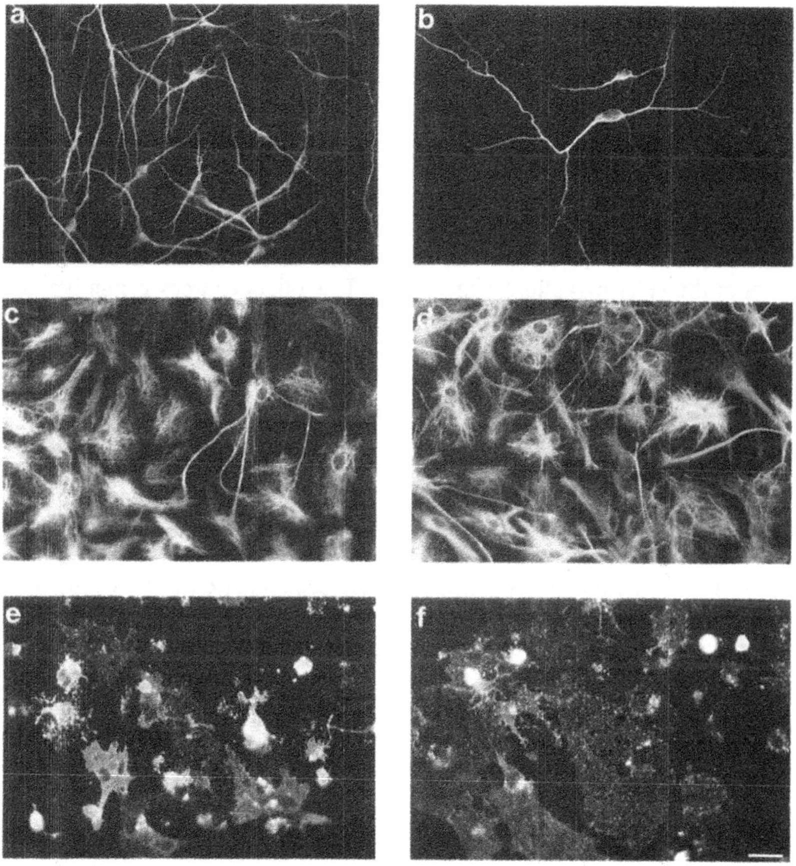

Figure 7-3. Immunocytochemical characterization at 17 dpp. Striatal (A, C and E) and mesencephalic (B, D and F) cells immunostained for ß-tubulin isotype III (A, B), GFAP (C, D) and the O1 antibody (E, F). At 17 dpp, cells have developed well differentiated morphologies. Most astrocytes have changed from a filamentous into a protoplasmic morphology, with a few ones extending long processes. Oligodendrocytes spread their citoplasms, covering large areas of the culture surface. Scale bar: 25 μm.

In striatal derived cultures, some cells became positive for ß-tubuline III (fig. 7-1D), a specific marker for neurons, and others were positive for GFAP (fig. 7-1E), indicating the presence of astrocytes. In our working conditions, ß-tubuline III^+ or $GFAP^+$ cells were undetectable in cultures from mesencephalon at 3 dpp, whereas $O1^+$ cells (oligodendrocytes) could be found in both striatal and mesencephalic cultures (fig. 7-1F) At 10 dpp, many cells were ß-tubuline III^+ in striatum (fig. 7-2A), extending long processes, showing typical neuronal morphology, and often arranged in clusters. A few ß-$tubuline^+$ cells were observed in mesencephalon (fig. 7-2B). At this time, a large number of astrocytes were present both in striatal and mesencephalic cultures. Their morphology was predominantly filamentous, with their processes intermingled in an amorphus network (fig. 7-2C, D). Oligodendrocytes were numerous, but the $O1^+$ staining just displayed a spheric cytoplasm with many short thin processes (fig. 7-2E, F). By 17 dpp neurons had developed very long processes in both types of cultures (fig. 7-3A, B). Mesencephalic neurons currently had two or three main processes steming form elongated bodies. Glial fibrillary acidic protein staining had changed from filamentous into protoplasmic (fig. 7-3C, D). Immunocytochemistry to O1 stained wide oligodendrocyte processes covering a large part of the culture surface (fig. 7-3E, F), while the cell bodies were often unstained. Cell counts indicated that neurons constituted 4-8% of striatal cells, while mesencephalic neurons were only 1-3% of the cells in culture (fig. 7-4A). The oligodendrocyte number increased with time in both types of cultures (fig. 7-4B).

PHARMACOLOGICAL CHARACTERIZATION

Effects of pharmacological agents on cyclic AMP and phosphoinositide (IP1) in neural stem cells

Since adenylate cyclase and phospholipase C have been reported to be coupled to a variety of neurotransmitter receptors, we examined selected agonists for their possible activity during NSC differentiation. Intracellular levels of cAMP and inositol-monophosfate (IP1) were determined as previously described by Morin et al., (1991), in brain slices, with minor modifications. Briefly, cells on plastic dishes were labeled simultaneously with 0.5-1μCi/ml [^{14}C]-adenine (84Ci/mmol, Amersham) and 1,5μCi/ml [^{3}H]-inositol (21Ci/mmol, New England Nuclear) for 24 hours in 500μl of a defined medium (D2), composed of Dulbecco's modified Eagle's medium, L-glutamine (2mM), sodium piruvate (1 mM), glucose (0.6%), and N_2 components in an atmosphere of 5% CO_2-95% air. The cells were rinsed and 1 ml of Krebs-Henseleit-20 mM HEPES (pH 7.4) buffer was added for 30 min at 37°C in the presence of 0.5mM isobutyl-methyl-xantine (IBMX) and 10mM LiCl. Glutamate, NA, DA, and the cholinergic muscarinic agonist carbachol (Cch) were added and the cells incubated at 37°C for 15 min. The reaction was stopped by aspiration of the buffer and the addition of 1ml of cold methyl alcohol (MeOH), and 0.12M HCl (1:1). Plates were allowed to sit in the freezer at

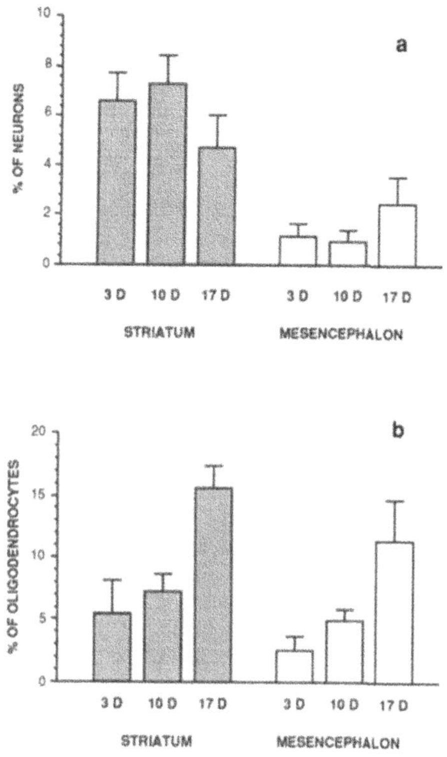

Figure 7-4. Cellular type composition of cultures from striatal and mesenephalic neural stem cells. Results are expressed as percentages of neurons (A) or oligodendrocytes (B) versus the total number of cells. The histograms represent the mean± S.E.M. from 7 to 9 coverslips. The presence of neurons and oligodendrocytes was detected by anti ß-tubulin isotype III and anti O1 immunostaining, respectively. Staining of nuclei with bis-Benzimide was used to count the total number of cells. (D = days post plating).

-20 °C for 1 hour, and aliquots (0.75ml) of the MeOH-HCl were transferred to tubes containing 3 ml of water. The fractions containing cAMP and IP1 were isolated by ion exchange chromatography as previously described (Strada et al., 1990; Johnson et al., 1990), and collected in scintillation vials. Fourteen ml of aquasol (NEN) was added and radioactivity was counted with a liquid scintillation spectrophotometer (LKB model 1219). MeOH-HCl remaining on the plates was aspirated and 0.5ml 0.4N NaOH was added to solubilize the cell particulate matter, for protein determination (Staufer, 1975).

As shown in figure 7-5, NA and DA increased cAMP levels in both types of cultures at the times analyzed. One hundred micromolar NA, maximally increased the levels of cAMP at 3 dpp (29.86±8 and 17±4 fold above basal for striatum and mesencephalon, respectively). One hundred micromolar DA increased the amount of cAMP in both striatum and mesencephalon showing the greatest increase in mesencephalon at 17 dpp (8.57±0.89 fold above basal).

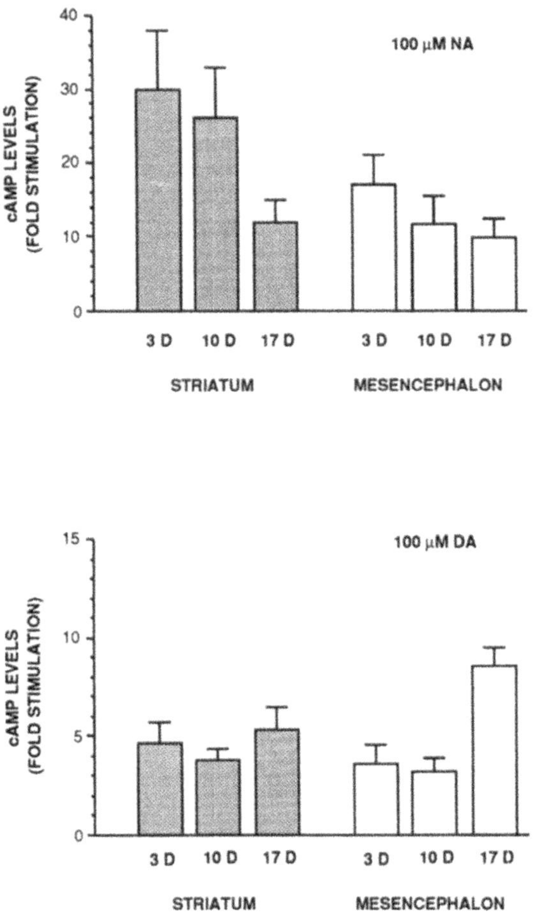

***Figure** 7-5*. Effect of noradrenaline and dopamine on cAMP accumulation during stiatal and mesencephalic neural stem cell differentiation.
Results are expressed as fold stimulation from basal (unstimulated) accumulation. Histograms represent the mean ± S.E.M. of 3 to 4 experiments done in triplicate. In striatal cultures, basal accumulation of cAMP (as DPM/mg protein) from a representative experiment ranged between 1443±143 (3dpp), 1876±24 (10dpp) and 2766±26 (17dpp). In mesencephalic cultures, basal values were 2620±760 (3dpp), 6187±861 (10dpp) and 2881±219 (17dpp). (D = days post plating).

Glutamate, NA, and Cch, stimulated PI lipid hydrolysis, as measured by the accumulation of IP1 (fig. 7-6). The maximum stimulation of phosphoinositide (PI) lipid hydrolysis with 1mM Cch was observed at 3 dpp decreasing with time in culture to reach minimal values at 17 dpp. One hundred micromolar Glu, presented a similar pattern to Cch, reaching maximal values of stimulation at 3 dpp (5.78±1.3 and 2.76±1.13 fold above basal for striatum and mesencephalon, respectively). At 17 dpp, no stimulation was observed with Glu in striatum (1.08±0.26 fold above basal) although the stimulation in mesencephalon remained at about 70% of basal levels (1.73±0.16 fold above basal). One hundred micromolar NA increased PI lipid hydrolisis with time in culture, showing maximal values at 17 dpp (2.32±0.9 and 7.95±2.3 fold above basal for striatum and mesencephalon, respectively).

ENDOGENOUS FREE AMINO ACIDS

The endogenous content of GABA, glycine (Gly) taurine (Tau) and Glu, which are amino acids with a presumptive neurotrophic activity other than their metabolic role, was studied at 3, 10 and 17 dpp. (fig. 7-7). Cells on plastic dishes were washed three times with 2ml of PBS and scraped at 4°C in 250µl of 0.2N perchloric acid with 0.2mM ethylenediaminetetraacetic acid (EDTA) and 0.5mM sodium meta-bisulfate ($Na_2S_2O_5$). Cell suspensions were sonicated three times for 15 sec and centrifuged at 10000g for 5 min. The supernatants were kept at -80°C and the pellets were used for protein evaluation. The supernatants were analyzed for amino acid content using the reverse-phase high pressure liquid chromatography (HPLC) procedure with O-phthaldialdehide (OPA) precolumn derivatization as described by Jones et al., (1981), except that, 3-mercaptopropionic acid was used instead of 2-mercaptoethanol in the derivatization reaction. This change improved the sensitivity of the analysis two-fold.

During *in vitro* NSC differentiation, the pattern of GABA was similar in both strial and mesencephalic cultures. It presented its lowest concetration at 3 dpp, with just trace values in mesencephalon (0.06nmoles/mg protein), and the highest at 17 dpp. Note, GABA levels were four times more concentrated in striatal than in mesencephalic cultures.Glycine maximal levels were measured at 3 dpp, being notably lower at 10 and 17 dpp. In striatal cultures, by contrast, Gly levels remained high at 10 dpp.

Taurine appeared in large amounts in both types of cultures, although there were differences between them in the concentration pattern. In striatal cultures Tau levels were high at 3 and 10 dpp, then decreasing to less than half at 17 dpp; on the other hand, mesencephalic cultures presented an inverted pattern, where lowest levels were found at 3dpp and steadily high concentrations were measured at 10 and 17 dpp. Mesencephalic Glu concentration was not modified over time. However, in striatal cultures Glu was found at high concentrations at 3 and 10 dpp, with a pattern that was similar to that of Tau.

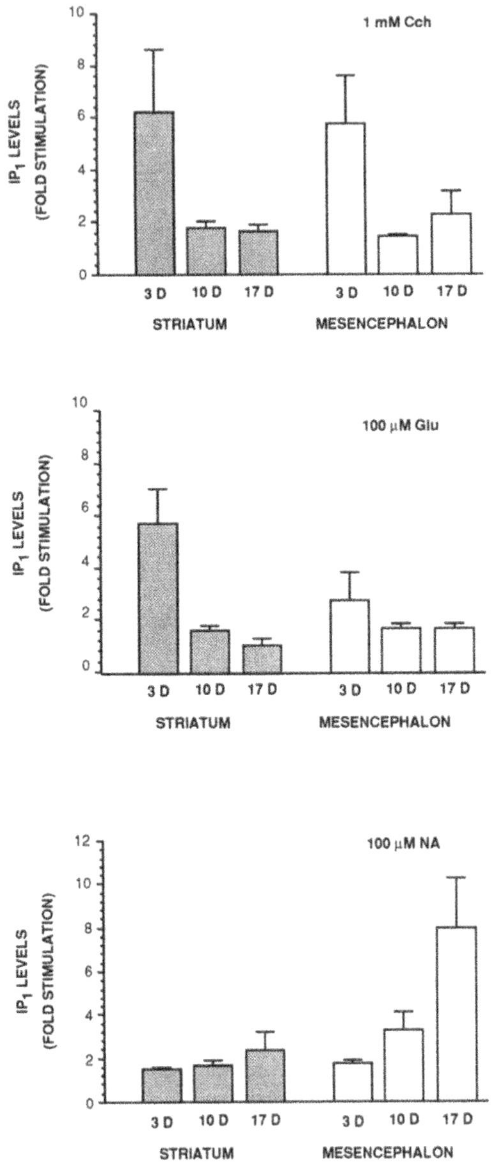

Figure 7-6. Effect of carbachol, glutamate, and noradrenaline on IP1 accumaltion during stiatal and mesencephalic neural stem cell differentiation. Results are expressed as fold stimulation from basal (unstimulated) accumulation. Histograms represent the mean±S.E.M. of 3 to 4 experiments done in triplicate. In striatum, basal accumulation of IP1 (as DPM/mg protein) from a representative experiment ranged between 7138± 327 (3dpp), 41619±12000 (10dpp), and 21165± 3519 (17dpp). In mesencephalon, basal values were 5562±665 (3dpp), 51054±2604 (10dpp), and 11681±700 (17dpp). (D= days post plating).

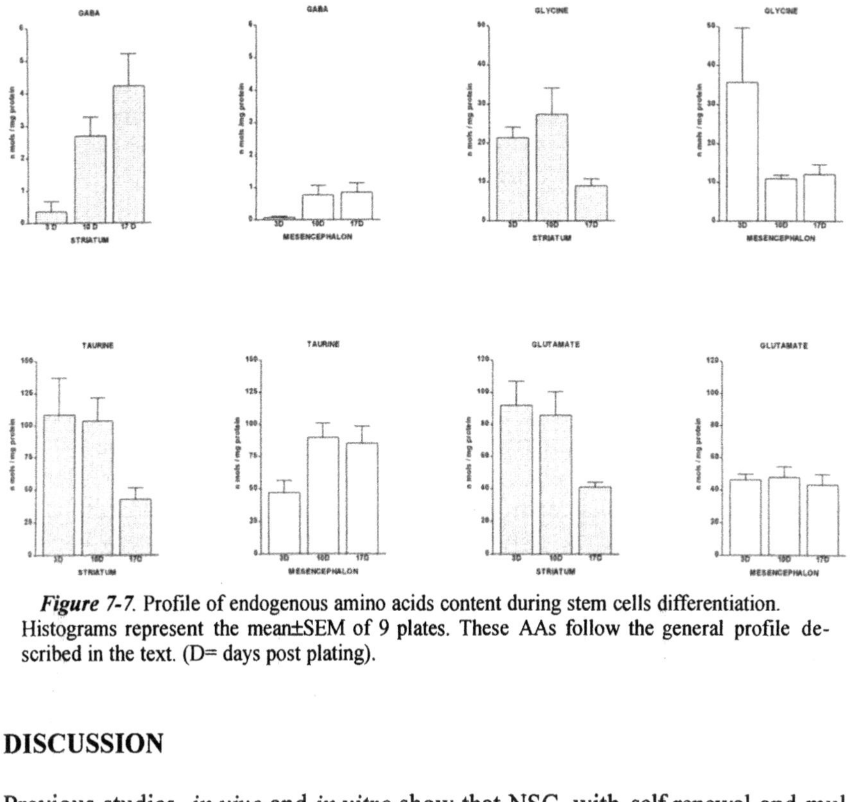

***Figure** 7-7.* Profile of endogenous amino acids content during stem cells differentiation. Histograms represent the mean±SEM of 9 plates. These AAs follow the general profile described in the text. (D= days post plating).

DISCUSSION

Previous studies, *in vivo* and *in vitro* show that NSC, with self-renewal and multilineage potential, are present in the embryonic through the adult mammalian CNS; however, many of these studies have been focused on mice (Reynolds, et al., 1992a, b; Weiss, et al., 1996; Suhonen, et al., 1996) rather than rats. In this study we show the presence of multipotential precursor cells in developing rat striatum and mesencephalon. These cells proliferate *in vitro* in response to EGF, and they differentiate into neurons, astrocytes and oligodendrocytes. The pattern of differentiation for glial cells (astrocytes and oligodendrocytes) was similar for both striatum and mesencephalon. At 17 dpp, oligodendrocytes constituted 12-15% of the cells and 80% were astrocytes. However, the percentage of neurons was higher (4-8%) in striatum than in mesencephalon (1-3%) at any given time. Our results agree with those obtained by O'Connor et al., (1996), who found that mouse EGF-responsive precursor cells from more rostral regions generate more neuronal progeny. Svendsen et al., (1995), reported that both rat striatal and mesencephalic precursor cells matured into a mixed culture rendering a 35% content of neurons. The differences between Svendsen's and our results, could be explained by differences in culture conditions. In contrast with us, Svendsen used the B27 supplemented medium, and an anti MAP2 as neuronal marker. The B27 supplement (Brewer et al., 1994) contains, among other components, antioxidant and differentiating agents, which could account for some of their neuronal yield. On the other hand, anti MAP2 can stain astrocytes in some conditions, as reported by others (Geisert et al., 1990) and observed by us.

In vitro differentiation of striatal and mesencephalic NSC into neurons or glial cells is accompanied by important changes in the functionality of NT receptors for cathecholamines, acetylcholine and glutamate, as well as in the content of the endogenous free AAs that have been implicated in brain development.

The presence of NA and DA receptors coupled to the cAMP system in neuronal and non-neuronal cells is well documented (Lauder, 1993). We found noradrenergic and dopaminergic receptors positively coupled to the cAMP system in NSC from striatum and mesencephalon. Neurotrophic actions have been reported for NA (Schwartz, 1992). Dopamine regulates the neurogenesis of its synaptic targets throughout activation of D1 receptors, positively coupled to adenylate cyclase (Lauder, 1993). However, we failed to detect intracellular levels of catecholamines measured by HPLC/ED (Mena, et al., 1995), and tyrosine hydroxylase immunostaining was not observed in our cultures (data not shown). These results suggest that even when functional catecholamine cAMP linked receptors are present in these cultures, their activation is not necessarily required for NSC *in vitro* differentiation.

Phosphoinositide lipid hydrolisis following stimulation with Ach, Glu, and NA has been shown to occur in the rat brain (Balduini, et al., 1991), as well as in cultured neurons (Weiss, et al., 1988), and astrocytes (Pearce and Murphy, 1988). Glutamate and Ach, have been involved in neuronal differentiation and astroglial proliferation through their activation of the PI pathway (Weiss, et al., 1988; Ashkenazi, et al., 1989). In our cultures of striatal and mesencephalic NSC, both Glu and Cch showed the highest stimulation of PI lipid hydrolisis at 3 dpp, and a reduction of this parameter was observed at longer periods of time in culture, which is in agreement with some data reported on brain slices (Heacok, et al., 1987; Palmer, et al., 1990; Balduini, et al., 1991). The authors found that muscarinic and glutamatergic responses were enhanced during the intrauterine and neonatal period, decreasing in adulthood. In NSC from striatum and mesencephalon, we found high levels of Glu, as well as the presence of choline acetyl transferase by western immunoblot, at two dpp (data not shown). These data suggest that Glu and Ach could be synthesized and released by either undifferentiated and differentiating NSC, interacting with functional receptors linked to PI lipid hydrolysis. Thus, these NTs may play a role during the process of NSC differentiation.

A sharp increase of PI lipid hydrolisis mediated by NA was found at 17 dpp in NSC from striatum and mesencepahalon (fig. 7-6). These results are parallel to those obtained in cerebral cortex and hippocampus (Roney and Nahorski, 1987; Balduini, et al., 1991). The authors found an increase in PI lipid hydrolisis from postnatal day one to adulthood. However, the role of NA during NSC differentiation is not well understood.

It is noteworthy to underline the presence of GABA and Tau, as endogenous AAs, when they were not supplied in the feeding medium. The importance of Tau in the developing brain is well known, presenting maximal concentrations in fetuses and decreasing in the adult brain (Sturman, 1993); and the role of GABA as a trophic neurotransmitter during brain development is well established (Barbin, et al., 1993; La Mantia, 1995). A previous study of NSC (Bazán, et al., 1996) showed that GABA was present at a very low concentration (1.31 nmol/mg prot) at two hours post plating, and was not detected again until 10 dpp, when many neurons were present. Similar results have been found in this study, although the concentration in striatum was four times higher than in mesencephalon. The presence of GABA neurons in striatum and mesencephalon is well documented (Kopin, 1994), and immunoreactivity for GABA has been shown in neurons derived from murine and rat stem cells (Reynolds et al., 1992a, Svendsen et al., 1995). Our results suggest that the GABA found in our cultures originate from gabaergic neurons generated from striatal and mesencephalic NSC. On the other hand, they also show that, at least, a population of cells derived from striatal and mesencephalic NSC have the machinery needed to syntethize GABA and Tau. Glycine, an inhibitory NT in the adult brain, showed a particular pattern, with a maximum peak at 3 dpp in both striatal and mesencephalic cultures. The characteristic peak at 3 dpp may indicate that glycine could be involved in the initial steps of NSC differentiation.

CONCLUSIONS

We conclude that neural stem cells from different embryonic rat regions proliferate and differentiate *in vitro* into neurons, astrocytes and oligodendrocytes. This pattern of differentiation is accompanied by important changes in the levels of endogenous AAs, and in the functionality of NT receptors coupled to second messenger systems. These results, open the possibility to manipulate *in vitro* the differentiation process of NSC by acting on functional NT receptors, and they suggest that the culture neural stem cells can be used as a potent tool to study the molecular mechanisms involved during CNS differentiation.

ACKNOWLEDGMENTS

We thank Ms. Ana Carazo, Ms. M. J. Asensio and Ms. M. Gómez-Calcerrada for their excellent technical help. This work was partially supported by grants from the Fondo de Investigaciones Sanitarias (FIS 94/0494) and DIGCYT (PB94/0338), and The International Institute for Paraplegia (P27/96), Zurich.

REFERENCES

Askhenazi, A., Ramachandran, J. and Capon, D.J. (1989). Acetylcholine analoge stimulate DNA synthesis in brain-derived cells via specific muscarinic receptor subtypes. Nature *340*: 146-150.

Balduini,W., Cadura, S.M. and Costa, L.G. (1991). Regional development of carbachol-, glutamate-, norepinephrine-, and serotonine-stimulated phosphoinositide metabolism in rat brain. Develop.Brain Res., *62*: 115-120.

Barbin,G., Pollard, H., Ga•arsa, J.L. and Ben-Ari, Y. (1993). Involvement of GABAa receptors in the outgrowth of cultured hippocampal neurons. Neurosci. Lett *152*:150-154.

Bazán, E., López-Toledano, M.A., Mena, M.A., Martín del Rio, R., Paíno C.L. and Herranz, A.S. (1996). Endogenous amino acid profile during in vitro differentiation of neural stem cells. In Gary Fiskum (ed): Neurodegenerative Diseases. New York, New York: Plenum Publishing Corporation, pp 225-234.

Berridge, M.J. (1993). Inositol trisphosphate and calcium signaling. Nature *361*: 315-325.

Brewer, G.J., Torricelli, J.R., Evege, E.K. and Price, P.J. (1993). Optimized survival of hippocampal neurons in B27-supplemented neurobasal, a new serum-free medium combination. J. Neurosci. Res. *35*: 567-576.

Furuichi, T. and Mikoshiba, K. (1995). Inositol 1,4,5-trisphosphate receptor-mediated Ca2+ signaling in the brain. J. Neurochem. *64*: 953-960.

Gage, F.H., Ray, J. and Fisher, L.J. (1995). Isolation, characterization, and use of stem cells from CNS. Annu. Rev. Neurosci. *18*: 159-92.

Geisert, E.E., Johnson, H.G. and Binder, L.I. (1990). Expression of microtubule-associated protein 2 by reactive astrocytes. Proc. Natl. Acad. Sci. USA *87*: 3967-71.

Gritti, A., Cova, L., Parati, E.A., Galli, R. and Vescovi. A.L. (1995). Basic fibroblast growth factor supports the proliferation of epidermal growth factor-generated neuronal precursor cells of the adult mouse CNS. Neurosci. Lett. *185*: 151-154.

Gritti, A., Parati, E.A., Cova, L., Frolichsthal, P., Galli, R., Wanke, E., Faravelli, L., Morassutti, D.J., Roisen, F., Nickel, D.D. and Vescovi, A.L. (1996). Multipotential stem cells from the adult mouse brain proliferate and self-renew in response to basic fibroblast growth factor. J. Neurosci. *16*: 1091-1100.

Hamberger, A., Chiang, G., Nylén, E., Scheff, S.W. and Cotman, C.W. (1978). Stimulus evoked increase in the biosynthesis of the putative neurotransmitter glutamate in the hippocampus. Brain Res. *143*: 549-555.

Johnson, C.L., Johnson, C.G., Bazán, E., Garver, D., Gruenstein, E. and Ahluwalia, M. (1990). Histamine receptors in human fibroblasts: inositol phosphates, Ca2+,and cell growth. Am. J. Physiol. *258* (Cell Physiol. *27*): C533-C543.

Jones, B.N., Paabo, S. and Stein, S. (1981). Amino acid analysis and enzymatic sequence determination of peptides by an improved O-phthaldialdehyde precolumn labeling procedure, J. Liquid Chrmatogr. *4*: 565-586.

Kopin, I.J. (1994). Neurotransmitter and diosorders of the basal ganglia. In G.J. Siegel et al (eds): Basic Neurochemistry: Molecular, Cellular, and Medical Aspects, 5th Ed. New York: Raven Press, pp 899-933.

La Mantia, A.S. (1995). The usual suspects: GABA and glutamate may regulate proliferation in the neocortex. Neuron *15*: 1223-1225.

Lauder, J.M. (1993). Neurotransmitters as growth regulatory signals: role of receptors and second messengers. TINS *16*: 233-240.

Legido, A. (1995). Neurotransmisores y desarrollo cerebral: efectos fisiológicos y fisiopatológicos. Rev. Neurol. *23*: S269-S277.

Leslie, F.M. (1993). Neurotransmitters as neurotrophic factors. In S. E. Laughlin and J.H. Fallon (eds): Neurotrophic Factors. San Diego, CA: Academic Press Inc., pp 565-598.

Mattson, M.P. (1990). Second messengers in neural growth and degeneration. In N.N. Osborne (ed): Current Aspects of Neuroscience Vol. 2. The Macmillan Press, pp 1-48.

Mena, M.A., Casarejos, M.J., Bonin, A., Ramos, J.A. and de Yébenes, J.G. (1995). Effects of dibutyryl cyclic AMP and retinoic acid on the differentiation of dopamine neurons: Prevention of cell death by dibutyryl cyclic AMP. J. Neurochem. *65*: 1-9.

Morin, D., Zini, R., Querol-Ferrer, V., Sapena, R and Tillement, J.P. (1991). A method to measure simultaneously cyclic AMP and inositol phosphate accumulation in brain slices. J. Neurochem. *56*:114-1120.

O'Connor, T.J., Vescovi, A.L., Hettiaratchi, P. and Reynold, B.A. (1996). Epidermal growth factor-responsive neural stem cells from various regions of murine embryonic CNS: differences in potential for neuronal differentiation. Soc. for Neurosci. *22*: 522 (Abstract).

Pierce, B. and Murphy, S. (1988). Neurotransmitter receptors coupled to inositol phospholipid turn-over and Ca++ flux: consequences for astrocyte function. In H. K. Kimelberg (Ed.): Glial Cell Receptors. New York: Raven Press, pp. 197-221.

Reynolds, B.A., Tetzlaff, W. and Weiss, S. (1992 a). A multipotent EGF responsive striatal embryonic progenitor cell produces neurons and astrocytes, J. Neurosci. *12*: 4565-4574.

Reynolds, B.A. and Weiss, S. (1992 b). Generation of neurons and astrocytes from isolated cells of the adult mammalian central nervous system. Science *255*: 1707-1710.

Roney, T.A. and Nahorski, S.R. (1987). Postnatal ontogeny of agonist and depolarization-induced phosphoinositide hydrolysis in rat cerebral cortex. J. Pharmacol. Exp. Ther. *243*: 333-341.

Schwartz, J.P. (1992). Neurotransmitters as neurotrophic factors: a new set of functions. Int. Rev. Neurobiol. *34*: 1-23

Sommer, I. and Schachner, M. (1981). Monoclonal antibodies (O1 to O4) to oligodendrocyte cell surfaces: An immunocytological study in the central nervous system. Dev Biol. *83*: 311-327.

Stauffer, C.E. (1975). A linear standard curve for the Folin Lowry determination of protein, Anal. Biochem. *69*: 646-648.

Strada, S.J., Duman, R.S. and Enna. S.J. (1990). Analysis of neurotransmitter receptor coupled cyclic nucleotide systems. In Henry I. Yamamura et al., (eds): Methods in Neurotransmitter Receptor Analysis. New York: Raven Press, pp 89-110.

Sturman, J.A. (1993). Taurine in development. Physiol. Rev. *73*: 119-147.

Suhonen, J.O., Peterson, D.A., Ray, J. and Gage, F.G. (1996). Differentiation of adult hippocampus-derived progenitors into olfactory neurons in vivo. Nature *383*: 624-627.

Svendsen, C.N., Fawcett, J.W., Bentlage, C. and Dunnet., S.B. (1995). Increased survival rat EGF-generated CNS precursor cells using B27 supplemented medium. Exp. Brain Res. *102*: 407-414.

Svendsen, C.N., Clarke, D.J., Roser, A.E. and Dunnet, S.B. (1996). Survival and differentiation of rat and human epidermal growth factor-responsive precursor cells following grafting into the lesioned adult central nervous system. Exp. Neurol. *137*: 376-388.

Tohyama, T., Lee, V.M.Y., Rorke, L.B., Marvin, M., McKay, R.D.G. and Trojanowski, J.Q. (1992). Nestin expression in embryonic human neuroepithelium and in human neuroepithelial tumor cells. Lab. Invest. *66* : 303-313.

Weiss, S., Schmidt, B.H., Sebben, M., Kemp, D.E., Beckaert, J. and Sladeczek, F. (1988). Neurotransmitter-induced inositol phosphate formation in neurons in primary culture. J. Neurochem. *50*: 1425-1433.

Weiss, S., Reynolds, B.A., Vescovi, A.L., Morshead, C., Craig, C.G. and van der Kooy, D. (1996). Is there a neural stem cell in the mammalian forebrain?. TINS *19*: 387-393.

8

EXPRESSION OF NEUROTROPHINS AND THEIR RECEPTORS WITHIN THE GLIAL CELLS OF RETINA AND OPTIC NERVE

Elena Vecino, Elena Caminos, Elena Becker,
Dionisio Martín-Zanca and Neville N. Osborne

The purpose of this chapter is to review some of our studies on the localisation of neurotrophins and their receptors in glial cells of the optic nerve and retina of the fish (tench) and rat. The fish optic nerve has the capacity to regenerate after damage and the retina grows throughout life. These characteristics are not associated with the same tissues of the rat. Neurotrophins are thought to be involved in development and regeneration so a difference in the distribution of neurotrophins and their receptors in retina/optic nerve in rat and fish may relate to such functions. At least five different-types of neurotrophin molecules exist of which nerve growth factor (NGF), brain-derived neurotrophic factor (BDNF), neurotrophin-3 (NT-3) and neurotrophin-4 (NT-4) have been best studied. Neurotrophin receptors of which there are two kinds, the low affinity receptor (p75) and the high affinity tyrosine kinase receptors (Trk) exists at varying states and have varying affinities for the different type of neurotrophins. The functional role of the neurotrophins and their receptors and ways of studying these molecules in the retina are also discussed in this overview.

A large body of evidence shows that neurotrophins and growth factors are involved in the survival, differentiation and development of neurones. Glial cells also have the important function of maintaining the survival of neurones. In contrast the mediators involved in the development and survival in the glial cells are less well defined. Glial cells like astrocytes and microglia are known to synthesise neurotrophic factors and these can mediate in direct or indirect neurotrophic actions (Hatten et al., 1988; Engele and Bohn, 1991; Yoshida and Gage, 1991; Elkabes et al., 1995; 1996). In the retina the major type of glial cell is the Müller cell although astrocytes, oligodendrocytes and microglial also exist. In this article we will summarise some of our studies on the distribution of neurotrophins and their receptors in retinal glial cells of the rat and fish. Although the fish and rat retina have a similar general morphology the fish retina grows throughout life (Müller, 1952; Johns, 1977; Meyer, 1978) and the optic nerve has the capacity to regenerate after lesion (Sperry, 1948; Jacobson and Gaze, 1965; Sharma, 1972; Murray, 1976). Neither of these characteristics is associated with the rat retina and optic nerve. Growth or cell proliferation and regeneration are thus properties associated with the fish retina and optic nerve and these involve glial cell functions (Bray et al., 1981; Silver et al.,1982).

RETINA AND OPTIC NERVE GLIAL CELLS

Müller cells

Müller (radial glial) cells are the main glial cells of the vertebrate retina. The retina originates from an evagination of the primitive neuroepithelium, which, like the brain and spinal cord, differentiates into neurones and radial glia. In contrast to the brain and spinal cord, radial glia, from the retina which are the Müller cells, maintain a radial appearance throughout adult life. The Müller cells are therefore the major supportive glia for neurones in the adult retina and perform many of the functions subserved by oligodendrocytes, astrocytes and ependymal cells in other regions of the CNS (Reichenbach and Wohlrab, 1986; Holländer et al., 1991; Bignami, 1995; Newman and Reichenbach, 1996).

The development of Müller cells, rod photoreceptor cells, bipolar cells, and a subset of amacrine cells from common precursor cells occurs during a late proliferative phase. A single Müller cell and a species-specific number of neurones form a “columnar” unit (Bignami, 1995). In contrast, "extracolumnar” neurones (ganglion cells, cone photoreceptor cells, horizontal cells, and another subset of amacrine cells) exist and begin to differentiate before the development of most Müller cells. These “extracolumnar” neurones may therefore have developed metabolic processes to support their own survival, whereas ("columnar") neurones may be more dependent on support from the Müller cell (Reichenbach et al., 1993).

In adult retina Müller cells have their somata within the inner nuclear layer (where they may form a distinct median sublayer in some species). They constitute the predominant glia of the retina and span its entire thickness. Although Müller cells from different species vary considerably in shape, some features are fairly universal. At the level of the outer limiting membrane, Müller cells extend apical microvilli into the subretinal space between the inner segments of photoreceptor cells. In some species, Müller cells also possess a cilium (Ennis and Kunz, 1986). Apicolaterally, Müller cells are connected to their neighbouring Müller and photoreceptor cells by specialised junctions to form the outer limiting membrane. In rat these junctions are *zonulae adherents* while in fish they are tight junctions (Reichenbach and Robinson, 1995). Müller cells send side branches into the two plexiform layers of the retina where they form sheaths around neuronal processes and synapses (Reichenbach et al., 1989). Extended smooth and sometimes rather long processes are sent from the somata of the Müller cells through the nerve fibre layer ending in a basal end-feet, which lies adjacent to the inner limiting membrane. This membrane is a basal lamina, at least partly produced by Müller cells.

Like other glial cells, Müller cells express a wide variety of voltage-gated ion channels and also express many types of neurotransmitter receptors, including a GABA receptor and several types of glutamate receptors. Studies on Müller cells have provided some of the clearest examples of glial-cell control of the neuronal microenvironment. It is known that Müller cells have the capacity to modulate neuronal activity by controlling the concentration of neuroactive substances in the extracellular fluid bathing the retinal cells. The concentrations of neurotransmitters and potassium, for example are regulated by glial cell homeostatic mechanisms. Müller cells possess high-affinity uptake carriers for many transmitters and are believed to regulate extracellular transmitter levels in the retina. This uptake is essential for terminating synaptic transmission as well as preventing the spread of transmitters away from the synaptic cleft (see Newman and Reichenbach, 1996).

Astrocytes and oligodendrocytes

Besides Müller cells, there are other types of glia in the retina, including astrocytes and in some cases (e.g. rabbit and fish retinas) oligodendrocytes in the nerve fibre layer, which form the optic nerve. The astrocytes in the retina do not originate from the retinal epithelium, thus it is generally believed that astrocytes migrate from the optic nerve and probably enter the retina together with blood vessels (Turner and Cepko, 1987; Watanable and Raff, 1988; Rungger et al., 1993). In the optic nerve, like in other white matter tracts, most astrocytes and oligodendrocytes are derive from local mitoses at the time of "myelination gliosis" (Skoff, 1990). The ventricular cavity disappears in the evagination which forms the retina and the optic nerve. It is believed that the mitotic precursors of the oligodendrocytes and astrocytes migrate into the optic nerve from the subependymal germinal layer in the brain (Bigmani, 1995). There are species differences in the extent of migration of the oligodendrocyte precursors as judged by the extent of myelination. Thus, in human and rat, myelination stops at the lamina cribosa, while in rabbit and dog, myelination extends beyond the lamina cribosa to reach the optic nerve head. In fish, where the lamina cribosa is absent, the myelination extends into the retina within the optic fibre layer (Wolburg, 1981).

Microglial cells

Microglial cells are the other type of non neuronal cells present within the retina as resident or resting microglia that can be activated in response to injury, infection and inflammation (Thanos, 1991). In the normal retina, microglia are located in the nerve fibre layer, inner plexiform layer and sparsely in the outer plexiform layer (Humphrey and Moore, 1996). Activated microglia proliferate exhibiting phagocytic activity disposing of degenerating elements (Thanos et al., 1992, 1993). Recent *in vivo* studies present evidence for enhanced motility of microglial cells during degenerative diseases or after neurotoxic lesions. This property enables the cells to migrate towards sites of lesion inflammation. The factors or conditions that serve to mobilise microglia to these sites are not well characterised (Thanos and Richter, 1993).

STRUCTURE AND FUNCTION OF THE NEUROTROPHINS AND THEIR RECEPTORS

Neurotrophins

Neurotrophins are polypeptide molecules that regulate the development and maintenance of specific functions of different populations of nervous cells in the peripheral (PNS) and central nervous systems (CNS). This family includes four closely related factors: the prototype nerve growth factor (NGF) (Levi-Montalchini and Angeletti, 1968; Levi-Montalcini, 1987), brain-derived neurotrophic factor (BDNF) (Barde et al., 1982; Leibrock et al., 1989), neurotrophin-3 (NT-3) (Hohn et al., 1990; Maisonpierre et al., 1990; Rosenthal et al., 1990) and neurotrophins-4/5 (NT4/5) (Berkemeier et al., 1991; Hallböök et al., 1991). These four factors share more than a 30% sequence homology and are relatively well conserved in all mammals. The most recent neurotrophin discovered is NT-6 whose receptor remains to be defined, although there is evidence for NT-6 binding to proteoglycans on the cell surface and/or extracellular matrix (Götz et al., 1994). Originally it was thought that neurotrophins were released from target cells as signals for appropriate connections to take place in the development of the peripheral and central nervous system (Ernfors et al., 1990). The mass of recent studies on neurotrophic factors has made it necessary to revise this view. It is now generally accepted that neurones might derive trophic support not only from innervated cells (retrograde mechanism), but also from afferent neurones (anterograde influence), axon ensheathing glial cells, or even from within themselves (autocrine mechanism). In addition, any given neurotrophic factor may affect many neuronal types and any given neuronal type may be influenced by more than one type of neurotrophic factor (Korsching, 1993). The biological functions proposed for neurotrophins within the PNS and the CNS are extensive and have been extensively described (for review, Barde, 1989; Persson and Ibañez, 1993; Johnson and Oppenheim, 1994; Thoenen, 1995).

Neurotrophin receptors

Neurotrophins exert their biological activities by activation of two unrelated types of transmembrane receptors. The first type is the low-affinity NGF receptor, also known as low-affinity neurotrophin receptor (LANR, $p75^{NGFR}$) (Radeke et al.,

1987) which bind neurotrophins at varying affinity (Rodriguez-Tébar et al., 1990; Ernfors et al., 1990) and does not possess intrinsic tyrosine kinase activity (Radeke et al., 1987). The second type of receptors has specificity for individual neurotrophins binding and is conferred by the Trk family of tyrosine kinases knowns as TrkA, TrkB, TrkC. The human trk proto-oncogene product gp140 encodes a glycoprotein designated TrkA (Martín-Zanca et al., 1989), which is the functional high-affinity NGF receptor (Kaplan et al., 1991, Klein et al., 1991a). It is also known that TrkA is a receptor for NT-3 (Cordon-Cardo et al., 1991; Davies et al., 1995) and NT4/5 (Kuman et al., 1993). TrkB is the high-affinity receptor for BDNF (Middlemas et al., 1991), NT-3 (Soppet et al., 1991; Klein et al., 1991b; Squinto et al., 1991) and NT4/5 (Berkemeier et al., 1991; Klein et al., 1992). TrkC is high-affinity receptor only for NT-3 (Lamballe et al., 1991). Each of these trk genes is transcribed into multiple mRNAs. In addition to proteins that display variants in their kinase domains, TrkB and TrkC receptors also exist as truncated proteins which lack intracellular kinase domains. The neurotrophins exist naturally as homodimers (and possibly heterodimers) and, as with ligands for other receptor tyrosine kinases, activate signal transduction by inducing dimerization and auto-phosphorylation of the appropriate Trk receptors (Lindsay et al., 1994).

Some of the reasons why neurotrophins have both high and low affinity receptors may be as follows. The co-expression of p75 with Trk family members could increase the affinity of the binding of the neurotrophin for receptors when competing for limited amounts of trophic factors, participate in the internalisation of ligand after binding, participate in the transport of NGF intracellularly during retrograde transport and possibly discriminate between different, but closely related neurotrophic factors (for review Chao and Hempstead, 1995, Greene and Kaplan, 1995). Several other roles for p75 have been postulated. These include an involvement in apoptosis (Rabizadeh et al., 1993; Barrett and Bartlett, 1994; Hartzopoulos et al., 1994; Curtis et al., 1995; Frade, 1996) and specific signalling mechanism for the neurotrophins.

NGF in particular has a signalling mechanism, through the p75 receptor, to stimulate the formation of ceramide, which is a potential mediator involved in cell growth, differentiation, programmed cell death and protein-phosphorylation (Casaccia-Bonnefil et al., 1996). The spectrum of neurotrophic factors, their receptors, their localization, their direct and indirect effects on the cells, and the timing of their expression generate a large variety of combinations in space and time. This potential number of effects may be compared with the highly specific and punctual actions required for the formation and maintenance of a complicated structure like the retina.

The types of responses that could arise from the interaction of neurotrophins and their receptors are potentially enormous and how they relate to the overall function of the retina remains to be ellucidated.

PREVIOUS STUDIES ON THE DISTRIBUTION OF NEUROTROPHINS AND THEIR RECEPTORS WITHIN THE RETINA AND OPTIC NERVE

Although the importance of neurotrophins and their receptors in the development, maintenance and survival of the retina has been extensively investigated (Turner, 1985; Johnson et al., 1986; Rodriguez-Tébar et al., 1989; Henderson, 1991; Mey and Thanos, 1993; De la Rosa et al., 1994; Jelsma and Aguayo, 1994; Mansour-Roubaey et al., 1994; Panni et al., 1994; Meyer-Franke et al., 1995; Peinado-Ramón et al., 1996), little is known about the cellular localisation of the neurotrophins and their receptors within the retina. Most of the investigations directed to study the neurotrophins within the retina have been carried out using binding assays (Rodriguez-Tébar et al., 1993), Northern blot analysis (Takahashi et al., 1993) or *in situ* hybridisation (Ernfors et al., 1992; Herzog et al., 1994; Perez and Caminos, 1995). The last technique reveals a cellular localisation of the mRNA and generally gives little information about the cellular morphology of the labelled cells. More extensive studies have been carried out on the topographical distribution of the neurotrophin receptors. The presence of p75 has been reported to occur in ganglion cells and Müller cells in both developing and adult retinas of several species (Richardson et al., 1986; Schatteman et al., 1988; Yan and Johnson, 1988; Carmignoto et al., 1991; Zanellato et al., 1993). The trkB and trkC mRNAs have been found in the inner nuclear layer and in the ganglion cell layer of the rat retina (Ernfors et al., 1992; Jelsma et al., 1993; Perez and Caminos, 1994; Rickman and Brecha, 1995) and trkC mRNA has also been shown to be present in nuclear layers of the chick retina (Bovolenta et al., 1996; Hallböök et al., 1996). Sandell et al. (1994) using a Trk antibody that cross-reacts with mammalian TrkA, B and C (pan-trk), described a pattern of immunoreactivity in developing zebrafish retina. In another study on fish retina Martin et al. (1995), using *in situ* hybridisation, demonstrated an expression of trkB and trkC mRNA in the zebrafish retina.

Recent studies on the rat optic nerve have demonstrated the presence of mRNAs for all the neurotrophins and trks (Jelsma et al., 1993; Elkabes et al., 1995). No studies have been reported to date on the cellular localisation of neurotrophins or their receptors in optic nerve of the rat or fish.

RECENT STUDIES ON THE NEUROTROPHINS WITHIN THE GLIAL CELLS OF FISH AND RAT RETINAS AND OPTIC NERVE

Technical approaches

The very few studies conducted so far on the distribution of the neurotrophins and their receptor levels in the retina have been due the lack of suitable antibodies with specificity for individual neurotrophins or their receptors. We have partially resolved this difficulty, particularly for the Trk specific receptors, by using a combination of four methods. Initial studies involved the use of reverse transcriptionpolymerase chain reaction (RT-PCR) to demonstrate the existence of trkA, trkB and trkC mRNAs (data non shown). The second approach was to incubate isolated retinas with NGF, BDNF or NT-3 so that phosphorylation of Trk receptors could take place and allow activity of the specific high-affinity receptors TrkA, TrkB and TrkC, respectively to be identified. This was done by electrophoresis/western blotting using an antiphosphotyrosine monoclonal antibody (fig. 8-1).

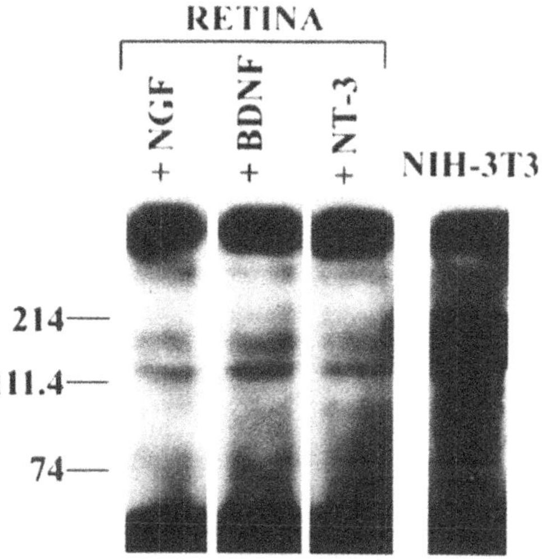

Figure 8-1. Electrophoresis/western blotting of Trk proteins following stimulation of fish (tench) retina with different neurtrophins. Retinas were cut into small pieces and incubated in phosphate buffer containing a cocktail of protease inhibitors (0.15U/ml aprotinin, 20μM leupeptine, 1mM PMSF, 100mM orthovanadate) and NGF, BDNF or NT-3 (50 ng/ml of PBS) for 5 minutes, at 37°C. Following a brief centrifugation (1.000xg for 5 min) the retinal pieces were placed in 1ml of ice-cold lysis buffer (20mM TrisHCl pH 8, 137mM NaCl, 10% glycerol, 1% NP-40, 2mM EDTA) containing protease inhibitors and a homogenous suspension produced by repeated pipetting with syringe containing an 18-gauge needle. After centrifugation (13.000xg) for 15 min. at 4°C and supernatants were immunoprecipitated with pan-trk. Phosphorylation products were fractionated by electrophoresis, transfered for western blotting and the high-affinity receptors of TrkA, TrkB or TrkC identified by use of an antiphospho-tyrosine monoclonal antibody. The right hand lane shows a mouse NIH-3T3 fibroblast line which is overexpressed for Trk (molecular weight between 111-214). The other lanes show expression of Trk following incubation with NGF, BDNF or NT-3. In each case two distinct phosphorylated proteins corresponding in molecular weight between 111.4 and 214 appear and these are thought to be Trks.

The third approach has been to use a variety of antibodies each recognising different parts of a receptor. To localise the TrkA receptor, for example, we have used three different antibodies, one recognising the extracellular domain, and the other two recognising different parts of the intracellular domain of the receptor molecule. The fourth approach is *in situ* hybridisation which allows for the topographical localisation of the genes related to neurotrophins and their receptors. However, as already discussed, *in situ* hybridisation provides limited information as to the type of cell labelled. This is especially difficult when the labelling is located to perykarya within nuclear layers of the retina, and not to the cell processes. This is a major problem when studying the Müller cells of the retina as none of the *in situ* hybridisation information can be related to these cells with certainty. For this reason we generally rely more on immunohistochemistry for studies on retinal glial cells.

In the described results no attempt will be made to define the characteristics of the antibodies used. Details of methodology used to determine neuronal distribution of neurotrophins and their receptors in the retina of rat and fish using immunohistochemistry and *in situ* hybridisation are described elsewhere (Caminos et al., 1994, 1995; Vecino et al., 1997). This article is mainly concerned with the localization of neurotrophins and their receptors to retina glial cells.

Neurotrophins

BDNF is a predominant neurotrophin within the fish retina, and is probably synthesised in Müller cells. As shown in figure 8-2B, BDNF is strongly localised in Müller cells with the labelling very similar to the localisation of S-100 immunoreactivity (Vecino et al., 1996). Some Müller cells also appear to contain very faint NGF and NT-4 specially located at the end-feet (fig. 8-2A, D). Moreover NT-3 is present within end-feets and perikarya of certain Müller cells (fig. 8-2C). Optic fibre layer has no labelling for any of the neurotrophins. However, the fish optic nerve reveals presence of oligodendrocyte-like cells containing BDNF and NT-3. After optic nerve crush more cell-like structures in the optic nerve show positive stain for BDNF and NT-3, some also express NT-4 and NGF.

Colocalisation experiments of neurotrophins in the optic nerve with the lectin from *Lycopersicon esculentum*, known to label macrophages in tench (Velasco et al., 1995), demonstrated that at least part of the population of neurotrophin positive cells that appear after optic nerve crush, are macrophages (data not shown).

Analysis of the rat retina shows that Müller cells do not contain the neurotrophins NGF, BDNF or NT-3, whereas a number of neurones do stain for all neurotrophins in the retina. Recent findings by Elkabes et al. (1995) using RT-PCR show that mRNA for NGF, BDNF, NT-3 and NT-4 exist in the rat optic nerve.

Neurotrophin receptors

Immunohistochemistry shows that neurotrophin receptors are associated with the Müller cell of the fish retina (fig. 8-2E-H). Within the optic fibre layer an antibody against the TrkA juxtamembrane domain labelled a small cluster of cells and the bipolar morphology and position of the cells within the retina suggest that they may be oligodendrocytes (fig. 8-2E; 8-4A, B). Use of antibodies to TrkB and the low affinity receptor p75 show clear staining within the end-feet of the Müller cells, but it was not possible to conclude whether the Müller cell perikarya were also labelled because many neuronal somata within the inner nuclear layer stained positively (fig. 8-2F, H). TrkC is localised in a distinctive way within the Müller cells, and is associated with the outer part of the Müller cells, in many ways resembling the localisation of GFAP (fig. 8-2G).

In the fish optic nerve TrkA and TrkB are associated within oligodendrocyte-like cells and TrkC appears within a few resident microglial cell, however only antibodies that recognise the intracellular domain recognises these cells (fig. 8-4D). After nerve crush, macrophages stain positively to p75. Moreover antibodies for TrkA and TrkB, stain not only oligodendrocytes but also macrophages (fig. 8-4C). In addition antibodies for TrkC stain macrophages (fig. 8-4E). The antibodies for anti-TrkA juxtamembrane or anti-TrkC intracellular do not "stain" these macrophages, instead the antibody for TrkC intracellular domain stain microglial cells and a few isolated astrocyte-like cells (fig. 8-4D). The positive labelling of the macrophages for the extracellular domain of TrkC but not for the intracellular domain implies that macrophages express a truncated TrkC receptor. This is in agreement with the results found of Elkabes et al., (1995) in which the truncated TrkC (where the intracellular domain of the molecule is absent) was the most abundant isoform in the rat optic nerve. On the other hand, microglial cells are immunoreactive to anti-TrkC intracellular domain, implying the presence of full-length TrkC receptor within these cells. The biological function of the various TrkC splice variants, remains to be clarified. Truncated TrkC may decrease neurotrophin signal transduction by forming inactive heterodimers within the full-length receptor or, alternatively, by competing for ligands, as has been suggested for the truncated TrkB (Knüsel et al., 1994).

Within the rat retina p75 immunoreactivity is associated with the whole of the Müller cell, staining its perikarya and branches and forming two distinct layers within the inner plexiform layer of the retina (fig. 8-3A). This clear and distinctive immunoreactive pattern contrasts with the weaker staining of the same cells in the fish retina (fig. 8-2H). Previous studies have shown that rat Müller cells express p75 (Richardson et al., 1986; Schatteman et al., 1988; Yan and Johnson, 1988; Carmignoto et al., 1991; Zanellato et al., 1993) but the distribution has not been compared with any other species. The p75 immunoreactivity of the rat Müller cells shows branching at two levels of the inner plexiform layer and appears to be altered after ischaemia (unpublished observations). The clear p75 low affinity immunoreactivity in the Müller cells contrasts with the weak expression of TrkB high-affinity receptor in these cells (fig. 8-3B, C). Within the rat optic nerve *trkA*, *trkB* and *trkC* mRNAs expression has recently been reported (Jelsma et al., 1993; Elkabes et al., 1995).

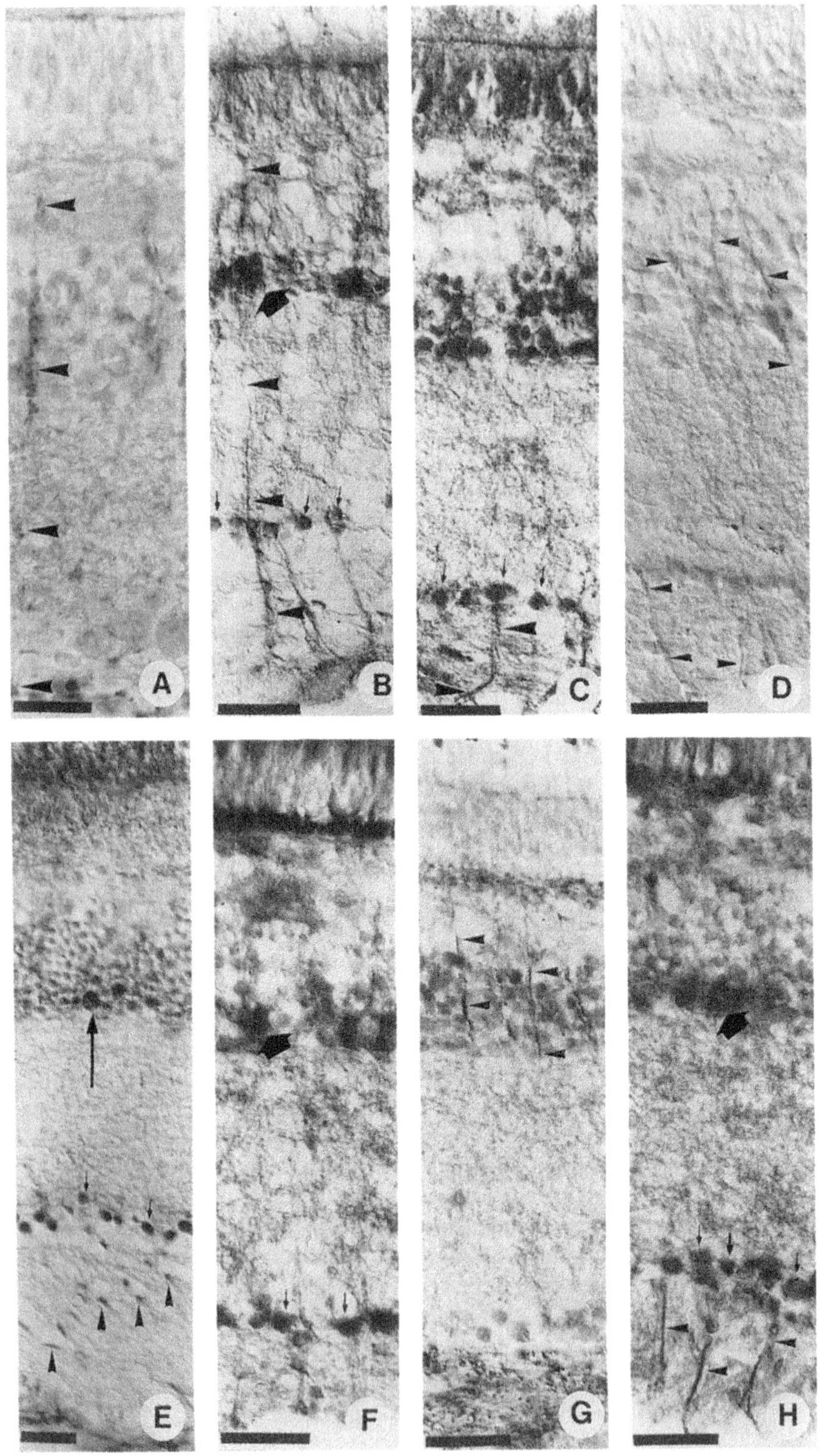
A
B
C
D
E
F
G
H

POSSIBLE ROLE OF RETINAL GLIAL CELLS AS SOURCE OF NEUROTROPHINS AND RECEPTORS

Neurotrophic factors were originally thought to originated from non-neuronal target cells and the nerve sheath. In the present chapter we have provided evidence that within the retina, the source of neurotrophins may be both neuronal and glial. Furthermore, the location of certain neurotrophins appears to be restricted to particular regions of the cell, as seen in Müller cells. The significance of this remains to be clarified.

The strong BDNF immunoreactivity associated with fish Müller cells contrasts with the absence of this neurotrophin from rat Müller cells, even though some neurones in the rat retina are BDNF positive. Moreover, TrkB receptors are present within most of the end-feet of the rat Müller cells (fig. 8-3C), while in the fish retina TrkB receptors are only present in few Müller end-feet (fig. 8-2F). Since it is known that BDNF enhances the survival and regrowth of the retinal ganglion cells (Johnson et al., 1986; Rodriguez-Tébar et al., 1989; Panni et al., 1994; Meyer-Franke et al., 1995) and that intravitreal injection of BDNF protects against ganglion cell death (Mansour-Robaey et al., 1994; Peinado-Ramón et al., 1996) it may be concluded that fish retina Müller cells generate a source of BDNF, to support ganglion cell survival and regeneration after optic nerve crush. In the rat retina this appears not to be the case; instead, BDNF has a neuronal origin. Since rat Müller cell end-feet have p75 and TrkB receptors, which are low and high-affinity receptors of BDNF, it is possible that when BDNF is injected intravitreously into the eye, the neurotrophins act on Müller cells.

An autocrine mode of action has been proposed for neurotrophins in which some types of neurones not only synthesise neurotrophic factors but also express the cognate receptors and respond to neurotrophic factors (Lindvall et al., 1994). Such a system now also appears to apply to glial cells, in particular Müller cells, since they express both neurotrophins and their high-affinity receptors. The possible autocrine mode of action of neurotrophins in the Müller cells *in vivo* is very

Figure 8-2. Immunocytochemical localisation of neurotrophins and their low and high affinity receptors in fish retina. Weak immunostaining for nerve growth factor (A) and neurotrophin-4 (D) are associated with some Müller cell process and end-feet (arrow heads). Brain derived neurotrophic factor (B) is also associated with Müller cell processes (arrow heads), and perikarya (large arrows) as well as with ganglion cells (small arrows). Neurotrophin-3 (C) is also associated with ganglion cells (small arrows) and end-feet of the Müller cells (arrow heads).
TrkA juxtamembrane immunoreactivity is associated with oligodendrocytes (arrow heads) in the nerve fibre layer and cell bodies in the ganglion (small arrows) and inner nuclear (big arrows) layers (E). TrkB (F) and p75 (H) immunoreactivities are very similar with staining being associated with Müller cell end-feet (arrow heads), possibly their perikarya (large arrows) and ganglion cells (small arrows). TrkC immunoreactivity is primarily associated with Müller cell processes (arrow heads) (G). Scale bars = 20μm.

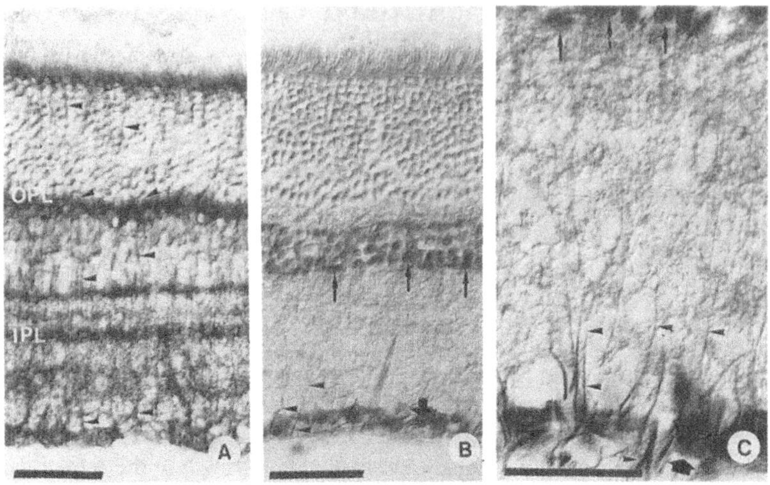

Figure 8-3. Rat retinas immunolabelled for the localisation of p75 (A) and TrkB extracellular domain (B and C). It can be seen that p75 is localised in Müller cells (arrowheads) and in laminae associated with the inner plexiform (IPL) and outer plexiform (OPL) layers. TrkB extracellular domain immunoreactivity is primarily located to Müller cells (arrow heads), ganglion cells (large arrow) and inner nuclear layer (small arrows). Scale bars = 20μm.

important, since it has been demonstrated *in vitro* that neurotrophic factors stimulate the synthesis of ganglioside levels in Müller cells (Hicks et al., 1996). Gangliosides are important signalling molecules involved in neural differentiation.

The presence of neurotrophins and in particular NT-3 within the optic nerve could be related with the proposed role for NT-3 within oligodendrocytes. It has been demonstrated that NT-3 participates in the proliferation of oligodendrocyte precursor cells *in vivo* and is thus required for normal oligodendrocyte development (Barres et al., 1994). In contrast, NGF and BDNF do not affect oligodendrocyte survival (Barres et al., 1993).

It is known that NGF promotes cell death in specific glial cell types through a ligand-dependent signalling mechanism involving the p75 neurotrophin receptor (Casaccia-Bonnefil et al., 1996). This mechanism may take place within the fish optic nerve since after optic nerve crushing, a small subpopulation of macrophages appears to express p75 and NGF as well as neurotrophins and growth factors.

CONCLUSIONS

In the present chapter we have summarised the distribution of neurotrophins and their receptors within the fish and rat retina and optic nerve using selected

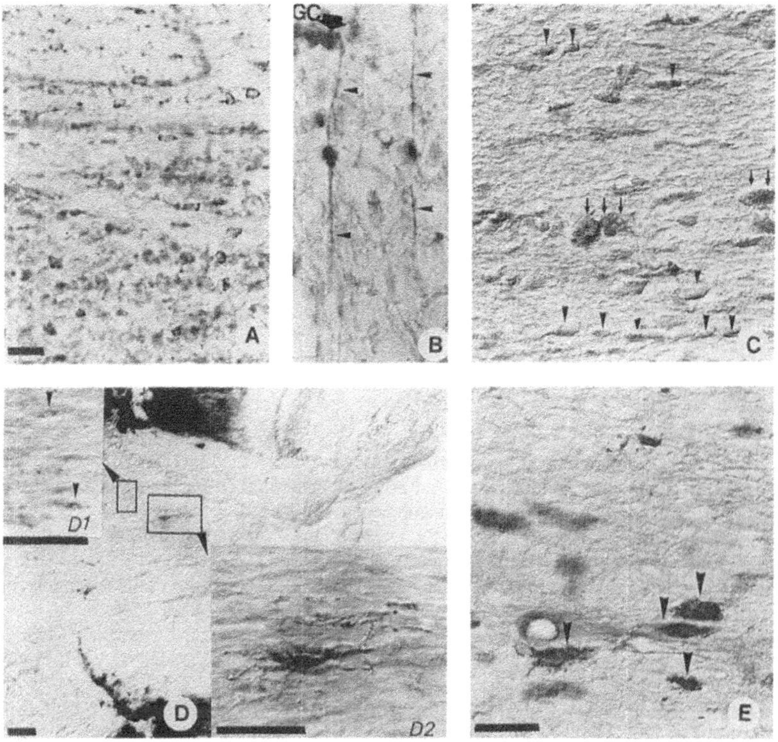

Figure 8-4. A shows TrkA-juxtamembrane positive oligodendrocytes in the nerve fibre layer of whole mount fish retina. B shows transverse sections of the nerve fibre layer shows the bipolar-like structure of these oligodendrocytes (arrowheads) some of which come into close contact with ganglion cells (small arrows). C is a longitudinal section of the fish optic nerve 7 days after optic nerve crush showing the presence of oligodendrocytes (arrow heads) and macrophages (big arrows) immunolabelled with antibodies directed to TrkB extracellular domain. D is a section through the optic nerve head showing TrkC intracellular domain positive microglial cells (seen more clearly in the magnified insert, D1) and astrocyte-like cells (seen more clearly in magnified insert, D2). E shows the presence of TrkC extracellular domain positive macrophages in fish optic nerve 7 days after optic nerve crush; note the location of these cells are close to blood vessels. Scale bars=20μm.

The regional overlap in the expression of neurotrophins and their receptors in the end-feet of the fish Müller cells suggests that neurotrophins might act locally via an autocrine mechanism. Müller cells of the fish stain particularly well for BDNF while in the rat the Müller cells are particularly well labelled for p75. In fish optic nerve, BDNF, NT-3, TrkA and TrkB are located to oligodendrocytes and following optic nerve crush all neurotrophins except NT3, and receptors appears to be associated additionally with macrophages.

Moreover, resident microglia and astrocyte-like cells stain with antibodies recognising TrkC intracellular domain. It is concluded that neurotrophins and their receptors play an important role in glial cells present of the retina and optic nerve.

ACKNOWLEDGMENTS

The authors wish to thank Cristina Otamendi for her photographic assistance. This work was supported by grants form DGICYT PB94-1104 and E.C. BMH2-ct96-0010 to D.M.Z. and E.C. Human Capital and Mobility grant CHRX-CT9-940435 grant to N.N.O and E.V.

REFERENCES

Barde, Y.A. (1989) Trophic factors and neuronal survival. Neuron *2*:1525-1534.

Barres, B.A., R. Schmid, M. Sendnter, and M. Raff (1993) Multiple extracellular signals are required for long-term oligodendrocyte suvival. Development *118*: 283-295.

Barres, B.A., M.C. Raff, F. Gaese, I. Bartke, G. Dechant, and Y.A. Barde (1994) A crucial role for neurotrophin-3 in oligodendrocyte development. Nature *367*: 371-5

Barrett, G.L., and P.F. Bartlett (1994) The p75 receptor mediates survival of death depending on the stage of sensory neurone development. Proc. Natl. Acad. Sci. USA *91*:6501-6505.

Bignami, A. (1995) Glial types, gliogenesis, and extracellular matrix in mammalian CNS. In A. Vernardakis and B.I. Roots (eds): Neuron-Glia Interactions During Phylogeny. Vol.I. Phylogeny and Ontogeny of Glial Cells. Humana Press, pp.3-39.

Bovolenta, P., J.M. Frade, E. Marti, M.A. Rodriguez-Pena, Y.A. Barde and A. Rodriguez-Tébar (1996) Neurotrophin-3 antibodies disrupt the normal development of the chick retina. J. Neurosci. *16*: 4402-4410.

Bray, G.M., M. Rasminsky, and A.J. Aguayo (1981) Interactions between axons and their sheath cells. An. Rev. Neurosci. *4*:127-162.

Berkemeier, L.R., J.W. Winslow, D.R. Kaplan, K. Nikolics, D.V. Goeddel, and A. Rosenthal (1991) Neurotrophin-5: a novel neurotrophic factor that activates trk and trkB. Neuron *7*:857-866.

Caminos, E., E. Vecino, E. Becker, A. Velasco, D. Martín-Zanca, and J. Aijón (1995) Immunohistochemical localization of neurotrophin receptors in the retina of the fish *Tinca tinca L.* and rat. Inves. Ophthalmol. Vis. Sci. Vol. *36*, p1165.

Caminos, E., E. Becker, A. Velasco, D. Martín-Zanca, J. Lara and E. Vecino (1995) Immunolocalization of the NGF high-affinity receptor (TrkA) in the normal tench retina and after optic nerve crush. Rev. Neurol (Barc), *23*(121):600.

Carmignoto, G., M.C. Comelli, P. Candeo, L. Cavichioli, Q. Yan, A. Merighi, and L. Maffei (1991) Expression of NGF receptor mRNA in the developing and adult rat retina. Exp. Neurol. *111*:302-311.

Casaccia-Bonnefil, P., B.D. Carter, R.T. Dobrowsky, and M.V.Chao (1996) Death of oligodendrocytes mediated by the interaction of nerve growth factor with its receptor p75. Nature *383*:716-719.

Chao, M.V., and L. Hempstead (1995) p75 and Trk: a two-receptor system. Trends Neurosci. *18*:321-326.

Cordon-Cardo, C., P. Tapley, S. Jing, V. Nanduri, E. O'Rourke, F. Lamballe, K. Kovary, R. Kelein, K.R. Jones, L.F. Reichard, and M. Barbacid (1991) The trk tyrosine protein kinase mediates the mitogenic proterties of nerve growth factor and neurotrophin-3. Cell *66*:173-183.

Curtis, R., K.M. Adryan, J.S. Patark, D.L. Compton, G. Weskamp, L.J. Huber, V.M. Chao, R. Jaenish, K.F. Lee, R.M. Lindsay, and P.S. DiStefano (1995) Differential role of the low affinity neurotrophin receptor (p75) in retrograde axonal transport of the neurotrophins. Neuron *14*:1201-1211.

Davies, A.M., L. Minichiello, and R. Klein (1995) Developmental changes in NT3 signalling via TrkA and TrkB in embryonic neurons. EMBO J. *14*:4482-4489.
De la Rosa, E.J., A. Arribas, J.M. Frade, and A., Rodriguez-Tébar(1994) Role of neurotrophins in the control of neural development: Neurotrophin-3 promotes both neuron differentiation and survival of cultured chick retinal cells. Neurosci. *58*:347-352.
Elkabes, S., D.G. Schaar, C.F. Dreyfus, and I.B. Black (1995) Developmental regulation of neurotrophin-3 and trkC splice variants in optic nerve glia *in vivo*. Neurosci. *66*:879-889.
Elkabes, S., E.M. Di Cicco-Bloom, and I.B. Black (1996) Brain microglia macrophages express neurotrophins that selectively regulate microglial proliferation and function. J. Neurosci. *16*: 2508-2521.
Engele, J., and M.C. Bohn (1991) The neurotrophic effects of fibroblast growth factors on dopaminergic neurons in vitro are mediated by mesencephalic glia. J. Neurosci. *11*:3070-3078.
Ennis, S., and Y.W. Kunz (1986) Differentiated retinal Müller glia are ciliated-ultrastuctural evidence in the teleost *Poecilia reticulata* P. Cell. Biol. Int. Rep. *10*:611-622.
Ernfors, P., C. Wetmore, L. Olson, and H. Persson (1990) Identification of cells in rat brain and peripheral tissues expressing mRNA for members of the nerve growth factor family. Neuron *5*:511-526.
Ernfors, P., J.P. Merlio, and H. Persson (1992) Cells expressing mRNA for neurotropnins and their receptors during embryonic rat development. Eur. J. Neurosci. *4*:1140-1158.
Götz, R., R. Köster, C. Winkler, F. Raulf, F. Lottspeich, M. Schartl, and H. Thoenen (1994) Neurotrophin-6 is a new member of the nerve growth factor family. Nature *372*:266-269.
Frade, J.M., A. Rodriguez-Tébar, and Y.A. Barde (1996) Induction of cell death by endogenous nerve growth factor through its p75 receptor. Nature *383*:166-168.
Greene, LL.A., and D.R. Kaplan (1995) Early events in neurotrophin signalling via Trk and p75 receptors. Curr. Opinion Neurobiol. *5*:579-587.
Hallböök, F., C.F. Ibañez, and H. Persson (1991) Evolutionary studies of the nerve growth factor family reveal a novel member abundantly expressed in *Xenopus* ovary. Neuron *6*:845-858.
Hallböök, F., A. Bäckström, K.,Kullander, T. Ebendal, and N. Carri (1996) Expression of neurotrophins and their receptors in the avian retina. J. Comp. Neurol. *364*:664-676.
Hatten, M.E., M. Lynch, R.E. Rydel, J. Sanchez, J. Joseph-Silverstein, D. Moscatelli, and D.B. Rifkin (1988) In vitro nerurite exptension by granule neurons is dependent upon astroglial-derived fibroblast growth factor. Dev. Biol. *125*:280-289.
Hartzopooulos, P.A., Ch. Suri, D.J. Glass, M.P. Goldfarb, and D. Yancopoulos (1994) The low affinity NGF receptor, p75, can collaborate with each of the Trks to potentiate functional responses to the neurotrophins. Neuron *13*:187-201.
Henderson, Z. (1991) Is nerve growth factor requiered for the survival of retinal ganglion cells during development? Brain. Res. *549*:322-326.
Herzog, K.H., K. Bailey, and Y.A. Barde (1994) Expression of the BDNF gene in the developing visual system of the chick. Development *120*:1643-1649.
Hicks, D., B. Guerold, and H. Dreyfus (1996) Stimulation of endogenous ganglioside metabolism by neurotrophic growth factors in cultured retinal Müller glia. Glia *16*:316-324.
Hohn, A., J. Leibrock, K. Bailey, and Y.A. Barde (1990) Identification and characterization of a novel member of the nerve growth factor/brain-derived neurtrophic factor family. Nature *344*:339-341.
Holländer, H. F. Makarov, Z. Dreher, D. Van Driel, T. Chan-Ling and J. Stone (1991) Structure of the macroglia of the retina: sharing and division of labour between astrocytes and Müller cells. J. Comp. Neurol. *313*:587-603.
Humphrey, M.F. and S.R. Moore (1996) Microglial responses to focal lesions of the rabbit retina: correlation with neural and macroglial reactions. Glia *16*:325-341.
Jacobson, M., and R.M. Gaze (1965) Selection of appropriate tectal connections by regenerating optic nerve fibres in adult goldfish. Exp. Neurol. *13*:418-430.
Jelsma, T.N., H.H. Friedman, M. Berkelaar, G.M. Bray, and A. Aguayo (1993) Different forms of the neurotrophin receptor trkB mRNA predominate in rat retina and optic nerve. J. Neurobiol. *24*:1207-1214.
Jelsma, T.N. and A.J. Aguayo (1994) Trophic factors. Curr. Opinion in Neurobiol. *24*:1207-1214.
Johns, P. R. (1977) Crowth of the adult goldfish eye. III. Source of the new retinal cells. J. Comp. Neurol.*176*: 343-357.

Johnson, J.E., Y.A. Barde, M. Schwab, and H. Thoenen (1986) Brain derived neurotrophic factor supports the survival of cultured rat retinal ganglion cells. J. Neurosci. *6*:3031-3038.
Johnson, J., and R. Oppenheim (1994) Keeping track of changing neurotrophic theory. Curr. Biology *4*:662-665.
Kaplan, D.Rl, B. L. Hampstead, D. Martín-Zanca, M.V. Chao, and L.F. Parada (1991) The trk proto-oncogene product: a signal transducing receptor for nerve growth factor. Science *252*:554-557.
Klein, R, S. Jing, V. Nanduri, E. O'Rourke, and M. Barbacid (1991a) The trk proto-oncogene encodes a receptor for nerve growth factor. Cell *65*:189-197.
Klein, R., J.S. Nanduri, F. Lambelle, P. Tapley, S. Bryant, C. Cordon-Cardo, K.R. Jones, L.F. Reichard, and M. Barbacid (1991b) The trkB tyrosine protein kinase is a receptor for brain-derived neurotrophic factor and neurotrophin-3. Cell *66*:395-403.
Klein, R., F., Lamballe, S. Bryant, and M. Barbacid (1992) The trkB tyrosine protein kinase is a receptor for neurotrophin-4. Neuron *8*:947-956.
Korsching, S. (1993) The neurotrophic factor concept: a reexamination. J. Neruosci. *13*:2739-2748.
Kummar, S., L.A. Peña and J. Vellis (1993) CNS glial cells express neurotrophin receptors whole levels are regulated by NGF. Mol. Brain. Res. *7*:163-168.
Knüsel, B., S. Rabin, F. Hefti, and D. Kaplan (1994) Regulated neurotrophin receptor responsiveness during neuronal migration and early differentiation (1994) J. Neurosci. *14*:1542-1554.
Lamballe, F., R., Klein, and M. Barbacid (1991) trkC, a new member of the trk family of tyrosine protein kinases, is a receptor for neurotrophin-3. Cell *66*:967-979.
Leibrock, J., F. Lottspeich, A. Hohn, M. Hofer, B. Henferer, P. Masiakowski, H. Thoenen, and Y.A. Barde (1989) Molecular cloning and expression of brain-derived neurotrophic factor. Nature *341*:149-152.
Levi-Montalcini, R., and P.U. Angeletti (1968) Nerve growth factor. Physiol. Rev. *48*:534-569.
Levi-Montalcini, R. (1987) the nerve growth factor, 35 years later. Sience *237*:1154-1162.
Lindvall, O., Z. Kokaia, J. Bengzon, E. Elmér, and M. Kokaia (1994) Neurotrophins and brain insults. Trends. Neurosci. *17*:490-496.
Lindsay, R.M., S.J. Wiegand, C. A. Altar and P.S. DiStefano (1994) Neurotrophic factors: from molecule to man. Trends Neurosci. *17*:182-190.
Maisonpierre, P.C., L. Belluscio, S. Squinto, M.E. Furth, R.M. Lindsay and B. D. Yancopoulos (1990) NT-3, BDNF, and NGF in the developing rat nervous system: parallel as well as reciprocal patterns of expression. Neuron *5*:501-509.
Mansour-Robaey, S., D.B. Clarke, Y.C. Wang, G.M. Bray, and A.J. Aguayo (1994) Effects of ocular injury and administration of brain-derived neurotrophic factor on survival and regrowth of axotomized retinal ganglion cells. Proc. Natl. Acad. Sci. USA. *91*: 1632-1636.
Martin, S.C, G. Marazzi, J.H. Sandell, and G. Heinrich (1995) Five Trk receptors in the zebrafish. Dev. Biol. *169*: 745-58
Martín-Zanca, D. R. Oskam, B. Mitra, T. Copeland, and M. Barbacid (1989) Molecular and biochemical characterization of the human trk proto-oncogene. Mol. Cell. Biol. *9*:24-33.
Mey, J., and S. Thanos (1993) Intravitreal injections of neurotrophic factors support the survival of axotomized retinal ganglion cells in adult rats in vivo. Brain. Res. *602*:304-317.
Meyer, R.L. (1978) Evidence from thymidine labeling for continuing growth of retina and tectum in juvenile goldfish. Exp. Neurol. *59*: 99-111.
Meyer-Franke, A., M.R. Kaplan, F.W. Pfrieger and B.A. Barres (1995) Characterization of the signaling interactions that promote the survival and growth of developing retinal ganglion cells in culture. Neuron *15*:805-819.
Middlemas, D.S., R.A., Lindberg and T. Hunter (1991) trkB, a neural receptor protein-tyrosine kinase: evidene for a full-length and two truncated receptors. Mol. Cell Biol. *11*:143-153.
Müller, H. (1952) Bau and Wachstum der Netzhaut des Guppy *(Lebistes reticulatus)*. Zool. Jb. *63*:275-324.
Murray, M. (1976) Regeneration of retinal axons into the goldfish optic tectum. J. Comp. Neurol. *168* :175-196.
Newman, E. and Reichenbach, A. (1996) The Müller cell: a functional element of the retina. Trends Neurosci. *19*:307-312.
Panni, M.K., J. Atkinson and R.D. Lund (1994) Evidence for tropic role of brain-derived neurotrophic factor in transplanted embryonic retinae. *81*:325-327.

Peinado-Ramón, P., M. Salvador, M.P. Villegas-Pérez, and M.Vidal-Sanz (1996) Effects of axotomy and intraocular administration of NT-4, NT-3, and brain derived neurotrophic factor on survival of adult rat retinal ganglion cells. A quantitative in vivo study. Invest. Ophtalmol. Vis. Sci. *37*:489-500.

Perez, M.T. and E. Caminos (1995) Expression of brain-derived neurotrophic factor and of its functional receptor in neonatal and adult rat retina. Neurosci. Lett. *183*:96-99.

Persson, H., and C.F. Ibáñez (1993) Role and expression of neurotrophins and the trk family of tyrosine kinase receptors in neural growth and rescue after injury. Curr. Opinion Neurobiol. *6*:11-18.

Rabizadeh, S. Oh. J., L-T. Zhong, J. Yand, C.M. Bitler, Ll. Burcher and D.E. Bredesen (1993) Induction of apoptosis by the low-affinity NGF receptor. Sciene *261*:345-348.

Radeke, M.J., T.P. Misko, C. Hsu, L.A. Herzengerg, and E.M. Shooter (1987) Gene transfer and molecular cloning of the rat nerve growth factor receptor. Nature *325*:593-597.

Reichenbach, A., and F. Wohlrab (1986) Morphometric parameters of Müller (glial) cells dependent on their topographic localization in the nonmyelinated part of the rabbit retina. A consideration of functional aspects of radial glia, J. Neurocytol. *15*:451-459.

Reichenbach, A., H. Schneider, L. Leibnitz, W. Reichelt, P. Schaaf, and R. Schümann (1989) The structure of rabbit retinal Müller (glial) cells is adapted to the surrounding retinal layers. Anat. Embryol. *180*:71-79.

Reichenbach, A., J.U., Stolzenburg, W., Eberhardt, T.I. Chao, D. Dettmer, and L. (1993) What do retinal Müller (glial) cells do for their neuronal "small siblings"?. J. Chem. Neuroanat. *6*: 201-213.

Reichenbach, A., and S. R. Robinson (1995) Ependymoglia and ependymoglia-like cells. In H. Kettenmann and B.R. Ranson (eds): Neuroglia. N.Y. Oxford Univ. Press, pp. 58-85.

Richardson, P.M., V.M.K. Issa, and R.J. Riopelle (1986) Distribution of neuronal receptors for nerve growth factor in the rat. J. Neurosci. *6*:2312-2321.

Rickman, D., and N.C. Brecha (1995) Expression of the proto-oncogene, trk, receptors in the developing rat retina. *12*:215-222.

Rodriguez-Tébar,A., P.L. Jeffrey, H. Thoenen and Y.A. Barde (1989) The survival of chick retinal ganglion cells in response to brain derived neurotrophic factor depends on their embryonic age. Dev. Biol. *136*:296-303.

Rodriguez-Tébar, A., G., Dechant, and Y.A. Barde (1990) Binding of brain-derived neurotrophic factor to the nerve growth factor receptor. Neuron *4*:487-492.

Rodriguez-Tébar, A., E. De la Rosa, and A. Arribas (1993) Neurotrophin-3 receptors in the developing chicken retina. Eur. J. Biochem. *211*:789-794.

Rosenthal, A., D.V. Goeddel, T. Nguyen, M. Lewis, A. Shih, G.R. Laramee, K. Nikolics. and J.W. Winslow (1990) Primary structure and biological activity of a novel human neurotrophic factor. Neuron *4*:767-773.

Rungger-Brandle, E., J.M. Messerli, G. Niemeyer, and H.M. Eppenberger (1993) Confocal microscopy and computer-assisted image reconstruction of astrocytes in the mammalian retina. Eur. J. Neurosci. *5*:1093-106.

Sandell, J.H., S.C. Martin and G. Heinrich (1994) The development of neurotrophin receptor Trk immunoreactivity in the retina of the zebrafish *(Brachydario rerio)*. Devel. Brain. Res. *81*:192-200.

Schatteman, G.C., L. Bibbs, A.A. Lahahan, P. Claude, and M. Bothwell (1988) Expression of NGF receptor in developing and adult primate central nervous system. J. Neurosci. *8*:860-873.

Sharma, S. (1972) The reformation of retinotectal projections after various tectal ablations in adult goldfish. Exp. Neurol. *34*: 171-182.

Silver, J., S.E., Lorenz, Wahlsten, D., and J. Coughlin (1982) Axonal guidance during development of the great cerebral commisures: descriptive and experimental studies, *in vivo*, on the role of preformed glial pathways. J. Comp. Neurol. *210*:10-29.

Skoff, R.P. (1990) Gliogenesis in rat optic nerve: Astrocytes are generated in a single wave before oligodendrocytes. Develop. Biol. *139*: 149-168.

Soppet, D., E. Escandon, J. Maragos, D.S. Middlemas, S.W. Reid, J. Blair, L.E. Burton, B.R. Stnaton, D. R. Kaplan, T. Hunter, K. Kidolics. and L.F. Parada (1991) The neurotrophic factors brain-derivd neurotrophic factor and neurotrophin-3 are ligands for the trkB tyrosine kinase receptor. Cell *65*:895-903.

Squinto, S.P., T.N. Stitt, T.H. Aldrich, S. Davis, S.M. Bianco, C. Radziejewski, D.J. Glass, P. Masiakowski, M.E. Furth, K.M. Valenzuela, P.S. Distefano and G.D. Yancopoulos (1991) trkB encodes a functional receptor for brain-derived neurotrophic factor and neurotrophin-3 but not for the nerve growth factor. Cell *65*:885-893.

Sperry, R.W. (1948) Patterning of central synapses in regeneration of the optic nerve in teleosts. Physiol. Zool. *21*: 351-361.

Takahashi, J.B., M. Hoshimaru, K.Harunhiko, and H. Masakazu (1993) Extension of optic nerve fibres on genetically modiefied cells producing brain-derived neurotrophic factor. Neurosci. Lett. *149*:83-86.

Thanos, S. (1991) The relationship of microglial cells to dying neurons during natural neuronal cell death and axotomy-induced degeneration of the rat retina. Eur. J. Neurosci. *3*:1189-1207.

Thanos, S., C. Pavlidis, J. Mey, and H.J. Thiel (1992) Specific transcellular staining of microglia in the adult rat after traumatic degeneration of carbocyanine-filled retinal ganglion cells. Exp. Eye. Res. *55*:101-117.

Thanos, S., and W. Richter (1993) The migratory potential of vitally labelled microglial cells within the retina of rats with hereditary photoreceptor dystrophy. In. J. Devel. Neurosci. *11*:671-680.

Thanos, S., J. Mey, and M. Wild (1993) Treatment of the adult retina with microglia-suppressing factors retards axotomy-induced neuronal degradation and enhances axonal regeneration in vivo and in vitro. J. Neurosci. *13*:455-466.

Thoenen, H. (1995) Neurotrophins and neuronal plasticity. Science *270*:593-598.

Turner, J.E. (1985) Promotion of neurite outgrowth and survival in dissociated fetal rat retinal cultures by a fraction derived from a brain extract. Dev. Brain Res. *18*:265-274.

Turner, D.L., and C.L. Cepko (1987) A common progenitor for neurons and glia persist in rat retina leate in development. Nature *328*:131-136.

Velasco, A., E. Caminos, E. Vecino, J. Lara, and J. Aijón (1995) Microglia in normal and regenerating visual pathways of the tench (*Tinca tinca L.*, 1758; Teleost): a study with tomato lectin. Brain Res. *705*:315-324.

Vecino, E., A. Velasco, E. Caminos, and J. Aijón (1997) Distribtion of S100 immunoreactivity in the retina and optic nerve head of the Teleost *Tinca tinca L.* Micros. Res. Tech.*36*:17-25.

Vecino, E., E. Caminos, M. Ugarte, D. Martín-Zanca, and N. Osborne (1997) Immunohistochemical distribution of neurotrophins and their receptors in the rat retina and influence of ischemia and reperfusion. Gen. Pharmacol. (in press).

Watanable, T., and M.C. Raff (1988) Retinal astrocytes are immigrants from the optic nerve. Nature *332*:834-837.

Wolburg, H. (1981) Myelination and remyelination in the regenerating visual system of the goldfish. Exp. Brain Res. *43*:199-206.

Yan, Q. and E.M. Johnson (1988) An immunohistochemical study of the nerve growth factor receptor in developing rats. J. Neurosci. *8*:3481-3498.

Yoshida, K., and F.H. Gage (1991) Fibroblast growth factors stimulate nerve growth factor synthesis and secretion by astrocytes. Brain Res. *538*:118-126.

Zanellato, A., M.C. Comelli, R. Dal Toso and G. Carmignoto (1993) Developing rat retinal ganglion cells express the fucntional NGF receptor p140 trkA. Dev. Biol. *159*:105-113.

9

EXPRESSION AND FUNCTION OF NEUROTRANSMITTER RECEPTORS IN GLIAL CELLS OF THE CENTRAL NERVOUS SYSTEM

Carlos Matute, David J. Fogarty, José María García-Barcina, Miroslav Gottlieb, María José Morán and María Victoria Sánchez-Gómez

Over the last decade, it has been clearly established that glial cells express a plethora of neurotransmitter and neuropeptide receptors. However, little information is still available about the functions mediated by these glial receptors. Since the mainstream hypothesis about the roles of glia considers these cells as providing structural and trophic support to neurons, it is likely that the receptors expressed by glial cells are key mediators of these phenomena. We summarize in this chapter recent work carried out in our laboratory illustrating the expression of glutamate, angiotensin II and ATP receptors, their molecular nature and some of the putative functions mediated by them. Overall, we have observed that these receptors are expressed in a highly heterogenous manner in glial cells and that they may participate in regulating the size of the glial cell population as well as in the production of trophic support relevant to brain repair.

GLUTAMATE RECEPTORS IN NORMAL GLIAL CELLS

Glutamate is the most widely distributed excitatory neurotransmitter in the mammalian brain. Signal transmission at glutamatergic synapses is mediated by ionotropic and metabotropic receptors that have now been molecularly characterized (Hollmann and Heinemann, 1994). According to molecular, pharmacological, and electrophysiological criteria, ionotropic glutamate receptors are classified as a-amino-3-hydroxy-5-methyl-isoxazolpropionic acid (AMPA), kainate, and *N*-methyl-D-aspartate (NMDA) subtypes. Non-NMDA ionotropic glutamate receptors are oligomeric ligand-gated ion channels with fast kinetics that are permeable to Na^+, K^+, and also to low levels of Ca^{2+} (for a review see Hollmann and Heinemann, 1994). In contrast, NMDA receptors are highly permeable to Ca^{2+} and exhibit voltage-dependence (Hollmann and Heinemann, 1994).

Glutamate receptors have also been found in glial cells in vitro, in brain slices from immature animals and in the adult brain (for a recent review see Steinhäuser and Gallo, 1996). Activation of these receptors on glial cells produces a large variety of responses, including membrane depolarization, and the release of neurotransmitters and growth factors, suggesting that they may have an active role in brain signaling and repair. However, the pattern of expression of receptor subtypes shows regional variations and does not always correlate with observations obtained in culture. The findings illustrated below were obtained by a combination of molecular and histological techniques and are restricted to the analysis of glutamate receptors present in glial cells of the adult brain white matter in situ. A summary of the results is shown in Table 9-1.

Table 9-1. *Summary of glutamate receptors subunits expressed by glial cells*

	RAT BRAIN		BOVINE WHITE MATTER			
	in situ	*in vitro*	*in situ*	*in vitro*	Cellular localization	Coexpression
GluR1	X	X	X	X	Astrocytes	GluR5/6/7
GluR2	X	X	X	X	Astrocytes	N.D.
GluR3	X	X	X	X		
GluR4	X	X	-	-	Oligodendroglial lineage cells	N.D.
GluR5	X	X	++	X	Astrocytes	GluR1 (Astrocytes)
GluR6	++	X	+	X	and	KA2 (Astrocytes and
GluR7	+	X	++	-	Oligodendrocytes	Oligodendrocytes)
KA1	X	X	++	X	N.D.	N.D.
KA2	+++	X	+++	X	Astrocytes and Oligodendrocytes	GluR5/6/7

Expression studies were carried out by RT-PCR (X, presence; -, absence) and/or by Northern blotting (+, weak; ++, moderate; +++ high). Rat brain tissues were used to assess the suitablility of the reagents used (primers and probes) to detect the different subunits. In vitro studies were done in glial cultures. Localization and co-expression analysis was performed by immunohistochemistry. N.D. not detected. The table summarizes results described in JMGB's PhD work, Matute and Miledi, 1993; Matute et al., 1994a,b; García-Barcina and Matute, 1996.

Different AMPA receptor subunits are expressed by different types of glial cells

AMPA preferring glutamate receptors are composed of subunits GluR1-4 (Hollmann and Heinemann, 1994). We found that GluR1-3 were the main subunits expressed in the corpus callosum, fornix and optic nerve of the adult bovine brain.

The GluR1-4 subunits are expressed as two alternative splice variants known as flip and flop, two isoforms which differ in a 115 bp exonic sequence located near the extracellular side of the transmembrane domain TMIV (Sommer et al., 1990). The flop isoform is predominant in the adult brain (Sommer et al., 1990). Interestingly, we found that the flip version appeared to be the most abundant in the adult white matter, suggesting that AMPA receptor genes may be processed differently in glia as compared to neurons.

The location of the AMPA receptors expressed in the white matter was investigated with antibodies to GluR1, to GluR2/3/4c and to GluR4 subunits (Petralia and Wenthold, 1992). The GluR1 subunit was exclusively present in the majority of fibrous astrocytes as recognized by the distribution of immunolabel throughout most of the cell surface as well as the comparable appearance of these cells to those revealed with antibodies to GFAP. Thus, it appeared that glutamate receptors including the GluR1 subunit may be located in the astrocyte cell body and the numerous radial processes which often were found to contact blood vessels and form labeled end feet. A similar distribution of label was observed with antibodies to GluR2/3/4c. However, the intensity of labeling in the cell bodies and proximal parts of the radial processes was lower with these antibodies whereas, in contrast, the processes surrounding the blood vessels and the endfeet were intensely labeled.

The GluR4 subunit was exclusively found in small cells with a rounded soma bearing one, two or three branches. These cells were randomly scattered throughout the entire extent of the corpus callosum and did not show any particular alignment in relation to the nerve fibers. To identify these cells, we carried out double labeling experiments with antibodies to the GluR4 subunit and subsequently with antibodies to gelsolin or CNPase. The results indicated that the GluR4 subunit was expressed in a small subpopulation of CNPase+ cells, in some gelsolin+ cells and in cells which were CNPase- and gelsolin-. This unidentified cell population often had a morphology similar to that described for progenitor cells in the adult brain (Fulton et al., 1992) and therefore, it may correspond to this cell class.

In summary, the AMPA receptor GluR1-3 subunits are present in astrocytes while GluR4 is expressed by cells of the oligodendroglial lineage. Thus, it appears that there is a certain segregation in the expression of the various AMPA selective glutamate receptor subunits in terms of cell types expressing the subunits.

All kainate receptor subunits are expressed in a large subpopulation of astrocytes and oligodendrocytes

Kainate selective glutamate receptors are composed of GluR5-7 and KA1-2 subunits (Hollmann and Heinemann, 1994). Transcripts for each of the five subunits were detected by PCR in the white matter of the adult bovine brain.

RNA editing has been proposed as a source of molecular heterogeneity in glutamate receptors (reviewed in Seeburg, 1993). Thus, pre-mRNAs of the AMPA subunit GluR2 and the kainate subunits GluR5 and GluR6 possess several sites which can potentially undergo editing. This is the case for the Q/R site which is located in the second transmembrane domain of these subunits and controls the Ca2+ permeability of the channel they form (Köhler et al, 1993). Restriction analysis of the GluR6 subunit expressed in glial cells of the adult bovine white matter revealed that both edited and unedited forms at the Q/R site coexist to a similar extent. In contrast, the GluR6 edited form was predominant in the adult rat brain, in agreement with other reports (Köhler et al, 1993). The more limited extent of editing observed in the white matter is probably due to different processing of the GluR6 transcript in glial cells as compared to whole brain preparations where most of the glutamate receptors are located in neurons. This data suggests that cells might form kainate receptors with biophysical properties distinct from their counterparts in neurons. The differences in the extent of GluR6 editing between glia and neurons may have important implications for the physiology of glial kainate receptors.

Immunohistochemical experiments revealed that the GluR5-7 and KA2 subunits appeared in about 50% of astrocytes and 40% of oligodendrocytes. In astrocytes the receptor subunits are located throughout the cell body and processes, whereas in oligodendrocytes they are mostly restricted to the soma. Double labeling studies demonstrated that in the white matter areas studied the GluR5-7 and KA2 subunits are coexpressed in the majority of macroglial cells endowed with kainate receptors. In addition, astrocytes, but not oligodendrocytes, expressing kainate selective receptors were found to express the GluR1 subunit.

Taken together, these results provide evidence for the presence and molecular diversity of kainate selective receptor subunits in glial cells of the adult white matter and for their coexpression with AMPA receptor subunits.

GLUTAMATE RECEPTORS IN REACTIVE GLIAL CELLS

One of the most outstanding features of glial cells is their ability to react to injury and that this response is accompanied by an up-regulation of many proteins (for reviews see Eddleston and Mucke, 1993; Kreutzberg, 1996). Because of that, we assayed the levels of ionotropic glutamate receptors present in glial cells of the ischemia-sensitive CA1 region.

Postischemic reactive astrocytes and microglia express glutamate receptors

Ischemia-induced neuronal damage involves a disturbance in calcium homeostasis, an excessive release of glutamate, and an increase in oxygen free radical formation. It is possible that changes in the extracellular glutamate concentration occurring during ischemia (Benveniste et al., 1984, 1989) and the subsequent neuronal cell death occurring in this pathological condition, may have a profound effect on the expression of GluRs in glial cells in the resulting damaged area.

To test this hypothesis, we examined by immunohistochemistry the expression of ionotropic glutamate receptor subunits in glial cells of the rat dorsal hippocampus, 3 days to 28 days after transient forebrain ischemia. In general, the expression of GluRs at all time points studied underwent a drastic reduction that was primarily restricted to the CA1 region. In addition to the disappearance of GluRs due to neuronal cell death, we observed their expression in reactive glial cells (Table 9-2). The time course of expression and the subunits involved were different for astrocytes and microglia. Reactive astrocytes exhibited kainate, GluR5-7, and NMDA, NR2A/B, receptor subunits, both of which were maximally expressed approximately 4 weeks post-ischemia. In contrast, reactive microglia expressed the GluR4 and NR1 subunits, AMPA and NMDA receptor subtypes respectively, with maximal expression being observed between 3 and 7 days post-ischemia. These results demonstrate that specific types of GluRs are expressed in reactive glial cells after ischemia and that overall, their expression levels peak around or after the periods of maximal astrogliosis (Petito et al., 1990) and microgliosis (Morioka et al., 1991). Thus, modulation of GluR expression may be one of the molecular components accompanying the gliotic process.

Table 9-2. *Glutamate receptor subunit expression in glial cells of postischemic CA1*

	Reactive Astrocytes	Reactive Microglia
Glu R1	weak	-
GluR2/3	weak	-
GluR4	-	+
GluR5/6/7	+	-
NR1	-	+
NR2A/B	+	-

The expression of nerve growth factor and glutamate receptors in postischemic astrocytes has a similar temporal profile

Delayed neuronal death subsequent to ischemia can be prevented at least in part, by exogenously applied NGF (e.g., Cheng and Mattson, 1991). Therefore, it is critical to understand the regulation of NGF expression in the areas of the brain such as the CA1 region, which are most sensitive to ischemic damage. To that end, we investigated by immunohistochemistry the expression of NGF in that hippocampal area at several stages of reperfusion following transient forebrain ischemia.

Ischemic damage induced the expression of NGF in reactive astrocytes located in all CA1 layers, as early as 3 days postlesion. NGF levels increased later and reached a peak at 28 days after reperfusion (fig. 9-1). Interestingly, the time course of NGF expression parallels that of kainate glutamate receptors in post-ischemic astrocytes observed in a previous study (see above; Gottlieb and Matute, 1997). This raises the possibility of a cause-effect relationship.

This idea is supported by evidence indicating that NGF production is up-regulated by the activation of glutamate receptors both in neurons (Zafra et al., 1990) and in astrocytes (Pechan et al., 1993). Thus, it is conceivable that the transient increase in extracellular glutamate occurring after ischemia (Benveniste et al., 1989) along with the increased expression of glutamate receptors in glial cells, results in an efficient signaling mechanism by which NGF production is enhanced. Consistent with this possibility, purines, a family of well established neurotransmitters and neuromodulators which activate specific receptors in glial cells, can trigger the production of cytokines and trophic factors in astrocytes and thus, minimize neuronal loss subsequent to brain ischemia and trauma (Neary et al., 1996). The trophic activity provided by NGF surplus may contribute to attenuate neurodegeneration caused by ischemic damage, particularly that occurring at the penumbra zone of the ischemic area, and to partially reverse the induced damage in the case of a brief blood supply deficit.

GLUTAMATE RECEPTOR ACTIVATION IMPAIRS OLIGODENDROCYTE VIABILITY

Oligodendrocytes myelinate axons in the Central Nervous System. This cell type mainly develops shortly after the majority of neurons are generated and therefore, it is likely that neurons play an important role in controlling oligodendrocyte development. Thus, it is known that the proliferation and survival of oligodendrocytes depends on electrical activity in axons and on the reciprocal contacts they establish, respectively (for a recent review, see Barres and Raff, 1996). Axon to oligodendrocyte signaling underlies the generation of the precise number of oligodendrocytes necessary to myelinate the entire extent of a given population of axons. Unfortunately, little information is available about the molecules participating in this signaling process.

We tested whether glutamate is involved in the shaping of the oligodendrocyte population, by studying the effects of glutamate agonists in oligodendrocyte cultures of the developing rat optic nerve. First, we investigated by immunocytochemistry the types of ionotropic glutamate receptors expressed by oligodendrocytes in vitro.

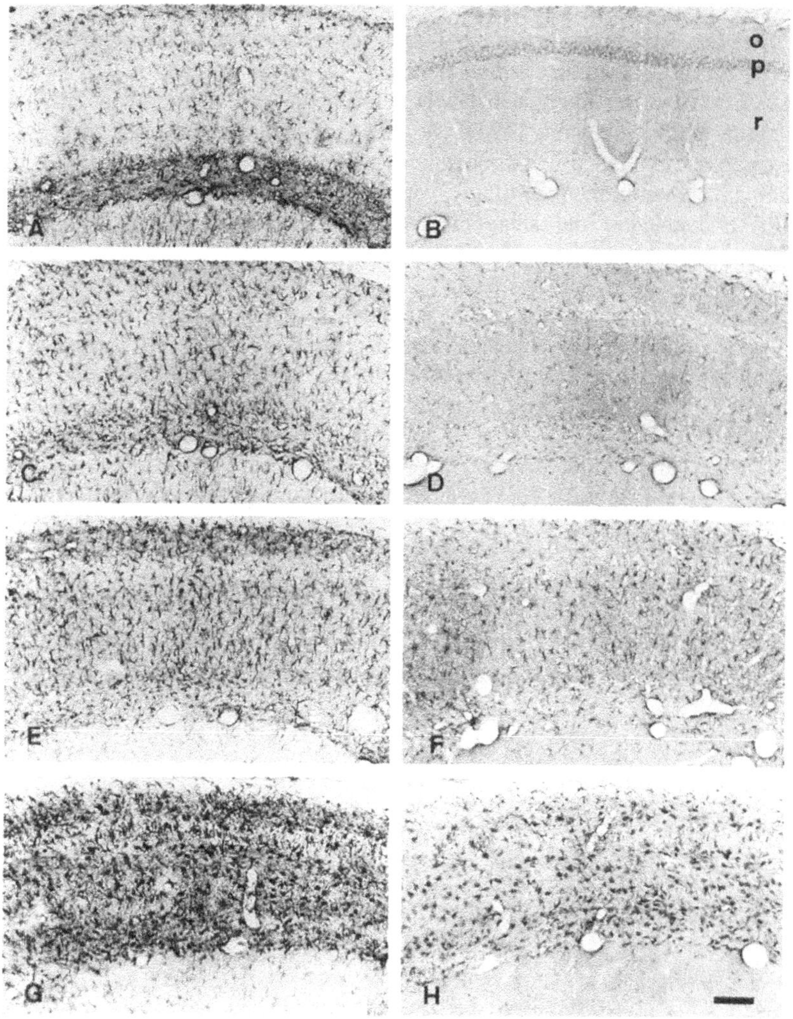

Figure 9-1. Immunolabeling in the CA1 region with antibodies to GFAP (left column) and NGF (right column) in control animals (A and B), and 3 (C and D), 7 (E and F) and 28 days after ischemia (G and H). Note the gradual increase in GFAP expression in all layers as compared to control animals as well as the increase of NGF expression in astrocytes with post-lesion time. o, stratum oriens; p, stratum pyramidale; r, stratum radiatum. Bar= 100 μm.

Cultures were derived from P7 rat optic nerves and grown in a chemically-defined medium (Barres et al., 1992). After 5 days, these cultures were highly enriched in oligodendrocytes (>95%) as assessed by antibodies to galactocerebroside C which were used for immunolabeling following a previously described procedure (Puchalski et al., 1994). The use of specific subunit antibodies indicated that the glutamate receptors expressed in the oligodendrocyte cultures consisted of AMPA subunits (GluR2-4), kainate subunits (GluR5-7, KA2) and the NMDA subunit NMDAR1. Because functional NMDA receptors are heteromeric in nature (Hollmann and Heinemann, 1994), we infer that the principal ionotropic glutamate receptors expressed in oligodendrocytes are those belonging to the AMPA and kainate subfamilies. Accordingly, we assayed the effects of their main agonists (AMPA and kainate) and antagonists (CNQX, 6-cyano-7-nitroquinoxaline-2,3-dione) on the viability of mature oligodendrocytes in vitro.

Cell viability and cell death was measured by counting cells which were labeled with fluorescein diacetate (FDA) and propidium iodide (PI), respectively. Application of AMPA (100 μM) resulted in an approximately 20% reduction in the number of viable cells which was evident only after 24 hours and which was antagonised by CNQX (fig. 9-2). In contrast, kainate, both at 10 μM and 100 μM, damaged up to about 30% and 60% respectively, of oligodendrocytes present in these cultures (fig. 9-2). This cytotoxic effect was prevented by using the non-NMDA antagonist CNQX (30 μM).

Figure 9-2. *Glutamate receptor-mediated toxicity in oligodendrocytes*

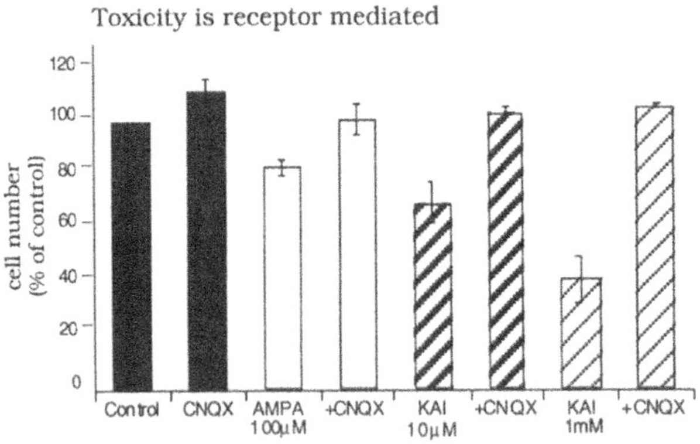

Effects of glutamate receptor agonists on the viability of oligodendrocytes in vitro. Optic nerve oligodendrocytes from P7-12 rat pups were cultured during 3-5 days on 12 mm diameter coverslips and then, agonists and/or antagonists were added to cultures. After an exposure of 24 h cells were labeled with FDA and PI and counted using a inverted microscope with a 20x objective. Ten fields/ coverslip were counted in four to six independent experiments. Averages ± SEM (n≥40) are shown.

The cytotoxic levels were not affected by cell density (data not shown), indicating that cytotoxicity is directly mediated by the receptor and not by substances released upon receptor activation. The fact that AMPA (100 μM) and low concentrations of kainate (10 μM) are toxic to oligodendrocytes suggests that the effects are partially mediated by both AMPA- and kainate- selective receptors (see Lerma et al., 1997). Intriguingly, 1mM glutamate, the endogenous agonist of glutamate receptors, did not affect oligodendroglial cell viability, an event which might be due to its possible instability as well as to its removal by excitatory amino acid transporters possibly present in oligodendrocytes.

These findings indicate that the ionotropic glutamate receptors present in oligodendrocytes negatively regulate the survival of these cells in vitro. Although we do not know whether this is the case in vivo, it is possible that glutamate released from Ranvier nodes following electrical activity in axons (Weinreich and Hammerschlag, 1975) may be involved in the shaping of the oligodendrocyte population in a fashion similar to that observed in vitro.

ANGIOTENSIN II RECEPTORS

Angiotensin II, the effector peptide of the renin-angiotensin system, is an important regulator of blood pressure and body fluid homeostasis. In the brain, stimulation of angiotensin II receptors leads to a complex pattern of cardiovascular and endocrine effects and to changes in behavior. Receptor autoradiography has shown relatively high levels of angiotensin II binding sites in various subcortical areas, such as the hypothalamic region, and also that both neuron and astrocyte cultures from newborn rat brain contain specific angiotensin II binding sites with distinct pharmacological properties (Summers et al, 1991).

Of the two families of angiotensin II receptors known, AT1 and AT2, we investigated expression and properties of AT1 receptors since it is the main receptor type expressed by glial cells (Summers et al., 1991). Part of the information mentioned below has been previously described in detail (Matute et al., 1994c).

Glial cells of the white matter express AT1 receptors in situ and in vitro

Northern blot analysis with an AT1 receptor probe corresponding to a gene previously isolated from the rat adrenal gland, revealed hybridization signals from several mRNA species of the adult bovine corpus callosum. The probe bound to bands of about 3.3, 2.0 and 1.0 kb, a fact that could reflect unique receptor subtypes or alternative splicing of the AT1 gene. The cells expressing the receptors were numerous and scattered throughout the extent of the corpus callosum.

In bovine and rat glial cultures, mRNA encoding angiotensin II receptors was detected in most cells. In addition, we observed that the pharmacology of the receptors expressed by cultured glial cells corresponded to that of AT1 receptors.

Thus, angiotensin II and Losartan competed strongly with 125I-angiotensin II specific binding with IC50 values of about 4 nM and 40 nM respectively. Angiotensin III was also a strong competitor (IC50 ≈ 10 nM), while angiotensin I and the AT2 receptor-selective ligand PD123319 displaced 125I-angiotensin II specific binding only at relatively high concentrations (IC50 > 10 μM).

After assessing the nature of the receptors, we proceeded to study the putative roles mediated by them.

Angiotensin II induces cell proliferation, aldosterone secretion and c-fos expression in astrocyte cultures

Incubation of cultured bovine astrocytes with angiotensin II induced a dose-dependent proliferative effect as measured by the incorporation of tritiated thymidine. The effect was observed at concentrations of angiotensin II as low as 10 nM and reached a maximum at 10 μM (EC50 ≈ 200 nM).

A similar effect was observed in U373 astrocytoma cells (fig. 9-3), a convenient cell line which we are currently using to delineate the molecular cascade triggered by AT1 receptor activation and resulting in proliferation.

Figure 9-3. *Mitogenic response to angiotensin II of U373 human glioma cells*

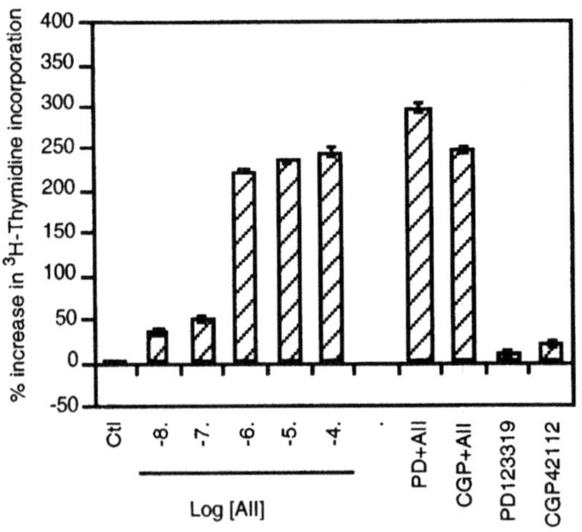

Angiotensin II promotes proliferation of the U373 human astrocytoma, an effect which is maximal at around 20 μM agonist. The response is not blocked by AT2 receptor antagonists PD123319 and CGP42112 (both at 20 μM), and therefore it appears to be mediated by AT1 angiotensin II receptors.

The effect of angiotensin II on the protein levels of the immediate early gene c-fos in cultured astrocytes was assayed by immunocytochemical and immunoblot analysis using an anti-synthetic peptide antibody. Under the culture conditions used in this study, serum-deprived astrocytes already expressed low amounts of c-fos protein but this was clearly increased after incubation with angiotensin II. Induction of c-fos was noticeable about 30 min after angiotensin II exposure and was still high 8 h later. Surprisingly, the proliferative effects of angiotensin II in astrocytes derived from c-fos knockout mice were comparable to those observed in wild type mice (fig. 9-4).

Figure 9-4. *Mitogenic response to angiotensin II of astrocytes from c-fos knockout mice*

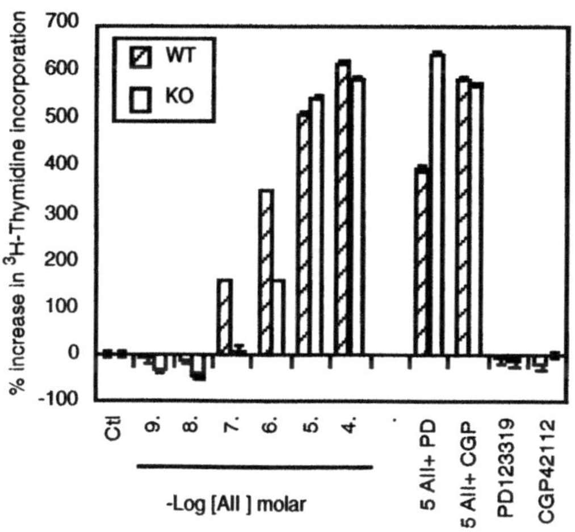

The mitogenic response to angiotensin II in astrocytes from *c-fos* knock out (KO) mice are similar to those observed in wild type mice (WT). The AT2 receptor antagonists CGP42112 and PD123319 had little effect on the angiotensin II response, suggesting the involvement of the AT1 receptor. In WT and KO mice EC_{50} values are approximately an order of magnitude different, suggesting that changes have occurred in signal transduction downstream of the receptor.

It is therefore, likely that in addition to c-fos, other transcription factors are also involved in the transactivation of the genes leading to astrocyte proliferation upon AT1 receptor activation.

It has been observed that several neurotransmitters and neuropeptides acting on glial cell receptors trigger the release of more neurotransmitters, peptides and cytokines (Martin, 1992; Levi and Gallo, 1995) via mechanisms not yet elucidated. We found that in astrocytes, angiotensin II induced the release of aldosterone, a mineralocorticoid essential for body fluid and electrolyte homeostasis

(Philips, 1987). Aldosterone has a major role in the maintenance of the sodium and potassium balance by promoting readsorption of the former and excretion of the latter. In blood, an increase in angiotensin II-induced aldosterone production can be inhibited by the AT1-selective antagonist Losartan but not by the AT2-selective antagonist PD123177 (Wong et al., 1990). Similarly, our findings indicate that aldosterone production in astrocytes is driven by AT1 receptors.

In summary, we have shown that astrocytes express AT1 receptors and provide evidence of several physiological functions mediated by those receptors in vitro. It will be important to elucidate the cascade of biochemical events linking AT1 receptor activation to such functions.

ATP RECEPTORS

Purines have recently been discovered to activate a plethora of membrane receptors. Pharmacological evidence has suggested the existence of multiple purinergic receptor subtypes based on their relative sensitivities to endogenous and synthetic adenosine derivatives.

P1 and P2 receptors were defined originally on the basis of their reactivity to adenosine and ATP respectively. Molecular cloning has led to the identification of two classes of P2 receptors: the P2X receptors consisting of ionotropic, ligand-gated ion channels (seven subtypes) and P2Y metabotropic receptors (seven subtypes) coupled to G-proteins (Burnstock and Wood, 1996). At present, these P2 receptor subclasses are pharmacologically distinguished by their agonist sensitivity profiles.

Thus, P2X: $\alpha\beta$ MeATP > ATP $\approx$ 2MeSATP
P2Y: 2MeSATP > ATP > αb MeATP

AMP-PNP is also used as an agonist to discriminate between P2X and P2Y receptors, having greater activity at P2Y receptors. Traditional P2 antagonists include: suramin (P2), PPADS (P2X) and Reactive Blue 2 (P2Y), (Dalziel and Wetfall, 1994).

Purinoceptors have been found to be expressed abundantly in the brain with characteristic distribution patterns. The most abundant P2X subtypes are X4 and X6. X1 and X2 mRNAs are present at low levels throughout the adult rat brain while in the neonate brain, they are somewhat more abundant, suggesting an important role in development (Kidd et al., 1995). The density of P2Y receptors in brain tissue is particularly high (Simon et al., 1995). Whereas P1 receptors have been found to play a regulatory role, modulating responses to classical neurotransmitters such as noradrenaline and acetylcholine, P2 receptors are coupled to the regulation of the level of intracellular calcium and to processes associated with cell growth and differentiation, both in development and in adulthood.

P2 purinoceptors are heterogeneously expressed in glial cells

The presence of P2 receptors in glial cells has been suggested on the basis of various responses of these cells to purinergic agonists. Thus it appears that cortical astrocytes express P2Y receptors both in vitro and in vivo, since the application of purinergic agonists has led to a variety of responses characteristic of G-protein coupled receptors (King et al., 1996). These include enhanced proliferation and morphological changes including stellation and an apparent up-regulation of GFAP expression (Neary and Zhu, 1994; Neary et al., 1996) which are characteristic of the gliotic process seen in vivo as a result of cerebral assaults such as trauma, ischemia, and some neurodegenerative diseases. Thus it is possible that the P2 receptor family may participate in regulating reactive gliosis and the outcome of neurological conditions in patients. More recently, attempts have been made to dissect the signal transduction pathway associated with these glial cell responses to ATP. In particular, ATP has been shown to act through G-protein coupled P2Y receptors to stimulate protein kinase C resulting in the activation of Mitogen-Activated Protein Kinase (Neary and Zhu, 1994). The latter kinase is a key molecule which can activate primary response genes such as fos and jun, which in their turn can activate the transcription of other genes or sets of genes such as GFAP, thus providing a molecular basis for nucleotide mediated gliosis. In addition, several lines of evidence point to the possibility that nucleotides may act as endogenous regulators of other glial cell types such as microglia, which have been demonstrated to express X7 (Surprenant et al., 1996) and oligodendroglia, by modulating calcium related biochemical processes (e.g., Kirischuk et al., 1995).

Due to the uncertainty about the exact receptor subtypes which may mediate these processes, we set out to characterize the expression pattern of several P2X and two P2Y subtypes in both normal and tumor glial cells. We used RT-PCR and subtype specific primers based on all available cloned sequences. The results are summarized in Table 9-3. It is immediately evident that combinations of P2X and P2Y receptors are expressed by normal glial cells both in vitro and in vivo. This contrasts with the findings of other authors who reported the absence of P2X transcripts in white matter employing in situ hybridization (Collo et al. 1996). This discrepancy is probably due to the greater sensitivity of the PCR technique. Interestingly, the loss of expression of P2X receptors seems to be associated with the tumor phenotype (see below). This rich expression of multiple P2X and P2Y receptors in glial cells suggests that ATP mediated events are more prevalent in these cells than previously thought and that these receptors may mediate many important physiological and pathological processes.

P2 purinoceptors regulate the proliferation of astrocytes and gliomas in vitro

We are currently comparing some of the long term trophic effects of P2 receptor activation in normal and tumor glial cells. Due to the fact that nucleotide molecules are rapidly hydrolyzed by ectonucleotidases present on cell membranes, into nucleosides and their metabolites, which can act on P1 receptors, we chose stable, non-hydrolyzable analogues to study the mitogenic effects of P2 stimulation. In normal cortical rat astrocytes, we found that P2Y activation by AMP-PNP (10μM) gave rise to a mitogenic response whereas P2X stimulation by αb MeATP (10μM) resulted in a partial inhibition of mitogenesis (-40%). This would suggest that P2X receptor stimulation may tend to inhibit proliferation since these cells express many of the P2X receptor subtypes examined.

Table 9-3. *P2 Receptor expression in glia*

Receptor subtype	X1	X2	X3	X4	X5	X6	X7	Y1	Y2	Y6
IN SITU										
Optic nerve	+	+	+	+	+	+	+	+	+	+
Rat Brain	+	+	+	+	+	+	+	+	+	+
IN VITRO										
NORMAL										
Astrocytes	+*	+*	+	+	+	+	+	+	+	
Microglia			+	+	+		+	+	+	
Oligodendrocytes	+*	+	+	+	+	+	+	+	+	
TUMOR										
C6					+	+		+	+	
U373								+		
SW1088								+	+	
T98				+				+	+	
U118				+				+	+	

* The amplified bands are multiplets suggesting the presence of alternative splice variants. Normal glial cells in vitro were obtained from primary cultures of rat cortical astrocytes. Oligodendrocytes in vitro were prepared from the optic nerves of P7 mice. C6 is a rat glioblastoma cell line; the other tumor glial cell lines are of human origin.

To simplify our studies, we chose to continue experiments with the human astrocytoma cell line U373, which expresses the Y1 subtype, in the absence of P2X receptors. We have observed a strong stimulation of proliferation induced with 20μM AMP-PNP, a preferential P2Y agonist (fig. 9-5). In contrast, the preferential P2X agonist αb Me ATP (20μM) did not exert any significant effect.

The antagonism of this response was somewhat unusual, with PPADS (20μM) acting as a partial antagonist and Reactive Blue (20μM) being ineffective.

Figure 9-5. *Mitogenic response to ATP receptor agonists of U373 human Glioma cells*

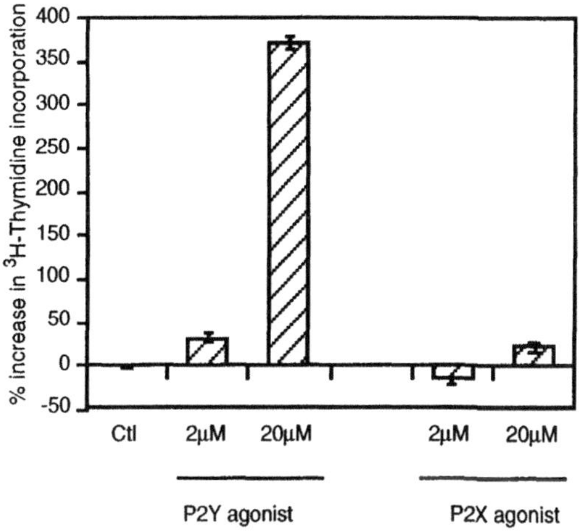

Mitogenic responses mediated by P2 purino-receptors in the U373 glioma cell line, are induced by the P2Y receptor agonist ß,γ-imido-adenosine 5′-triphosphate (AMP-PNP), but not by the P2X receptor agonist α,ß-methylene ATP (αß-Me ATP).

Thus this Y1 receptor shows a different pharmacological profile to that observed in other cell systems (Schachter et al., 1996). At present, we are examining the possibility that the ionotropic, physiological effects (eg. calcium entry) of P2X stimulation by ATP (the endogenous ligand) may modulate the trophic effects associated with simultaneous P2Y activation and that the absence of P2X effects leads to exaggerated trophic responses.

The reduced presence of P2X receptor subtypes in transformed glial cells (see Table 9-3) may be related to the tumorigenic process. If this is a general feature in gliomas, the development of P2Y receptor antagonists may have a great anti-tumor therapeutical value.

CONCLUSIONS

Glial cells of the central nervous system express a large variety of receptors to neurotransmitters and neuropeptides. It seems at present that most of the receptors to neuroactive substances, found in neurons are also present in glial cells. We have illustrated here that the receptors to glutamate, angiotensin II and ATP are present in most glial cell types including microglia and that they have a heterogeneous molecular nature. In addition, we have provided evidence indicating that these receptors may participate in shaping the size of the glial population and in generation of trophic activity after injury.

ACKNOWLEDGMENTS

This work was supported by the Gobierno Vasco (PI94-106), University of País Vasco (EC237/95) and Ministerio de Educación (PB94-1377).

REFERENCES

Barres, B.A., Hart, I.K., Coles, H.S.R, Burne, J.F., Voyvodic, J.T., Richardson, W.D., and M.C. Raff (1992) Cell death and control of cell survival in the oligodendrocyte lineage. Cell *70*:31-46.

Barres, B.A., and M.C. Raff (1996) Axonal control of oligodendrocyte development. In: Glial cell development. Basic principles and clinical revelance. Jessen, K.R. and Richardson, W.D. (eds.). BIOS Scientific Publishers Ltd, Oxford, UK, pp. 71-84.

Benveniste, H., Brejer, J., Schousboe, A., and N.H. Diemer (1984) Elevation of the extracellular concentrations of glutamate and aspartate in rat hippocampus during transient cerebral ischemia monitored by intracerebral microdialysis. J. Neurochem. *43*:1369-1374.

Benveniste, H., Jorgensen, M.B., Sandberg, M., Christensen, T., Hagberg, H., and N.H. Diemer (1989) Ischemic damage in hippocampal CA1 is dependent on glutamate release and intact innervation from CA3. J. Cereb. Blood Flow Metab. *9*:629-639.

Burnstock, G., and J.N. Wood (1996) Purinergic receptors: their role in nociception and primary afferent neurotransmission. Curr. Opin. Neurobiol. *6*:526-532.

Cheng, B., and M.P. Mattson (1991) NGF and bFGF protect rat hippocampal and human cortical neurons against hypoglycemic damage by stabilizing calcium homeostasis. Neuron *7*:1031-1041.

Collo, G., North, R.A., Kawashima, E., Merlo-Pich, E., Neidhart, S., and A. Surprenant (1996) Cloning of $P2X_5$ and $P2X_6$ receptors and the distribution and protperties of an extended family of ATP-gated ion channels. J. Neurosci. *16*:495-2507.

Dalziel, H.H., and D.P. Westfall (1994) Receptors for adenine nucleotides and nucleosides: subclassification, distribution, and molecular characterization. Pharmacological reviews. *46*:449-466.

Eddleston, M., and L. Mucke (1993) Molecular profile of reactive astrocytes- Implications for their role in neurologic disease. Neuroscience. *54*:15-36.

Fulton, B.P., Burne, J.F., and M.C. Raff. (1992) Visualization of O-2A progenitor cells in developing and adult rat optic nerve by quisqualate-stimulated cobalt uptake. J. Neurosci. *12*:4816-4833.

García-Barcina, J.M., and C. Matute (1996) Expression of kainate-selective glutamate receptor subunits in glial cells of the adult bovine white matter. European J. Neurosci. *8*:2379-2387.

Gottlieb, M., and C. Matute (1997) Expression of ionotropic glutamate receptor subunits in glial cells of the hippocampal CA1 area following transient forebrain ischemia. J. Cereb. Blood Flow Metab. *17*:290-300.

Hollmann, M., and S. Heinemann (1994) Cloned glutamate receptors. Annu. Rev. Neurosci. *17*:31-108.

Kidd, E.J., Grahames, C., Simon, J., Michel, A., Barnard, E., and P.P.A. Humphrey (1995) Localization of P2X purinoceptor transcripts in the rat nervous system. Mol. Pharmacol. *48*:569-573.

King, B.F., Neary, J.T., Zhu, Q., Wang, S., Norenberg, M.D., and G. Burnstock (1996) P2 purinoceptors in rat cortical astrocytes: expression, calcium-imaging and signalling studies. Neuroscience. *74*:1187-1196.

Kirischuk, S., Scherer, J., Kettenmann, H., and A. Verhratsky (1995) Activation of P2-purinoceptors triggers Ca^{2+} release from InsP3-sensitive internal stores in mammalian oligodendrocytes. J. Physiol. *483*:41-57.

Köhler, M., Burnashev, N., Sakmann, B., and Seeburg, P. H. (1993) Determinants of Ca2+ permeability in both TM1 and TM2 of high affinity kainate receptor channels: diversity by RNA editing. Neuron *10*:491-500.

Kreutzberg, G.W. (1996) Microglia: a sensor for pathological events in the CNS. Trends Neurosci. *19*:312-318.

Lerma, J., Morales, M., Vicente, M.A., amd O. Herreras (1997) Glutamate receptors of the kainate type and synaptic transmission. Trends Neurosci. *20*: 9-12.

Levi, G. and Gallo, V. (1995) Release of neuroactive amino acids from glia. In: Neuroglia. Kettenmann, H. and Ransom, B. R. (eds.), Oxford Univ. Press, New York, pp. 815-826.

Martin, D.L. (1992) Synthesis and release of neuroactive substances by glial cells. Glia. *5*:81-94.

Matute, C., García-Barcina, J.M., and R. Miledi (1994a) Expression of neurotransmitter receptors and Ca^{2+} channels in the adult fornix and optic nerve. NeuroReport. *5*:1457-1460.

Matute, C., Gutiérrez-Igarza, K., Río, C., and R. Miledi (1994b) Glutamate receptors in astrocytic end-feet. NeuroReport. *5*:1205-1208.

Matute, C., and R. Miledi (1993). Neurotransmitter receptors and voltage-dependent Ca2+ channels encoded by mRNA from the adult corpus callosum. Proc. Natl. Acad. Sci. USA. *90*:3270-3274.

Matute, C., Pulakat, L., Río, C., Valcárcel, C., and R. Miledi (1994c). Properties of angiotensin II receptors in glial cells from the adult corpus callosum. Proc. Natl. Acad. Sci. USA. *91*:3774-3778.

Morioka, T., Kalehua, A.N., and W.J. Streit (1991) The microglial reaction in the rat dorsal hippocampus following transient forebrain ischemia. J. Cereb. Blood Flow Metab. *11*:966-973.

Neary, J.T., and Q. Zhu (1994) Signalling by ATP receptors in astrocytes. NeuroReport. *5*: 1617-1620.

Neary, J.T., Rathbone, M.P., Cattabeni, F., Abbracchio, M.P., and G. Burnstock (1996) Trophic actions of extracellular nucleotides and nucleosides on glial and neuronal cells. Trends Neurosci. *19*:13-18.

Pechan, P.A., Chowdhury, K., Gerdes, W., and W. Seifert (1993) Glutamate induces the growth factors NGF, bFGF, the receptor FGF-R1 and c-fos mRNA expression in rat astrocyte culture. Neurosci. Lett. *153*:111-114.

Petito, C.K., Morgello, S., Felix, J.C., and M.L. Lesser (1990) The two patterns of reactive astrocytosis in postischemic rat brain. J. Cereb. Blood Flow Metab. *10*:850-859.

Petralia, R.S., and R.J. Wenthold (1992) Light and electron immunocytochemical localization of AMPA-selective glutamate receptors in the rat brain. J. Comp. Neurol. *318*:329-354.

Philips, M.I. (1987) Functions of angiotensin in the central nervous system. Annu. Review Physiol. *49*:413-435.

Puchalski, R.B., Louis, J.C., Brose, N., Traynelis, S.F., Egebjerg, J., Kukekov, V., Wenthold, R.J., Rogers, S.W., Lin, F., Moran, T., et al (1994) Selective RNA editing and subunit assembly of native glutamate receptors. Neuron.*13*:131-147.

Schachter, J.B., Li, Q., Boyer, J.L., Nicholas, R.A., and T.K. Harden (1996) Second messenger cascade specificity and pharmacological selectivity of the human $P2Y_1$-purinoceptor. Br. J. Pharmacol. *118*:167-173.

Seeburg, P. H. (1993) The molecular biology of mammalian glutamate receptor channels. Trends Neurosci. *16*:359-365.

Simon, J., Webb, T.E., and E.A. Barnard (1995) Characterization of a P2Y purinoceptor in the brain. Pharmacol. Toxicol. *76*:302-307.

Sommer, B., Keinänen, K., Verdoorn, T.A., Wisden, W., Burnashev, N., Herb, A., Köhler, M., Takagi, T., Sakmann, B. and P.H. Seeburg (1990) Flip and flop: a cell-specific functional switch in glutamate-operated channels of the CNS. Science. *249*:1580-1585.

Steinhäuser, C., and V. Gallo (1996) News on glutamate receptors in glial cells. Trends Neurosci. *19*:339-345.

Summers, C., Tang, W., Zelezna, B., and M.K. Raizada (1991) Angiotensin II receptor subtypes are coupled with distinct signal transduction mechanims in cultured neurons and astroglia from rat brain. Proc. Natl. Acad. Sci. USA. *88*:7567-7571.

Surprenant, A., Rassendren, F., Kawashima, E., North, RA., and G. Buell (1996) The cytolitic P2Z receptor for extracellular ATP identified as a P2X receptor ($P2X_7$). Science. *272*: 735-738.

Weinrich, D., and R. Hammerschlag (1975) Nerve impulse enhanced release of amino acids from non-synaptic regions of peripheral and central nerve trunks of bullfrog. Brain Res. *84*:137-142.

Wong, P.C., Hart, S.D., Zaspel, A.M., Chiu, A.T., Ardecky, R.J., Smith, R.D. and Timmermans, P.B. (1990) Nonpeptide angiotensin II receptor antagonists. VIII. Characterization of functional antagonism displayed DuP 753, an orally active antihypertensive agent. J. Pharm. Exp. Ther. *255*:584-592.

Zafra, F., Hengerer, B., Leibrock, J., Thoenen, H., and D. Lindholm (1990) Activity dependent regulation of BDNF and NGF mRNAs in the rat hippocampus is mediated by non-NMDA glutamate receptors. EMBO J. *9*:3545-3550.

10

ROLE OF ASTROGLIA IN THE NEURAL EFFECTS OF SEX HORMONES AND NEUROACTIVE STEROIDS

María del Carmen Fernández-Galaz, Julie Ann Chowen and Luis Miguel García-Segura

The development and function of the nervous tissue is affected by hormonal steroids produced by the adrenal glands and gonads and by endogenous steroids, known as neurosteroids, synthesized directly from cholesterol in the brain, mainly by glial cells. In this chapter we review the present knowledge about the involvement of astroglia in the effects of sex steroids and neurosteroids on the central nervous system. The roles play by astroglia in developmental and plastic effects of neuroactive steroids, including effects on neural tissue response to injury, are analyzed. Particular consideration is given to the role of astrocytes and tanycytes in synaptic plastic events induced by sex steroids in the hypothalamus. Current evidence indicating that tanycytes may be a source of growth factors that could act as neuroendocrine regulators is also discussed.

Circulating steroids produced by the adrenal glands and gonads affect the development, structure and function of the central nervous system. The mechanism of action of hormonal steroids in the brain involves their binding to specific intranuclear receptors that then act as transcription factors for specific genes (Beato, 1989). Furthermore, steroid hormones may modulate neuronal excitability by having rapid non-genomic actions (McEwen, 1991). Neural activity is also modulated by local steroids (neurosteroids) that are synthesized de novo in the brain from cholesterol (Baulieu and Robel, 1990). Neurosteroids affect neuronal excitability by modulating calcium channel currents and by acting on ionotrophic neurotransmitter receptors (Majewska, 1992).

It is now clearly established that in addition to their effects on neurons, circulating steroids and neurosteroids may affect brain function by acting on glial cells as well (Garcia-Segura et al., 1996). All types of glial cells, microglia, Schwann cells, oligodendroglia and astroglia, have been shown to be modulated by steroids (Garcia-Segura et al., 1996). Steroid hormones and neurosteroids affect myelination by acting on oligodendrocytes and Schwann cells and modulate the response of nerve tissue to pathological insults by acting on microglia and astroglia. In addition, glial cells participate in the metabolism of hormonal steroids and in the synthesis of neurosteroids (Garcia-Segura et al., 1996). Here we focus on a subtype of glial cells, the astroglia, and examine recent studies that indicate that these cells participate in plastic neural responses promoted by sex steroids. We also review evidence that these cells may mediate sex steroid effects on neurons through the release of growth factors.

ASTROGLIA ARE AFFECTED BY SEX HORMONES AND NEUROSTEROIDS

Astroglia in several brain areas are affected by sex hormones. For instance, administration of estradiol to adult ovariectomized rats, castration of newborn males and testosterone administration to newborn females result in significant changes in the distribution of the astrocytic marker glial fibrillary acidic protein (GFAP) in the striatum and hippocampus (Garcia-Segura et al., 1996), while castration of adult male rats results in elevated levels of GFAP mRNA in the hippocampus (Day et al., 1993). In adult females, the surface density of GFAP-immunoreactive cells in the hilus of the dentate gyrus fluctuates during the estrous cycle following the physiological variations in the circulatory levels of ovarian hormones, decreases after removal of circulating gonadal steroids by ovariectomy and increases after the pharmacological administration of either 17ß-estradiol or progesterone to ovariectomized animals (Luquín et al., 1993). In contrast to the significant modifications in the surface density of GFAP-immunoreactive cells, the number of GFAP-immunoreactive astrocytes is not affected by experimental manipulations of gonadal steroid levels and does not change during the estrous cycle (Luquín et al., 1993). This suggests that differences in the extent of branching of astrocytic processes are the most likely source for the observed modifications in the surface density of GFAP-immunoreactive cells. This is also supported by studies on hippocampal explant cultures showing that 17ß-estradiol and testosterone induce the growth of GFAP-immunoreactive processes (Del Cerro et al., 1995).

In these cultures it was also shown that neurosteroids affect astrocyte morphology: pregnenolone and its sulfated derivative induced the growth of GFAP-immunoreactive processes. Thus, the effect of this neurosteroid on hippocampal astroglia was qualitatively similar to that exerted by the gonadal hormones 17ß-estradiol and testosterone. In contrast, pregnenolone oleate was without effect on hippocampal astroglia. Another neurosteroid, dehydroepiandrosterone, did not affect the extension of GFAP-immunoreactive processes, but induced the formation of hypertrophied astrocytes, highly immunoreactive for GFAP and with an appearance similar to reactive astroglia. These findings indicate that astroglia are not only affected by peripheral steroid hormones, but also by endogenous steroids. This opens the possibility for interactions between the effects of circulating hormones and endogenous steroids on astroglia. For instance, progesterone is among the hormonal steroids that induce plastic astroglia changes and is also a neurosteroid produced by glial cells from cholesterol (Jung-Testas et al., 1989; Koenig et al., 1995). Progesterone and estradiol have complex interactions on astroglia in vivo when both steroids are administered simultaneously to ovariectomized rats. In the hippocampus an initial additive effect during the first 24 hours after steroid administration is followed by an antagonistic effect within 48 hours (Luquín et al., 1993). In contrast, in the hypothalamus progesterone completely blocks the effect of estradiol on astroglia (Garcia-Segura et al., 1994b).

The differing effects of progesterone on hippocampal and hypothalamic astroglia suggest a heterogeneity in the response of these cells to steroids. This is also supported by the fact that astroglia from the granular layer of the cerebellar cortex of adult female rats fail to respond to natural and experimental changes in gonadal steroids (Luquín et al., 1993). Furthermore, Day et al. (1993) have detected regional differences in the responses of astroglia to castration and sex hormone replacement in adult male rats. These regional differences in the effects of steroids on astroglia may reflect the existence of astroglia subpopulations with differing degrees of steroid sensitivity, may be the consequence of differences in neuron to glia interactions, or may result from possible differences in the local production of neurosteroids.

SEX HORMONES AND NEUROSTEROIDS MODULATE ASTROGLIA PLASTICITY AFTER BRAIN INJURY

Gonadal steroids and neurosteroids not only modulate physiological astroglia plasticity, but also affect astroglia responses in pathological conditions (Day et al., 1990; Garcia-Estrada et al., 1993). The effect of steroids on astroglia has been studied in rats after a penetrating injury of the cerebral cortex and hippocampus. Such a lesion results in the proliferation and hypertrophy of astrocytes in the proximity of the wound. Castration of rats of both sexes increases the proliferation of astrocytes after injury. In contrast, administration of high physiological levels of estradiol or progesterone to ovariectomized females, or the administration of testosterone to castrated males, decreases the proliferation and hypertrophy of astrocytes in the injured brain. This effect was observed either after systemic administration of the hormones or after their intracerebroventricular administration. Intracerebroventricular administration of neurosteroids, such as pregnenolone of dehydroepiandrosterone, also decreased astrocyte proliferation (Garcia-Estrada et al., 1993 and unpublished results).

The mechanisms involved in the effect of steroids on astroglia proliferation after brain injury are yet unknown. Recent studies on cultured hippocampal slices (Del Cerro et al., 1996) have revealed that the response of astrocytes to neurosteroids and gonadal hormones is different depending on the extracellular concentration of K^+. Since extracellular K^+ levels rise in the brain parenchyma after injury, this may be one of the reasons why androgens affect astroglia proliferation after injury, but do not affect astroglia proliferation during normal physiological conditions, at least during hypothalamic development (Garcia-Segura et al., 1995b).

ASTROGLIA ARE INVOLVED IN SYNAPTIC PLASTICITY INDUCED BY SEX STEROIDS AND IN THE SEXUAL DIFFERENTIATION OF SYNAPTIC CONNECTIVITY

Cellular mechanisms mediating steroid actions on synapses and glial cells have been extensively studied in the rat arcuate nucleus, a hypothalamic center involved in the feed-back regulation of gonadotropins. Perinatal androgens, probably after its conversion to estrogens in the brain, induce the sexual differentiation of synaptic connectivity in the arcuate nucleus (Garcia-Segura et al., 1995b; Matsumoto, 1991). Furthermore, in postpubertal females, there is an estrogen-induced transient disconnection of inhibitory GABAergic inputs to the somas of arcuate neurons during the preovulatory and ovulatory phases of the estrous cycle (Garcia-Segura et al., 1994a). This synaptic remodeling is blocked by progesterone and begins with the onset of female puberty (Garcia-Segura et al., 1994a). Astrocytes appear to play a significant role in these synaptic changes (Garcia-Segura et al., 1994a, b). In the afternoon of proestrus, estradiol induces synthesis of GFAP (Garcia-Segura et al., 1994a,b; Kohama et al., 1995) and the growth of glial processes which ensheathe the neuronal membrane and displace synaptic terminals. Glial processes retract and synapses reform in the afternoon of estrus (Garcia-Segura et al., 1994a,b). Similar changes are observed after the administration of estradiol to ovariectomized rats. Progesterone, in contrast, blocks the effect of estradiol on glial cells and synaptic plasticity (Garcia-Segura et al., 1994a,b).

Astrocytes also may be involved in the organizational effects of sex steroids on brain synaptic connectivity (Chowen et al., 1995; Garcia-Segura et al., 1995b). Perinatal steroids appear to regulate the number of synaptic inputs to arcuate neurons by affecting the expression of cytoskeletal components of astrocytes and the amount of neuronal membrane covered by glial processes and, therefore, the amount of membrane available for the formation of synaptic contacts (Chowen et al., 1995; Garcia-Segura et al., 1995b).

NEURON TO ASTROCYTE SIGNALING MEDIATED BY ADHESION MOLECULES MAY BE IMPLICATED IN THE EFFECT OF STEROIDS

The signaling mechanisms involved in the steroid effects on neurons and glia have been investigated in hypothalamic cultures (Garcia-Segura et al., 1994a, 1995a). Estradiol induces the growth of GFAP immunoreactive processes in astrocytes in serum-free mixed neuronal-glial primary rat hypothalamic cultures. This effect is detected as early as 30 minutes after the addition of the hormone to the cultures, is dose-dependent, reversible and specific to hypothalamic cells.

The estradiol-induced growth of astrocyte processes in vitro depends on the presence of neurons since it is not observed in pure glial cultures. Moreover, this effect is dependent on pre-existing direct membrane to membrane contact between living neurons and glia and appears to involve specific hypothalamic neuro-glia interactions, since it is not observed in co-cultures of hypothalamic neurons with astrocytes from other brain areas (Garcia-Segura et al., 1994a, 1995a). Furthermore, the expression of a specific form of the neural cell adhesion molecule (N-CAM) on the neuronal membranes is essential for the plastic glial changes induced by estradiol.

N-CAM exists in a variety of isoforms differing in the length of their cytoplasmic domain and/or their carbohydrate content. In embryonic brain, N-CAM is enriched in polysialic acid (PSA). This PSA-N-CAM isoform is gradually replaced in most brain areas during the perinatal and early postnatal periods by less polysialylated N-CAM isoforms. However, PSA-N-CAM expression persists in areas of the adult brain that can undergo neuronal-glial changes, such as the hypothalamo-neurohypophysial system, arcuate nucleus and median eminence. In hypothalamic monolayer cultures, PSA-N-CAM is expressed on the surface of neurons, while astrocytes are not PSA-immunoreactive (Garcia-Segura et al., 1994a, 1995a).

Enzymatic removal of PSA from neuronal surfaces prevents the effect of estradiol on astrocyte shape, but not the effect of other signals such as fibroblast growth factor which act directly on astrocytes (Garcia-Segura 1994a, 1995a). This indicates that PSA-N-CAM on the neuronal surface is necessary for the manifestation of the hormonal effect on astrocytes and suggests that a neuron to glia signaling pathway, mediated by adhesion molecules, participates in the plastic effects of gonadal hormones. The role of PSA-N-CAM on astrocyte shape changes may be related, in part, to modulation of cell adhesion properties. PSA can exert broad steric effects at the cell surface and may modulate not only the homophilic binding of N-CAM to other N-CAM molecules, but the interaction of other cell surface molecules as well. In addition, the role of PSA may be related to the activation of second messenger pathways (Garcia-Segura, 1995a).

TROPHIC FACTORS RELEASED BY ASTROGLIA MAY BE INVOLVED IN THE EFFECTS OF SEX HORMONES

In addition to direct neuron to glia signaling mediated by adhesion molecules, soluble factors may also be involved in steroid effects on brain cells. Glial cells may release different growth factors that could alter neuronal function and that may be involved in the regulation of synaptic connectivity. For instance, astroglia are a source of trophic factors, such as transforming growth factor (TGF) α and β and insulin-like growth factor I (IGF-I), substances that may mediate some of the neuroendocrine effects of gonadal hormones (Hiney et al., 1996; Ma et al., 1992; Melcangi et al., 1995; Ojeda et al., 1990).

IGF-I is involved in the feedback regulation of growth hormone by affecting the synthesis or the release of growth hormone-releasing hormone and somatostatin by hypothalamic neurons (Berelowitz et al., 1991; Tannenbaum et al., 1983). IGF-I may also affect the reproductive axis by modulating the secretion of gonadotrophin releasing hormone by hypothalamic cells and, therefore, the release of gonadotrophins (Hiney et al., 1996).

Several studies suggest an interaction between IGF-I and gonadal steroids and an active participation of glial cells in this interaction. Tanycytes, a specific form of astroglia present in the arcuate nucleus and median eminence, are IGF-I immunoreactive (Dueñas et al., 1994). IGF-I immunoreactive levels in tanycytes of the arcuate nucleus show sex differences, with adult females showing significantly lower IGF-I levels than males of the same age. This sex difference is abolished by early postnatal androgenization of females (Dueñas et al., 1994), suggesting that it may be dependent on the perinatal burst of androgen production by the testis of developing male rats. Furthermore, IGF-I levels in tanycytes increase in both male and female rats at the time of puberty. Females show an abrupt increase in IGF-I immunoreactive levels in tanycytes between the morning and afternoon of the first proestrus. Henceforth, IGF-I immunoreactivity fluctuates according to the different stages of the estrus cycle. IGF-I immunoreactive levels are high in the afternoon of proestrus, after the peak of estrogen in plasma, remain increased on the morning of the following day and then decrease to basal conditions by the morning of metestrus (Dueñas et al., 1994). In addition, IGF-I levels decrease in tanycytes when gonadal steroid levels are reduced by ovariectomy and increase in a dose-dependent manner when ovariectomized rats are injected with 17ß-estradiol (Dueñas et al., 1994).

Tanycytes are a specialized form of ependymocyte whose function is still unclear. These cells are considered to be phylogenetical precursors of astrocytes (Suárez et al., 1995). The ultrastructural and immunological (expression of GFAP and vimentin) characteristics of tanycytes correspond to those of astroglia. For many years it has been proposed that tanycytes in the arcuate nucleus and median eminence may be involved in endocrine regulation (Garcia-Segura et al., 1996; Ma et al., 1992; McQueen, 1994) and may accumulate substances from the cerebrospinal fluid or from the capillaries of the median eminence. In agreement with this possibility, previous studies have shown that tanycytes are able to take up ß-endorphin from the cerebrospinal fluid (Bjelke and Fuxe, 1993). Furthermore, tanycyte processes are closely associated with neurosecretory terminals in the median eminence (Kozolowski and Coates, 1985; Oota et al., 1974) regulating neuronal contact with portal capillaries (King and Letourneau, 1994; King and Rubin, 1994; Kozolowski and Coates, 1985; Meister et al., 1988) and may also result in the focal concentration of factors, such as IGF-I, that may affect neurohormone release. In addition to IGF-I, other factors may be involved in the endocrine effects of tanycytes. Ma et al. (1992) have reported a marked increase in TGF-α gene expression in tanycytes during the initiation of puberty in the rat. TGF-α mRNA levels increase gradually after the anestrous phase of puberty, reaching peak values on the afternoon of the first proestrus.

An interesting question in regard to the hormone-induced changes in IGF-I immunoreactivity in tanycytes is the source of IGF-I. Since tanycytes do not express mRNA for IGF-I (Dueñas et al., 1994), the modifications in IGF-I levels may be the result of the hormonal modulation of its uptake from extrahypothalamic areas.

To test this possibility, IGF-I was labeled with digoxigenin and injected in the lateral cerebral ventricle. The labeled IGF-I was accumulated by tanycytes (Fernandez-Galaz et al., 1996). This accumulation was specific since it was substantially decreased by the administration of unlabeled IGF-I or unlabeled insulin, which also acts on IGF-I receptors, and was blocked by a specific IGF-I receptor antagonist (Fernandez-Galaz et al., 1996).

In 1974 Brawer et al. reported ultrastructural modifications in the ventricular surface of tanycytes in the arcuate nucleus of female rats at different stages of the estrous cycle. The number of microvilli increases during proestrus, remains elevated during estrus and decreases in metestrus. The significance of these changes is still unknown, but may well be related to the uptake of substances from the cerebrospinal fluid.

By using colloidal gold ultrastructural immunolocalization techniques we have recently observed that the IGF-I receptor is enriched in the microvilli of tanycytes (Garcia-Segura et al., unpublished). Therefore, changes in the extension of microvilli may be related to changes in the number of IGF-I receptors exposed in the lumen of the third ventricle and this, in turn, may influence the uptake of IGF-I by tanycytes. In order to test whether IGF-I uptake by tanycytes fluctuates during the different stages of the estrous cycle, we injected IGF-I labeled with digoxigenin into the lateral cerebral ventricle of cycling female rats. Animals were killed 1 hour after injection and the number of digoxigenin labeled tanycytes was counted (Fernandez-Galaz et al., 1996). The number of tanycytes labeled with digoxigenin showed prominent changes according to the different phases of the estrous cycle. Compared with diestrus and metestrus, the number of labeled tanycytes showed a significant decrease in the afternoon of proestrus.

A recovery in IGF-I accumulation was observed by the morning of estrus, and this was followed by a significant increase by the afternoon of estrus. The decrease in the number of tanycytes that accumulate digoxigenin-labeled IGF-I in the afternoon of proestrus was unexpected, since at this stage of the estrous cycle there is a significant increase in IGF-I immunoreactivity in these cells (Dueñas et al., 1994). This apparent discrepancy may reflect the competition between exogenous IGF-I and increased endogenous IGF-I levels in the afternoon of proestrus.

Further studies are necessary to determine the cause of the differences observed in the number of tanycytes that accumulate IGF-I after its intracerebroventricular injection. Nevertheless, these findings suggest that tanycytes may play a key role in the neural effects of gonadal steroids by regulating the availability of IGF-I to hypothalamic neurons.

CONCLUSIONS

In summary, the data examined in this chapter indicate that glial cells participate in mediating the effects of sex steroids and neurosteroids in the central nervous system under physiological and pathological conditions. Gonadal steroids and neurosteroids promote astroglia plasticity in several areas of the central nervous system, including the hypothalamus, the striatum and the hippocampus and modulate astroglia proliferation and the formation of reactive astroglia after brain injury. Under physiological conditions, the effects of steroids on astroglia may be involved in the sexual differentiation of synaptic connectivity and in plastic synaptic events during adult life. Recent evidence indicates that the effects of neuroactive steroids in brain may be mediated in part by growth factors, such as IGF-I. Astroglia also may play a crucial role by regulating the availability of such factors to neurons.

ACKNOWLEDGMENTS

This work was supported by Fundación Ramón Areces, DGICYT, and Fundación Endocrinología y Nutrición.

REFERENCES

Baulieu, E.E. and P. Robel (1990) Neurosteroids: a new brain function? J. Steroid Biochem. Mol. Biol. *37*:395-403

Beato, M. (1989) Gene regulation by steroid hormones. Cell 56:335-344

Berelowitz, M., M. Szabo, L.A. Frohman, S. Firestone, L. Chu, and R.L. Hintz (1981) Somatomedin C mediates growth hormone negative feedback by effects on both the hypothalamus and the pituitary. Science *212*:1279-1281

Bjelke, B., and K. Fuxe (1993) Intraventricular ß-endorphin accumulates in DARPP-32 immunoreactive tanycytes. NeuroReport *5*:265-268

Brawer, J.R., P.S. Lin, and C. Sonnenschein (1974) Morphological plasticity in the wall of the third ventricle during the estrous cycle in the rat: a scanning electron microscopic study. Anat. Rec. *179*:481-489

Chowen, J.A., S. Busiguina, and L.M. Garcia-Segura (1995) Sexual dimorphism and sex steroid modulation of glial fibrillary acidic protein messenger RNA and immunoreactive levels in the rat hypothalamus. Neuroscience *69*:519-532

Day, J.R., N.J. Laping, T.H. McNeil, S.S. Schreiber, G. Pasinetti, and C.E. Finch (1990) Castration enhances expression of glial fibrillary acidic protein and sulfated glycoprotein-2 in the intact and lesion-altered hippocampus of the adult male rat. Mol. Endocrinol. *4*:1995-2002

Day, J.R., N.J. Laping, M. Lampert-Etchells, S.A. Brown, J.P.O'Callaghan, T.H. McNeill, and C.E. Finch (1993) Gonadal steroids regulate the expression of glial fibrillary acidic protein in the adult male rat hippocampus. Neuroscience *55*:435-443

Del Cerro, S., J. Garcia-Estrada, and L.M. Garcia-Segura (1995) Neuroactive steroids regulate astroglia morphology in hippocampal cultures from adult rats. Glia *14*:65-71

Del Cerro, S., J. Garcia-Estrada, and L.M. Garcia-Segura (1996) Neurosteroids modulate the reaction of astroglia to high extracellular potassium levels. Glia *18*:293-305

Dueñas, M., S. Luquin, J.A. Chowen, I. Torres-Aleman, F. Naftolin, and L.M. Garcia-Segura (1994) Gonadal hormone regulation of insulin-like growth factor-I-like immunoreactivity in hypothalamic astroglia of developing and adult rats. Neuroendocrinology *59*:528-538

Fernandez-Galaz, M.C., I. Torres-Aleman, and L.M. Garcia-Segura (1996) Endocrine-dependent accumulation of IGF-I by hypothalamic glia. Neuroreport *8*:373-377.

Garcia-Estrada, J., J.A. Del Rio, S. Luquin, E. Soriano, and L.M. Garcia-Segura (1993) Gonadal hormones down-regulate reactive gliosis and astrocyte proliferation after a penetrating brain injury. Brain Res. *628*:271-278

Garcia-Segura, L.M., J.A. Chowen, A. Parducz, and F. Naftolin (1994a) Gonadal hormones as promoters of structural synaptic plasticity: cellular mechanisms. Prog. Neurobiol. *44*: 279-307

Garcia-Segura, L.M., S. Luquin, A. Parducz, and F. Naftolin (1994b) Gonadal hormone regulation of glial fibrillary acidic protein immunoreactivity and glial ultrastructure in the rat neuroendocrine hypothalamus. Glia *10*:59-69

Garcia-Segura, L.M., B. Cañas, A. Parducz, G. Rougon, D. Theodosis, F. Naftolin, and I. Torres-Aleman (1995a) Estradiol promotion of changes in the morphology of astroglia growing in culture depends on the expression of polysialic acid on neuronal membranes. Glia *13*:209-216

Garcia-Segura, L.M., M. Dueñas, S. Busiguina, F. Naftolin, and J.A. Chowen (1995b) Gonadal hormone regulation of neuronal-glial interactions in the developing neuroendocrine hypothalamus. J Steroid Biochem. Molec. Biol. *53*:293-298

Garcia-Segura, L.M., J.A. Chowen, and F. Naftolin (1996) Endocrine glia: roles of glial cells in the brain actions of steroid and thyroid hormones and in the regulation of hormone secretion. Front. Neuroendocrinol. *17*:180-211

Hiney, J.K., V. Srivasta, C.L. Nyberg, S.R. Ojeda, and W. Les Dees (1996) Insulin-like growth factor I of peripheral origin acts centrally to accelerate the initiation of female puberty. Endocrinology *137*: 3717-3728

Jung-Testas, I., Z.Y. Hu, E.E. Baulieu, and P. Robel (1989) Neurosteroids: Biosynthesis of pregnenolone and progesterone in primary cultures of rat glial cells. Endocrinology *125*:2083-2091

King, J.C., and R.J. Letourneau (1994) Luteinizing hormone-releasing hormone terminals in the median eminence of rats undergo dramatic changes after gonadectomy, as revealed by electron microscopic image analysis. Endocrinology *134*:1340-1351.

King, J.C., and B.S. Rubin (1994) Dynamic changes in LHRH neurovascular terminals with various endocrine conditions in adults. Horm. Behav. *28*:349-356

Koenig, H.L., M. Schumacher, B. Ferzaz, A.N. Do Thi, A. Ressouches, R. Guennoun, I. Jung-Testas, P. Robel, Y. Akwa, and E.E. Baulieu (1995) Progesterone synthesis and myelin formation by Schwann cells. Science *268*:1500-1503

Kohama, S.G., J.R. Goss, T.H. McNeill, and C.E. Finch (1995) Glial fibrillary acidic protein mRNA increases at proestrus in the arcuate nucleus of mice. Neurosci. Lett. *183*:164-166

Kozolowski, G,P., and P.W. Coates (1985) Ependymoneuronal specializations between LHRH fibers and cells of the cerebroventricular system. Cell. Tiss. Res. *242*:301-311

Luquin, S., F. Naftolin, and L.M. Garcia-Segura (1993) Natural fluctuation and gonadal hormone regulation of astrocyte immunoreactivity in dentate gyrus. J. Neurobiol. *24*:913-924

Ma, Y.J., M.P. Junier, M.E. Costa, and S.R. Ojeda (1992) Transforming growth factor-α gene expression in the hypothalamus is developmentally regulated and linked to sexual maturation. Neuron *9*: 657-670

Majewska, M.D. (1992) Neurosteroids: endogenous bimodal modulators of the $GABA_A$ receptor. Mechanism of action and physiological significance. Prog. Neurobiol. *38*:379-395

Matsumoto, A. (1991) Synaptogenic action of sex steroids in developing and adult neuroendocrine brain. Psychoneuroendocrinology *16*: 25-40

McEwen, B.S. (1991) Non-genomic and genomic effects of steroids on neural activity. Trends Pharmacol. Sci. *12*:141-147

McQueen, J.K. (1994) Glial cells and neuroendocrine function. J. Endocrinol. *143*:411-415

Meister, B., T. Hökfelt, Y. Tsuruo, H. Hemmings, C. Ouimet, P. Greengard, and M. Goldstein (1988) DARP-32, a dopamine- and cyclic AMP-regulated phosphoprotein in tanycytes of the mediobasal hypothalamus: distribution and relation to dopamine and luteinizing hormone-releasing hormone neurons and other glial elements. Neuroscience *27*:607-622

Melcangi, R.C., M. Galbiati, E. Messi, F. Piva, L. Martini, and M. Motta (1995) Type 1 astrocytes influence luteinizing hormone-releasing hormone release from the hypothalamic cell line GT1-1: is transforming growth factor ß the principle involved? Endocrinology *136*:679-686

Ojeda, S.R., H.F. Urbanski, M.E. Costa, D.F. Hill, and M. Moholt-Siebertdd (1990) Involvement of transforming growth factor α in the release of luteinizing-hormone releasing hormone from the developing female rat hypothalamus. Proc. Natl. Acad. Sci. USA *87*:9698-9702

Oota, Y., H. Kobayashi, R.S. Nishioka, and H.A. Bern (1974) Relationship between neurosecretory axon and ependymal terminals on capillary walls in the median eminence of several vertebrates. Neuroendocrinology *16*:127-136

Sohrabji, F., L.A. Greene, R.C. Miranda, and D. Toran-Allerand (1994a) Reciprocal regulation of estrogen and NGF receptors by their ligands in PC12 cells. J. Neurobiol. *25*:974-988

Sohrabji, F., C. Miranda, and C.D. Toran-Allerand (1994b) Estrogen differentially regulates estrogen and nerve growth factor receptor mRNA in adult sensory neurons. J. Neurosci. *14*:459-471

Suárez, I., G. Bodega, M. Rubio, and B. Fernández (1995) Evolution of astrocytes in the vertebrate CNS. In A. Vernadakis and B. Roots (eds): Neuron-glia interactions during phylogeny. I. Phylogeny and ontogeny of glial cells. Totowa, N.J.: Humana Press, pp. 41-57

Tannenbaum, G.S., H.J. Guyda, and B.I. Posner (1983) Insulin-like growth factors: A role in growth hormone negative feedback and body weight regulation via brain. Science *220*:77-79

11

CYTOKINES IN ASTROGLIAL CELLS: FUNCTIONS AND MECHANISMS OF ACTION

Francisco Molina-Holgado, Eduardo Molina-Holgado, Alberto Lledó and Carmen Guaza

The pathological mechanisms that cause central nervous system (CNS) dysfunction in most neurological diseases are not well known. In recent years, there has been increasing evidence that soluble mediators such as cytokines from activated T-lymphocytes and macrophages are able to modulate the function of cells within the CNS. Glial cells upon activation can secrete immunoregulatory factors that influence lymphoid/mononuclear cells as well as the glial cells themselves. Thus, the potential exists for bi-directional communication between lymphoid cells and glial cells within the CNS, which in part is mediated via cytokines. In neurological disorders with an immunological component, glial cells, microglia and astrocytes, may contribute to inflammation and immune events occurring in the CNS. The cytokines in the CNS mainly implicated in contributing to inflammatory and immune responses include interleukin-1 (IL-1α and β), interleukin-6 (IL-6), tumor necrosis factor-α (TNF-α), and interferon γ (IFN-γ). We review here the production of immunoinflammatory mediators like nitric oxide and prostaglandins as well as cytokines such as IL-6 and TNF-α by cultured astrocytes upon IL-1α or β stimulation. Emphasis will be drawn on mechanisms of action of IL-1 on astrocytes, showing that PKC, isoform-α, is involved on the signal transduction pathways mediating astrocyte PGE_2 production.

Changes in astrocyte function occur in response to a variety of neuropathological processes, including trauma, degeneration, infection or inflammation. The precise molecular signals involved in astrocyte reactivity remain unclear, but evidence derived mainly from *in vitro* studies suggests that cytokines may play a critical role. Cytokines are polypeptidic soluble factors that control the growth and differentiation of cells involved in immune reactivity and are pleitropic, complementary and counteractive. Moreover, production of several cytokines is controlled by other cytokines in stimulatory or inhibitory manners and in some cases, acts as a cascade (Tadamitsu et al.,1994). These complex cytokine actions are referred as cytokine network. Microglia, macrophage like cells in the brain and astrocytes are the most important cells in the CNS cytokine network (Sawada et al.,1994; Fabry et al.,1994). Astrocytes are the most numerous of the glial cells, and in the adult nervous system, astrocytes, unlike neurones, retain the ability to divide and multiply, and not only when the CNS is injured, but also in neurological alterations and neurodegenerative processes, astrocytes respond by becoming reactive (see Eddelston and Mucke, 1993).

ROLE OF GLIAL CELLS IN CNS CYTOKINE NETWORK

Studies on the effects of immune cell-derived cytokines on glial cells were initially directed toward investigating whether factors from immune cells may be contributing to astrocytic activation seen in the region of inflammatory infiltrates in diseases such as MS (Fontana et al., 1982). Cytokines, like interleukin-1 (IL-1), interleukin-6 (IL-6), tumor necrosis factor (TNF-α) and interferon γ (IFN-γ) are present in the CNS during neurological and neurodegenerative disorders (see Benveniste, 1992). TNF-α is localized within astrocytes in the multiple sclerosis plaque (Hoffman et al.,1989; Selmaj et al.,1991). IL-1 is present also in brain from patients with Down's syndrome in Alzheimer 's disease and appears to be localized in microglia (Griffin et al.,1995). IFN-γ and IL-6 has been detected in cerebrospinal fluid during viral meningitis and encephalitis (Frei et al., 1989) and increased cerebrospinal fluid levels of IL-6 and TNF-α have been documented in patients with MS and other inflammatory neurological diseases. Local production of cytokines within the CNS may come from not only infiltrating mononuclear such as T cells and macrophages but also by endogenous glial cells such as the astrocytes and microglia. It is currently a matter of debate whether elevated cytokine levels are the consequence or the reason of the pathological alterations of the above neurological disorders.

Interleukin-1 is a pro-inflammatory cytokine produced in response to injury in the periphery and in the CNS (Rothwell and Stribjos, 1995; Rothwell and Hopkins, 1995). Although the exact origin of IL-1 in the CNS remains unclear, microglia is the major source of this cytokine (Giulian et al.,1986; Nieto Sampedro, 1987) and astrocytes have been considered to be both, a target for and a possible source of IL-1. Interleukin-1 is known to induce astrocytes to express other cytokines, such IL-6 and TNF-α as well as other inflammatory mediating substances that may play critical roles in the response to injury (Aloisi et al.,1992; Norris et al.,1994; Lafortune, 1996). There are two forms of IL-1, IL-1α and IL-1ß which are the products of different genes, and the IL-1 receptor antagonist (IL-1RA) which acts as a true receptor antagonist. The two forms of IL-1 agonists have 29% structural homology to one another, but they bind to the same surface receptor and have essentially identical biological activities. Two receptors for IL-1 have been described, named type I and type II, but IL-1R type II acts as a decoy receptor modulating IL-1 levels extracellularly. Features of the IL-1 receptor included its high affinity for IL-1 and the relatively low number of receptors per cell

(Dinarello, 1994). Interestingly, mouse astrocytes have been described to present functional IL-1 receptors (IL-1R) in their cytoplasmic membrane (Ban et al., 1993).

INDUCTION OF NITRIC OXIDE IN GLIAL CELLS

Nitric oxide (NO), a free radical gas synthesized by the enzyme nitric oxyde synthase (NOS) in a variety of tissues including CNS, participates in multiple physiological and pathophysiological processes (Moncada et al., 1991; Hooper et al., 1995).

To date, three distinct forms of NOS have been described. The constitutive rat and human brain forms (NOSI) and the bovine and human constitutive endothelial cell forms (NOSIII) share a 60% aminoacid sequence homology being both calcium and caldmoduline dependent (Bredt and Snyder, 1990). The inducible form of the enzyme, NOSII, binds calmodulin with very high affinity and has a minimal requirement of calcium (Nathan and Xie, 1994). Several reports postulate that NO under determined conditions may play a role in neurotoxicity (Dawson et al., 1991; Merrill et al., 1993; Chao et al., 1996), Elevated levels of nitrites/nitrates have been reported to occur in CSF of patients with several neurological disorders (Milstein et al., 1994) and mRNA for iNOS is upregulated in the brain during infection with rabies virus and borna disease virus (Koprowski et al.,1993; Zheng et al.,1993; Hooper et al.,1995; Van Dam et al., 1995). Although there is strong evidence for the induction of iNOS in astrocytes after lipopolyssacharide (LPS) or cytokine stimulation, mainly IL-1 and IFNγ plus IL-1 (Galea et al., 1992; Simmons and Murphy, 1992), several authors claimed for microglia as the major source of NO following brain injury and inflammation (Boje et al., 1992; Vincent et al., 1996).

EICOSANOIDS AS INFLAMMATORY MEDIATORS IN THE CNS

Other potent inflammatory agents like prostaglandins (PG) are produced by astrocytes in response to IL-1 and LPS and a link between NO production and cyclooxygenase activation has been reported not only in peripheral macrophages but also in astrocytes (Salvemini et al., 1993; Molina-Holgado et al., 1995). In pathological situations PG production increases dramatically. Elevated levels of PG have been observed in multiple sclerosis and neurodegenerative processes (Fretland, 1992; Griffin et al., 1994). Some information is available concerning the ability of IL-1 to stimulate eicosanoids production by astrocyte cultures (Hartung et al., 1989; Katsuura et al.,1989) but the nature of the intracellular signalling mechanisms responsible for IL-1 induced PGE_2 production in astrocytes remain unclear at this time. In fact, there is increasing interest in delineating the signal transduction pathways mediating astrocyte IL-1 actions. So far, several studies have suggested that expression of both IL-1ß-induced IL-6 and TNF-α genes in astrocytes is mediated by protein kinase C (Bethea et al., 1992; Norris et al., 1994), whereas protein kinase A is not involved. However, there are other second messenger systems as well as transcription factors such as nuclear factor-kß (Sparacio et al., 1992) and AP-1 (Daffada et al., 1994) that might be considered as mediators of IL-1 effects (O'Neill, 1995).

In this review about actions of cytokines on astroglial cells, we will present our recent work showing that IL-1 induced PGE_2 release by astrocytes is dependent upon protein kinase C activation (PKC-α isoform). We also summarize some of the data concerning the ability of cultured astrocytes to produce NO, TNF-α and IL-6 upon IL-1 stimulation.

EXPERIMENTAL PROCEDURES

Astrocyte Cultures

Primary astrocyte cultures were prepared from the cerebral cortex of 1-day-old neonatal mice (Balb/c and SJL/J, Instituto Cajal, Madrid, Spain) according to well established protocols (McCarthy and de Vellis, 1980) and grown in T150 flasks for at least 14 days in Dulbecco's modified Eagles medium (DMEN) and 10% heat-inactivated fetal calf serum (Whittaker, France), 20mM glutamine (INC Biomedicals, CA), and gentamicin with the medium being changed twice per week. On reaching the confluency (2 weeks) the cells were shaken at 240rpm overnight at 37°C to dislodge cells adhering to the astrocyte layer (primarily oligodendrocytes and microglia). The media were replaced and the cells were allowed to recover for 2-3 days before experiments. The resulting cultures consist of $\geq$95% astrocytes as determined by Glial Fibrilar Acidic Protein (GFAP) staining, and less than 2% of microglia as indicated by OX42 positive cells.

Cell treatments

For stimulation, astrocyte monolayers were incubated during different periods of time (up to 18 hours) with different doses of human recombinant IL-1α (generous gift of Hoffman La Roche), hrIL-1ß (Genzyme) or recombinant murine IFNγ (Genzyme) plus IL-1α or IL-1ß. In some experiments we used lipopolysacharide (LPS, serotipe 06:26B, Sigma Chemical Co. St Louis) with an incubation period of 18 hours. Specific incubation times with PMA or TPA are specified in the text in each experiment.

Nitric oxide assay

NOS activity was assessed by measurement of NO_2^- (a break down product of NO) accumulation in the cell culture media. A 100µl aliquot of the culture media was mixed with 100µl of Griess reagent (Green et al., 1982) and incubated at room temperature for 10min, and then the absorvance was measured at 570 nm in a microlate reader (Titertek Multiskam MCC). Fresh culture medium served as the blank in all experiments. Solutions of sodium nitrite (Na NO_2^-) diluted in culture media were used as standards. The detection limit for NO_2^- was about 0.5mM.

Eicosanoid production

The cell free supernatants were collected and stored at -70°C until measurement of PGE_2. The amount of PGE_2 was determined by radioimmunoassay (RIA) (NEN Research Products). PGE_2 standard curves range from 1 to 100pg/ml. The sensitivity of the assay was 0.44pg/tube. The protein content was determined by a BCA Protein Assay Reagent with bovine serum albumin as the standard.

Cytokine assays

The presence of TNF-α and IL-6 in the harvested supernatants from astrocyte cultures was performed using enzyme-linked immunoadsorbent assays (Intertest-6XTM mouse IL-6 ELISA kit and Factor-test XTM mouse TNF-α ELISA kit, Genzyme, Cambridge, MA). For TNF-α the detection limit was of 15pg/ml and for IL-6 the detection limit was of 8.8pg/ml. Coefficients of variation intra-assay were for TNF-α of 2.5% and for IL-6 of 8,1%. Coefficients of variation interassay were 8.7% for TNF-α and 11,9% for IL-6.

Western analysis

Cells were cultured for varying lengths of time with or without TPA and IL-1ß. At the end of the incubation period, the cells were first washed with PBS and then were lysed with lysis buffer (20mM Tris-HCl, NP-40 1% SDS0 1%, glicerol 10%, 137mM NaCl, 0,5mM EDTA, 1mM PMSF, 1mg/ml aprotinin and 1mM sodium ortovanate).

The lysate was boiled for 10min, and was then microfuged for 15min. The proteins separated by standards SDS-PAGE (Bio-Rad) were electrophoretically transferred at 4°C to nitro-cellulose membrane (Bio-Rad). The membrane was then immersed for 1 hour at 25°C in PBS containing 5% nonfat dry milk, washed three times with PBS, and incubated for 2 hours at 25°C with rabbit antisera directed againtsts various PKC isoenzymes (1:3000) in the absence or presence of corresponding blocking Ag peptides.

The membranes were then washed three times with PBS containing 0.05% Tween 20, twice with PBS and incubated for 1hour at 25°C with the second antibody (HRP-Goat anti-rabbit (1: 10.000, Santa Cruz). The membrane was then washed three times in PBS containing 0.5% Tween 20, and two times more with PBS and leave in PBS and developed with ECL (Amersham).

INTERLEUKIN-1 IN THE REGULATION OF ASTROGLIAL FUNCTION

IL-1 stimulates prostaglandin production in astrocyte cultures: Involvement of PKC activity

We first examined the potential for PKC activation to stimulate PGE_2 production by using the phorbol ester, PMA. As shown in Table 11-1 PMA (1, 10, 100ng/ml; 18hr) treatment of glial cells enhanced dose relatedly the release of PGE_2 from astrocytes. Table 11-1, also shows that incubating PMA-stimulated astrocytes in the presence of indomethacin, a known inhibitor of the cyclooxygenase enzyme which converts arachidonic acid into prostaglandins, totally abolished the stimulatory action of PMA in PGE_2 production indicating that cellular prostaglandin production is due to cyclooxygenase activity. The inactive phorbol ester steroisomer, 4α phorbol (100 ng/ml) failed to increase PGE_2 release over basal levels. As expected LPS (1μg/ml;18hr) treatment induced a potent increase in PGE_2 production. The effect of a medium dose of PMA (10 ng/ml) was inhibited by 1μM dexamethasone and completely prevented by the presence of the protein inhibitor cycloheximide and the transcriptional inhibitor actinomycin D (Table 11-1).

Table 11-1. *PMA increases PGE_2 release from astroglial cells in a dose related manner. Effects of indomethacine, dexamethasone and actinomycin.*

Addition	PGE_2 (pg/mg protein)
None	9.28 ± 3.1
PMA (1ng/ml)	18.31 ± 4.6*
PMA (10ng/ml)	51.35 ± 4.8***
PMA (100ng/ml)	221.0 ± 21.5***
Inactive 4-a-phorbol (100ng/ml)	11.21 ± 4.1
PMA (10ng/ml) + INDO (5ng/ml)	9.71 ± 3.4
PMA (10ng/ml) + DEX (1mM)	16.1 ± 3.6*
PMA (10ng/ml) + Actinomycin (5 mg/ml)	10.3 ± 2.9
PMA (10ng/ml) + Cycloheximide (10mg/ml)	2.1 + 1.8
LPS (1 mg/ml)	92.3 ± 8.6***

Values are means ± SEM of three assays in duplicate. Statistical analysis were performed by ANOVA followed by the Student-Newman-Keuls t-test to establish differences between means. *P<0.01 vs none; ***P<0.0001 vs none.

Addition of IL-1α to astrocyte cultures increased PGE_2 levels in the culture medium in a dose-related manner (fig. 11-1). Dexamethasone at a dose of 0,1μM reduced the stimulatory action of IL-1α on PGE_2 release by a 67% and a maximal dose of 1μM dexamethasone almost completely prevented increased PGE_2 release by IL-1α. Addition of IL-1ß to astrocyte cultures stimulated, dose related, the production of PGE_2 too (data not shown).

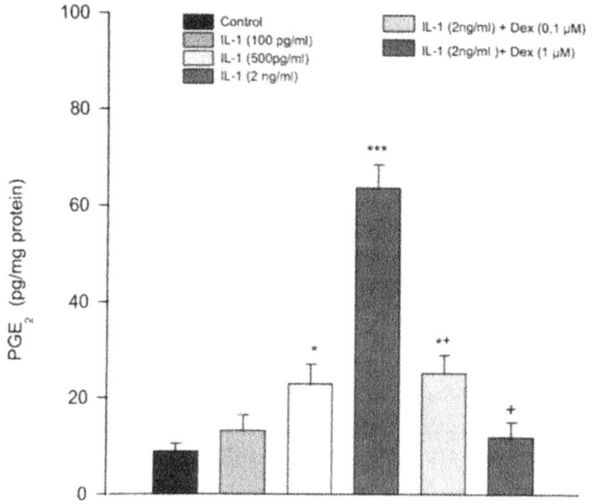

Figure 11-1. Effect of different doses of IL-1α (100pg/ml; 500pg/ml or 2ng/ml) on PGE_2 release from astrocyte cultures in the absence or presence of dexamethasone (0,1μM and 1μM). The data shown are mean ± SEM of triplicate assays in three separate experiments. Statistical significances were performed by ANOVA followed by the posteriori test of Newman-Keuls-t test to establish differences between means. ***P<0.0001 vs control (none addition); *P<0.001 vs control; $^{+}$P<0.0001 vs IL-1α (2ng/ml).

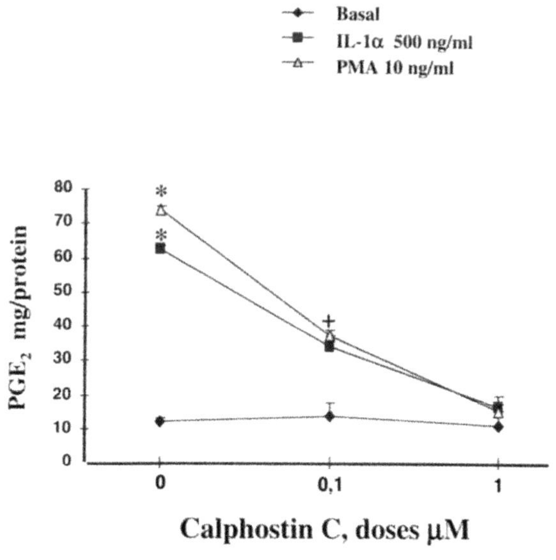

Figure 11-2. Effects of Calphostin C on PGE_2 production by astrocytes. Calphostin C inhibited, as a function of dose, the effects of IL-1α (500 pg/ml) or PMA (10ng/ml) -induced PGE_2 release by astrocytes. The data shown are mean ± SEM of triplicate assays in three separate experiments. Statistical significances were performed by ANOVA followed by the posteriori test of Newman-Keuls-t test to establish differences between means. *P< 0.0001 vs control; $^{+}$P<0.01 vs control.

Pretreatment of cells with H-7 (50 μM), a compound which inhibits PKC and PKA, or staurosporine (10nM) prior to the stimulation by IL-1α resulted in a clear reduction in PGE_2 release (Table 11-2). To confirm a role of PKC in PGE_2 production by IL-1α, we used a more specific PKC inhibitor, such as calphostin C, at known non toxic effective doses to inhibit PKC, 0.1 and 1μM. Under these conditions, the increased production of PGE_2 by astrocytes stimulated with IL-1α was markedly blunted with the dose of 0.1μM and almost completely inhibited by the dose of 1μM (fig. 11-3).

Table 11-2. *Effects of H7, Staurosporine and on IL-1α induced PGE_2 production.*

Addition	PGE_2 (pg/mg protein)
None	11.5 ± 2.90
IL-1α (500pg/ml)	48.3 ± 5.10 ***
H7 (50mM)	9.6 ± 2.30
Staurosporine (10nM)	10.3± 2.78
IL-1α (500pg/ml) + H7 (50mM)	23.5 ± 3.60 *
IL-1α (500pg/ml) + Staurosporine (10nM)	16.4 ± 3.90 *

Values are means ± SEM of three assays in duplicate. Statistical analysis were performed by ANOVA followed by the Student-Newman-Keuls-test to establish differences between means. *p<0.01 vs none, ***p<0.0001 vs none.

PKC depletion studies are another method by which the involvement of PKC as a second messenger can be examined. Prolonged exposure to PMA results in down-regulation of PKC by PKC protein degradation or by inhibition of translation of PKC mRNA, thus attenuating or abolishing PKC-mediated responses. In our study, a PMA pretreatment of 4 hours attenuated significantly the increased production of PGE_2 by subsequent stimulation with PMA or IL-1 (data not shown). As illustrated in figure 11-4, the PMA and IL-1α stimulatory actions on PGE_2 production were completely inhibited following cell pretreatment with PMA for 24 hours. These results confirm that the increased PGE_2 production by IL-1 treatment is due to a conventional PKC isoform, i.e. phorbol ester dependent. Neonatal rat astrocytes express the PKC α, δ, ε and ζ isoforms but not the γ isoform (Ballestas and Benveniste, 1995). Western blot analysis using a panel of isoform-specific antibodies showed that neonatal mouse astrocytes also failed to express γPKC isoform (data not shown).

We then examined PKCα immunoreactivity in astrocytes subjected to PKC depletion by pretreatment with TPA for 18 hours. As shown in figure 11-5, PKCα immunoreactivity was almost completely eliminated following TPA pretreatment. Astrocytes subjected to PKC depletion were stimulated with IL-1ß or TPA for 4 hours and the results we obtained were consistent with the involvement of PKC-α isoform on IL-1 actions in astrocyte cell cultures.

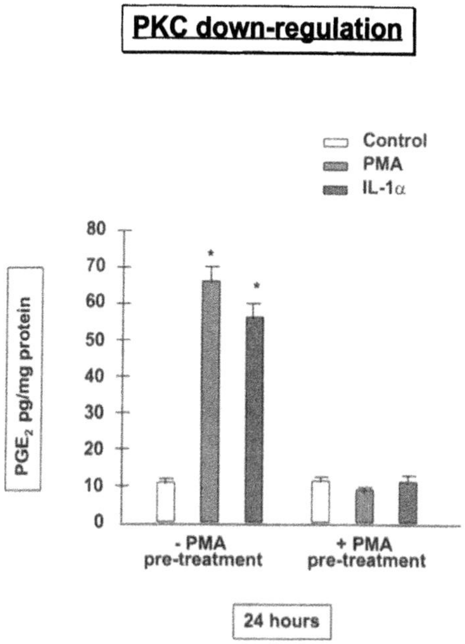

Figure 11-3. Effects of prolonged exposure to phorbol ester on IL-1a and phorbol ester-induced PGE_2 release by astrocyte cultures. Astrocytes were incubated for 24 hours with or without the phorbol ester PMA (100ng/ml). Cells were then washed with DMEM and added fresh DMEM medium containing either PMA (10 ng/ml) or IL-1a (2ng/ml) for a 4 hour incubation. The data shown are mean ± SEM of triplicate assays in three separate experiments.

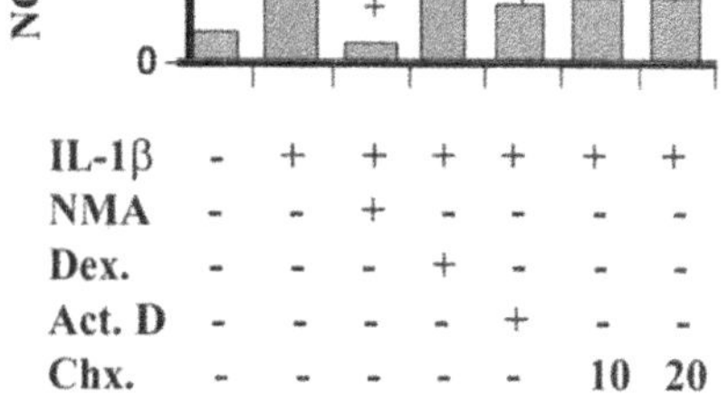

Figure 11-4. Effects of IL-1ß (2ng/ml) stimulation on NO_2^- accumulation in culture supernatants. Astrocytes were cultured for 18 hrs in the presence or absence of an inhibitor of NOS, NMA, dexamethasone, actinomicyn or cycloheximide. Results are mean ±SEM of three individual experiments done in triplicate. *P<0.001 vs control; $^+$P<0.001 vs IL-1ß.

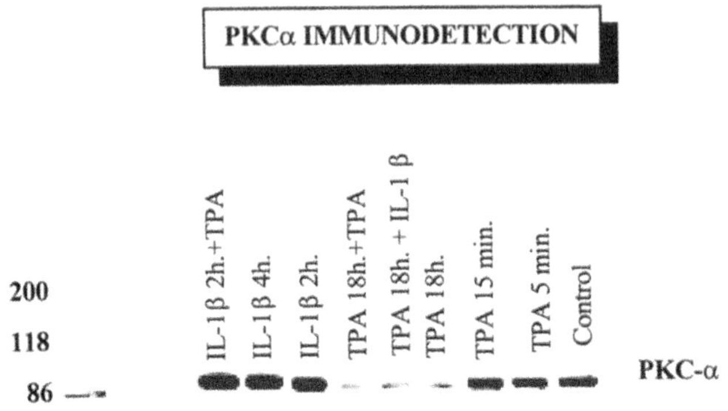

Figure 11-5. Expression of PKC-α isoenzyme in astrocyte cultures. Effects of prolonged exposition to TPA (100nM) on PKC-α expression after the following treatments: IL-1ß (5ng/ml), TPA, or a combination of IL-1ß+TPA. Cells were prepared for Western analysis with anti PKC-α.

Astrocyte NO production following stimulation with IL-1α and IL-ß

IL-1 and IFNγ + IL-1 have been shown to transcriptionally increase iNOS gene expression in various rodent cell types. In astrocyte cultures induced with LPS, NOS activity peaks at 12 hours and is maintained for at least 36 hours (Simmons and Murphy, 1992).

Table 11-3. *Nitric oxide production in astrocyte cultures subjected to different doses of IL-1α or β*

Addition	NO_2^- (nmol/mg protein)
None	7.90± 3.5
IL-1α (100 pg/ml)	12. 80 ±3.1
IL-1 α (500 pg/ml)	23.75 ±3.8*
IL-1 α (2 ng/ml)	31.60 ±4.1**
IL-1ß (100 pg/ml)	14.30±2.9
IL-1ß (500 pg/ml)	22.90±3.5*
IL-1ß (2 ng/ml)	35.40±3.8**

Results are the means ±SEM of four experiments done in triplicate. Statistical analysis were performed by ANOVA followed by the Student-Newman-Keuls-test to establish differences between means. *P<0.05; **P<0.001 vs none addition.

When induced with IL-1ß/IFNγ, NOS activity peaks at 4 hours and decreases steadily (Simmons and Murphy 1993). The induced NOS activity is reflected by measurement of NO_2^- accumulation. The time course of nitrites accumulation following IL-1α or ß over a period of time of 24 hours indicated us that an adequate incubation time for measuring a reliable amount of nitrites in astrocyte supernatants was 18 hours (data not shown). In cultured mouse astrocytes the exposure to both IL-1α or ß (18 hours), increased NO_2^- accumulation in the supernatants, in a dose response fashion (Table 11-3). Control cultures and cultures treated with other cytokines, such as IL-6 or TNFα alone did not produce detectable amounts of NO_2^- (data not shown). In figure 11-4 it is shown that increased nitrites accumulation in astrocyte supernatants following IL-1ß stimulation (2ng/ml) is counteracted by the presence the inhibitor of nitric oxide synthase, monomethyl arginine (NMA). The presence of dexamethasone (1µM) inhibited IL-1ß induced NO production by 60%. The addition of the transcriptional inhibitor actinomicynD (2mg/ml) or the protein synthesis inhibitor, cycloheximide (10 and 20 mg/ml), significantly blocked nitrites accumulation occurred by IL-1ß.

Interleukin-6 and tumor necrosis factor-α production by astrocytes stimulated with IL-1 ß

Interleukin-1 has been shown to activate IL-6 and TNF-α gene expression in astrocytes. We show that mouse astrocyte cultures enhance significantly the production of the proteins IL-6 and TNF-α as measured in culture supernatants over a time period of 18 hours following IL-1ß or IFNγ+IL-1ß stimulation (Table 11-4). The level of IL-6 induced by IL-1ß occurs in a dose-dependent manner, while in the case of TNF-α the previous exposure to IFN-γ seems to be a requisite for IL-1ß induction of this cytokine in a substantial amount. The basal medium content of IL-6 and TNF-α for unstimulated cells was undetectable in >50% of samples and averaged 37,8±2,89 pg/ml for IL-6 and 42,30±4,50 pg/ml for TNFα.

Table 11-4. *Interleukin-1ß induction of IL-6 and TNF-α production by astrocytes.*

Addition	IL-6 pg/ml	TNF-α pg/ml
None	37.8±2.89	42.3±8,5
IL-1ß (100pg/ml)	51.2± 4.55*	46.7±6,9
IL-1ß (500 pg/ml)	89.6±9.50**	51.4±7.3
IL-1ß (2 ng/ml)	124.3±13.1***	76.5±10.2**
IFNγ (50U) + IL-1ß (500pg/ml)	295.7±18.6***	219.3±20.5**

Mouse astrocytes were incubated with control medium, IL-1ß, and IFN-γ+IL-1ß for 18 hours, supernatants harvested and assayed for IL-6 and TNF-α as described in methods. Results are the mean±SEM of three experiments in duplicate. Statistical analysis were performed by ANOVA followed by the Student-Newman-Keuls-test to establish differences between means. *P<0.05; **P<0.01; ***P<0.001 vs none addition.

ROLE OF ASTROCYTES AS TARGET CELLS FOR IL-1 ACTIONS IN INFLAMMATORY PROCESSES IN THE BRAIN

We have analysed PGE_2 and NO responsiveness, together with IL-6 and TNF-α production following IL-1 exposure in a primary cell type, the astrocyte. We showed that stimulation of astrocytes with IL-1 results in: i) increased production of PGE_2; ii) accumulation of nitrites; and iii) enhanced release of IL-6 and TNF-α to the culture medium.

IL-1 actions on astrocytes. signal transduction mechanisms

Both cytokines, IL-1α or IL-1ß enhanced PGE_2 production by mouse astrocyte cultures. This is in agreement with previous studies in which IL-1α and IL-1ß elicited PG production in cultured astrocytes (Hartung et al.,1989; Katsuura et al.,1989; Nam et al.,1995). However, the second messengers activated via IL-1 receptors to increase PG production by astrocytes have not been fully delineated. The PKC family of serine/threonine kinases are implicated in a variety of cell surface receptors-mediated signal transduction pathways, including those used by IL-1 in several cell types (Brook and Mizel, 1994). Evidence for the involvement of protein kinases on IL-1 mediated PGE_2 production was obtained by our results using H7 and staurosporine, since these protein kinase inhibitors (Hidaka and Kobayashi, 1992) attenuated the increased PGE_2 production and a specific PKC inhibitor, calphostin reproduced this effect. Calphostin C interacts with the regulatory domain of PKC and competes with the DAG phorbol ester binding site located in the C1 domain. Therefore, calphostin C may not inhibit the activity of PKC isoforms which lack DAG/phorbol ester binding sites suggesting the involvement of a conventional PKC isoform. The expression of various PKC isoenzymes has been shown to differ depending on the tissue and cell types and individual PKC isoenzymes may play cell type-specific roles in various cellular responses (Nishizuka 1988). The results of Western analysis showed that mouse astrocytes constitutively express PKC-α protein, a conventional PKC isoenzyme and that the pretreatment of astrocytes with TPA results in a downregulation of PKC isoform-α following IL-1ß stimulation, suggesting the involvement of this isoenzyme on IL-1 actions. Activation of PKC has been previously involved in astrocyte function (Aloisi et al., 1994; Ballestas and Benveniste,1995; Simmons and Murphy, 1994) and it is an important step in the initiation of arachidonic acid synthesis and release in astrocytes which are the major source of cyclooxygenase products in the brain. Cellular PG production is due to cyclooxygenase activity as well as the availability of arachidonic acid substrate, usually controlled by enzymes such as phospholipase A_2 (PLA_2). Previous studies (Oka and Arita, 1991) have shown that enhanced PGE_2 production by IL-1ß in rat astrocytes is subsequent to activation of PLA_2 and it is known that activation of PKC by PMA stimulates the release of arachidonic acid, due to increased PLA_2 activity which involved also MAP kinase activation (Lin et al., 1993). Thus, it is possible that enhanced PGE_2 release by PMA-stimulated astrocytes also involved PLA_2 activation.

Interestingly, PKC activation has been reported to be essential for other actions of IL-1 in astroglial cells, such as IL-1 mediated induction of IL-6 (Norris et al.,1994) or induction of leukemia inhibitory factor (Aloisi et al.,1994). Induction of cyclooxygenase activity (PGHS-2 or COX-2) has been recently described to occur in astrocytes in response to IL-1ß and the PKC activator TPA, suggesting that increased expression of PGHS-2 can also contribute to PG production (O'Banion et al., 1996). Using primary astrocyte cultures we have showed that NO activates cyclooxygenase pathway stimulating PG synthesis (Molina-Holgado et al.,1995) and PKC activation has been reported to induce NOS in microglia (Merrill et al.,1993), macrophages (Severn et al.,1992) and partially also in astrocytes (Simmons and Murphy, 1994). As described in other cell types, dexamethasone significantly blunted the production of PGE_2 by astrocytes upon L-1 or PMA treatment. This is in agreement with the important antiinflammatory properties of glucocorticoids by inhibiting PG production, both at PLA_2 (Flower, 1988; Ling et al.,1992) and/or COX-2 level (Masferrer et al.,1992; O'Banion et al., 1992).

Potential relevance of IL-1 induced nitric oxide by astrocytes in mediating neurodegeneration

Transcriptional induction of iNOS is initiated in astrocytes in vitro by LPS or cytokines (rev: Murphy et al., 1993) that activate signalling pathways involving protein kinases (Feinstein et al., 1994; Simmons and Murphy, 1994). As previously described, here we showed that both IL-1α or IL-1ß stimulation caused NOS induction in astrocyte cultures (Simons and Murphy, 1992, 1993; Galea et al., 1992; Lee et al., 1993). In addition iNOS can be induced in C6 cells and in a human astrocytoma cell line (Simons and Murphy, 1992; Vigne et al., 1993; Feinstein et al., 1993). Although microglia is considered to be the major source of NO (Merril et al., 1993), astrocytes are the majority cell type in the brain and then understanding the regulation of induced NO production from these cells is an important issue. In neuropathological conditions, as NO can be cytotoxic when produced in substantial amount and for periods of time, it would be of great importance the existence of efficient mechanisms to regulate or terminate NO production. In fact, there is increasing interest to delineate the regulatory mechanisms involved in iNOS expression at the transcriptional, translational and posttranslational levels. Studies with astrocytes in vitro have revealed that iNOS expression can be inhibited at the transcriptional level by corticosteroids, anti-inflammatory cytokines, and NO itself (Park et al., 1994). The potential contribution of astrocyte- derived NO by IL-1 to cell damage remains to be established, but during immunoinflammatory disorders the production of IL-1 by activated microglia may lead to an exaggerated production of NO by astrocytes which are the main target for IL-1 action (Ban et al., 1993). The above results support the hypothesis that IL-1 plays a critical role in cell injury during inflammatory disease of CNS.

Astrocytes as source of cytokines. Influence on the cytokine CNS network and neuropathological actions

The production of cytokines such as IL-6 and TNF-α by astrocytes has now been documented in a number of conditions like the stimulation with IL-1 (Aloisi et al., 1992; Norris et al., 1994, Lieb et al., 1996). IL-6 can be detected in Alzheimer disease brains but not in brains of normal control persons, particularly in very early stages of the disease (Huell et al., 1995). Elevated levels of IL-6 in the CNS of EAE diseased animals but not in their spleens or sera are suggestive of a local production of IL-6 within the brain (Gijbels et al., 1990). IL-6 has multiple effects on astrocytes and may function in an autocrine manner as glioblastoma, astrocytoma cell lines, and primary astrocyte cultures express specific high affinity receptors for IL-6 (Taga et al., 1987; Sawada at al., 1995). In spite of these observations, the role of IL-6 in neurological disorders remains unclear. Although overproduction of IL-6 in brain of transgenic mice results in important deficits including tremors and astrogliosis (Campbell et al., 1993), astrocytes can respond to IL-6 by secreting nerve growth factor (Frei et al., 1989) and IL-6 has been shown to suppress demyelination in a viral model of multiple sclerosis (Rodriguez et al., 1994). By contrast, the biologic significance of TNF-α production in astrocytes upon IL-1 stimulation is supported by the recognition of TNF-α as an important effector in neurotoxicity specially in immune-mediated demyelination by damaging oligodendrocytes (Selmaj and Reine, 1988). TNF-α is predominantly localized in astrocytes of active lesions of multiple sclerosis brains (Hoffman et al., 1989; Selmaj et al., 1991) and its level of appearance correlates with the demyelinating activity in humans and experimental animals (Sharief and Hentges, 1991).

The release of inflammatory mediators such as prostanoids, reactive oxygen species and pro-inflammatory cytokines like IL-6 and TNF-α by astrocytes upon IL-1 stimulation point out the potential contribution of these glial cell to the pathogenesis of inflammatory brain diseases. Inflammatory mechanisms have been also involved in degenerative brain diseases like AD (Aisen and Davis, 1994), AIDS dementia (Merrill, 1992; Griffin et al., 1994), multiple sclerosis (Merrill 1992; Woodroofe and Cuzner, 1993) or prion diseases (Betmouni et al., 1996). The importance of microglial cells, first line of defense against inflammation, as the main source of IL-1 in the CNS and of astrocytes the main target cell for IL-1 actions in promoting or preventing damage following CNS injury or during neurodegenerative processes is likely to be promptly established.

CONCLUDING REMARKS

The activation of cells by IL-1 may turn in a number of signal transduction pathways which may diverge, adding complexity. Furthermore, different pathways may be activated in different cell types. Here, we showed the involvement of PKCα in PGE_2-stimulated production by IL-1. It seems clear that a protein kinase cascade involving PKC, MAP kinases and /or novel protein kinases would allow for the amplification needed to generate a signal from the small number of receptors triggered by IL-1. These kinases are also likely to activate transcription factors such as NFkB, altering gene expression. The exact nature of these kinases and how they modify gene expression in astrocytes has yet to be determined.

ACKNOWLEDGMENTS

We would like to thank C. García for skillful technical assistance. The experimental work from the authors'laboratory was funded by DGICYT (PM 92-0014 and PB 94-0098) and Fundación Salud 2000.

REFERENCES

Aisen, P.S., and K.L Davis (1994) Inflammatory mechanisms in Alzheimer's disease: implications for therapy. Am. J. Psychiatry. *151*: 1105-1133.

Aloisi, F., A.Care, G.Borsellino, P.Gallo, S.Rosa, A.Cabibbo, U. Testa, G. Levi, and C. Peschle (1992) Production of hemolymphopoietic cytokines (IL-6, IL-8, colony-stimulating factor) by normal human astrocytes in response to IL-1ß and tumor necrosis factor. J. Immunol. *149*:2358-2366

Aloisi, F., S. Rosa, U. Testa, P. Bonsi, G. Russo, C. Peschle, and G. Levi (1994) Regulation of leukemia inhibitory factor synthesis in cultures human astrocytes. J. Immunol. *152*:5022-5031.

Ballestas, M.E., and E.N. Benveniste (1995) Interleukin-1ß and tumor necrosis factor α mediated regulation of ICAM-1 gene expression in astrocytes requires protein kinase C activity. Glia *14*:267-278.

Ban, E., L.L. Sarlieve, and F.G. Haour (1993) Interleukin-1 binding sites on astrocytes. Neuroscience. *3*:725-733.

Barnes, P.J.,and I. Adcock (1993) Anti-inflammatory actions of steroids: molecular mechanisms. Trends Pharmacol. Sci. *14*:436-441.

Benveniste, E.N. (1992) Inflammatory cytokines within the central nervous system: sources, function and mechanisms of action. Am. J. Physiol. *263*:C1-C16.

Benveniste, E.N. (1995) The role of cytokines in multiple sclerosis/autoimmune encephalitis and other neurological disorders.In Human Cytokines: Their Role in Research and Therapy (eds): Blackwell Scientific Publications. Boston: B. Aggarawal and R. Puri pp. 195-216.

Bethea, J.R., G.Y. Gillespie, and E.N. Benveniste (1992) Interleukin-1ß induction of TNF-α gene expression: involvement of protein kinase C. J. Cell Physiol. *152*:264-273.

Betmouni S., V.H. Perry, and J.L. Gordon (1996) Evidence for an early inflammatory response in the central nervous system of mice with scrapie. Neuroscience *74*:1-5.

Boje, K.M., and P.K. Arora (1992) Microglial-produced nitric oxide and reactive nitrogen oxides mediates neuronal cell death. Brain Res. *587*:250-256.

Bredt, D. S., and S. H. Snyder (1990) Isolation of nitric oxide synthase, a calmodulin-requiring enzyme. Proc. Natl. Acad. Sci. USA. *87*:682-685.

Brooks, J.W. and S.B. Mizel (1994) Interleukin-1 signal transduction Eur Cytokine Netw 5: 547-561.

Campbell, I.L., C. R.Abraham, E.Masliah, P.Kemper, J.D.Inglis, M.B.A. Oldtone, and L. Mucke (1993) Neurologic disease induced in transgenic mice by the cerebral overexpression of interleukin-6 in transgenic mice. Proc. Natl. Acad. Sci. USA. *90*:10061-10065.

Chao, C.C., S.Hu, W.S.Sheng, D.Bu, M.I. Bukrinsky, and P.K. Petterson (1996) Cytokines-stimulated astrocytes damage human neurons via nitric oxide mechanism. GLIA. *16*:276-284.

Daffada A. A. Murray E.J. and Young S.P. (1994) Control of activator protein -1 and nuclear factor kappa B activity by interleukin-1, interleukin-6 and metals in HEPG2 cells. Biochem. Biophys. Acta 1222, 234-240.

Dawson V.L. Dawson T.M., Bartley,D.A., Uhl, G.R. and Snyder S.H. (1993) Mechanisms of nitric oxide mediated neurotoxicity in primary brain cultures J. Neurosci,13, 2651-2661.

Dinarello, C.A. (1994) The interleukin-1 family: 10 years of discovery. FASEB J., *8*:1314-1325

Di Rosa, M., M.Radomski, R.Carnuccio, and S. Moncada (1990) Glucocorticoids inhibit the induction of nitric oxide synthase in macrophages. Biochem. Biophys. Res. Commun.. *172*:1246-1252.

Eddleston M., and L. Mucke (1993) Molecular profile of reactive astrocytes: Implications for their role in neurologica disease. Neuroscience *54*:15-36.

Endöh, M., K. Maiese, and J.A. Wagner (1994) Expression of the inducible form of nitric oxide synthase by reactive astrocytes after transient global ischemia. Brain Res. *651*: 91-100.

Fabry, Z., C.S. Raine, and M.N. Hart (1994) Nervous system as an immune compartment: the dialect of the immune response in the CNS. Immunol Today. *15*: 218-224.

Feinstein, D., E. Galea, S. Roberts, H.Berquist, H. Wang, and D.J. Reis (1993)Induction of nitric oxide synthase in rat C6 glioma cells. J. Neurochem.*62*:315-321.

Flower, R.J. (1988) Lipocortin and the mechanism of action of glucocorticoids. Br. J. Pharmacol. *98*:1287-1295.

Frei, K., U.V.Malipiero, T.P.Leist, R.M.Zinkernagel, M.E. Shwab, and A. Fontana (1989) On the cellular source and function of interleukin-6 produced in the central nervous system in viral diseases. Eur. J. Immunol. *19*:689-694.

Fontana, A., F. Kristensen, R. Duls, D.Gemsa, and Weber E (1982) Production of prostaglandin E and an interleukin-1 like factor by cultures astrocytes and C6 glioma cells. J. Immunol. *126*:2413-2419.

Fretland, D.J. (1992), Potential role of prostaglandin's and leukotrienes in multiple sclerosis and experimental allergic encephalomyelitis. Prostaglandin's Leukot. Essent. Fatty Acids. *45*: 249-257.

Galea, E., D.L. Feinstein, and D.J. Reis (1992) Induction of calcium independent nitric oxide synthase activity in primary cultures of astrocytes. Proc Natl. Acad. Sci. USA. *89*:10945-10949.

Gebicke-Haerter, P.J., A. Schobert, and G. Hertting (1991) Pertussis and cholera toxins inhibit prostaglandin synthesis in rat astrocyte cultures at distinct metabolic steps. Biochem Pharmacol. *42*:1267.

Gijbels, K., J.Van Damme, P.Proost, W.Put, H.Carton, and, A. Billiau (1990) Interleukin-6 production in the central nervous system during experimental autoimmune encephalomyelitis. Eur. J. Immunol. *20*:233-235.

Giulian, D., T.J. Baker, L.H. Shich, and L.B. Lachman (1986) Interleukin-1 of the central nervous system is produced by ameboid microglia. J. Exp. Med. *164*:594-604.

Green, L. C., D.A.Wagner, J.Glowgowski, P.L.Skipper, J.S. Wishnock, and S.R. Tannenbaum (1982) Analysis of nitrate, nitrite and ^{15}N nitrate in biological fluids. Anal. Biochem. *126*: 131-138.

Griffin D.E., S.L. Wesselingh, and J.C. Mc Arthur (1994) Elevated central nervous system prostaglandin's in human immunodeficiency virus-associated dementia. Ann. Neurol.*35*: 592-597.

Griffin W.S.T., J.G. Sheng, G.W. Roberts, and R.E. Mrak (1995) Interleukin-1 expression in different plaque types in Alzheimer's disease: signicance in plaque evolution. J. Neuropathol. Exp. Neurol. *54*:276-281.

Hartung H.P., B.Schäfer, K. Heininger, and K.V. Toyka (1989) Recombinant interleukin-1ß stimulates eicosanoid production in rat primary culture astrocytes.Brain Res. *489*:113-119.

Hidaka, H., and R. Kobayashi (1992) Pharmacology of protein kinase inhibitors. Annu Rev. Pharmacol. Toxicol.. *32*:377.

Hoffman, F.M., D.R. Hinton, K. Johnson, and J.E. Merrill (1989) Tumor necrosis factor identified in multiple sclerosis brains. J.Exp. Med. *170*:607-612.

Hooper, D. G., T. Ohnishi, R. Kean, Y. Numagami, B. Dietzschold, and H. Koprowski (1995) Local nitric oxide production in viral and autoimmune diseases of the central nervous system. Proc. Natl. Acad. Sci. USA, *92*:5312-5316.

Katsuura, G., P.E. Gottschall, R. R. Dahl, and, A. Arimura (1989) Interlukin-1ß increases prostaglandin E_2 in rat astrocyte cultures: modulatory effect of neuropeptides. Endocrinology *124*: 3125-3127.

Koprowski, H., Y. M. Zheng, E.Hebert-Katz, N.Fraser, L.Rorke, F.Fang, C. Hanlon, and B. Dietzshol (1993) In vivo expression of inducible nitric oxide synthase in experimentally induced neurological diseases. Proc. Natl. Acad. Sci. USA. *90*:3024-3027.

Lafortune L., J. Nalbantoglu and J. P. Antel (1996) Expression of tumor necrosis factor α (TNF α) and interleukin-6 (IL-6) mRNA in adult human astrocytes: comparison with adult microglia and fetal astrocytes. J. Neuropathol. Exp. Neurol. 55: 515-521

Lee, S.C., D.W.Dickson, W. Liu, and C.F. Brosnan (1993) Induction of nitric oxide synthase activity in human astrocytes by interleukin-1ß and interferon-γ. J Neuroimmunol. *46*:19-24.

Lieb K., C. Kaltschmidt, B.Kaltschmidt, P.A. Bauerle, M. Berger, J. Bauer, and B.L. Fiebich (1996) Interleukin-1ß uses common and distinct signaling pathways for induction of the interleukin-6 and tumor necrosis factor α genes in the human astrocytoma cell line U373. J. Neurochem. *66*: 1496-1503.

Lin L.L., M. Wartman, A. Y. Lin, J.L. Knopf, A. Seth and R.J. Davis (1993) $cPLA_2$ is phosphorilated and activated by MAP kinase, Cell *7*:, 269-278.

Masferrer, J.L., K. Seibert, B. Zweifel, and P. Needleman (1992) Endogenous glucocorticoids regulate an inducible cyclooxygenase enzyme. Proc. Natl. Aca. Sci. USA *89*:3917-1921

McCarthy, K.D. and, J. de Vellis (1980) Preparation of separate astroglial and oligodendroglial cell cultures from rat cerebral tissue. J. Cell. Biol.. *85*: 890-902.

Merrill J.E. (1992) Proinflammatory and antiinflammatory cytokines in multiple sclerosis and central nervous system acquired immunodeficiency syndrome. J. Immunother. *12*: 167-170.

Merril, J.E., L.J.Ignarro, M.P.Sherman, J.Melinek, and T.E. Lane, (1993) Microglial cell cytotoxicity of oligodendrocytes is mediated through nitric oxide. J. Immunol.. *151*: 2131-2141.

Milstein, S., N.Sakai, B.J.Brew, C.Krieger, J.H.Vickers, K.Saito, and M.P. Heyes (1994) Cerebrospinal fluid nitrite/nitrate levels in neurological diseases. J. Neurochem. *3*:1178-1180.

Molina-Holgado, F., A.Lledó, and C. Guaza (1995) Evidence for cyclooxygenase activation by nitric oxide in astrocytes. GLIA. *15*:167-172.

Moncada, S., R.M.J. Palmer, E.A. Higgs (1991) Nitric oxide: physiology, pathophysiology, and pharmacology. Pharmacol. Rev. *43*:109-142.

Nam, M.J., C. Thore, D. Busija (1995) Rapid induction of prostaglandin synthesis in piglet astroglial cells by interleukin-1α. Brain Res. Bull.. *3*:215-218.

Nathan, C., and Q.W. Xie (1994) Regulation and biosynthesis of nitric oxide J. Biol. Chem. *269*:13725-13728.

Navikas,V., D.Matusevicius, M.Söderstrom, S. Fredrikson, P.Kivisäkk, A.Ljungdahl, B. Höjeberg, and H. Link (1996) Increased interleukin-6 mRNA expression in blood and cerebrospinal fluid mononuclear cells in multiple sclerosis. J. Neuroimmunol. *64*: 63-69.

Neary, J.T., L.O.B. Noremberg, M.D. Noremberg (1988) Protein Kinase C in primary astrocyte cultures : Cytoplasmic localization and translocation by a phorbol ester. J. Neurochem. *50*:1179.

Nieto Sampedro M., and M.A. Berman (1987) Interleukin-1 activity in rat brain: sources targets and the effects of injury J. Neurosci Res. *17*:214-219.

Nishizuka Y. (1988) The molecular heterogeneity of protein kinase C and its implications for cellular regulation. Nature, *334*:661-665

Norris, J.G., L.P. Tang, S.M. Sparacio, and E.N. Benveniste (1994) Signal transduction pathways mediating astrocyte IL-6 induction by IL-1ß and tumor necrosis factor-α. J. Immunol. *152*:841-850.

O'Banion, M.K., J.C. Miller, J.W. Chang, M.D. Kaplan, and P.D. Coleman (1996) Interleukin-1ß induces prostaglandin G/H Synthase-2 (Cyclooxy-genase-2) in primary murine astrocytes cultures. J. Neurochem. *66*, 2532.

O'Banion, M.K., V.D. Winn, and D.A. Young (1992) cDNA cloning and functional activity of a glucocorticoid regulated inflammatory cyclooxygenase. Proc. Natl. Acad. Sci. USA. *89*: 4888.

Oka, S., and H. Arita (1991) Inflammatory factors stimulate expression of group II phospholipase A_2 in rat cultures astrocyte. J.Biol. Chem. *266*:9956.

O'Neill L.A.J (1995) Interleukin-1 signal transduction. Int. J. Clin. Lab. Res. *25*:169-177.

Park S.K., H.L. Lin, and S. Murphy (1994) Nitric oxide limits transcriptional induction of nitric oxide synthase expression in CNS glial cells. Biochem. Biophys. Res. Commun.*201*:762-768.

Perry V.H., H.C. Bell Brownand, and M.K. Matyszak (1995) Inflammation in the nervous system. Curr Op. Neurobiol. *5*:636-641.

Rodriguez, M., K.D. Pavelko, and W. C. Mc Kinney (1994) Recombinant human IL-6 suppresses demyelination in a viral model of multiple sclerosis. J. Immunol.. *153*:3811-3821.

Rothwell N.J., and J.L.M. Stribjos (1995) Cytokines in neurodegeneration and repair. Int. J. Devl. Neuroscience. *13*:179-185.

Rothwell, N.J., and S.J. Hopkins (1995) Cytokines and the nervous system II: actions and mechanism of action. TINS *18*:130-136.

Salvemini, D., T.P. Misko, J.L.Masferrer, K., Seibert, M.G. Currie, and, P. Needleman (1993) Nitric oxide activates cyclooxygenase enzymes. Proc. Natl. Acad. Sci. USA. *90*:7240-7244.

Sawada, M., A.Suzumura, and T. Marunouchi (1995) Cytokine netwok in the central nervous system and its roles in growth and differentiation of glial and neuronal cells. Int. J. Devl. Neuroscience. *13*:253-264.

Selmaj, K.W., and C.S. Raine (1988) Tumor necrosis factor mediated myelin and oligodendrocyte damage in vitro. Ann Neurol. *23*:339-346.

Selmaj K.W., C.S. Raine, B. Canella, and C.F. Brosnan (1991) Identification of lymphotoxin and tumor necrosis factor in multiple sclerosis lesions. J. clin. Invest. *87*:949-954

Severn, A., M.J.O. Makelam, and F.Y. Liew (1992) The role of protein kinase C in the induction of nitric oxide synthase by murine macrophages. Biochem Biophys. Res. Commun. *188*: 997.

Sharief, M.K. and M.D. Hentges (1991) Association between tumor necrosis factor-α and disease progressionin patients with multiple sclerosis, N. England J. Med. 325:467-451

Simmons, M.L., and S. Murphy (1992) Induction of nitric oxide synthase in glial cells. J. Neurochem. *59*:897-905.

Simmons, M. L., S. Murphy (1994) Roles for protein kinases in the induction of nitric oxide synthase in astrocytes. Glia. *11*: 227-235.

Sparacio S.M., Y.Zhang, J.Vilcek, and E.N.Benveniste (1992) Cytokines regulation of interleukin-6 gene expression in astrocytes involves activation of NFkB like nuclear protein. J.Neuroimmunol. *39*:231-242.

Tadamitsu K., T. Tetsuya, and S. Akira (1994) Cytokine signal transduction. Cell *76*:253-262.

Taga, T., Y.Kawanishi, R.R.Hardy, T. Hirano, and, T. Kishimoto (1987) Receptors for B cell stimulatory factor 2. Quantification, specificity, distribution, and regulation of their expression. J. Exp. Med.*166*:967-981.

Vam Dam, A.M., J.Bauer, A.Man, W.K.H. Hing, C.Marquette, F.J.H. Tilders, and F. Berkenbosch (1995) Appearance of inducible nitric oxide synthase in the rat central nervous system after rabies virus infection and during experimental allergic encephalomyelitis but not after peripheral administration of endotoxin. J. Neurosci. Res.*40*:251-260.

Vigne P., C.Damais, and C. Frelin (1993) IL-1 and TNF-α induce cGMP formation in C6 astrocytomna cells via the nitridergic pathway. Brain Res. *606*:332-334.

Woodroofe M.N., and M.L. Cuzber (1993) Cytokines mRNA expression in inflammatory multiple sclerosis lesions: detection by non radiactive in situ hibridization, Cytokine *5*:583-588

Zheng, Y.M., M.H. H.Schafer, E. Ehihe, H. Sheng, S.Corisdeo, Z. F.Fu, H. Koprowski, and B. Dietzschold (1993) Severity of neurological signs and degree of inflammatory lesions in the brains of rats with Borna disease correlate with the induction of nitric oxide synthase. J. Virol. *67*:5786-5791.

12

THE CRITICAL FACTOR FOR L-DOPA TOXICITY ON DOPAMINE NEURONS IS GLIA

Maria Angeles Mena, Maria José Casarejos,
Alberto Alcazar, Antonio Herranz,
Carlos Luis Paíno and Justo García de Yébenes

L-DOPA is toxic for dopamine neurons in culture but its toxicity has not been proven in animals or in patients with Parkinson's disease. Since most experiments in vitro showing L-DOPA toxicity were performed in neurons cultured in the absence of glia we hypothesized that the discrepancy between the effects of L-DOPA in vivo and in vitro may be related to the presence or absence of glia, respectively. Fetal midbrain neuronal cultures were treated with L-DOPA, 200 μM, in the presence or absence of mesencephalic glia conditioned medium (GCM). In the absence of GCM, L-DOPA greatly reduced the number of tyrosine hydroxylase (TH) immunoreactive neurons and increased the levels of quinones in the medium; GCM prevented these effects of L-DOPA and increased the length and arborization of neurites of the TH immunoreactive cells. In order to characterize the compounds produced by glia and responsible for the protection of dopamine neurons the GCM was fractionated in two samples through filters of molecular size of 10 kD. Both fractions were protective against L-DOPA toxicity, through the fraction <10 kD was more effective than the fraction >10 kD. Candidate neuroprotective agents, including growth factors (BDNF, FGF, GDNF, NGF) and antioxidants (ascorbic, glutathione) were tested and found to provide protection against L-DOPA toxicity. In conclusion, glia prevents L-DOPA toxicity on dopamine neurons through the release of multiple soluble compounds to the extracellular space.

INTRODUCTION

L-DOPA, the most effective compound for the symptomatic treatment of Parkinson´s Disease (PD), is toxic for dopamine (DA) neurons in vitro (Melamed, 1986; Mena et al., 1992, Mena et al., 1993; Mytilineou, 1993; Pardo et al.,

1995) and induced cell death in neuronal cells is mediated by induction of apoptosis (Walkinshaw and Waters, 1994 and 1995) but its effect on DA neurons of healthy animals (Hefti et al., 1981, Perry et al., 1984, Yurek et al., 1991) and humans (Quinn et al., 1986) is debated. In animals, L-DOPA is toxic for DA neurons only in those previously treated with other DA neurotoxic agents (Blunt et al., 1993). Regarding the influence of L-DOPA on the progression of PD, though some authors reported an increase in the prevalence of fluctuations and dyskinesia related to the duration of the L-DOPA treatment (Lesser et al., 1979) the majority of authors described no deleterious effect of L-DOPA or improvement by this compound on life expectancy and no evidence of negative impact on the progression of the disease (Blint et al., 1988; Markham and Diamond, 1986; Uitti et al., 1993).

L-DOPA toxicity on DA neurons in vitro occurs in the absence of glia while in intact animals DA neurons are surrounded and protected by glia. We have shown that midbrain glia conditioned medium (GCM) protects midbrain dopamine neurons from L-DOPA neurotoxicity (Mena et al., 1996). Here we investigated the putative protective factors produced by mesencephalic glia and responsible for the protection of DA neurons from L-DOPA toxicity.

EXPERIMENTAL STUDIES IN VITRO

Neuronal cultures

Neuronal-enriched cultures from embryonic rat (crown-rump length 10-12mm) midbrain were obtained and prepared as previously described (Mena et al., 1993). The cells were seeded in Dulbecco modified Eagle's medium with 15% fetal calf serum (DMEM-FCS) at a density of 10^5 cells/cm^2 on 16mm-diameter multiwells or glass coverslips previously coated with poly-D-lysine, 4.5µg/cm^2, in 0.1M borate buffer, pH 8.4. The cultures were kept in a humidified chamber at 37°C in a 5% CO_2 atmosphere. Twenty-four hours after plating, the cells were changed to serum-free defined medium (EF12) as previously reported (Mena et al., 1993). EF12 consisted of a 1:1 (v/v) EMEM and nutrient mixture of Ham's F-12, supplemented with D-glucose (6mg/ml), insulin (25µg/ml), putrescine (60µM), progesterone (20nM) and sodium selenite (30nM).

Protocols of treatment

The cells, randomly allocated to the different experimental groups, after 6 days in culture, received, for 24 hours, the following treatments: C=Control group: vehicle; D=L-DOPA group: L-DOPA, 200x10^{-6}M; GCM=GCM group: mesencephalic GCM, as described below; L-DOPA+GCM=L-DOPA, 200x10^{-6}M+ GCM; L-DOPA+LMwF=L-DOPA + large molecular weight fraction, >10kD, of GCM; L-DOPA+ SMwF=L-DOPA + small molecular weight fraction, <10kD, of

GCM. Fractions of GCM were obtained by filtration with Centripep 10, (Amicon, Inc, Beverly, MA.). The effect of candidate neuroprotective agents was tested in other experimental groups. Neurotrophic factors were investigated in cells treated with L-DOPA, $200x10^{-6}$M. Nerve growth factor (NGF) was added to a final concentration of 100ng/ ml, and brain derived neurotrophic factor (BDNF), glia derived neurotrophic factor (GDNF) or basic fibroblast growth factor (bFGF) to a final concentration of 50ng/ml for each compound (human recombinant proteins, from PeproTech EC LTD, London). These doses of neurotrophic factors and L-DOPA were chosen according to those found effective in previous experiments (Mena et al., 1995). Other midbrain cultures were treated with glutathione (GSH) 200 or $400x10^{-6}$M, and ascorbic acid (AA), 3 to 2000μM, in the presence or absence of L-DOPA, $200x10^{-6}$M.

Methods of evaluation

Glia-enriched mesencephalic cultures were obtained from sibling cells kept in culture for 7 days in DMEM-FCS (Levison and McCarthy, 1991). Then the DMEM-FCS medium was discarded, the cells were washed out three times with Leibovitz's L-15 and then cultured in serum-free defined medium. The conditioned medium was collected from cells cultured for 24 hours. In these cultures the astrocytes were identified by their positive staining with glial fibrillary acidic protein (GFAP) antibodies (Dako, Denmark) After seven days in culture, the number of astrocytes was around 80-90% of total cells.

Neuronal-enriched cultures were characterized by immunostaining with mouse anti-ß-tubulin isotype III (Sigma, St. Louis, Banerjee et al.,. 1990), mouse anti-nestin (DSHB, Iowa, Hockfield and McKay, 1985) and the monoclonal antibodies A2B5 (Eisenbarth et al., 1979, Raff et al., 1983) and O1 (Sommer and Schachner, 1981) and with rabbit anti-GFAP. Dopamine (DA) neurons were immunostained with a monoclonal mouse anti-tyrosine hydroxilase (TH) antibody (Boehringer-Mannheim). Nuclei were stained with bisbenzimide (Hoechst 33342, Hilwing and Gropp, 1975), and apoptotic cells were identified by chromatin condensation.

After 7 days in culture, the number of viable cells was $9.5\text{-}12.8x10^4$cells/cm^2. The percentage of neurons (ß-tubuline immunoreactive cells) was 63±2.5%, GFAP positive cells were only 2.33±0.58%, and oligodendrocytes accounted for 9.38±1.76% of cells, as assessed by the galactocerebroside staining with the O1 antibody. A small population of bipotential glial precursors (O-2A) was present in the cultures, since 14.3±1.56 of cells were stained by the A2B5 antibody and didn't possess the morphology of neurons. Most of the remaining cells were pluripotential neural progenitors and stained for nestin intermediate filament. Figures 12-1 and 12-2 show the immunocytochemical characterization of cultures. At this time in culture, the number of DA neurons, TH-positive ranged between 2.1 and 5.4% of the total (3.57 ± 0.14%).

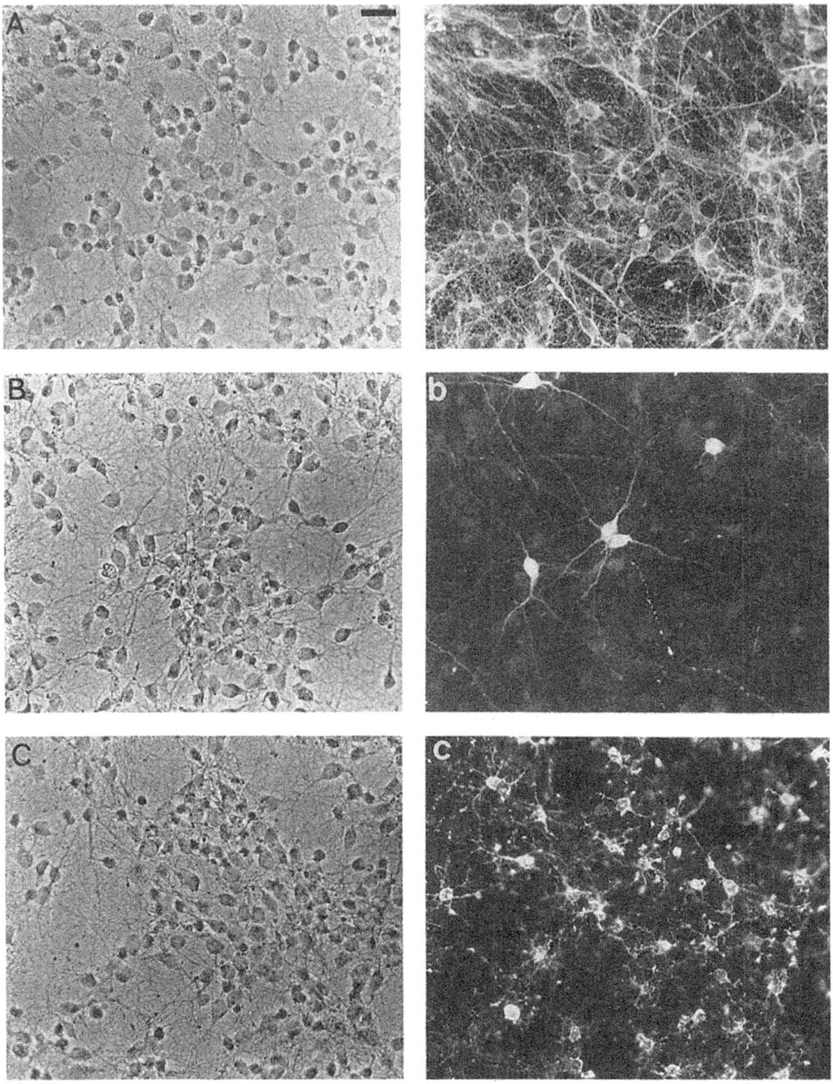

Figure 12-1. Cellular characterization of neuronal-enriched cultures from rat fetal (E14) midbrain cultures. Immunocytochemical characterization (a-c) of neuronal-enriched cultures after 7 days in vitro, with their corresponding phase contrast images (A-C). a: ß-tubulin isotype III positive cells (neurons); b: TH-positive cells (dopamine neurons). c: A2B5-stained cells (most of them are bipotential glial precursors, O-2A, and a few are maturing oligodendrocytes or type II astrocytes). Scale bar: 25 μm. (see also figure 12-2).

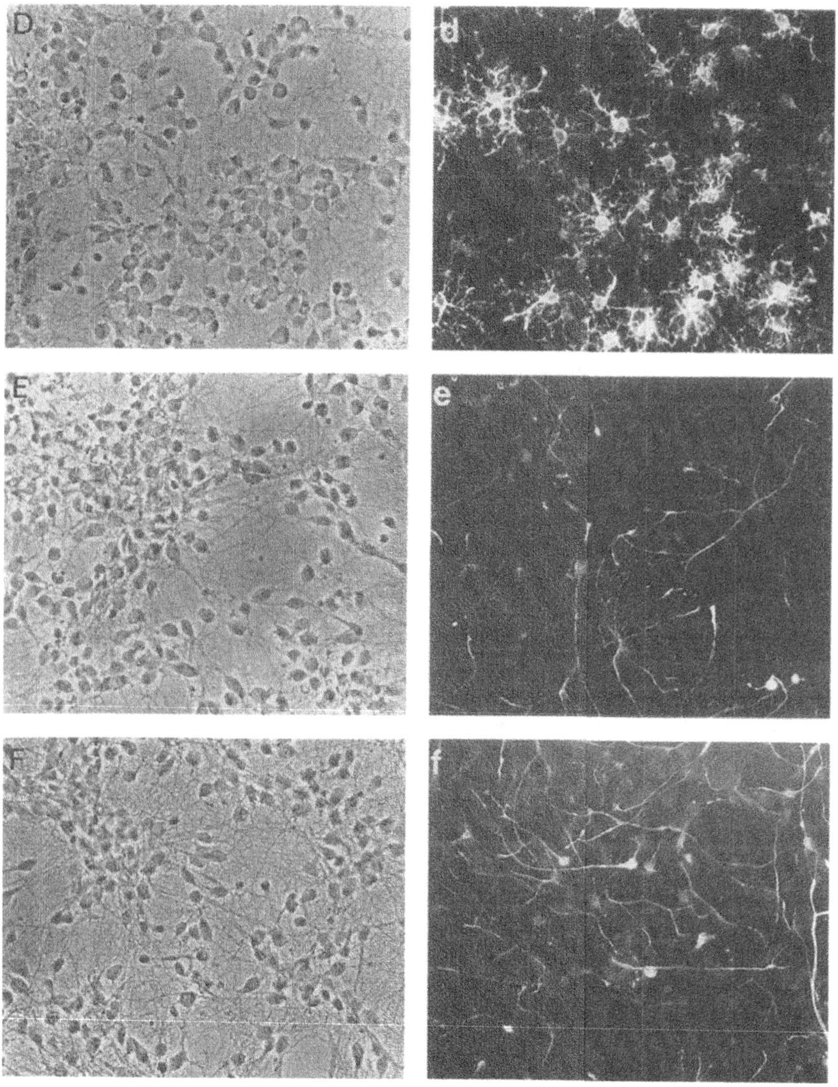

Figure 12-2. Cellular characterization of neuronal-enriched cultures from rat fetal (E14) midbrain cultures. Immunocytochemical characterization (d-f) of neuronal-enriched cultures after 7 days in vitro, with their corresponding phase contrast images (D-F). d: O1-stained cells (mature oligodendrocytes); e: GFAP-positive cells (astrocytes); f: Nestin-positive cells, which stains pluripotential neural progenitors, as well as maturing astrocytes (including radial glia). Scale bar: 25 μm.

^{3}H-DA uptake

It was measured after incubation of the cells with ^{3}H-DA, 10^{-8}M, in the presence of pargyline, 10^{-5}M, and AA, 10^{-3}M, at 37°C for 30min. Non-specific uptake/binding was calculated at 0° C and/or in the presence of 10^{-5}M mazindol (Beart and McDonald, 1980). Proteins were measured according to Bradford (1976). Quinones were estimated by spectrophotometry at 490 nm (Mena et al., 1993).

Ascorbic acid (AA) was measured by high performance liquid chromatography with electrochemical detection (HPLC/ED) according to previously described techniques (Washko et al., 1989).

Glutathione measurements

GSH was measured according to Hyland and Bottiglieri (1992), with minor modifications. The chromatographic system consisted of two Beckman-112 pumps, a 231 XL-Gilson autosampler, and a 5µm Beckman ODS (15cm x 4.6mm I.D.) reversed phase column. The OPA-derivatized GSH was achieved using a Perkin Elmer LS4 spectrofluorimeter. Excitation and emission wave lengths were set at 365 and 455 nm. Respectively.

Mobil phase A was 0.05M sodium acetate pH 6.8, and phase B was acetonitrile. The flow rate was 1.3ml/min. and a linear gradient from 8% to 25% B in 5 min. was performed, 20µl of sample (culture media plus perchloric acid to a 0.4M final concentration) could be injected every 10min.

Gel Electrophoresis of proteins in glia conditioned medium and defined medium

One-dimensional sodium dodecyl sulfate (SDS)-polyacrylamide gel electrophoresis (PAGE) resolution of proteins was performed on 5.5-cm-long minigels containing 15% acrylamide and 0.375% N,N'-methylenebi-sacrylamide according to the technique of Hames (1986). Gels were stained with silver reagent. Stained bands were scanned and compared with an image analyser equipped with the software package Diversity One (PDI, New York).

MECHANISMS OF ACTION

Effects of glia conditioned medium (GCM) on L-DOPA induced apoptosis

The effect of GCM was mediated, at least in part, by inhibition of apoptosis of DA neurons. The percentage of apoptotic cells, which was 13.84% in control cultures, rose to 27.52% in cultures treated with L-DOPA; returned to control levels, 15.53% in cultures treated with L-DOPA+GCM, and was below control levels, 7.25%, in cultures treated with GCM (fig. 12-3).

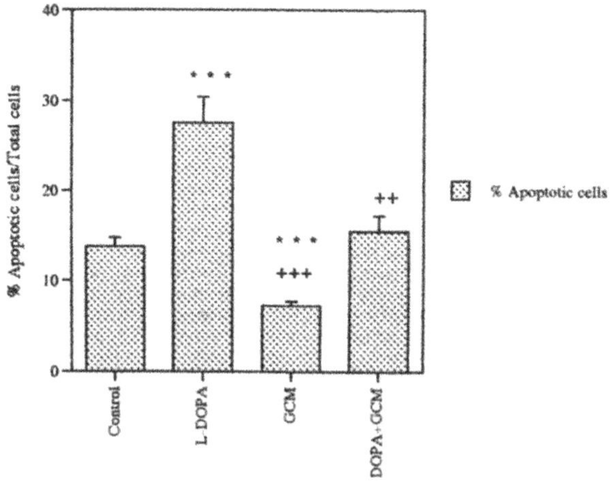

Figure 12-3. Effects of midbrain glia conditioned medium (GCM) on apoptotic cell death in L-DOPA-treated fetal midbrain cultures. Cells were treated at 5 days post-plating with L-DOPA or vehicle and/or GCM. The numbers represent the mean±s.e.m. of 6 samples, expressed in % of apoptotic cells versus total cell number. Statistical analysis was performed by analysis of variance followed by student's t test. Differences were considered significant when $p<0.05$. ***=$p<0.001$, with respect to the control; ++=$p<0.01$, +++= $p<0.001$, with respect to the group treated with L-DOPA. The concentration of L-DOPA was allways 200 x 10^{-6} M. Control group treated with defined medium.

Effects of glia conditioned medium (GCM) and its different fractions on L-DOPA toxicity in dopamine neurons

GCM protected from L-DOPA-toxicity in fetal mesencephalic cultures (Table 12-1, Fig. 12-4 and 12-5) and this protective effect was retained by the fractions of molecular size greater and smaller than 10kD, but mostly by the small molecular size fraction. L-DOPA reduced the number of viable cells, their neuritic arborization, the high affinity ^{3}H-DA uptake and the number of TH immunoreactive cells and increased the levels of quinones. At the doses of L-DOPA used in these experiments, 200x10^{-6} M, the number of surviving DA neurons, as evaluated by the number of TH immunoreactive cells, was reduced to 27% of the controls. The neurotrophic effect of GCM was very marked since it increased the number of TH immunoreactive cells and the number of high affinity ^{3}H-DA uptake sites to 207 and 216% of controls, respectively (Table 12-1). The effects of mesencephalic GCM was selective for DA neurons since it did not modify the total number of nuclei but only the number of TH positive cells. Co-treatment with GCM not only prevented completely the effects of L-DOPA on the number of viable cells, their neuritic arborization, their ^{3}H-DA uptake to control values and levels of quinones, but even increased the number of TH positive cells to 171% of the average values for vehicle treated cells (Table 12-1). Regarding the cell morphology, cotreatment with GCM prevented the damage of the neuritic network induced by L-DOPA (fig. 12-4). TH immunoreactivity, which virtually disappeared from cultures treated with L-DOPA (Fig. 12-5B), was greatly enhanced, above control levels, in cultures treated with L-DOPA+GCM (fig. 12-5D). The neurotrophic effects of midbrain GCM are retained by its fractions of molecular size larger and smaller than 10kD, though the former was less effective than the latter (Table 12-1, fig. 12-5E and 12-5F).

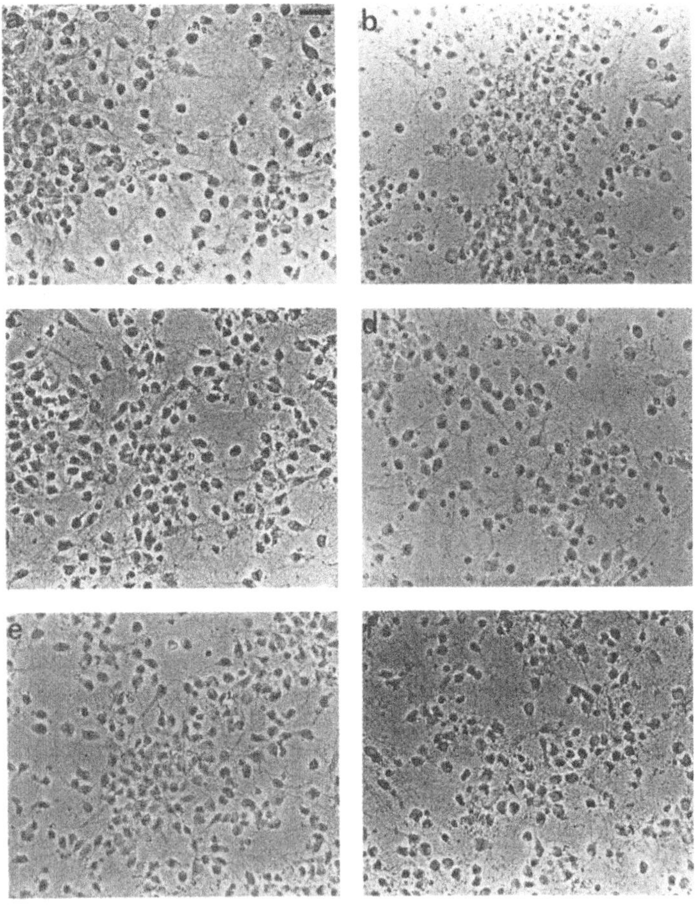

Figure 12-4. Effects of midbrain GCM and its fractions on cell morphology in L-DOPA-treated fetal midbrain cultures. Phase contrast microphotographs. Cells were treated at 5 days post-plating with: a: Control group treated with defined medium; b: Group treated with defined medium + L-DOPA, $200x10^{-6}$ M; c: Group treated with GCM; d: Group treated with GCM + L-DOPA, $200x10^{-6}$ M; e: Group treated with the fraction <10kD of GCM + L-DOPA, $200x10^{-6}$ M; f: Group treated with the fraction >10kD of GCM + L-DOPA, $200x10^{-6}$ M. Scale bar= 25μm.

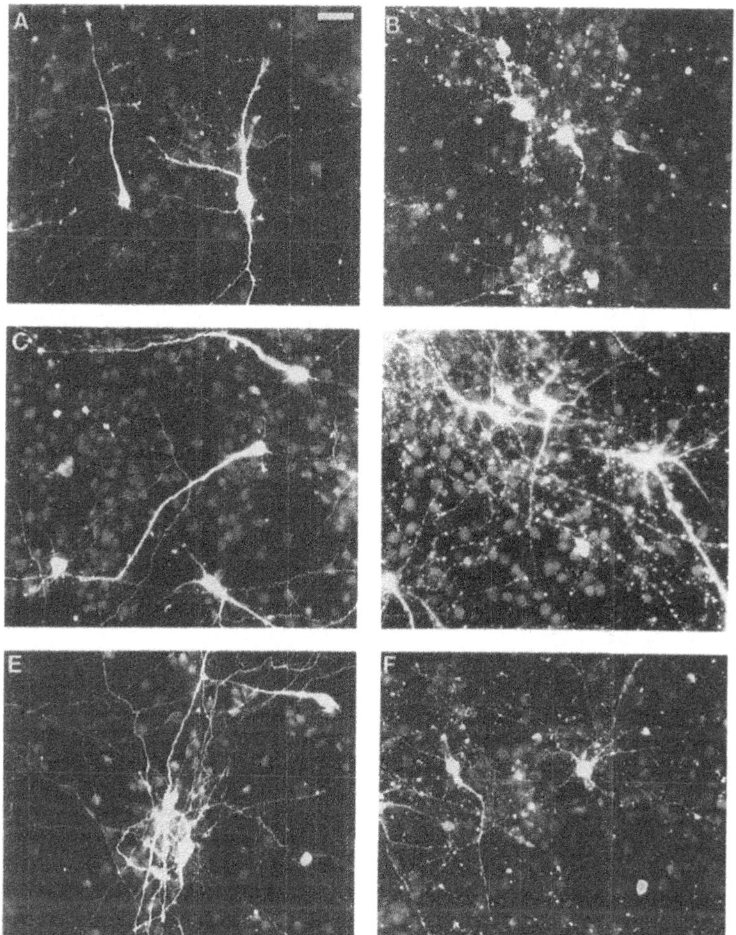

Figure 12-5. Effects of midbrain GCM and its fractions on TH immunohistochemistry in L-DOPA-treated fetal midbrain cultures.Cells were treated at 5 days post-plating with: A: Control group treated with defined medium; B: Group treated with defined medium+L-DOPA, $200x10^{-6}$M; C: Group treated with GCM; D: Group treated with GCM +L-DOPA, $200x10^{-6}$M; E: Group treated with the fraction <10kD of GCM+L-DOPA, $200x10^{-6}$M; F: Group treated with the fraction >10 kD of GCM+L-DOPA, $200x10^{-6}$M. Scale bar= 25μm.

Protein gel electrophoresis of glia conditioned medium and defined medium

Aliquots of 5μg protein from glia conditioned medium (A) and defined medium (B), were subjected to SDS-PAGE and densitometric scanning analysis. The silver staining of the gel (left) and its densitometric scanning (right) are shown (fig. 12-6). The numbers 1 to 7 (fig. 12-6A) indicate the seven new most prominent bands present on GCM, with relative molecular weight 181, 158, 143, 55, 42, 35, and 33 (x1000), respectively. The arrows indicate the migration position of relative molecular weight markers (Mr x 1000) (fig. 12-6).

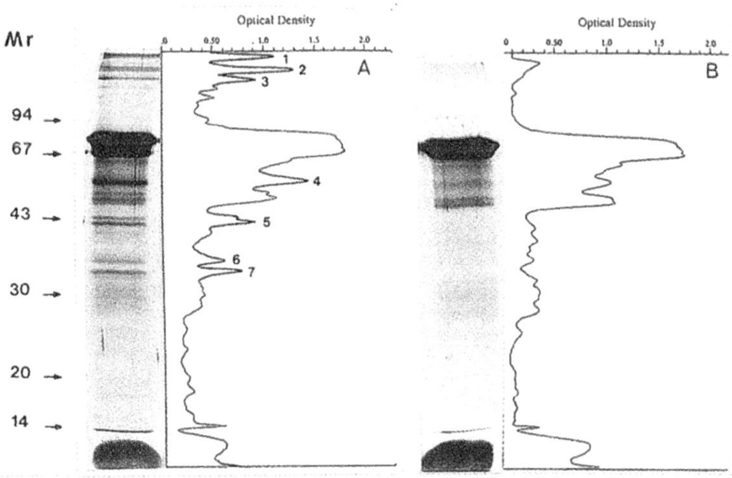

Figure 12-6. Gel electrophoresis of glia conditioned medium (GCM) and defined medium. SDS-PAGE and densitometric scanning of glia conditioned medium (A) and defined medium (B). 5μg of protein were analyzed by SDS-PAGE. The silver staining of the gel (left) and its densitometric scanning (right) are shown. Numbers 1 to 7 on figure (A) indicate the seven news and majorities bands presents on GCM, with relative molecular weight 181, 158, 143, 55, 42, 35, and 33 (x1000), respectively. The arrows indicate the migration position of relative molecular weight markers (Mr, x1000).

Effect of neuroprotective agents produced by glia on L-DOPA toxicity

Putative neuroprotective agents produced by glia were subdivided into two categories, neurotrophic factors of peptidic nature and free radical scavenging agents. NGF was the only neurotrophic factor that significantly reduced the elevation of quinones induced by L-DOPA (Table 12-1); however, the effect of NGF on TH immunoreactive cells was modest. The treatment with NGF increased the number of TH immunoreactive cells in L-DOPA+NGF to 61% of control from 27% of TH immunoreactive cells in L-DOPA treated group; bFGF increased the number of TH immunoreactive cells to 80% of control in L-DOPA treated cultures and BDNF, and GDNF to 124% and 111%, respectively, of control in L-DOPA treated group (Table 12-1). This discrepancy between quinone quenching effect and preservation of TH immunoreactive cells suggest that the protection of midbrain dopamine neurons by mesencephalic GCM may be related to several mechanisms, some related to the suppression of quinones and others unrelated.

Table 12-1. *Effects of GCM and its fractions, neurotrophic factors and glutathione on L-DOPA toxicity in rat midbrain fetal neurons.*

Treatment	N	Nuclei (well)	TH+cells (well)	3HDA upt (cpm/well)	Quinones (O.D. 490nm)
Control	22	28137±2739	1005±43	20552±450	0.147±0.001
% of controls		100±10	100±4.2	100±2.1	100±1
L-DOPA	20	83±8	27±9[c]	35±2[c]	259±13[c]
GCM	20	116±7	207±30[c]	216±19[c]	106±26
L-DOPA+GCM	22	106±7	171±12[c,z]	124±12[z]	108±1[z]
L-DOPA+LMwF	15		130±10[a,z]	50±2[c]	210±5[c,y]
L-DOPA+SMwF	15		170±11[c,z]	85±8[z]	150±2[c,z]
L-DOPA+NGF	6		61±23		180±9[c,z]
L-DOPA+BDNF	6		124±27[z]		253±4[c]
L-DOPA+bFGF	6		80±35		260±21[c]
L-DOPA+GDNF	6		111±17[z]		226±20[c]
GSH 200	6	115±10	108±9	93±4	102±2
GSH 400	6	122±8	179±24[c]	117±9	98±2
L-DOPA+GSH200	7	107±6	89±5[z]	39±5[c]	132±14[b,z]
L-DOPA+GSH400	7	114±14	100±8[z]	46±3[c,x]	101±1[z]

The numbers represent the mean±s.e.m. of N samples, expressed in % of the corresponding controls included in each experiment. Statistical analysis was performed by analysis of variances followed by Student's t test. Differences were considered significant when $p<0.05$. a=$p<0.05$, b=$p<0.01$, c=$p<0.001$, with respect to the control; x=$p<0.05$, y=$p<0.01$, z=$p<0.001$, with respect to the group treated with L-DOPA. The concentration of L-DOPA was allways 200×10^{-6} M. GCM and its small (SMwF, <10kD) and large (LMwF, >10kD) molecular weight fractions were added in the same volume. We estimate that during the process of fractionation the conditioned medium was concentrated 2.5times. BDNF, bFGF and GNDF were added to a final concentration of 50ng/ml; NGF was added to a final concentration of 100ng/ml. GSH was added to a final concentration of 200 and 400×10^{-6} M.

Glutathione induced a dose dependent increase of TH positive cells and ^{3}H-DA uptake and a dose dependent reversal of L-DOPA-induced elevation of quinones (Table 12-1). In control cultures GSH, 400×10^{-6} M, increased the number of TH immunoreactive cells, to 179% of the control value (Table 12-1). Since GSH didn't significantly modify the total number of cells in the culture, this increase in TH immunoreactive cells suggested that GSH had a selective neurotrophic effect on DA neurons. In cultures treated with L-DOPA, GSH, 400 x 10^{-6} M totally prevented the elevation of quinones and the disappearance of TH immunoreactive cells but not the reduction of ^{3}H-DA uptake induced by L-DOPA (Table 12-1). Since ^{3}H-DA uptake is considered an index of arborization, GSH is more effective for the prevention of cell death than for the damage to the terminals of the DA neurons.

The effect of GCM was mediated, at least in part, by the elevation of glutathione on DA cultures. The levels of GSH, which was 2.25±0.22μM (100%) in the medium of control cultures, was below control levels, 1.23±0.18 μM (55%) in cultures treated with L-DOPA, and rose to 25.6±0.28μM (1138%) in cultures treated with GCM, and to 27.4±0.36μM (1218%) in cultures treated with L-DOPA+GCM (fig. 12-7). The levels of ascorbic acid (AA) were higher in mesencephalic GCM (22μM) than in defined medium (8μM). Concerning the different components of glia, the low molecular weight fraction had higher concentration of AA (17μM) than the large molecular weight fraction (5μM). AA prevented L-DOPA toxicity on midbrain DA neurons. but the dose required for full protection is greater (from 50 to 100μM) than the concentration found in GCM. AA at concentrations found in GCM, from 12 to 25μM, rose the percentage of surviving DA neurons after exposure to L-DOPA, 200μM, from 52 to 71 and 87% of the controls, respectively, and decreased the levels of quinones from 283 to 143 and 115% of controls (Table 12-2). Therefore, though AA accounts for part of the protective effects of GCM, other factors should be considered.

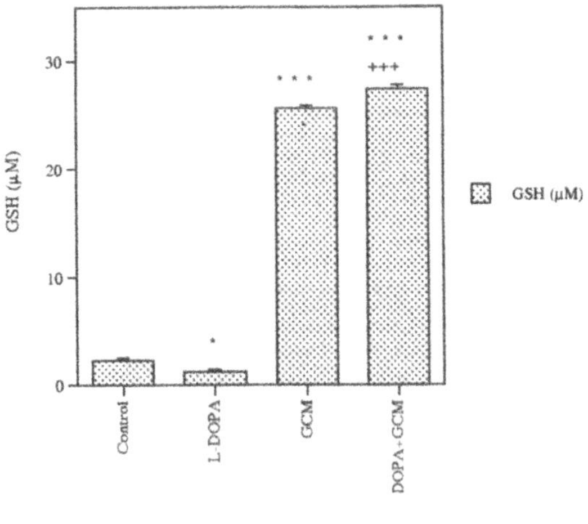

Figure 12-7. Effects of GCM on GSH levels in the medium of fetal midbrain cultures. Cells were treated at 5 days post-plating with L-DOPA or vehicle and/or GCM. The numbers represent the mean±s.e.m. of 6 samples. Statistical analysis was performed by analysis of variance followed by student's t test. Differences were considered significant when $p<0.05$; *=$p<0.05$; ***=$p<0.001$, with respect to the control; +++= $p<0.001$, with respect to the group treated with L-DOPA. The concentration of L-DOPA was allways 200 x 10^{-6} M. Control group treated with defined medium.

***Table 2.** Effect of ascorbic acid on L-DOPA toxicity in rat midbrain fetal neurons.*

AA (μM)	L-DOPA (μM)	TH+(cells/well)	Quinones (O.D.490nm)
0	0	1283±89 (100%)	0.122±0.02 (100%)
0	200	664±92[c] (52%)	0.345±0.03[c] (283%)
3	200	812±92[b] (63%)	0.209±0.03[c,x] (171%)
6	200	843±87[b] (66%)	0.177±0.03[b,y] (145%)
12	200	913±71[a] (71%)	0.174±0.01[c,y] (143%)
25	200	1113±104[y] (87%)	0.140±0.001[c,z] (115%)
50	200	1126±77[y] (88%)	0.130±0.001[b,z] (106%)
100	200	1438±125[z] (112%)	0.128±0.002[z] (105%)
500	200	1238±131[y] (96%)	0.124±0.001[z] (102%)
1000	200	1166±104[y] (91%)	0.123±0.001[z] (101%)
2000	200	378±60[c] (29%)	0.121±0.001[z] (99%)

The numbers represent the mean±s.e.m. of (n= 5-9) samples. Statistical analysis was performed by analysis of variance followed by Student's t test. Differences were considered significant when $p<0.05$. a=$p<0.05$, b=$p<0.01$, c=$p<0.001$, with respect to the control; x=$p<0.05$, y=$p<0.01$, z=$p<0.001$, with respect to the group treated with L-DOPA.

CONCLUSIONS

Our studies reveal that the toxicity of L-DOPA in DA neurons in vitro is prevented by soluble factors secreted by midbrain glial cells. Therefore, these experiments provide an explanation for the discrepancies between experimental studies on L-DOPA toxicity in vitro and results obtained in experiments in vivo in animals and clinical data in patients with PD.

Glia produce neurotrophic and neurite-promoting agents (Engele et al., 1991; Nagata et al., 1993; Hoffer et al., 1994; Takeshima et al., 1994a and 1994b; Muller et al., 1995) that enhance development, survival, neurite extension and resistance to neurotoxins of DA neurons. Confluent, type 1, astrocytes and oligodendrocyte, type 2 astrocyte (O-2A) progenitors, and conditioned medium from the O-2A progenitor cultures, increase the survival of DA neurons after serum deprivation (Takeshima et al., 1994a and b). Microglial conditioned medium promotes the survival and development of cultured embryonic midbrain neurons (Nagata et al., 1993). Glial neurotrophic factors, including fibroblast growth factor (FGF) (Mena et al., 1995) and glial derived neurotrophic factor (GDNF) (Hoffer et al., 1994) are effective on DA neurons. GDNF reverses 6-OHDA-induced injury to midbrain DA neurons in vivo (Hoffer et al., 1994).

In this study, the protective effect of GCM on L-DOPA toxicity was associated to a reduction of quinones in the medium. Several neurotrophic factors increase the activity of important enzymes of the free radical scavenging system. NGF induces superoxide dismutase (Nistico et al., 1992). Therefore, GCM may decrease L-DOPA toxicity by reducing pro-oxidants in culture. NGF and BDNF are expressed at high levels in cultured glia (Spranger et al., 1990) though they have been detected only in brain neurons in vivo. GDNF was isolated as a DA selective neurotrophic factor produced by glia (Lin et al., 1993) and bFGF is expressed by both glia and neurons (Gall et al., 1994).

The neuroprotective effect of mesencephalic GCM on L-DOPA toxicity is mainly retained by the filtrate fraction smaller than 10kD. Most neurotrophic factors appear to be retained in the higher molecular weight fraction, although the possibility of the presence of inhibitors in either fraction has not been examined. The fraction smaller than 10kD contains small molecules with antioxidant properties including AA and GSH. The levels of AA were three times higher in midbrain GCM than in defined medium This difference of concentration suggested that glia accumulated AA. We have shown that AA at concentrations of $200 x 10^{-6}$ M, protects midbrain dopamine neurons from L-DOPA toxicity (Mena et al., 1993, Pardo et al., 1995). In this study we show that the levels of AA found in GCM could provide a partial, but not a complete protection against L-DOPA toxicity.

GSH is a major component of the mechanisms of cell protection against oxidative stress (Meister, 1988). GSH was decreased in the substantia nigra of patients with idiopathic PD (Riederer et al., 1989; Sian et al., 1994) and that reduction is considered relevant to the pathophysiology of this disease. The nigrostriatal DA neurons, the most important disease target in patients with PD, are normally surrounded by a low density of glia (Damier et al., 1993; Makar et al., 1994; Sagara et al., 1993) suggesting that these neurons are less protected from and thereby more susceptible to oxidative stress. GSH peroxidase, the major protective enzyme against peroxide toxicity, is exclusively located in glia in human midbrain (Damier et al., 1993). Astrocytes play an important role in anti oxidative processes in the brain (Makar et al., 1994). Astrocytes, but not neurons, are rich in GSH peroxidase, the enzyme which produces GSH in the brain (Slivka et al., 1987) and in culture (Raps et al., 1989, Sagara et al., 1996, Siushansian et al., 1996). Therefore, GSH is an important small antioxidant produced by glia, with neurotrophic properties on midbrain DA neurons. Our studies show that GSH increases the number of TH+ cells and protects from L-DOPA toxicity in midbrain cultures.

In PD, the density of glia GSH peroxidase immunoreactive cells surrounding the surviving DA neurons is increased, suggesting a compensatory proliferation of glia that protects surviving neurons against pathological death (Damier et al., 1993). The physiological effect of L-DOPA and catecholamines (CA) on development and survival of the nervous system in general, and CA producing neurons in particular, is unknown. Though L-DOPA is toxic for CA neurons in vitro (Melamed et al., 1986; Mena et al., 1992 and 1993; Pardo et al., 1995) CA are needed for the development of the nervous system and survival in mammals (Zhou et al., 1995; Thomas et al., 1995). Therefore, it is likely that L-DOPA and CA are critical factors for the survival and differentiation of DA neurons in developmental stages and adulthood. The effect of L-DOPA on DA cells in vivo, might be neurotrophic or neurotoxic, depending on several environmental elements produced by glia. In this case L-DOPA and CA may promote, during development, differentiation or apoptosis of CA neurons, depending on the balance between their concentration and that of neuroprotective elements produced by glia. This possibility requires further investigation.

ACKNOWLEDGMENTS

This work was supported by CICYT, SAF 96-1099 and CAM97-85/19. The authors thank to Ana Carazo for expert technical assistance.

REFERENCES

Banerjee A, M.C. Roach, P. Trcka and R.F. Ludueña (1990) Increased microtubule assembly in bovine brain tubulin lacking the type III isotype of beta-tubulin. J. Biol. Chem. *265*:1794-1799.

Beart P.M. and D. McDonald (1980) Neurochemical studies of the mesolimbic dopaminergic pathway: somatodendritic mechanisms and GABAergic neurons in rat ventral tegmentum. J. Neurochem. *34*: 1622-1629.

Blint J, A.M. Bonnet and Y. Agid (1988) Does L-DOPA aggravate Parkinson's disease?. Neurology *38*: 1410-1416.

Blunt S.B., P. Jenner and C.D. Marsden (1993) Suppressive effect of L-DOPA on dopamine cells remaining in the ventral tegmental area of rats previously exposed to the neurotoxin 6-hydroxydopamine. Mov. Disord. *8*: 129-133.

Bradford M.M. (1976) A rapid and sensitive method for the quantification of microgram quantities of protein utilizing the principle of protein-dye-binding. Anal. Biochem. *72*: 248-254.

Damier P., E.C. Hirsch, F. Javoy-Agid, P. Zhang and Y. Agid (1993) Protective role of glutathione peroxidase against neuronal death in Parkinson's disease. Neuroscience *52*: 1-6.

Eisenbarth, G.S., F.S. Walsh and N. Niremberg (1979) Monoclonal antibody to a plasma membrane antigen of neurons. P.N.A.S. USA *76*:4913-4917.

Engele J., D. Schubert, M.C. Bohn (1991) Conditioned media derived from glial cell lines promote survival and differentiation of dopaminergic neurons in vitro: role of mesencephalic glia. J. Neurosci. Res. *30*: 359-371.

Gall C.M., R. Berschauer, P.J. Isackson (1994) Seizures increase basic fibroblast growth factor messenger mRNA in adult rat forebrain neurons and glia. Mol. Brain Res. *21*: 190-205.

Hames B.D. (1986) An introduction to polyacrylamide gel electrophoresis. In: Gel electrophoresis of proteins. (Hames BD and Rickwood D., eds), pp 1-92. Oxford, IRL press.

Hefti F., E. Melamed, J. Bhawan and R.J. Wurtman (1981) Long-term administration of L-DOPA does not damage dopaminergic neurons in the mouse. Neurology *31*: 1194-1195

Hilwig I. and A. Gropp (1975) pH dependent fluorescence of DNA and RNA in cytologic staining with "33342 Hoechst". Exp. Cell Res. *91*:457-460.
Hockfield S. and R.D. McKay (1985). Identification of major cell classes in the developing mammalian nervous system. J. Neuroscience *5*:3310-3328.
Hoffer B.J., A. Hofman, K. Bowenkamp, P. Huettl, J. Hudson, D. Martin, L.F.H. Lin, G.A. Gerhardt (1994) Glial cell line-derived neurotrophic factor reverses toxin-induced injury to midbrain dopaminergic neurons in vivo. Neurosci. Lett. *182*: 107-111
Hyland K. and T. Bottiglieri (1992) Measurerment of total plasma and cerobrospinal fluid homocysteine by fluorescence following high-performance liquid chromatography and precolumn derivatization with O-phthaldialdehyde. J. Chromatogr. *570*: 55-62
Lesser R.P., S. Fahn, S.R. Snider, L.J. Cote, W.P. Isgreen and R.E. Barret (1979) Analysis of the clinical problems in parkinsonisms and the complications of long-term L-DOPA therapy. Neurology *29*: 1253-1260
Levison S.W. and K.D. McCarthy (1991) Astroglia in culture. In: G Banker and K Goslin (eds) Culturing Nerve cells, pp 309-336. The MIT Press, Cambridge, Massachusetts
Lin L.F.H., D.H. Doherty, J.D. Lile, S. Bektosh, F. Collins (1993) GDNF: a glial cell line-derived neurotrophic factor for midbrain dopaminergic neurons. Science *260*: 1130-1132.
Makar T.K., M. Nedergaard, A. Preuss, A.S. Gelbard, A.S. Perumal, and A.J. Cooper (1994) Vitamin E, ascorbate, glutathione, glutathione disulfide and enzymes of glutathione metabolism in cultures of chick astrocytes play an important role in antioxidative processes in the brain. J. Neurochem. *62*: 45-53.
Markham C.H. and S.G. Diamond (1986) Long term follow up of early DOPA treatment in Parkinson's Disease. Ann. Neurol. *19*: 365-372.
Meister A. (1988) Glutathione metabolism and its selective modification. J. Biol. Chem. *263*: 17205-17208.
Melamed E. (1986) Initiation of L-DOPA therapy in parkinsonian patiens should be delayed until the advanced stages of the disease. Arch. Neurol. *43*: 402-405.
Mena M.A., B. Pardo, M.J. Casarejos, S. Fahn, J. G. de Yebenes (1992) Neurotoxicity of L-DOPA on catecholamine-rich neurons. Mov. Disord. *7*: 441-445
Mena MA, B Pardo, CL Paino, J G de Yebenes (1993) L-DOPA toxicity in foetal rat midbrain neurones in culture: modulation by ascorbic acid. NeuroReport *4*: 438-440.
Mena M.A., M.J. Casarejos, G. Gimenez-Gallego, G. de Yebenes (1995) Fibroblast growth factors: structure-activity on dopamine neurons in vivo. J. Neural. Transm. *9*: 1-14
Mena M.A., M.J. Casarejos, C.L. Paino, J. G. de Yebenes (1996) Glial conditioned medium protects fetal rat midbrain neurons in culture from L-DOPA toxicity. NeuroReport 7: 441-445
Muller H.W., U. Junghans, J. Kappler (1995) Astroglial neurotrophic and neurite-promoting factors. Pharmac. Ther. *65*: 1-18.
Mytilineou C., S. Han and G. Cohen (1993) Toxic and protective effects of L-DOPA on mesencephalic cell cultures. J. Neurochem. *61*: 1470-1478
Nagata K., N. Takei, K. Nakajima, H. Saito, S. Kohsaka (1993) Microglial conditioned medium promotes survival and development of cultured mesencephalic neurons from embryonic rat brain. J. Neurosci. Res. *34*: 357-363.
Nistico G., M.R. Ciriolo, K. Fiskin, M. Iannone, A. De Martino and G. Rotillo (1992) NGF restores decrease in catalase activity and increase superoxide dismutase and glutatione peroxidase activity in the brain of the aged rats. Free Radic. Biol. Med. *12*: 177-181.
Pardo B., M.A. Mena, M.J. Casarejos, C.L. Paino, J.G. de Yebenes (1995) Toxic effects of L-DOPA on mesencephalic cell cultures: protection with antioxidants. Brain Res. *682*: 133-143.
Perry T.L., V. Wee Yong, M. Ito, J.G. Foulks, R.A. Wall, D.V. Godin and R.M. Clavier (1984). Nigrostriatal dopamine neurons remain undamaged in rats given high doses of L-DOPA and carbidopa chronically. J. Neurochem. *43*: 990-993.
Quinn N., D. Parker, I. Janota and C.D. Marsden (1986) Preservation of the substantia nigra and locus coeruleus in a patient receiving L-DOPA (2Kg) plus decarboxylase inhibitor over a four-year period. Mov. Disord *1*: 65-68.
Raff, M., E.R. Abney, J.Cohen, R.Lindsay and M. Noble (1983) Two types of astrocytes in cultures of developing rat white matter: differences in morphology, surface gangliosides and growth characteristics. J. Neurosci. *3*:1289-1300.

Raps S.P., J. Lai, I. Hertz, A.J.L. Cooper (1989) Glutahione is present in cultured astrocytes but not in cultured neurons. Brain Res. *493*: 398-401.

Riederer P., E. Sofic, W.D. Rausch, B. Schmidt, G.P. Reynolds, K. Jellinger, M.B.H.Youdim (1989) Transition metals, ferritin, glutathione, and ascorbic acid in parkinsonian brains. J. Neurochem. *52*: 515-520.

Sagara J., K. Miura and S. Bannai (1993) Maintenance of neuronal glutathione by glial cell. J. Neurochem. *61*: 1672-1676.

Sagara J., N. Makino, and S. Bannai (1996) Glutathione efflux from cultured astrocytes. J. Neurochem. *66*: 1876-1881.

Sian J., D.T. Dexter, A.J. Lees, S.E. Daniel, Y. Agid, F. Javoy-Agid, P. Jenner, C.D. Marsden (1994) Alterations in glutathione levels in Parkinson's disease and other neurodegenerative disorders affecting basal ganglia. Ann. Neurol. *36*: 348-355.

Siushansian R., S.J. Dixon, and J.X. Wilson (1996) Osmotic swelling stimulates ascorbate efflux from cerebral astrocytes. J. Neurochem. *66*: 1227-1223.

Slivka A., C. Mytilineou, G. Cohen (1987) Histochemical evaluation of glutahione in brain. Brain Res. *409*: 275-284.

Sommer, I. and M. Schachner (1981) Monoclonal antibodies (O1 to O4) to oligodendrocyte cell surfaces: An immunocytological study in the central nervous system. Dev. Biol. *83*:311-327.

Spranger M., D. Lindholm, C. Bandtlow, R. Heuman, H. Gnahn, M. Naher Noe, H. Thoenen (1990) Regulation of nerve growth factor (NGF) synthesis in rat central nervous system: comparison between the effects of interleukin-1 and various growth factors in astrocyte culture and in vivo. Eur. J. Neurosci. *2*: 69-76.

Takeshima T., K. Shimoda, Y. Sauve, J.W. Commissiong (1994a) Astrocyte-dependent and independent phases of the development and survival of rat embryonic day 14 mesencephalic neurons in culture. Neuroscience *60*: 809-823.

Takeshima T., J.M. Johnston, J.W. Commissiong (1994b) Oligodendrocyte-type-2 astrocyte (O-2A) progenitors increase the survival of rat mesencephalic, dopaminergic neurons from death induced by serum deprivation. Neurosci. Lett. *166*: 178-182.

Thomas S.A., A.M. Matsumoto, R.D. Palmiter (1995) Noradrenaline is essential for mouse fetal development. Nature *374*: 643-646.

Uitti R.J. J.E. Ahlskog, D.M. Maraganore, M.D. Muentner, E.J. Atkinson, R.H. Cha, and P.C. O'Brien (1993) L-DOPA therapy and survival in idiopathic Parkinson's disease: Olmsted County Project. Neurology *43*: 1918-1926.

Walkinshaw G. and C.M. Waters. (1994) Neurotoxin-induced cell death in neuronal PC12 cells is mediated by induction of apoptosis. Neuroscience. *63*: 975-987.

Walkinshaw G. and C.M. Waters (1995) Induction of apoptosis in catecholaminergic PC12 cells by L-DOPA. Implications for the treatment of Parkinson's disease. J. Clin. Inv. *95*: 2458-2464.

Washko P.W., W.O. Hartzell and M. Levine (1989) Ascorbic acid analysis using High-performance liquid chromatography with coulometric electrochemical detection. Anal. Biochem. *181*: 276-282.

Yurek D.M., K. Steece-Collier, T.J. Collier and J.R. Sladek (1991) Chronic L-DOPA impairs the recovery of dopamine agonist-induced rotational behavior following neural grafting. Exp. Brain Res. *86*: 97-107.

Zhou Q.Y., C.J. Quaife, and R.D. Palmiter (1995) Target disruption of the tyrosine hydroxylase gene reveals that catecholamines are required for mouse fetal development. Nature *374*: 640-643.

III. INVOLVEMENT IN DAMAGE AND REGENERATION

13

ETHANOL EXPOSURE DURING BRAIN DEVELOPMENT ALTERS ASTROGLIOGENESIS AND ASTROCYTE FUNCTIONS

Jaime Renau-Piqueras, Luis Megías and Consuelo Guerri

One of the most severe consequences of maternal alcohol consumption during pregnancy is the damage to the developing central nervous system (CNS) which is manifested by mental retardation and other CNS dysfunctions. The extreme consequence of these effects is the fetal alcohol syndrome (FAS). Alterations in neuronal migration have been proposed as a mechanism involved in alcohol-induced CNS damage. Since astroglial cells are responsible for the guidance of migrating neuronal bodies, promotion of neurite outgrowth, and shaping of neuronal assemblies during development, we have postulated that alcohol-induced alterations of astroglial functions could be important in contributing to the effect of ethanol on the developing brain. Using an animal model which reproduces in the rat many of the alterations observed in children with FAS, we have demonstrated, both in vivo and in vitro (astrocytes in primary cultures), that prenatal alcohol exposure induces important alterations in the astrogliogenesis. In fact, several groups including ours have demonstrated that alcohol induces decreases in either DNA synthesis or cell proliferation, decreases in protein synthesis and content, decreases in serotonin uptake, alterations in cytoskeleton organization and in the acquisition of several cytoskeletal proteins, delays in morphological maturation, increases in oxidative stress and alterations in neurotrophins. The main conclusion which can be drawn from these results is that the effects of prenatal expsoure to alcohol on astrocytes could significantly contribute to ethanol-associated alterations in CNS development.

THE FETAL ALCOHOL SYNDROME

Alcohol consumption during pregnancy induces a wide spectrum of adverse effects in the offspring, the extreme consequence of which is the fetal alcohol syndrome

(FAS). However, although the teratogenicity of ethanol has been known since antiquity, the first, modern description of FAS was made by Lemoine et al. (1968) and it was defined by Jones et al. (1973) and by Jones and Smith (1975). The full syndrome has been defined as the presence of prenatal and postnatal growth retardation below the tenth percentile, central nervous system (CNS) dysfunction (neurologic abnormalities, developmental delay and/or intellectual deficits), and characteristic cranial/facial abnormalities, including microcephaly, short palpebral fissures, poorly developed philtrum, recessed maxillary area, etc. Infants with FAS may also manifest other congenital anomalies involving the heart, liver, kidneys, eyes, and musculoskeletal system among others (Jones and Smith, 1975).

Even in absence of the full FAS, the developing human CNS appears to be particularly sensitive to ethanol-induced growth suppression, and prenatal alcohol exposure has been found to result in a high incidence of mental retardation and/or behavioural deficits that do not respond well to therapy (Darby et al., 1981; Streissguth et al., 1980, 1986). Thus, the correlation of mental retardation with small brain size is strongest at high levels of maternal alcohol intake. Available evidence also indicates that the effects of moderate drinking on fetal development tend to be behavioural rather than physical (Streissguth et al., 1994), and that they are more subtle, variable and difficult to document systematically. Interestingly, intrauterine growth retardation as a result of alcohol exposure is found in all animal models of FAS and alcohol-related birth defects, including animals exposed to ethanol on a single day of gestation.

The incidence of FAS is estimated to be 1.9 per 1000 live births, and in Western societies alcohol consumption during pregnancy is considered to be the most prevalent cause of preventable mental retardation (Abel and Sokol, 1987).

NEUROANATOMICAL ALTERATIONS

Early ethanol exposure induces morphological alterations in the brain of children with FAS and in experimental animals (reviewed in Clarren, 1986 and Druse, 1992). These alterations include a reduction in the size of the cerebral hemisphere, agenesis of corpus callosum and olfactory bulbs, dysplasia of the hypophysis and anencephaly, and a reduction in the weight of the cerebrum and cerebellum.

Neocortex

The neocortex is a complex brain structure composed, in mammals, of at least six neocortical layers with 10 to 20 functional subdivisions. One of the basic functions of this structure is the analysis and representation of the relationship between the components of sensory and motor patterns that support functions as diverse as perception, motor programming, memory, language processing and reasoning (Singer, 1996). During brain development, the plasticity of cortical processing architectures is susceptible to experience-dependent modifications, but they become fixed in the adult. Therefore, ethanol-induced alterations in the formation of cortical architectures would be expected to have profound and long-term behavioural consequences. Prenatal exposure to ethanol induces a 13% reduction in this structure by significantly decreasing the number of neurons (33%) and glial cells (36% reduction) in the motor and somatosensory cortices, while the auditory cortex is less affected by ethanol (Miller, 1992).

At the cellular level ethanol affect many of the ontogenetic phases of neurogenesis, which include generation and proliferation of different types of neurons and glia cells, neuronal migration along radial glia fascicles, differentiation (which includes production of redundant dendritic and axonal branches and the formation of transient connections); and, finally, pruning of excessive growth and death of redundant neurons (Jacobson, 1991). The definitive pattern of cellular relationships is thus the result of several processes that are patterned in space and time.

Early ethanol exposure depresses the proliferation of cortical neurons. However, differential effects of ethanol on the ventricular and subventricular zones have been observed. For example, whereas the proliferative activity in the ventricular zone is depressed and, therefore, there is a decrease in early-generated neurons, ethanol stimulates proliferative activity in the subventricular zone, where the neurons are generated later (Miller, 1992). It has been suggested that this differential effect on the two proliferative zones is due to the different levels of ethanol that the cells are exposed to. Low levels, according to this hypothesis, would stimulate cell proliferation, whereas high levels would inhibit it (Miller, 1993).

Fetal alcohol exposure also induces abnormal neuronal migration. This effect has been observed in autopsies of children with FAS (Clarren, 1978) and animals following prenatal exposure to ethanol (Miller, 1992). Indeed, sheets or cluster of ectopic neurons, abnormal glial migration, leptomeningeal heterotopias and disruption in the lamination of the neocortex have been observed following prenatal exposure. The alterations in neuronal migration and in cell death may be the result of either changes in radial glia guiding fibbers (Miller, 1993) or an alteration in the production of glial factors which are essential to neuronal migration (i.e. neural adhesion molecules) or survival (i.e. growth factors). In fact, we have recently demonstrated that prenatal ethanol exposure affects the early stages of astrogliogenesis (fetal day 14 to fetal day 21), inducing a marked alteration in radial glia (precursor cell of astrocytes) morphology and significantly decreasing the expression of glial fibrillary acidic protein, as astrocyte marker (Vallés et al., 1996). Therefore, alterations in radial glia might be an underlying mechanism in the ethanol-induced alterations in neuronal migration, Finally, dysmorphic dendritic spines and altered synaptic development have also been observed in the cortices of rats exposed prenatally to ethanol.

In conclusion, although ethanol could interfere with may of the complex ontogenetic processes of neocortical organisation, disruptions in the proliferation of the stem cell population may lead to a reduction in the generation of both glial cells and neurons, and this, in turn, could lead to secondary effects in later development such as neurogenetic and astrogliogenetic deficits. Alteration in astroglial development could then influence neurogenesis, neural survival and differentiation. Impairment of the spinal projections and cortical output can explain some of the cognitive and neurobehavioural dysfunctions observed after prenatal exposure to ethanol (Streissguth et al., 1983, 1994).

Hippocampus

A large number of studies have demonstrated that the hippocampus is a target site for the toxic effects of ethanol during brain development. The main reason for researchers interest is this area is that memory formation has traditionally been associated with the hippocampus, and both clinical and experimental studies demonstrate that learning and anterograde and spatial memory deficits are common sequelae of prenatal ethanol exposure (Streissguth, 1990).

Morphological studies on hippocampal formation demonstrate that the effects of ethanol depend on the developmental stage of the hippocampal area studied and on the blood alcohol levels. Prenatal exposure to moderate ethanol levels decreases the total number of pyramidal neurons in hippocampus field CA1 generated during this period, but induces hypertrophy of CA3 mossy fibers. Moreover, exposure of neonatal rats to high blood alcohol levels (BALs) (~300 mg/dl) resulted in a marked reduction in the number of neurons in the CA1 region, while the cell populations in the hippocampal fields CA3, CA4 and the dentate gyrus were affected. These anatomical changes were found to be associated with alterations in the learning process.

One critical neurochemical process during hippocampal formation and the consolidation of the learning and memory processes is glutamate-mediated neurotransmission. Glutamate is known to play an important neurotrophic role in neuronal development and synaptic plasticity, which are involved in the long-term potentiation. Ethanol depresses glutamate release in fetal hippocampus, and prenatal exposure to ethanol has been shown to depress glutamate release, decrease glutamate binding, diminish long-term potentiation of synaptically-activated population spines in CA1 and in *N*-methyl-D-aspartate (NMDA) binding site density, assessed in adult rats at 45 days of age. Moreover, exposure to ethanol during hippocampal formation significantly decreases the number of specific NMDA receptors, with no change in their affinity (Vallés et al., 1995). A recent study demonstrated that the sensitivity of NMDA receptor-mediated activity to ethanol is greater in the immature than in the mature hippocampus.

Given the relevance of the hippocampus for normal behavioural functioning, the neurochemical, neurophysiological and anatomical changes observed after prenatal ethanol exposure may underlie the persistent memory and learning deficits commonly noted in children and animals prenatally exposed to ethanol (Streissguth et al., 1994).

Cerebellum

The cerebellar cortex is another area of the brain that has received particular attention because clinical and experimental data have consistently demonstrated that ethanol exposure during brain development damages the motor system, including the cerebellum. In fact, the role of cerebellum in motor coordination and in cognition processes is well demonstrated.

Children exposed to ethanol exhibit numerous symptoms associated with cerebellar damage, including delays in motor development, problems with fine motor tasks, ataxia and gait alterations (Streissguth et al., 1983, 1990). Malformations in the structure of the cerebellum, and complete agenesis of cerebellar vermis, have been noted in autopsies of FAS children. Animal models have also shown functional and structural damage related to the cerebellum following ethanol exposure during development. Behavioural deficits that could reflect cerebellar damage have been reported, including impairment of locomotion, coordination, fine motor performance and spatial cognitive processes (Meyer et al., 1990).

Experimental studies have also demonstrated that exposure to ethanol during gestation and/or the neonatal period leads to a reduction in the size of the cerebellum. However, this reduction is more pronounced when ethanol is administered during the brain growth spurt, and is directly related to the BALs reached in the pups. These changes in the cerebellar size are accompanied by a loss of cerebellar Purkinje cells, which are particularly affected by either prenatal or early postnatal ethanol exposure. Moreover, a significant correlation between Purkinje cell loss and the loss of granule cells has been observed and suggests that the reduction in granule cells is due to the loss of their targets, the Purkinje neurons. However, a recent reports demonstrates that ethanol can interact directly with cerebellar granule cells in culture, resulting in cell death, and these effects were reduced by stimulation of NMDA receptors. Therefore, a reduction in NMDA receptors may underlie the ethanol-induced granule cell death in cerebellum.

In conclusion, the above results support the idea that the stage of brain development at which the cerebellum is most vulnerable to the toxic effects of ethanol is the brain growth spurt, which corresponds to the neonatal period in the rat and to the third trimester in human gestation (although the maturation of the cerebellum continues until 2 years of life). Therefore, binge ethanol consumption during this period can induce structural and functional deficits in the cerebellum which may explain the long-lasting cognitive and motor deficits observed in the children of alcohol-drinking mothers (Larroque et al., 1995).

The next sections will highlight some of the major findings of several groups, including ours, on the effect of prenatal alcohol exposure on astrocytes and astrogliogenesis. Other reviews on alcohol and glial cells have recently been published (Miller, 1992; Guerri et al., 1993; Ledig and Tholey, 1994; Phillips, 1994; Kimelberg and Aschner, 1994; Guerri, 1996).

THE ANIMAL MODEL

The following four criteria have been considered as essential for establishing of a relevant animal model of alcoholism: i) The animal should voluntarily and selectively consume a solution of the drug to yield significant blood alcohol concentrations (BAC). ii) When alcohol is consumed for several weeks, the animals should exhibit tolerance to the effects of the drug. iii) Abrupt termination of alcohol administration should results in physical signs of drug withdrawal. iv) Finally, alterations in liver, brain and endocrine function, which occur in association with chronic alcohol ingestion in humans, should also occur in animals.

From the main rodent models the most satisfactory appears to be the "balanced liquid diet" model which is an attempt to eliminate nutritional variables. The main advantages of this model are: i) High alcohol consumption leading to high BAC. ii) No loss in body weight during treatment. iii) Tolerance and physical dependence develop. iv) Large numbers of animals can be treated at once.

To study effects of prenatal alcohol exposure on development, we use a rodent model, based on the "balanced liquid diet", which reproduces many of the alterations described in childrens with FAS (Sanchis et al., 1986). Female Wistar rats received a liquid diet (ethanol or control) containing either 5% (wt/vol) ethanol or isocallorically balanced with dextrin-maltose for pair-fed controls. The animals were treated for a minimum of 40 days before and during pregnancy. Mean blood alcohol levels reached by pregnant rats and their fetuses were 22.5±8.0 mmol/L and 23.5±5.0 mmol/L, respectively. We have demonstrated in previous studies that the progeny of chronic alcoholic rats reproduce some of the alterations observed in children affected by FAS (Sanchis et al., 1986, 1987). These alterations included a decrease in fetal viability, increase in neonatal mortality, reduced fetal and neonatal body and brain weights, and several hormonal, liver ultrastructural, and neurochemical alterations (Renau-Piqueras et al.,1987; Sanchis et al., 1986, 1987; Guerri et al., 1984).

EFFECTS OF PRENATAL EXPOSURE TO ALCOHOL ON ASTROCYTES AND ASTROGLIOGENESIS: THE "IN VITRO" MODEL

The fact that neuroglial heterotopies, and agenesis of the corpus callosum and anterior commisure (areas originally formed by glial cells) have been observed in post-mortem studies of FAS children (Clarren et al., 1978) has led some authors to suspect, as commented on above, that abnormalities in glial development contribute to the adverse effects of ethanol on the developing brain. In these studies ectopic clusters have been observed near the pial surface, and they are composed not only of neurons but also of glia, undifferentiated cells and pial cells. In addition, one of the developmental stages most vulnerable to ethanol-induced reduction in brain size in experimental animal is the brain growth spurt, a period of major development of glial and myelin structures, and this suggests a potential effect of ethanol on glial cells.

The complexity of mammalian CNS is an important disadvantage when it comes to elucidating the effects of prenatal exposure to ethanol on nervous cell types during development. The results obtained by several groups clearly indicate that primary cultures of astrocytes or astroglial precursors have biochemical, morphological and functional characteristics comparable to those of their "in vivo" counterparts. Among the advantages of these cultures is that they can serve as a model for studying the direct effect of ethanol on cell proliferation, differentiation and functions and, therefore, on astrogliogenesis.

Primary cultures of cortical astrocytes from 21-day-old fetuses are prepared from brain hemispheres (Renau-Piqueras, 1989; Gómez-Lechón, 1992). Fetuses are obtained from under sterile conditions from both control and chronic alcoholic rats. The purity of the cultures is determined using a monoclonal anti-GFAP and immunofluorescence. The GFAP-positive cells ranged from 85-90%, indicating that almost all the cells in the cultures are astrocytes (Renau-Piqueras, 1989).

Four different groups of astrocyte cultures are used: (i) cells obtained from control fetuses which are cultured in absence of ethanol (control or C cells); (ii) the same cells but cultured in presence of several concentrations of ethanol (25, 50 and 100 mM: 25-CA, 50-CA and 100-CA cells, respectively); (iii) astrocytes derived from fetuses prenatally exposed to ethanol but cultured without ethanol (prenatally exposed to ethanol or PEA); and (iv) PEA cells cultured in presence of 25mM ethanol (pre+ postnatal exposure). The concentration of ethanol in the medium is controlled by daily analysis by head-space gas chromatography.

Radial glia cells are cultured as previously described in detail. In brief, whole brains of 13-day rat fetuses (control and PEA) are removed under sterile conditions. Brains are pooled and mechanically dissociated, and isolated cells are plated (250-300 cells/mm^2) on poly-L-lysine-coated plates in Earle19s minimal essential medium supplemented with glucose, 10% heat-inactivated horse serum, and L-glutamine at 37°C in 5% CO_2. After two days in culture, cells are washed for 3 s with cold (4°C) medium to remove neurons, reefed, and returned to the incubator. Medium is changed every 2-3 days. The purity of cultures was determined using the monoclonal antibody Rat 401, which identifies a transient radial glia cell in the embryonic rat CNS.

CELL GROWTH: DNA AND PROTEIN SYNTHESIS.CELL CYCLE: DNA AND PROTEIN CONTENT

Astrocytes in primary cultures derived from control animals (C cells) showed a decrease in the content of both protein and DNA when they were exposed to 25mM ethanol (Renau-Piqueras et al., 1986; Guerri et al., 1990). A diminution in the content of protein induced by "in vitro" exposure of astrocytes to alcohol has also been reported by other groups (Ledig et al., 1991; Ledig and Tholey, 1994; Lokhorst and Druse, 1993). However, in contrast with our results, it has been published that the protein content was more affected than the DNA content in control astrocytes exposed to alcohol (Kennedy and Mukerji, 1986a, b; Lokhorst and Druse, 1993). The discrepancies may be due to different treatment schedules and ethanol doses. On the other hand, PEA cells showed a decrease in cell growth as well as in the DNA and protein content. Moreover, when 25mM ethanol was added to these PEA cultures, there was an important decrease in their ability to divide (Renau-Piqueras et al., 1986; Guerri et al., 1990).

A similar effect on cell division was found when control cells were incubated with alcohol (43, 108 and 217mM) during the logarithmic phase of growth (Davies and Cox, 1991). Moreover, a dose-dependent inhibition by ethanol of the growth of C_6 (morphologically astrocyte-like) cells was also reported (Isenberg et al., 1992).

DNA and protein synthesis. Cell cycle

Control astrocytes showed peaks of DNA and protein synthesis (determined by ^{3}H-thymidine and ^{3}H-leucine incorporation) at 8 and 15 days, respectively (Guerri et al., 1990). Prenatal exposure to alcohol clearly reduces the incorporation of both precursors, although the reduction was more pronounced for ^{3}H-thymidine (75% at 7 days). This effect on DNA synthesis was also observed when C astrocytes were exposed to ethanol (25 and 100mM) during the entire culture period. However, although decreases in DNA and proteins synthesis were observed during the proliferation (7 days) and differentiation (21 days) periods, this effect was more important in proliferative cells, and correlates with an accumulation of astrocytes in the G_0/G_1 phase of the cell cycle (Guerri et al., 1990).

Incubation of astrocytes with alcohol (25, 50, 100 or 200 mM) for 18 hr also results in dose-dependent changes in the rates of incorporation of tritiated thymidine, uridine and valine (Snyder et al., 1992). Since nucleoside transporters have been demonstrated in cultured cells and Wu et al. (1986) have described two saturable routes of nucleoside entry into cultured astrocytes, and as ethanol alters lipids or lipoproteins of the cell membrane (Tabakoff and Hoffman, 1987; Yeagle, 1988; Slomiany et al., 1992), the effect of ethanol on precursor uptake may be linked to alterations in astroglial membranes.

The decrease in protein synthesis in ethanol-exposed astrocytes was due to a diminution in the total RNA and mRNA in these cells, which suggest that ethanol interfere with protein synthesis at the transcription level. Indeed when the same amounts of RNA were used in "in vitro" protein synthesis experiments, no differences in the pattern and the amount of proteins synthesized by control and PEA cells were observed. This effect of ethanol on protein synthesis agrees with the results of Horbach et al. (1989) obtained in fetal hepatocytes.

ULTRASTRUCTURAL AND CYTOSKELETAL ALTERATIONS

When astrocytes were exposed to alcohol "in vivo" and "in vitro", several ultrastructural alterations were found. Astrocyte cultures showed a dose-dependent attenuation in both the depth and complexity of the cell layer (Davies and Cox, 1991). Moreover, prenatal exposure to ethanol induced qualitative and quantitative ultrastructural changes in proliferating and differentiated astrocytes in primary culture. Several components such as lysosomes, mitochondria, rough endoplasmic reticulum and Golgi apparatus, displayed important alterations. These appeared accompanied by alterations in the cytochemical and biochemical activity of several enzyme markers including acid phosphatase (lysosomes and trans-Golgi apparatus), uridine diphosphatase (endoplasmic reticulum and Golgi apparatus), thyamine pyrophosphatase (trans-Golgi apparatus), glucose-6-phosphatase (endoplasmic reticulum), 519-nucleotidase, (Na-K) ATPase (plasma membrane), etc. It is important to note that these effects of alcohol were more pronounced in proliferating cells that in differentiated ones (Mayordomo et al., 1992). Changes in the activity of several astrocyte enzymes (butylcholinesterase, glutamine syntethase) after alcohol exposure have been also described (Renau-Piqueras et al., 1988; Guerri et al., 1989; Ledig and Tholey 1994).

However, one of the most striking changes observed in astrocytes prenatally exposed to ethanol is a morphological disorganization of the cytoskeleton (CSK). Immunofluorescence studies using an anti-GFAP antibody revealed that in 1-day-old control astrocytes fluorescence was restricted to a small area surrounding the nucleus. In 4- and 7-day-old control cells the CSK showed a reticular pattern, whereas in 14- and 21-day-old control astrocytes intermediate filaments (IF) in most of the cells surrounded the nucleus and were arranged in parallel arrays, mainly in the processes. This pattern was found in mature multipolar astrocytes. In the PEA cells similar patterns were found, although 25% of the cells showed a reticular distribution of IF after 14 days in culture. This number was reduced in 21-day-old PEA astrocytes. These qualitative and quantitative results were also observed when cells were examined by electron microscopy and immunogold (Renau-Piqueras et al., 1989) (fig. 13-1). When cells were stained for tubulin and vimentin, morphological alterations resembling those described for GFAP were observed (Saez et al., 1991). In addition, using immunoblotting and immunofluorescence procedures we have demonstrated that prenatal alcohol exposure decreased the amount of GFAP, vimentin and tubulin, during the entire culture period (fig. 13-2). A similar effect on GFAP was observed when homogenates of brain were used (Saez et al., 1991). Moreover, recent results showed that GFAP and vimentin mRNA expression was also altered in prenatally exposed to alcohol astrocytes (Vallés, 1996) (fig. 13-3).

Since astrocytes exposed to ethanol show a decrease in morphological differentiation and the phosphorylation of IF proteins (GFAP and vimentin) seems to be implicated in this event (Goldman and Chiu, 1984; McCarthy et al., 1985), we investigated whether ethanol alters the phosphorylation of these IF proteins and whether this effect is mediated by protein kinases A (PKA) and C (PKC). We have demonstrated that in control cells, activators of both protein kinases (Harrison and Mobley, 1991) induce an increase in phosphorylation of GFAP and vimentin, which correlates with an increase in CSK morphological differentiation.

However, in astrocytes exposed to ethanol, decreases in phosphorylation of both CSK proteins and in the activation of PKA and PKC were observed, which suggests that ethanol induces changes in the levels of PKA and PKC (Guerri et al., 1993).

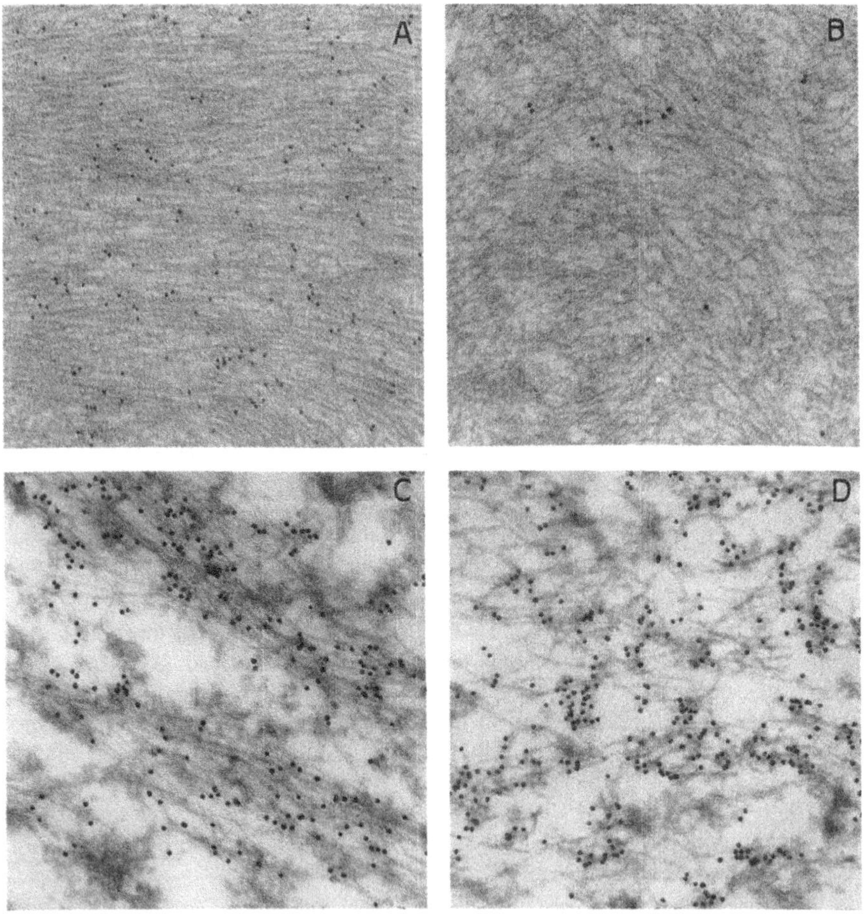

Figure 13-1. Ultrastructural localization of anti-GFAP binding sites in 14-day-old control (A,C) and prenatally alcohol-exposed (B,D) astrocytes. A and B post-embedding procedure, C and D pre-embedding procedure. In A and C, IF are distributed in parallel arrays, whereas in B and D they are disposed in a reticular network. A-D, x52700.

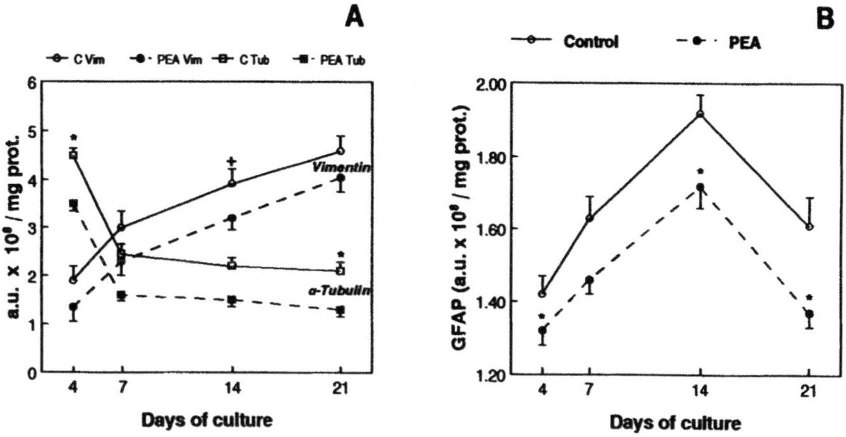

Figure 13-2. Changes in cytoskeletal vimentin and α-tubulin (A), and GFAP (B) content with time in control astrocytes (C) and astrocytes prenatally exposed to ethanol (PEA) in primary culture. Relative absorbance units (a.u.) were obtained from the densitometric analysis of the immunoblots with days of primary culture. Each point is the mean±SD of seven different cultures for each group. (*) Significantly different ($p \leq 0.01$) from control values; (+) significantly different ($p \leq 0.05$).

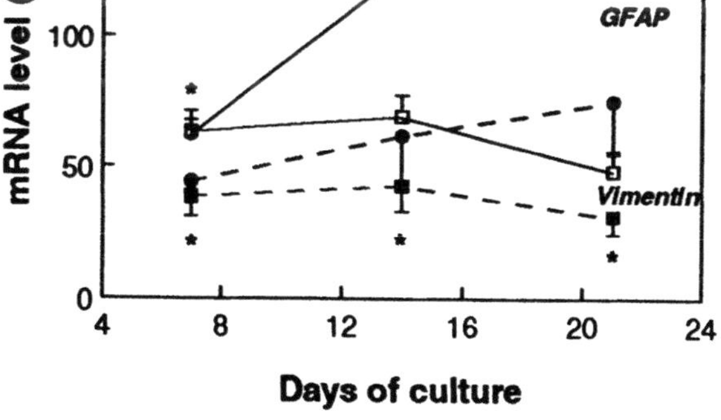

Figure 13-3. Densitometric analysis of Northern blot demonstrates GFAP and vimentin mRNA levels, in relative densitometric units (R.D.U.), from control (C) and ethanol-treated (PEA) astrocytes at different days of culture. A constant RNA concentration (10mg per lane) was used. Data are mean±SD (bars) values of four different cultures. (*) Significantly different, p(GFAP) ≤0.001, p(vimentin) ≤0.02, from control values.

These results suggest that some important CSK-dependent functions, including guidance of neurons during brain development and mitosis, may be altered by prenatal exposure to ethanol. Damage to the CSK could also be involved in alterations in macromolecular transport and secretion. In fact, alcohol-induced variations in the secretion and amount of several plasma membrane proteins such as NGF and its receptor, 5'-nucleotidase, (Na-K) ATPase, N-CAM, etc. have been found (Ledig et al., 1985; Renau-Piqueras et al., 1988, 1992; Guerri et al., 1989; Kentroti et al., 1995). Some of these proteins (NGF and N-CAM) play an important role in modulating neuronal migration (Lu, 1991; Birling and Price, 1995; Rutishauser 1996). Preliminary results obtained in our laboratory suggest that prenatal exposure to alcohol also alters other processes that depend on CSK integrity, such as endocytosis and translocation of macromolecules. However, alterations in glycoprotein secretion could also be due, as occurs in fetal and adult hepatocytes, to alcohol-induced damage to the Golgi apparatus structure and function (Renau-Piqueras, 1997). Since ethanol alters the lipids or lipoproteins of the cell membrane (Tabakoff and Hoffman, 1987; Yeagle, 1988; Slomiany et al., 1992), the effect of ethanol on the activity and/or amount of surface glycoproteins may also be linked to perturbations in astroglial membranes.

OXIDATIVE STRESS IN ASTROCYTES

Reactive free radical species have been shown to occur physiologically and can lead to toxicity (Halliwell,1987). Among free radical, the hydroxyl radicals show a strong ability to react with many biological molecules, thus inducing lipoperoxides and altering membrane structure. Lipid peroxidation (LP) may therefore be of major significance in cell injury produced by the free radical mechanism (Halliwell,1987). However, cells are equipped with an antioxidant defense against these toxic radicals. The defense mechanisms range from low-molecular-weight compounds (glutathione, vitamin E, etc.) to complex enzymatic systems (superoxide dismutase (SOD), catalase, etc.). A disturbance of the cellular prooxidant/ antioxidant balance in favour of the former may denote oxidative stress (Sies, 1985).

Evidence regarding the role of oxygen-derived free radicals and LP in the pathogenesis of alcohol liver injury has accumulated (Albano et al., 1993), and recent data indicate that ethanol can also induce LP in extrahepatic tissues with alcohol metabolism such as brain (Nordmann et al., 1990). This organ possesses characteristics that may make it especially prone to oxiradical injury. Its membranes are rich in polyunsaturated fatty acid side chains, and it is poor in catalase and SOD activities (Cohen, 1988). Moreover, the levels of antioxidant substances such as tocopherol and glutathion are relatively low (Sinet et al., 1980). We have demonstrated that ethanol induces a dose-dependent increase in lipid peroxidation in adult brain homogenates and that the peroxidative effects of alcohol seem to be related to both cytochrome P450 and the ethanol-inducible form of cytochrome P450 (CYP2E1) (Montoliu et al., 1994).

Using the formation of dichlorfluorescein (DCF), we also demonstrated that both "in vitro" and chronic alcohol exposure significantly increased the formation of reactive oxygen species (ROS) in synaptosomes. Chronic alcohol treatment also leads to an induction of cytochrome P450, NADP cytochrome c reductase, NADPH oxidation and CYP2E1 in brain microsomes. In addition, this treatment decreases the GSH/GSSG ratio in brain and increases the activity of SOD and catalase (Guerri et al., 1994; Montoliu et al., 1994).

These results suggest that chronic ethanol intake induces oxidative stress in brain, and that this effect is, in part, due to induction of CYP2E1 and its generation of oxygen radical species in this organ. However, due to the diversity of the nervous cells in the CNS, it remains to be determined whether there is selective expression of CYP2E1 in the different nervous cell types, and whether ethanol-induced ROS are related to the induction of this isoform and/or to the damage to the nervous cells. Astrocytes play important roles in brain and in the metabolism of many toxic compounds such as ammonia, glutamate, free radicals, xenobiotics and metals, thus protecting the brain from the toxicity of these agents. In fact, we have recently demonstrated the presence of CYP2E1 in astrocytes in primary culture, its induction by ethanol and the consequent generation of free radical species (Montoliu et al., 1995). Moreover, exposure of astrocytes to 25 or 50mM ethanol for 14 days causes an enhancement of cytochrome P450 and NADPH-cytochrome c reductase (Montoliu et al., 1995) (fig. 13-4).

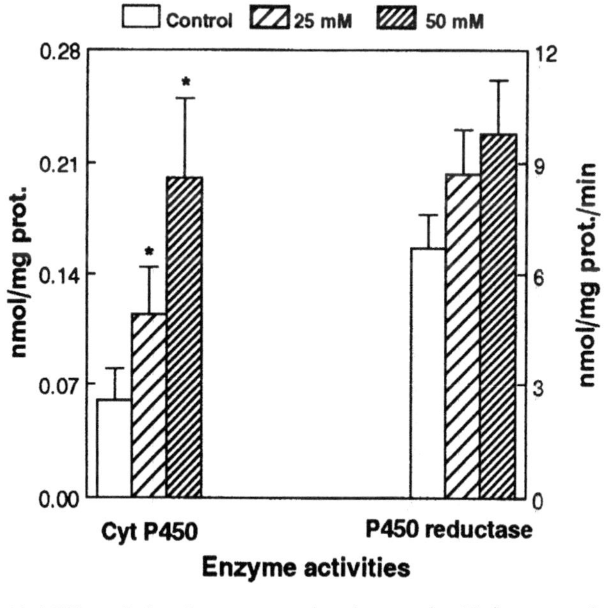

Figure 13-4. Effect of ethanol exposure on the microsomal oxidation system in astrocyte microsomes. Microsomes were isolated from control or ethanol-exposed (25 or 50mM) astrocytes (14 days of culture). Data are the means±SD of three different preparations per group. (*) Significantly different from control astrocytes ($p \leq 0.05$).

Double immunofluorescence using anti-CYP2E1 and anti-GFAP showed that this cytochrome was distributed throughout the cytoplasm and processes, although labelling was more evident over the nuclear membrane. This pattern was confirmed using immunogold label. Addition of ethanol to the astrocyte culture medium for 14 days resulted in an enhancement of CYP2E1, a dose-dependent increase in ROS, and malonaldehyde and hydroxynonenal formation (fig. 13-5). The formation of ROS, induced by ethanol, was partially prevented by incubating the astrocytes with anti-CYP2E1. The increase in free radical formation was accompanied by depletion of the GSH content (Montoliu et al., 1995) (fig. 13-6).

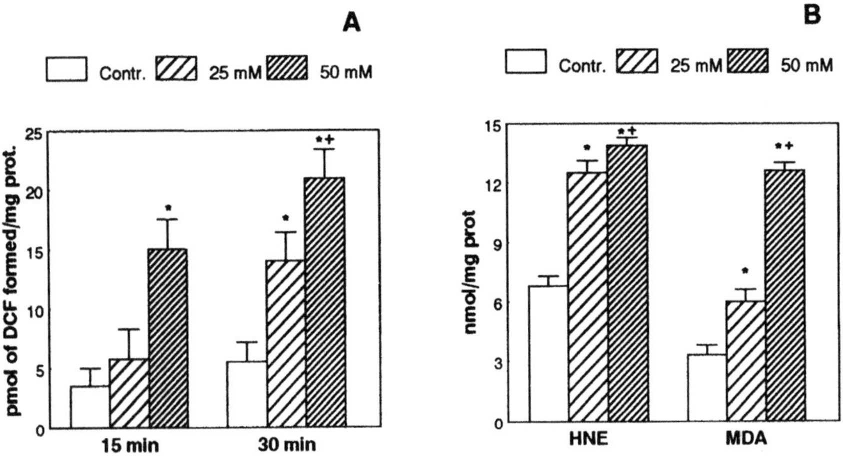

Figure 13-5. (A) Effect of ethanol on oxygen radical formation in astrocytes, measured as the amount of dichlorfluorescein (DCF) formed (pmol/mg of protein). Cells were cultured with (25 and 50mM) or without ethanol in the culture medium. At 14 days of culture, control and ethanol-exposed astrocytes were preloaded for 15 min with DCFH-DA, and the rate of DCF formed was measured at 15 and 30min. (B) 4-Hydroxynonenal (HNE) and Malonaldehyde (MDA) content in control and ethanol-exposed (25 and 50mM) astrocytes. Astrocytes in primary culture were grown in the presence or absence of 25 or 50mM ethanol in the culture medium. At 14 days after plating, the contents of HNE and MDA were determined. Data are mean±SD (bars) values from six different determinations. (*) Significantly different ($p \leq 0.001$) from control values, (+) significantly different ($p \leq 0.001$) from 25mM ethanol treatment.

Moreover, an increase of SOD activity was observed in astrocyte cultures in presence of ethanol (Ledig and Tholey, 1994). Finally, alcohol induces a decrease in the content of formaldehyde dehydrogenase (FALDH) or ADH-Class III (Iborra et al., 1992, 1993), an enzyme responsible for formaldehyde metabolism, one of the most toxic products of LP. The inhibition of FALDH synthesis could result in a lack of protection of the cell against formaldehyde. These results suggest that ethanol induces oxidative damage in astrocytes, which could explain some of the toxic effects of alcohol on these cells, such as CSK alterations. This assumption is supported by the fact that an increase in GSH content prevents the effect of alcohol on astrocyte CSK. Thus these results indicate the importance of oxidative stress as a mechanism involved in alcohol-induced brain damage.

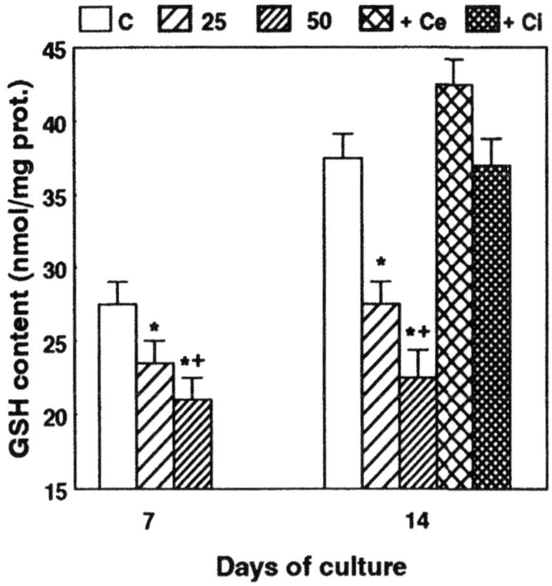

Figure 13-6. Changes in GSH content in control and ethanol-treated astrocytes. After 7 and 14 days in culture, the GSH levels were determined in control (C) and ethanol-exposed (25 or 50mM) astrocytes. In some experiments, cells were grown in a medium containing 50mM ethanol and L-cysteine (0.02mM; +Ce) or L-cystine (0.02mM; +Ci), and the GSH content in the cells was determined after 14 days in culture. Data are mean±SD (bars) values from five different experiments. (*) Significantly different ($p \leq 0.002$) from control values; (+) significantly different ($p \leq 0.002$) from 25mM ethanol treatment.

RADIAL GLIA. "IN VIVO" AND "IN VITRO" STUDY

The morphogenesis of the CNS is characterised by a series of complex processes including interaction between neurons and glial cells. Glia are involved in the rearrangement and elimination of synaptic junctions, and in the exchange of nutrients, neurotransmitters, second messengers, ions, and growth factors, all of which are critical processes in normal brain development (Rakic, 1991). During early embryonic development radial glia cells play integral roles in the guidance of axons and growth cones during the formation of neuronal circuits (Rakic, 1991). In mammals radial glia are transient, for shortly after neuronal migration is completed, they are transported into astrocytes. CSK characteristic is one of the criteria which have been used to understand how neuronal and glial morphology become specific during development. The maturation of astroglial cells includes changes in CSK components and morphology, and modifications in the expression and content of IF occur during this period. We have demonstrated (Sancho-

Tello et al., 1995) that although GFAP immunostaining appeared late (fetal day 21) in gestation, and at day 5 in radial glia cultures, GFAP mRNA expression was first detected on fetal day 15 (F15) and increased to F21. During postnatal development a striking increase in GFAP and its encoding messenger occurs. In contrast, the level of vimentin and its mRNA expression in brain were very high during the fetal stage (F15 to F21). Thereafter, vimentin expression declined during postnatal development until postnatal day 21 and remained constant at adult levels. In contrast, when radial glial in primary culture were used, an increase in vimentin expression was observed in glial cells throughout the entire culture period. From this apparently contradictory result, the question arises as to whether the increase of vimentin in astroglial cell cultures has a biological significance or represents an immature state of the astroglia cell in culture. Vimentin and GFAP can coexist in the same filament forming copolymers, and recent studies revealed that astrocytes and their precursor cells are composed of GFAP-vimentin heteropolymers or vimentin alone and that the ratio of GFAP to vimentin reflects the degree of differentiation and the functional state of the cell (Abd-el-Basset et al., 1992). Therefore, during the fetal stage or radial glia differentiation both "in vivo" and in primary culture, new GFAP molecules may be incorporated into preexisting vimentin filaments and new GFAP-vimentin heteropolymers could be formed during astroglial maturation. The high levels of vimentin expression observed during the fetal period "in vivo" may be due to the large proportion of cells of mesenchymal origin which express vimentin during the fetal period. At later stages of brain development, most of the vimentin positive cells lose the capacity to synthesize vimentin, while astroglial cells, like others, still contain vimentin in adult brain and, therefore, the total vimentin content in brain decreases (Sancho-Tello et al., 1995).

"In utero" alcohol exposure delays the appearance of GFAP and its mRNA and significantly decreases the GFAP expression in fetal brain and in primary culture of radial glia (Vallés et al., 1996) (fig. 13-7). In contrast, no changes were observed in vimentin. In addition, some morphological alterations were observed in PEA glial cells in culture. Alterations in radial glia have also been reported in rats exposed to ethanol from gestational days 7 (E7) to E21 (Miller and Robertson, 1993). Cells obtained from PEA fetuses showed some alterations throughout the culture interval: i) The long radial glia processes characteristic of these cells were shorter than in the control cell. ii) Whereas in control cultures most of the GFAP-positive radial glia cells had been transformed into branched stellate astrocytes by day 15, in PEA cells the number of astrocytes with a stellate configuration was smaller. iii) The appearance of GFAP by immunofluorescence was delayed $\approx$ 2 days (day 7 of culture). iv) Finally, PEA cells showed important alterations in their cytoskeleton such as changes in the intermediate filament organisation pattern (Renau-Piqueras, et al.,1989; Saez et al., 1991).

Overall, our results demonstrate that ethanol exposure during early embryogenesis alters the normal development of radial glia and significantly decreases GFAP expression in both fetal brains and radial glia in culture, which could result in a disturbance of neuronal migration. On the other hand, ethanol-induced alterations in radial glia may also underlie the defects in corpus callosum observed in children with FAS. It has been shown that during embryonic development, the corpus callosum, initially and transiently formed by fascicles of radial glia, supports the growth axons from one side of the brain to the other (Norris and Kalil, 1991). Furthermore, disruption of the radial glia "sling" at the cerebral midline, as in congenital mouse mutants lacking this structure, prevents growth of callosal axons across the midline, and the corpus callosum does not form (Silver and Ogawa, 1983). Therefore, an alteration in radial glia could be involved in the agenesis or abnormalities of the corpus callosum observed in children of alcoholic mothers (Riley et al., 1995).

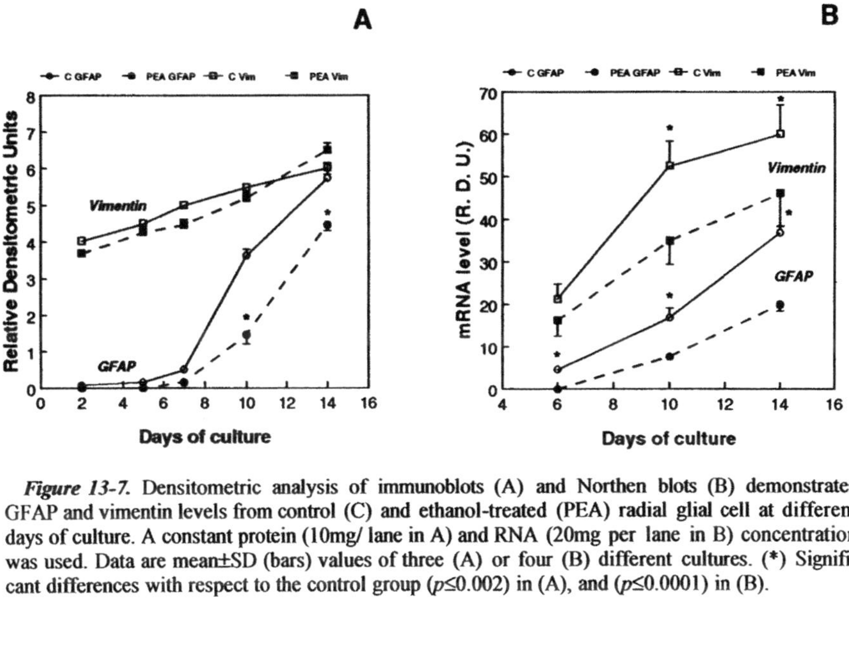

Figure 13-7. Densitometric analysis of immunoblots (A) and Northen blots (B) demonstrates GFAP and vimentin levels from control (C) and ethanol-treated (PEA) radial glial cell at different days of culture. A constant protein (10mg/ lane in A) and RNA (20mg per lane in B) concentration was used. Data are mean±SD (bars) values of three (A) or four (B) different cultures. (*) Significant differences with respect to the control group ($p \leq 0.002$) in (A), and ($p \leq 0.0001$) in (B).

CONCLUSIONS

The above results obtained using an animal model that reproduces in the rat many of the alterations observed in children with FAS support the conclusions that prenatal exposure to ethanol affects the proliferation and differentiation of astroglial cells. Astroglia damage would be likely to have profound effects on the glial functions of removing a metabolizing glutamate and other neurotransmitters from synaptic areas, and may influence neurogenesis, differentiation and neuronal migration. Therefore, ethanol-induced alterations in astrogliogenesis may influence the development of the nervous system.

ACKNOWLEDGMENTS

This work was supported by grants SAF 94-0065, SAF 96-0185 from the CICYT (Spain).

REFERENCES

Abel, E.L., and R.J. Sokol (1987) Incidence of fetal alcohol syndrome and economic impact of FAS-related anomalies. Drug Alcohol Depen. *19:*51-56.

Abd-el-Basset, E.M., I. Ahmed, V. Kalnins, and S. Fedoroff (1992) Immuno-electron microscopical localization of vimentin and glial fibrllary acidic protein in mouse astrocytes and their precursor cells in culture. Glia *6:*149-153.

Albano, A., P. Clot, M. Tabone, S. Arico, and M. Ingelman-Sundberg (1993) Oxidative damage and human alcoholic liver diseases. In G. Poli, E. Albano and M.U. Dianzani (eds): Free Radicals: From Basic Science to Medicine. Basel: Birkhäuser Verlag, pp. 310-322.

Birling, M.C., and J. Price (1995) Influence of growth factors on neuronal differentiation. Curr. Opin. Cell Biol. *7:*878-884.

Clarren, S.K., E.C. Alvord, S.M. Sumi, A.D. Streissguth, and D.W. Smith (1978) Brain malformations related to prenatal exposure to ethanol. J. Pediatr. *92:*64-70.

Clarren, S.K. (1986) Neurophatology in fetal alcohol syndrome. In J.R. West (ed): Alcohol and Brain Development. New York: Oxford University Press, pp. 158-169.

Cohen, G. (1988) Oxygen radicals and Parkinson19s disease. In B. Halliwell (ed): Oxygen Radicals and Tissue Injury. Bethesda: FASEB, pp. 130-135.

Darby, B., A.P. Streissguth, and D.W. Smith (1981) A preliminary follow-up of 8 children diagnosed fetal alcohol syndrome in infancy. Neurobehav. Toxicol. Teratol. *3:*157-159.

Davies, D.L., and W.E. Cox (1991) Delayed growth and maturation of astrocytic culture following exposure to ethanol: electron microscopic observations. Brain Res. *547:*53-61.

Druse, M.J. (1992) Effects of maternal alcohol consumption on the developing nervous system. In R.R. Watson (ed): Alcohol and Neurobiology. Brain Development and Hormone Regulation. Boca Raton: CRC Press, pp. 1-30.

Goldman, J.E., and F.-C. Chiu (1984) Dibutyryl cyclic AMP causes intermediate filament accumulation and actin reorganization in astrocytes. Brain Res. *306:*85-95.

Gómez-Lechón, M.J., F.J. Iborra, I. Azorín, C. Guerri, and J. Renau-Piqueras (1992) Cryopreservation of rat astrocytes from primary cultures. J. Tiss. Cult. Meth. *14:*73-78.

Guerri, C., A. Esquifino, R. Sanchis, and S. Grisolía (1984) Growth, enzymes and hormonal changes in offspring of alcohol fed rats. In: Mechanism of Alcohol Damage in Utero. Ciba Symposium 105. London: Pitman, pp. 85-89.

Guerri, C., A. Marqués, M. Sancho-Tello, and J. Renau-Piqueras (1989) Effect of prenatal exposure to alcohol on membrane-bound enzymes during astrocyte development "in vivo" and in primary culture. Int. J. Dev. Biol. *33:*239-244.

Guerri, C., R. Saez, M. Sancho-Tello, E. Martín de Aguilera, and J.Renau-Piqueras (1990) Ethanol alters astrocyte development: A study of critical periods using primary cultures. Neurochem. Res. *15:*559-565.

Guerri, C., R. Saez, M. Portolés, and J. Renau-Piqueras (1993) Derangement of astrogliogenesis as a possible mechanism involved in alcohol-induced alterations of central nervous system in development. Alcohol Alcoholism suppl. *2:* 203-208.

Guerri, C., C. Montoliu, and J. Renau-Piqueras (1994) Involvement of free radical mechanism in the toxic effects of alcohol: implications for fetal alcohol syndrome. In D. Armstrong (ed): Free Radicals in Diagnostic Medicine. New York: Plenum Press, pp. 291-305.

Guerri, C. (1996) Teratogenic effects of alcohol: current status on animals research and "in vitro" models. Arch. Toxicol. (in press).

Halliwell, B. (1987) Oxidants and human disease: some new concepts. FASB J. *1:*358-364.

Harrison, B.C., and P.L. Mobley (1991) Phorbol myristate acetate and 8-bromo-cyclic AMP-induced phosphorylation of glial fibrillary acidic protein and vimentin in astrocytes: comparison of phosphorylation sites. J. Neurochem. *56:*1723-1730.

Horbach, G.J.M.J., G.I. Henderson, and S. Schenker (1989) The effect of ethanol exposure on the transcriptional activity of fetal hepatocytes. Hepatology *10:*653-661.

Iborra, F.J., J. Renau-Piqueras, M. Portolés, M.D. Boleda, C. Guerri, and X. Parés (1992) Immunocytochemical and biochemical demonstration of formaldehyde dehydrogenase (class III alcohol dehydrogenase) in the nucleus. J. Histochem. Cytochem. *40:*1865-1878.

Iborra, F.J., J. Renau-Piqueras, X. Parés, and C. Guerri (1993) Effect of ethanol on GSH-formaldehyde dehydrogenase (ADH-class III) distribution in astrocytes. Alcohol Alcoholism *28:*231.

Isenberg, K., X. Zhou, and B.W. Moore (1992) Ethanol inhibits C_6 cell growth: fetal alcohol syndrome model. Alcohol. Clin. Exp. Res. *16:*695-699.

Jones , K.L., D.W. Smith, C.M. Ulleland, and A.P. Streissguth (1973) Patterns of malformation in offspring of chronic alcoholic mothers. Lancet *1:*1267-1271.

Jones , K.L., and D.W. Smith (1975) The fetal alcohol syndrome. Teratol. *12:*1-10.

Kentroti, S., H. Rahman, J. Grove, and A. Vernadakis (1995) Ethanol neuronotoxicity in the embryonic chick brain "in ovo" and in culture: interaction of the neural cell adhesion molecule (NCAM). Int. J. Dev. Neurosci. *13:*859-870.

Kennedy, L.A., and S. Mukerji (1986a) Ethanol neurotoxicity. 1. Direct effects on replicating astrocytes. Neurobehav. Toxicol. Teratol. *8:*11-15.

Kennedy, L.A., and S. Mukerji (1986b) Ethanol neurotoxicity. 2. Direct effects on differentiating astrocytes. Neurobehav. Toxicol. Teratol. *8:*17-21.

Kimelberg, H.K., and M. Aschner (1994) Astrocytes and their functions, past and present. In F.E. Lancaster (ed): Alcohol and Glial Cells. Bethesda: NIH, pp. 1-40.

Larroque, B., M. Kaminski, P. Dehaene, D. Subtil, M.-J. Delfosse, and D. Querleu (1995) Moderate prenatal alcohol exposure and psychomotor development at preschool age. Am. J. Public Healt *85:*1654-1661.

Ledig, M., P. Kopp, and P.Mandel (1985) Effect of ethanol on adenosine triphosphatase and enolase activity in rat brain and cultured nerve cells. Neurochem. Res. *10:*1311-1324.

Ledig, M., L. Megías-Megías, and G. Tholey (1991) Maternal alcohol exposure before and during pregnancy: effect on development of neurons and glial cell in culture. Alcohol Alcoholism *26:*169-176.

Ledig, M., and G. Tholey (1994) Fetal alcohol exposure and glial cell development. In F.E. Lancaster (ed): Alcohol and Glial Cells. Bethesda: NIH, pp. 117-132.

Lemoine, P., M. Harousseau, J.P. Borteyru, and J.C. Menuet (1968) Les enfants des parents alcooliques. Anomalies observees a propos de 127 cas. Quest. Med. *21:*476-482.

Lokhorst, D.K., and M.J. Druse (1993) Effects of ethanol on cultured fetal astroglia. Alcohol Clin. Exp. Res. *17:*810-815.

Lu, B., M. Yokohama, C.F. Reyfus, and I.B. Black (1991) NGF gene expression in actively growing brain glia. J. Neurosci. *11:*318-326.

McCarthy, K.D., J. Prime, T. Harmon, and R. Pollenz (1985) Receptor-mediated phosphorylation of astroglial intermediate filament proteins in cultured astroglia. J. Neurochem. *44:*723-730.

Mayordomo, F., J. Renau-Piqueras, L. Megías, C. Guerri, F.J. Iborra, Y. Azorín, and M. Ledig (1992) Cytochemical and stereological of rat cortical astrocytes during development in primary culture. Effect of prenatal exposure to ethanol. Int. J. Dev. Biol. *36:*311-321.

Meyer, L.S., L.E. Kotch, and E.P. Riley (1990) Neonatal ethanol exposure. Functional alterations associated with cerebellar growth retardation. Neurotoxicol. and Teratol. *12:15*-22.

Miller, M.W. (1986) Effects of alcohol on the generation and migration of cerebral cortical neurons. Science *233:*1308-1311.

Miller, M.W. (1987) Effects of prenatal exposure to alcohol on the distribution and time of origin of corticospinal neurons in the rat. J. Comp. Neurol. *257:*372-382.

Miller, M.W. (1989) Effect of prenatal exposure to ethanol on the development of cerebral cortex: II Cell proliferation in the ventricular and subventricular zones of the rat. J. Comp. Neurol. *287:*326-338.

Miller, M.W. (1992) Effects of prenatal exposure to ethanol on cell proliferation and neuronal migration. In M.W.Miller (ed): Development of the Central Nervous System: Effects of Alcohol and Opiates. New York: Wiley-Liss, pp. 47-69.

Miller, M.W., and S. Robertson (1993) Prenatal exposure to ethanol alters the postnatal development and transformation of radial glia to astrocytes in the cortex. J. Comp. Neurol. *337:*253-266.

Montoliu, C., S. Vallés, J. Renau-Piqueras, and C. Guerri (1994) Ethanol-induced oxygen radical formation and lipid peroxidation in rat brain: effect of chronic alcohol consumption. J. Neurochem. *63:* 1855-1862.

Montoliu, C., M. Sancho-Tello, I. Azorín, M. Burgal, S. Vallés, J. Renau-Piqueras, and C. Guerri (1995) Ethanol increases cytochrome P4502E1 and induces oxidative stress in astrocytes. J. Neurochem. *65:*2561-2570.

Nordmann, R., C. Riviere, and H. Rouach (1990) Ethanol-induced lipid peroxidation and oxidative stress in extrahepatic tissues. Alcohol Alcoholism *25:*231-237.

Norris, C.R., and K. Kalil (1991) Guidance of callosal axons by radial glia in the developing cerebral cortex. J. Neurosci. *11:*3481-3492.

Phillips, D.E. (1994) Effects of alcohol on glial cells development "in vivo": morphological studies. In F.E. Lancaster (ed): Alcohol and Glial Cells. Bethesda: NIH, pp. 69-91.

Rakic, P. (1991) Glial cells in development. "In vivo" and "in vitro" approaches. Ann. N. Y. Acad. Sci. *633:*96-99.

Renau-Piqueras, J., F. Miragall, C. Guerri, and R. Báguena-Cervellera (1987) Prenatal exposure to alcohol alters the Golgi apparatus of newborn rat hepatocyte: a cytochemical study. J. Histochem. Cytochem. *35:*221-228.

Renau-Piqueras, J., M. Sancho-Tello, R. Zaragoza, and C. Guerri (1988) Effect of ethanol on the development of astrocytes in primary cultures. Adv. Biosci. *71:*269-273.

Renau-Piqueras, J., R. Zaragoza, P. de Paz, R. Báguena-Cervellera, L. Megías, and C. Guerri (1989) Effects of prolonged ethanol exposure on the glial fibrillary acidic protein-containing intermediate filaments of astrocytes in primary culture: A quantitative immunofluorescence and immunogold electron microscopic study. J. Histochem. Cytochem. *37:*229-240.

Renau-Piqueras, J., C. Guerri, M. Burgal, P. de Paz, R. Saez , and F. Mayordomo (1992) Prenatal exposure to ethanol alters plasma membrane glycoproteins of astrocytes during development in primary culture as revealed by concanavalin A binding and 519-nucleotidase activity. Glia *5:*65-74.

Renau-Piqueras, J., R. Guasch, I. Azorín, J.M. Seguí, and C. Guerri (1997) Prenatal alcohol exposure affect galactosyltransferase activity and glycoconjugates in the Golgi apparatus of fetal rat hepatocytes. Hepatology (in press).

Riley, E.P., S.M. Mattson, E.R. Sowell, E.L. Jernigan, D.F. Sobel, and K.L. Jones (1995) Abnormalities of the corpus callosum in children prenatally exposed to alcohol. Alcohol Clin. Exp. Res. *19:*1198-1202.

Rutishauser, U. (1996) Polysialic acid and the regulation of cell interactions. Curr. Opin. Cell Biol. *8:*679-684.

Saez R., M. Burgal, J. Renau-Piqueras, A. Marqués, and C. Guerri (1991) Evolution of several cytoskeletal proteins of astrocytes in primary culture: effect of prenatal alcohol exposure. Neurochem. Res. *16:*737-747.

Sanchis, R., M. Sancho-Tello, and C. Guerri (1986) The effects of chronic alcohol consumption on pregnant rats and their offspring. Alcohol Alcoholism. *21:*295-301.

Sanchis, R., M. Sancho-Tello, M. Chirivella, and C. Guerri (1987) The role of maternal alcohol damage on the ethanol teratogenicity in the rat. Teratology *36:*199-208.

Sancho-Tello, M., S. Vallés, C. Montoliu, J. Renau-Piqueras, and C. Guerri (1995) Developmental pattern of GFAP and vimentin gene expression in rat brain and radial glial cultures. Glia *15:*157-166.

Sies, H. (1985) Oxidative Stress. Academic Press, London.

Silver, J., and M.Y. Ogawa (1983) Postnatally induced formation of the corpus callosum in acallosal mice on glia-coated cellulose bridges. Science *220:*1067-1069.

Sinet, P.M., R.E. Heikkila, and G. Cohen (1980) Hydrogen peroxide by rat brain "in vivo". J. Neurochem. *34:*1421-1428.

Singer, W. (1996) development and plasticity of cortical processing architectures. Science *270:*758-764.

Slomiany, A., M. Grabska, E. Grzekinska, K. Yamaki, C. Kasinathan, and B.L. Slomiany (1992) Membrane biogenesis in presence of ethanol. Alcohol Clin. Exp. Res. *16:*1152-1161.

Snyder, A.K., S.P.Singh, and S. Ehmann (1992) Effects of ethanol on DNA, RNA and protein synthesis in rat astrocyte cultures. Alcohol. Clin. Exp. Res. *16:*295-300.

Streissguth, A.P., S. Landesman-Dwyer, J.C. Martin, and D.W. Smith (1980) Teratogenic effects of alcohol in human and laboratory animals. Science *209:*353-361.

Streissguth, A.P., H. Barr, P.D. Sampson, J. Parrish-Johson, G. Kirchner, and J.C. Martin (1986) Attention, distraction and reaction time at age 7 years and prenatal alcohol exposure. Neurobehav. Toxicol. Teratol. *8:*717-725.

Streissguth, A.P., H. Barr, and P.D. Sampson (1990) Moderate prenatal alcohol exposure: effects on child IQ and learning problems at age 7/12 years. Alcohol Clin. Exp. Res. *14:*662-669.

Streissguth, A.P., P.D. Sampson, H.C. Olson, F.L. Bookstein, H.M. Barr, M. Scott, J. Feldman, and A.F. Mirsky (1994) Maternal drinking during pregnancy: Attention and short-term memory in 14-year-old offspring. A longitudinal prospective study. Alcohol Clin. Exp. Res. *18*:202-218.

Tabakoff, B., and P.L. Hoffman, (1987) Biochemical pharmacology of alcohol. In H.Y. Meltzer (ed): Psychopharmacology: The Third Generation of Process. New York: Raven, pp. 1521-1526.

Vallés, S., V. Felipo, C. Montoliu, and C. Guerri (1995) Alcohol exposure during brain development reduces 3H-MK-801 binding and enhances metabotropic-glutamate receptor stimulates phosphoinisitide hydrolysis in rat hippocampus. Life Sci. *56:* 1373-1383.

Vallés, S. (1996) Cambios en la expresión de genes de filamentos intermedios astrogliales durante el desarrollo del cerebro: efecto de la exposición al etanol. Ph. Dis. University of Valencia. Spain.

Vallés, S., M. Sancho-Tello, R. Miñana, E. Climent, J. Renau-Piqueras, and C. Guerri (1996) Glial fibrillary acidic protein expression in rat brain and in radial glia culture is delayed by prenatal ethanol exposure. J. Neurochem. *67:*2425-2433.

Wu, J.-S.R., R. Teoh, J.Vallance-Owen, and J.D. Young (1986) Nucleoside transport in primary cultures of differentiated mouse astrocytes. Neurosci. Lett. *25:*S49.

Yeagle, P.L. (1988) Lipid regulation of cell membrane structure and function. FASEB J. *3:*1833-1847.

14

NEURON-GLIA ENSEMBLES AND MAMMALIAN CNS LESION REPAIR

Manuel Nieto-Sampedro

Damage to the CNS breaks the physical and functional boundaries of the nervous tissue, disrupts the interactions between neural cells, and destroys both the neurons and the connections in which they are involved. At the beginning of the century, Cajal showed that damaged CNS did not regenerate spontaneously, an observation supported by subsequent morphological studies. However, at the same time, it was observed that adult CNS "rejuvenated" in response to injury. Mechanical trauma, stroke and many degenerative diseases, evoke developmental capabilities in the adult CNS, such as glial proliferation and differentiation, axonal sprouting and synaptogenesis. CNS response to injury is best understood as the response of neuron-glia ensembles. These ensembles have properties that, unaided, are capable of correcting minor CNS lesions. We believe that, adequately controlled, these properties could also permit to correct the consequences of major CNS injury. True regeneration, i.e. the restitution of the CNS to its pre-injury state, is clearly impossible in the absence of neuronal division. However, we think that functional repair can be achieved. The present chapter describes, within the above context, the work carried in our laboratory regarding: i) Brain inhibitors of astrocyte and astrocytoma proliferation. ii) Inhibitors of neurite outgrowth in reactive glia and iii) Enhancement of nerve fiber regeneration by olfactory bulb ensheathing glia.

INTRODUCTION

The nervous system of the vertebrates is in charge of integrating the function of the various organs of the body. Because it also controls and coordinates the interactions of the organism with its environment, nervous system lesions are a major survival problem. To deal with it, mammalian evolution has found two major solutions: prevention and repair. The first solution is used in the central nervous system (CNS); the second, in the peripheral nervous system (PNS). Regarding CNS lesions in mammals, evolutive emphasis seems directed towards prevention, by means of efficient mechanical protection. Major CNS lesions are incapacitating and the individuals affected have very low chances of survival. Hence, the possibility of their competitive reproduction is practically null and mutations conducive to efficient CNS lesion repair have practically no chances of conservation. Spontaneous repair of mammalian CNS lesions does not occur.

Accidents of various types are the main cause of mortality in children and young adults. The devastating individual effects of brain and spinal cord injuries often obscure the enormous social and economic importance of these lesions. Only cardiac disease and cancer surpass the number of deaths caused by accidents and, examining the number of potential work-years lost, lesions surpass all other problems. This is because most brain or spinal cord injuries cause chronic incapacity and the lesions most frequently occur to individuals under 45 years of age. Nowadays edema and other acute events can be efficiently treated and major CNS lesions are rarely mortal, though incurable. The final consequences of CNS injury depend on the area damaged and the extent of the lesion, but the best present therapies can offer is mere relief of the symptoms. Rehabilitation, afterwards, will teach the victims to adapt to their new situation and make the best possible use of their remaining functions.

The incurability of large injuries of brain or spinal cord was practically a dogma until a decade ago. Since then, there have been hopeful advances, first realizing that CNS structure and function are highly adaptable and then, describing the cellular and molecular bases of this adaptability (plasticity). Plasticity, a property of neuron-glia ensembles, includes self-repair as a typical trait and may be manipulated for the treatment of CNS lesions. Plasticity studies have permitted the interventive repair of experimental lesions that were considered incurable until a few years ago (Ramón-Cueto and Nieto-Sampedro, 1994; Cheng et al., 1996). These advances permit us to face the future of CNS lesion repair with a certain optimism, for we believe that its main features are understood and the remaining task is technical. However, given the complexity of the problem, this task will be still long and difficult.

CONTROL OF NEURAL ENVIRONMENT: NEURON-GLIA UNITS OF FUNCTION

Approximately half the mass of nervous tissue is constituted by glial cells, which outnumber neurons by 10-fold (Pope, 1978). Described by its discoverer tissue as nervous glue (Virchow, 1859), this static and distorted image of glia was maintained for more than one hundred years, primarily by neuroanatomists and neuropathologists. However, the last twenty years have seen a dramatic change and our present view of nervous function has enlarged to a new perspective that is no longer dominated by a neuronal point of view. We have reached the consensus that neural development, nervous activity, neural maintenance and neural pathology, must be considered from the point of view of neuron-glia ensembles. They are best understood in terms of dynamic *neuron-glia units* of function. The idea of neuron-glia units has been developed independently by a number of researchers during the last 15 years, but was formulated explicitly by Arenander and de Vellis (1983) and Nieto-Sampedro (1988*a*).

Astrocytes are intimately associated with neurons, as well as with the rest of the organism. They envelop all CNS synapses and also form the *glia limitans* (the frontier between CNS and other tissues, especially blood). The specialized high conductance regions of the end-feet of astrocytes surrounding blood vessels, can monitor nutrients, oxygen, vitamins and hormones in blood. Moreover, astrocytes are sensitive to ions and are able to bind and transport neurotransmitters. When depolarized by excitatory neurotransmitters, some of them are able to conduct action potentials. They are intimately related to the production and inactivation of the most abundant CNS excitatory neurotransmitter: glutamate. Astrocytes can generate glutamate through non-vesicular liberation (Parpura et al., 1994) and terminate its action through binding, uptake and amidation (refs. in Nieto-Sampedro, 1988). An exciting development in astrocyte-neuron intersignaling is the ability of astrocytes to propagate Ca^{2+} waves, initiated by neuronal activity (Cornell-Bell et al., 1990). Ca^{2+} waves, transmitted along astrocytes, can affect the intracellular Ca^{2+} concentration in neurons far away from the initiating site (Nedergaard, 1994). Astroglia also synthesize nitric oxide, a highly diffussible neuromodulador capable of strong effects on both neurons and glial cells (Garthwaite et al., 1991). Finally, astrocytes are in direct communication with each other and with neurons by means of gap-junctions (Sontheimer, 1995).

All these properties make astrocytes able to continuously monitor (receive, interpret and react according to) the information signaled by neurons and blood. They seem to function as 'environment-stats', the set point of which is the normal neural environment. Feed-back signals maintain an environmental composition that is appropriate to the normal pattern of neuronal activity. Any deviation from that composition will trigger a compensating glial response, for example, by removing excess excitatory molecules before they reach excitotoxic levels, or by expressing neurotrophic factors required to maintain Ca^{2+} levels preventing neuronal apoptosis. In summary, astrocytes seem to be the primary controllers of CNS environment, regarding ionic composition, neurotransmitter concentration and supply of neuronal growth factors.

In contrast to astrocytes, the information available on microglia physiology and cell biology is much more incomplete, although research is actively progressing (see Microglia, 1993). Microglia certainly play an essential role in neuroimmune interactions and, very likely, make a major contribution to the establishment of the boundaries of axonal territories.

In summary, nervous tissue is made of functional units, consisting of neurons associated to groups of glial cells. The main controllers of the microenvironment in these ensembles, regarding ion balance, neurotransmitter levels and growth factor supply, are the glial cells. Neuron-glia ensembles can communicate and synchronize with each other by means of Ca^{2+} waves, gap junctions and diffussible transmitters. To understand correctly the response of nervous tissue to perturbations, we have to consider the coordinated response of neurons and glia.

CENTRAL NERVOUS SYSTEM LESIONS

Lesions are the CNS perturbations that interest our research group. There are many types of lesion, each of them probably evoking a specific physiological response. Classic neuropathology, however, classifies lesions in two global types, according to their gross morphological effects: *anisomorphic* and *isomorphic* (Greenfield, 1958). These two lesion types differ considerably in their effects at the cellular and molecular levels.

Disruption of the glia limitans: anisomorphic and isomorphic lesions

The soma and, more frequently, the "end feet" of astrocytes, fully cover the surfaces of brain and spinal cord. Astrocytes also envelop CNS blood vessels, thus forming a boundary between CNS and the blood supply. Some properties of the astrocyte end feet, such as ionic conductivity and excitability, are different from the rest of the same cell. Additionally, blood vessels in the CNS are not fenestrated and the astrocyte-blood vessel association has special permeability properties, which limit free access to the CNS of cells and molecules originating from other tissues. This boundary glia forms a physical and functional frontier between the CNS and the remaining tissues, the so-called *glia limitans*.

Lesions that alter the gross morphology of the CNS are called *anisomorphic*. These lesions, typically caused by mechanical trauma, disrupt the *glia limitans* and the blood-CNS barrier at the injury site. Vascular breakage and vaso-spasm cause ischemia/ hypoxia/ hypoglycemia, together with invasion of the injury area by blood cells and serum proteins. Edema, resulting from both extracellular fluid accumulation and astrocyte swelling, is obvious 24 hr later (Klatzo, 1967). Ultrastructural axonal abnormalities can be observed in both, grey and white matter, immediately after a contusive injury. Uniform necrosis and myelin degeneration follow 8-24 hours later (Balentine, 1978). Many blood-derived macrophages accumulate 48 hr after an open wound and engulf degenerating myelin and other cell debris. A destructive phenomenon, called *secondary neuronal death*, begins one or two day(s) after the lesion. Groups of neurons close to the damaged area, or connected with it, begin to die a few days later. Gradually enlarging cysts develop and coalesce, forming cavities within the CNS parenchyma (Kao and Chang, 1977). The number of neurons lost as a consequence of secondary neuronal death is generally much larger than that of cells acutely lost (primary neuronal death). Therefore, secondary injury is probably the main cause of functional loss after CNS trauma.

Isomorphic lesions neither alter the gross morphology of the CNS (Greenfield, 1958), nor damage the *glia limitans*, although they partly alter the blood-CNS barrier. They are caused, for example, by degenerative disease, neurotoxins, cardiac arrest, local infarctions and wallerian degeneration. Microglia initiate the CNS response to isomorphic lesions, proliferating abundantly 3 to 5 days after the injury and then differentiating in reactive microglia and macrophages. Following the initial microglial response, reactive astrocytes appear around the microglial cells, as if avoiding them (Fernaud-Espinosa et al., 1993).

GLIAL RESPONSE TO LESIONS: RESTING AND REACTIVE GLIA

Normal astrocytes are called *"resting"* astrocytes. The term *reactive*, refers to cells that react to a serious perturbation, i.e. an epileptic episode, or a lesion. There are many types of perturbations and, correspondingly, many possible kinds of glial response. The use of the same name for very different phenomena and cells, generates the feeling that reactive glia is a single thing. In the following paragraphs, I will describe that this is not so.

Reactive astrocytes, glia limitans and glial scar

Astrocytes close to an anisomorphic lesion undergo two concomitant, but not necessarily related phenomena: i) proliferation and ii) differentiation into fibrous astrocytes. Fibrous astrocytes are larger and have processes thicker and more numerous than resting astrocytes.

The main constituent of their intermediate filaments, the glial fibrillary acidic protein (GFAP), undergoes a large increase in expression and the filaments associate. Cells with more abundant and larger processes define the most common morphological image of *reactive* astrocytes. Reactive and fibrous are synonyms for the same cell type, one with many, associated GFAP filaments. *Astrogliosis* reflects the appearance of many of these cells, usually after a deletereous event. The "glial scar" is of an accumulation of reactive astrocytes at a lesion surface, overlayed by fibroblasts and collagen. It probably represents the attempt of the CNS to reconstitute a new *glia limitans*. At least part of the astrocytes in it arise as the result of injury-induced proliferation and some authors also call these proliferating cells reactive astrocytes. However, the properties and appearance of proliferating astrocytes are very different from those of fibrous astrocytes.

Proliferating astrocytes in adult mammalian CNS may be either mature astrocytes that have undergone de-differentiation, or remaining astrocyte precursors, or both. Although this problem is clearly established, its definite answer needs further research. Viable astrocytes cannot be cultured from adult mammalian CNS, unless tissue is cultured adjacent to an injury site. The astrocytes that proliferate from injured adult tissue are cells that exhibit morphology and immunological properties similar to that of cultured type 1 astroblasts from neonatal brain. Such astroblasts promote neuritogenesis and are an excellent substrate for neurite extension. Conversely, the membranes of some fibrous astrocytes, contain proteoglycans that cause growth cone collapse, inhibit neurite outgrowth and repel growing neurites (Bovolenta et al., 1997). The ratio of fibrous astrocytes to astroblasts determines the neurite promoting or neurite inhibiting properties of the 'glial scar'. This ratio changes with post-lesion time and glial scar properties change with it. The glial scar separates neurons previously connected and because it either prevents the growth of regenerating sprouts, or supplies them with misguidance cues, it constitutes the most general obstacle to the re-establishment of lost connections. Unaided, regenerating axons cannot cross this barrier.

Astrocyte number and CNS inhibitors of astrocyte division

The number of astroglial cells in the mammalian CNS remains practically constant throughout adulthood, in spite of the existence in CNS of a variety of mitogens. Because division remains more a potentiality than a frequent event, we hypothesized that mitogen inhibitors coexisted with the mitogens and were crucial for maintaining astrocyte number stationary. Evidence of inhibitor presence in brain was obtained in the late 1980's (Nieto-Sampedro, 1988; Nieto-Sampedro and Broderick, 1989), but their purification and characterization was achieved only recently (Abad-Rodríguez and Nieto-Sampedro, 1993; Nieto-Sampedro and Abad-Rodríguez, 1995). Two families of low molecular weight inhibitory compounds were purified (from rodent and bovine brain), using a combination of conventional and high resolution chromatography. One group was composed of small acidic peptides, with amino acid composition and size reminiscent of those of "chalones" (immune system mitosis inhibitors). The other family of inhibitors consisted of complex glycolipids, with 2 or 3 sialic acids per molecule and chemical properties intermediate between GD1b and GT. Preliminary chemical, bi-dimensional

NMR and immunological results suggest that unique chemical features of the inhibitor would make it chemically different from known gangliosides.

Both types of inhibitor showed dose-dependent, saturable cytostatic activity on normal astroblasts and transformed glioma lines. The concentration that inhibited serum-promoted thymidine incorporation by 50% in these cells (ID50) was 210 to 270 nM. Although normal endothelial cells seemed similarly sensitive, thymidine incorporation in normal or transformed fibroblasts was not inhibited by glycolipid up to a concentration of 2 μM. Inhibitory peptides showed cytotoxicity at concentrations 4-fold over their ID50s, but the glycolipid was not cytotoxic, even at concentrations 10-fold above its ID50. The growth of experimental brain tumors, formed after transplatation of C6 glioma cells in adult rat cortex, was inhibited by both the glycolipid and a synthetic analogue of its carbohydrate moiety. In addition, treatment with the inhibitor evoked tumor destruction (Nieto-Sampedro et al., 1996). The receptors of the glycolipid are two Ca^{2+}-dependent, carbohydrate binding proteins of molecular weight 150 and 250 kDalton, respectively, and are probably selectin or integrin-type intercellular adhesion molecules. The well known (but not explained) phenomenon of contact inhibition of cell proliferation, may be mediated these or similar receptors and their glycolipid ligands. Relatively simple compounds, modeled on the glycolipid ligand and capable of crossing the blood-brain barrier, may head a new class of antimitotics/antimetastatics clinically useful for the destruction of brain tumors.

REACTIVE AND REGENERATIVE GROWTH

CNS lesions elicit, in the last instance, new synapse formation. The formation of new synaptic contacts occurs after the initiation and growth of new axon terminals or *axon sprouting*. Sprouting is independent of subsequent formation of new synapses and, in fact, CNS sprouts frequently degenerate without ever forming synapses. The term sprouting simply designates a type of growth response; it may or may not be the first step towards new synapse formation. When new synapses are formed by sprouts of axons undamaged by the lesion, the process is called *reactive synaptogenesis*. Here, the term *reactive* denotes synapse formation triggered by stimuli that are not part of the developmental program. The process is called *regenerative synaptogenesis* when the sprouts contributing to new synapse formation arise from damaged axons. Regenerative synaptogenesis involves long distance axon extension and is not observed in the mammalian CNS, except after transplantation.

The existence of reactive synaptogenesis, shows that in mature mammalian CNS new synapses can form, replace those lost by injury and mediate synaptic transmission (Cotman et al., 1981; Cotman and Nieto-Sampedro, 1984). Reactive synaptogenesis was, for a long time, synonymous to synaptic plasticity, i.e. CNS plasticity. Although the new synapses do not replace the original circuitry, they may have beneficial consequences in the case of small lesions. Reactive sprouts, by providing presynaptic input to deafferented neurons, may help to prevent dendritic atrophy associated with deafferentation. Furthermore, it may facilitate functional recovery, as in the hippocampus, where reactive synaptogenesis facilitated recovery of performance on a reinforced alternation task, after unilateral entorhinal cortex ablation. On the negative side, it may interfere with functional recovery by introducing abnormal connections, which compete with regenerative sprouts for available synaptic spaces.

Axonal sprouting and glial growth factors

New axon growth has two essential requirements: i) the presence of specific molecules responsible for sprout initiation and collectively denominated *neuritogenic factors,* and ii) the existence of a substrate appropriate for adhesion and elongation of the sprouts. Two further types of growth factor are of great importance for restoration of correct synapses: iii) *chimiotactic* or directional factors, that orient neurite growth and iv) differentiation factors, that intervene in synapse maturation, directing neurotransmitter choice. There is a certain confusion regarding the activity and identity of these factors because some of them, particularly the first and best known, NGF (Levi-Montalcini, 1982), possess several activities. NGF is neurotrophic (survival promoting), neuritogenic and chimiotactic for sympathetic neurons. Nonetheless, neurotrophic factors generally do not initiate neurite outgrowth. They are only permissive factors, i.e., molecules that permit neurons to be alive and available to receive instructions from specific factors. Instructive factors are then responsible for eliciting defined structural or functional modifications. For example, NGF is neurotrophic for some sensory neurons and the high molecular weight basal lamina protein, laminin, is capable of evoking neurite initiation both during development and after a lesion. Interaction of factors usually helps; NGF associated to laminin is much more efficient as a neurotropic and a neuritogenic factor for sensory neurons than either agent alone (Millaruelo et al., 1988).

Perhaps the most significant recent advancement in the elucidation of the mode of action of growth factors, has been the establishment of the connection between neuritogenic and neurotrophic factors. Walsh and his team found that the neuritogenic activity of cell adhesion proteins was explained by their ability to link to the receptor of a glial neurotrophic factor, fibroblast growth factor (FGF), and share its tyrosin-kinase domain (Williams et al., 1994). The protein phosphorylation cascade thus initiated, finally affected Ca^{2+} homeostasis (Mattson et el., 1993; Schulman, 1995), triggering neurite outgrowth. In this respect, it is suggestive that intracellular Ca^{2+} levels are the meeting point of long term potentiation (LTP; Malenka, 1995) and neurite initiation and elongation (i.e. filament protein polymerization and cytoskeleton reorganization; Hoffman, 1995). Glial cells are probably the major source of growth factors in the CNS (Nieto-Sampedro et al., 1983; Rudge, 1993).

Reactive glia, sprout inhibition and sprout misguidance

Functional loss after CNS lesions, is mainly due to circuit disruption subsequent to axonal damage. Lack of axon regeneration is not due to the inability of central neurons to produce axonal sprouts, or to their lack of long distance growth capability, as shown by Aguayo and collaborators (1987). Inhibition of axon regeneration is, rather, a consequence of the properties of reactive glia. What are these properties? Anisomorphic and isomorphic gliotic tissue differ in cell composition and mode of axon growth inhibition. In anisomorphic lesions, the glial scar is widely regarded as the most general obstacle both to regeneration and to transplant-host integration. Regenerating axons usually fail to grow across an injury zone or through the host-transplant boundary. Most frequently, they turn around and either degenerate or make inappropriate synaptic contacts on available sites proximal to their place of origin. Astrocytes become reactive less than one day after trauma. In the following days they migrate and proliferate, and their enlarged fibrous processes form a web that is not a simple mechanical barrier. Axons do not cross a transection zone even when the 'scar' is very thin or virtually absent (Gearhart, et al., 1979). However, they can cross thick collagen barriers (de la Torre, 1982) or dense implanted 'scar' tissue (Reier et al., 1983). The glial scar functions in much the same way as the normal *glia limitans*, in that axons crossing neither of them. Recent experiments in our laboratory, comparing *in vitro* the neurite promoting properties of plasma membranes from normal tissue and anisomorphic gliotic tissue (Bovolenta et al., 1992), suggested one mode of regeneration inhibition: *misguidance*. Proliferating anisomorphic lesion astrocytes have astroblast-like properties and neurites prefere to grow adhering to them, thereby being misguided away from their synaptic targets.

The alternative proposed by Ramón y Cajal (1914), i.e. that reactive astrocytes actively inhibit sprout outgrowth, is observed in the case of isomorphic gliosis. Plasma membranes from isomorphic gliotic tissue, totally inhibited neurite initiation and repelled and caused the collapse of neurites previously evoked by laminin (Bovolenta et al., 1992). The inhibitory molecule, a heparan-sulfate, chondroitin-sulfate proteoglycan (Bovolenta et al., 1993), was present in fibrous astrocytes (Bovolenta et al., 1997).

All these observations place the glia at the center stage of the regeneration problem and emphasize the importance of the cellular and molecular processes that control the number and type of the various glial populations.

REPAIR OF DAMAGED CNS

The major causes of functional loss after CNS lesions are the interruption of neuronal circuits (due to damage or even death of neurons) and the activation of a variable proportion of resting glia. Generally, damaged CNS is incapable of complete functional self-repair without external help. Clinical intervention is necessary to prevent secondary neuronal death, to replace lost neurons, to prevent inhibition of sprouting and to guide regenerative sprouts to grow accross gliotic tissue, towards their physiological targets. The problem of secondary neuronal death is being tackled with excitatory neurotransmitter blockers and by using embryonic tissue transplants to replace lost neurons. Axonal regeneration requires overcoming the inhibitory properties of reactive glia and creating an environment conducive to neurite growth. Two complementary approaches are being developped in our laboratory to solve the problem of axonal regeneration and navigation: i) blocking antibodies against glial neurite outgrowth inhibitors; ii) transplantation of olfactory bulb ensheathing cells, a type of growth-promoting macroglia.

Neurite outgrowth inhibitors. Blocking antibodies

Axonal navigation is controlled by the growing tip of the axon, the growth cone, which senses guidance cues distributed non-homogeneously in the external environment. Growth promoting and growth repellent or inhibitory signals are the most prominent of these cues. Injury-induced gliotic tissue, the classic barrier to CNS axon regeneration (Ramón y Cajal, 1914), may contain growth inhibitory molecules, but its molecular nature has not been determined, partly because reactive glia could not be cultured as such. Astrocytes cultured from gliotic tissue either die, or grow in a "blast-like" form. Further, a putative *in vitro* model of reactive astrocytes, type 1 astroblasts treated with cAMP, did not inhibit neurite outgrowth like gliotic tissue. Therefore, to characterize possible neurite inhibitory molecules in normal and reactive glia, we had to develop an *in vitro* model of gliotic tissue (Bovolenta et al., 1992).

Using that model, we found that cultured astrocytes, independently of their lineage, morphology, immunological type and treatment with differentiating agents, induced profuse neurite outgrowth from embryonic CNS neurons. The outgrowth was also elicited by isolated plasma membranes and was comparable to that promoted by laminin (Bovolenta et al., 1992). Conversely, membranes isolated from isomorphic gliotic tissue (deafferenting lesion, or excitotoxic injury), powerfully inhibited central neurite outgrowth. Anisomorphic gliotic tissue membranes had intermediate properties. They were neurite outgrowth permissive, but growth was more limited than over membranes of cultured astrocytes. When given a choice, growing neurites actively avoided membranes from isomorphic gliotic tissue, while they seemed to follow anisomorphic membrane boundaries and crossed unhindered over membranes of cultured astrocytes. In conclusion, reactive glia contained both, inhibitory and neurite promoting molecules, the proportion of which depended on the way gliosis was generated. For isomorphic gliotic tissue membranes, the balance favored inhibition of neurite outgrowth, while anisomorphic membranes probably expressed inhibitory components at lower levels and growth promoting activity predominated.

Using an inhibition assay based on the *in vitro* glial scar model, we isolated a neurite inhibitory proteoglycan from isomorphically injured adult brain membranes. The proteoglycan had an apparent molecular weight of 160-220 kDa and its GAG portion, which contained heparan sulphate and chondroitin sulphate, appeared necessary for inhibition when tested *in vitro*. The inhibitor, when presented as a surface bound molecule, prevented neurite initiation; when added in a soluble form to growing neurites, induced their retraction. This activity was greatly diminished by digestion with heparinase and, to a slightly lesser extent, with chondroitinase ABC, implying that both heparan sulphate and chondroitin sulphate chains were involved in the inhibitory activity (Bovolenta et al., 1993).

Proteoglycans and particularly their sugar moieties, the glycosaminoglycans, modulate neurite growth in either a positive or a negative way during development; their expression was low in the adult, being enhanced after injury or disease (Bovolenta et al., 1993; Levine, 1994).

We have developed monoclonal antibodies (Mabs) against the inhibitory proteoglycan, two of which were characterized and used to study the mode of action and localization of the inhibitor in normal and injured adult brain (Bovolenta et al., 1997). The antibodies recognized the 200 kDa inhibitor in Western blots and, in adult brain, reacted predominantly with neurons although, after a lesion, immunoreactivity was associated mainly with reactive astrocytes.

The localization of neurite outgrowth inhibitors in reactive glia, supports the idea that gliotic tissue is largely responsible for the failure of axonal regeneration in mammalian CNS. One of the Mabs blocked *in vitro* both the inhibition of neurite outgrowth and the growth cone collapse activity of the proteoglycan. Tests are in progress to examine whether the antibody also shows the same activity *in vivo*.

Ensheathing glia transplants promote, protect and guide axon growth

CNS neuronal environment consists essentially of glial cells, which have the greatest influence on axonal growth cone navigation. When a large enough proportion of glial cells becomes reactive, the microenvironment becomes hostile to both, sprouting and axon growth. Navigation in adult CNS is easy for embryonic axons, whereas adult sprouts are only capable of progressing in adult CNS when surrounded by the environment provided by peripheral nerve. It is known that olfactory bulb is the single mammalian CNS locus where damaged axons regenerate spontaneously. Olfactory axons regenerate with the help of a special type of macroglia, so-called ensheathing glia, that is present in both the peripheral and central components of the olfactory system.

Ensheathing glia accompany olfactory axons on their long and complex path from the olfactory mucose (PNS) to the olfactory bulb (CNS) and are, probably, largely responsible for the ability of olfactory axons to find their synaptic targets in the CNS. Ensheathing cells show survival and neuritogenesis promoting properties similar to those of Schwann cells but, like astrocytes, are capable of integration and migration in the CNS (Gudiño-Cabrera and Nieto-Sampedro, 1996). They envelope axons, isolating them from surrounding environment, an ability that has been used to promote regeneration of sprouts from transected sensory axons into the spinal cord (Ramón-Cueto and Nieto-Sampedro, 1994).

A spinal cord lesion that can be repaired: dorsal root avulsion

Neuropathic pain, a common sequel to traumatic injury in motorbike or labour accidents, is due to sensory root avulsion. Although the damaged central branch of adult dorsal root ganglion (DRG) axons regenerates, it fails to re-enter into the spinal cord (Ramón y Cajal, 1914; Carlstedt et al. 1989; Siegal et al. 1990). The failure is caused by the presence in the dorsal root transitional zone of a non-permissive environment, the isomorphic gliotic tissue formed after Wallerian degeneration of DRG afferents. In an attempt to promote the ingrowth of damaged dorsal root axons into the cord, several groups have tried to create artificial axonal growth supportive substrates in the dorsal root entry zone. However, although axons were able to bypass the transition zone, they only grew as far as the implanted permissive substrate. Beyond that, axons elongated for very short distances and failed to reinnervate the spinal cord.

The unique properties of olfactory bulb ensheathing glia suggested that these glial cells could be used to promote axonal regeneration in non-regenerating system. Ensheathing cells were purified away from other cell types by immunoaffinity, using anti-p75-NGF receptor and suspensions of pure ensheathing cells were used for the experiment. After laminectomy at the lower thoracic level, the spinal cord was exposed and one dorsal root (T10) was completely transected at the cord entry point. The root stump was microsurgically anastomosed to the cord and a suspension of immuno-pure ensheathing cells was transplanted in the spinal cord at the dorsal root entry zone. Three weeks after transplantation, using DiI and antibodies against CGRP and GAP-43, numerous regenerating dorsal root axons were observed re-entering the spinal cord. Regenerating sensory afferents invaded laminae 1, 2 and 3, grew through laminae 4 and 5 and reached the dorsal grey commissure and lamina 4 of the contralateral side. Transplanted ensheathing cells reached the same laminae as regenerating axons did and neither ensheathing cells nor regenerating axons invaded those laminae they did not inervate under normal circumstances. Additionally, neither regenerating axons nor ensheathing cells were observed within the ipsilateral ventral horn or the dorsal column. In conclusion, ingrowth of regenerating dorsal root axons into adult spinal cord was made possible by ensheathing glia transplants. Although observed reinnervation is incomplete, it may be sufficient for neuropathic pain treatment.

Present work is directed to promote the recovery of a well-defined physiological function: the control of urinary excretion. The urinary bladder of the rat is rendered neurogenic by selective bilateral transection of the relevant lumbar and sacral afferents. The complex and delicate surgery required involves laminectomy, laparotomy, selective root stimulation, transection and transplantation. The subsequent tracing at the optical and electron microscope levels make obvious the convenience of multidisciplinary collaboration, in this case of teams from Navarra Hospital and Navarra University (Drs. Pascual-Piedrola and Insausti, respectively) and the Cajal Institute (Gudiño-Cabrera and Nieto-Sampedro). Rapid transfer of positive results to clinical practice is thus assured. Transplantation of ensheathing cells into the dorsal root entry zone, permits regenerating sensory axons to cross unhindered thick layers of isomorphic reactive glia in the deafferented dorsal horn. Because ensheathing glia transplanted in the spinal cord seem to behave like they do in the olfactory system during development, we are also testing the possible general use of these cells to promote axonal regeneration in the CNS. Ensheathing cells can be purified in large amounts from the olfactory bulb of adult animals of several species, including humans, even long after death and show high viability after months of freeze storage (Gudiño and Nieto-Sampedro, 1996). Hence, there are no obvious scientific or ethical obstacles to their clinical use.

DISCUSSION AND PERSPECTIVES: TEMPORAL WINDOWS FOR CNS REPAIR

The study of the cellular and molecular basis of neural plasticity during the past 15 years has changed the traditionally pessimistic view of CNS damage. The view emerging from CNS plasticity studies and progressively accepted by researchers and neurologists is more optimistic. We believe that damaged CNS probably will never regererate, that is, perhaps will never be like it was before the lesion. However, we firmly believe that the detailed knowledge of the plasticity mechanisms will permit replacement of damaged circuits in a way that will make possible *functional repair* of damaged CNS. Yet, still is difficult to indicate the precise steps that this type of intervention will follow. Two clear objectives are prevention of secondary neuronal death and substitution of damaged circuits. Probably various succesive interventions will be needed, as suggested earlier (Nieto-Sampedro, 1988). Immediately after a lesion, paramedic treatment with excitatory neurotransmitter antagonists to prevent secondary neuronal death may become routine. With the same purpose, treatment with exogenous neurotrophic factors may be used, until that time (8-12 days postlesion) where endogenous production of neurotrophic factors by injured tissue is sufficient for neuronal maintenance. Exogenous neuritogenic factors may also help to promote sprouting, together with blocking antibodies to neurite outgrowth inhibitors present in both, gliotic tissue and central myelin. Another interventive tool, in this respect, could be transplants of ependymocytes, which would both promote sprouting and help regenerative sprouts to cross gliotic tissue. Embryonic neuron transplantation may help in disrupted circuit reconstruction, perhaps substituting monosinaptic paths by multisynaptic relays. Finally, the data available thus far suggest that rehabilitation may be directed to stimulate intrinsic CNS plasticity mechanisms, that would ensure the correctness and functionality of the new connections. Present optimistism concerning the possibility of functional repair of CNS lesions is given further support by the success in repair of experimental models of rhizotomy (Ramón-Cueto and Nieto-Sampedro, 1994) and spinal chord transection (Cheng et al., 1996; Li et al., 1997).

To summarize, our working hypothesis is that functional repair of CNS lesions is possible, applying some general principles: i) We should use to its maximum the spontaneous autorepair response of the CNS. ii) Because lesion damage evolves with time, lesion treatment will very likely require multiple interventions. iii) Interventions must be inserted in the autorepair process, in their appropriate temporal windows, and must be directed to manipulate the autorepair response. Different intervention types are required at different times postlesion and correct timing is critical. iv) If exogenous compounds and cells are needed, these should be, as much as possible, those normally used by the nervous system.

ACKNOWLEDGMENTS

This work was supported by the CSIC, the Science and Technology Interministerial Committee (CICYT), Health Research Funds (FIS), the Madrid Autonomous

Community, and Europharma S.A. I thank Miss Chi Bui for creative proofreading of the manuscript.

REFERENCES

Abad-Rodríguez, J. and Nieto-Sampedro, M.(1993) Purification from bovine brain of an inhibitor of astrocyte an C6 glioma cell division. Eur. J. Neurosci. supp. *6:* 23.

Aguayo, A.J., Vidal-Sanz, M., Villegas-Pérez, M.P., y Bray, G.M. (1987) Growth and connectivity of axotomized retinal neurons in adult rat with optic nerves substituted by PNS grafts linking the eyes and midbrain. Ann. N.Y. Acad.Sci. *495*: 1-9.

Arenander, A. y de Vellis, J. (1983) Frontiers of Glial Physiology. In The Clinical Neurosciences (R. Rosenberg, ed., section V). Churchill Livingstone, New York, pp.53-91.

Balentine, J.D. (1978) Pathology of experimental spinal cord trauma. II. Ultrastructure of axons and myelin. Lab. Invest., *39*: 254-266.

Bovolenta, P., Wandosell, F. y Nieto-Sampedro, M.(1992) CNS glial scar tissue: a source of molecules which inhibit central neurite outgrowth. Progress Brain Res. *94*: 367-379.

Bovolenta, P., Wandosell, F. and Nieto-Sampedro, M. (1993a) Characterization of a neurite outgrowth inhibitor expressed after CNS injury. Eur. J. Neurosci., *5:* 454-465.

Bovolenta, P., Fernaud-Espinosa, I., Méndez-Otero, R. and Nieto-Sampedro, M. (1997) Neurite outgrowth inhibitor of gliotic brain tissue. Mode of action and cellular localization studied with specific monoclonal antibodies. Eur. J. Neurosci., *in press.*

Carlstedt, T., S. Cullheim, M. Risling, and B. Ulfhake (1989) Nerve fibre regeneration across the PNS-CNS interface at the root-spinal cord junction. Brain Res. Bull. *22*: 93-102.

Cheng, H., Cao, Y. and Olson, L. (1996) Spinal cord repair in adult paraplegic rats: partial restoration of hind limb function. Science *273*: 510-513.

Cotman, C.W., Nieto-Sampedro, M. y Harris, E. W. (1981) Synapse replacement in the nervous system of adult vertebrates. Physiol. Rev. *61:* 684-784.

Cotman, C.W.and Nieto-Sampedro, M. (1984) Cell biology of synaptic plasticity. Science *225*: 1287-1294.

Cornell-Bell, A.H., Finkbeiner, S.M., Cooper, M.S. y Smith, S.J. (1990) Glutamate induces Ca2+ waves in cultured astrocytes: long range glial signalling. Science *247:* 470-473.

De la Torre, J.C. (1982) Catecholamine fiber regeneration accross a collagen bio-implant after spinal cord transection. Brain Res. Bull. *9:* 545-552.

Fernaud-Espinosa, I., Nieto-Sampedro, M. and Bovolenta, P. (1993) Differential activation of microglia and astrocytes in aniso- and iso-morphic gliotic tissue. Glia *8:* 277-291.

Garthwaite, J. (1991). Glutamate, nitric oxide and cell-cell signalling in the nervous system. Trends Neurosci. *14:* 60-67.

Gearhart, J., Oster-Granite, M.L., and Guth, L.(1979) Histological changes after transection of the spinal cord of fetal and neonatal mice. Exp. Neurol. *66*: 1-15.

Greenfield, J.G. (1958) General pathology of nerve cell and neuroglia. In: Neuropathology. (Greenfield, J.G., Blackwood, W., Meyer, A., McMenemey, W.H., Norman, R.M. eds.) Ed. Arnold, Ltd. London. pp 1-66.

Gudiño-Cabrera, G. and Nieto-Sampedro, M. (1996) Ensheathing cells: Large scale purification from adult olfactory bulb, freeze-preservation and migration of transplanted cells in adult brain. Restor. Neurol. Neurosci.: *10:* 25-34.

Hoffman, P.N. (1995) The synthesis, axonal transport and phosphorylation of neurofilaments determine axonalcaliber in myelinated nerve fibers.The Neuroscientist. *1:* 76-83.

Kao, C.C. and Chang, L.W. (1977) The mechanism of spinal cord cavitation following spinal cord transection. Part 1: A correlative histochemical study. J. Neurosurg. *46:* 197-206.

Klatzo, I. (1967) Neuropathological aspects of brain edema. J. Neuropathol. Exp. Neurol. *26*: 1-14.

Levi-Montalcini, R. (1982) Developmental neurobiology and the natural history of nerve growth factor. Annu. Rev. Neurosci. *5*: 341-362.

Levine, J. M. (1994) Increased expression of the NG2 chondroitin sulfate proteoglycan after brain injury. J. Neurosci. *14*: 4716-4730.

Li, Y., Field, P.M. and Raisman G. (1997) Repair of adult corticospinal tract by transplants of olfactory ensheathing cells. Science *277*: 2000-2002.

Malenka, R.C. (1995) LTP and LTD: Dynamic and interactive processes of synaptic plasticity. The Neuroscientist. *1:* 35-42.

Mattson, M.P., Rydel, R.E., Lieberburg, I., and Smith-Swintowsky, V.L. (1993) Altered calcium signaling and neuronal injury: stroke and Alzheimer's disease as examples. Ann. N.Y. Acad. Sci. *679*: 1-21.

Microglia: Special Issue. (1993) Glia *7(1)* pp.1-120.

Millaruelo, A. I., M. Nieto-Sampedro and C. W. Cotman (1988). Cooperation between NGF and laminin or fibronectin in promoting sensory neuron survival and neurite outgrowth. Dev. Brain Res. *38:* 219.228.

Nedergaard, M. (1994) Direct signalling from astrocytes to neurons in cultures of mammalian brain cells. Science *263*: 1768-1771.

Nieto-Sampedro, M., Manthorpe, M., Barbin, G., Varon, S. y Cotman, C.W. (1983) Injury-induced neuronotrophic activity in adult rat brain: correlation with survival of delayed implants in the wound cavity. J. Neurosci. *3:* 2219-2229.

Nieto-Sampedro, M. (1988a) Growth factor induction and order of events in CNS repair. In Pharmacological approaches to the treatment of brain and spinal cord injury. (eds. D.G.Stein and B.A. Sabel). Plenum Press, New York, pp.301-337.

Nieto-Sampedro, M. (1988b) Astrocyte mitogen inhibitor related to epidermal growth factor receptor. Science *240:* 1784-1786.

Nieto-Sampedro, M. and Broderick, J.T. (1989) A soluble brain molecule related to epidermal growth factor receptor is a mitogen inhibitor for astrocytes. J. Neurosci. Res., *22:*.28-35.

Nieto-Sampedro, M. and Abad-Rodríguez, J. (1995) Inhibidores de la division glial. Rev. Neurol. *23:* 605.

Nieto-Sampedro, M., Bailón, C., Fernandez Mayoralas, A., Martín-Lomas, M., Mellström, B. and Naranjo, J.R. (1996) Experimental brain glioma: growth arrest and destruction by a blood-group-related tetrasaccharide. J. Neuropathol. Exp. Neurol. *55:* 169-177.

Parpura, V., Basarsky, T.A., Liu, F., Jeftinija, K., Jeftinija, S. y Haydon, P.G. (1994) Glutamate-mediated astrocyte-neuron signalling. Nature *369*: 744-747.

Pope, A. (1978) Neuroglia: Quantitative aspects. En Dynamic properties of glial cells (E. Schoffeniels y F.G. Tower, eds.) New York, Pergamon Press, pp. 13-20.

Ramón y Cajal, S. (1914) Estudios sobre la degeneración y regeneración del sistema nervioso. Imprenta Hijos de Nicolás Moya, Madrid.

Ramón-Cueto, A. y Nieto-Sampedro, M. (1994) Regeneration into the spinal cord of transected dorsal root axons is promoted by ensheathing glia transplants. Exp. Neurol. *127:* 232-244.

Reier, P.J., Stensaas, L.J. and Guth, L. (1983) The astrocytic scar as an impediment to regeneration in the central nervous system. In: Spinal Cord Reconstruction (C.C. Kao, R.P.Bunge and P.J. Reier, eds.), Raven Press, New York, pp. 163-195.

Rudge, J.S. (1993) Astrocyte-derived neurotrophic factors. In: Astrocytes. Pharmacology and function (S. Murphy, ed.) Acad. Press, San Diego, pp.267-305.

Schulman, H. (1995) Protein phosphorylatiuon in neuronal plasticity and gene expression. Curr. Opinion Neurobiol. *5:* 375-381.

Siegal, J. D., M. Kliot, G. M. Smith, and J. Silver (1990). A comparison of the regeneration potential of dorsal root fibers into gray or white matter of the adult rat spinal cord. Exp. Neurol. *109*: 90-97.

Sontheimer, H. (1995) Glial influences on neuronal signalling. The Neuroscientist *1:* 123-126.

Williams, E.J., Furtness, J., Walsh, F.S., and Doherty, P. (1994) Activation of the FGF receptor underlies neurite outgrowth stimulated by L1, N-CAM and N-cadherin. Neuron *13*: 583-594.

15

GLIAL RESPONSE TO EXCITOTOXIC INJURY IN THE IMMATURE RAT BRAIN

Laia Acarin, Berta González,
Anthony J. Castro and Bernardo Castellano

Damage to the immature rat brain is often characterized by considerable anatomical reorganization and sprouting of target-depleted axons; but the injury associated glial reaction, playing a potential role in plasticity, is not well documented. It was classically thought that the glial response represented a negative aspect of the brain response to degeneration, but more recent studies have pointed out the participation of glial cells in the production of trophic factors important for central nervous system repair. This apparent neuronotrophic role of glial cells seems to be particularly important in the young brain, where the microglial response after injury follows a rapid timecourse, leading to a fast removal of neuronal debris, and astrocytes are assumed to be more growth supporting than in the adult. In our laboratory we are working on an experimental model based on thalamocortical connectivity in the young brain in order to study the microglia and astroglial response after different grades of neuronal affectation. In this chapter we describe different aspects of the microglial and astroglial response to developmental injury, including the interaction between these glial populations by means of cytokine and growth factor release.

MICROGLIA AND ASTROGLIAL CELLS

The brain response to injury characteristically involves a marked astroglial reaction occurring after microglial activation. In the central nervous system (CNS), microglial cells constitute aproximately 10% of the total glial cell population and they are the only glial cells of mesodermal origin. Microglia arise from monocytes that populate the CNS after it has been vascularized (Boya et al., 1987; Hickey and Kimura, 1988; Ling and Wong, 1993; Ling, 1994). Monocyte-derived ameboid microglial cells are already observed on the second embryonic week in the rat brain and still present during the first weeks of postnatal life (Ashwell, 1991; Dalmau et al., 1997). Ameboid microglial cells, rounded and devoid of any processes, show phagocytic activity and are believed to contribute to CNS development by removal of dead cells occurring by natural cell death. During early postnatal development, ameboid microglial cells transform into ramified microglia (Boya et al., 1991; Wu et al., 1993), the microglial cell type found in the adult brain. Characteristically, ramified microglial cells have an elliptical or elongated cell body with several processes branching out from the perinuclear cytoplasm. Processes are usually further divided into spindle-like thinner projections and distal parts often appear to end in proximity to blood vessels or in relation to neurons. The functional role of ramified microglial cells in the normal adult brain is still not known. Their characteristic distribution forming a network of cells covering the entire CNS territory and their reaction to injury by up regulating macrophage characteristics suggests that in their resting state they are down-regulated macrophages that form a system of immunocompetent cells in the CNS (Graeber and Streit, 1990; Thomas, 1992; Nakajima and Kohsaka, 1993b). Some authors have suggested that the pinocytic activity of ramified microglial cells represents a fluid cleansing system for the extracellular fluid of normal tissue (Thomas, 1990; Glenn et al., 1991). Moreover, the presence of specific enzymes for the cleavage of adenosine and inosine phosphoesters on microglial cells (Castellano et al., 1991; Vela et al., 1995) suggests a role of these cells in controlling extracellular concentrations of nucleotides and nucleosides, key factors in the control of proliferation and differentiation processes in the CNS.

The response of microglia to neural damage is by far the most studied aspect of microglial cell biology. As described by del Río Hortega more than 75 years ago (Del Río-Hortega, 1920a; Del Río-Hortega, 1920b), microglial cells characteristically react to injury in a graded manner, depending on the severity of the lesion or the developmental stage of the nervous system. Becoming apparent by rapid morphological, immunophenotypical and functional changes, microglial cell activation can imply proliferation and migration, and it may even involve active phagocytosis when neuronal death or axonal degeneration occurs (see Nakajima and Kohsaka, 1993b; and Kreutzberg, 1996 for review).

Astrocytes are the most abundant cell of the CNS, outnumbering neurons by about 10:1. Derived from the neuroectoderm of the neural tube, astrocytes are star-shaped cells whose processes extend to the surface of blood vessels (forming the so-called end feet processes) or to the surface of the brain to form the glial limiting membrane (glia limitans). In the normal CNS, astrocytes play multiple important functions that contribute to neural development and the integrity of normal brain function, including the structural support of the nervous system parenchyma, guidance for neuroblast migration or neurite outgrowth, and induction of blood brain barrier properties in endothelial cells (Moonen et al., 1990). Astrocytes are also involved in synaptic function, helping to terminate neurotransmitter action by uptake or by providing synaptic insulation, and they participate in ion metabolism and the regulation of cerebral blood flow by potassium siphoning (Hansson and Ronnback, 1994). Moreover, specific enzyme systems enable astrocytes to metabolize ammonia, glutamate, and free radicals, protecting the brain from the toxicity of these compounds (Norenberg, 1994).

In addition to these widely varied functions in the normal brain, astrocytes play crucial roles in response to neural damage. The molecular mechanisms that lead to reactive astrogliosis are not well established, although it is likely that contributing factors are derived from the necrosis or injury of neurons, myelin breakdown, edema formation and blood-brain barrier (BBB) disruption. The early microglial cell activation after injury and the consequent secretion of cytokines also seem to play a key role in triggering the astroglial response. In fact, accumulating evidence suggests that both glial cell populations react to injury in a synchronized way by means of the release of cytokines, mitogens and growth factors that modulate each other's reactivity (Frei et al., 1986; Giulian et al., 1988b; Benveniste, 1993).

After astrocytic swelling, which is the earliest astroglial response to injury and is characterized by swelling of astrocytic processes and glycogen accumulation, reactive astroglial cells undergo hypertrophy of the cytoplasm and processes due to an increase in the expression of glial fibrillary acidic protein (GFAP) and *de novo* expression of vimentin. Astroglial hyperplasia can also occur in this stage, especially after anisomorphic injury (when tissue integrity is damaged and BBB is disrupted) (Janeczko, 1988; Miyake et al., 1992; Schiffer et al., 1993). While astrocytic hypertrophy, hyperplasia and glial scar formation (the long term result of astroglial response) have long been thought to represent a negative factor for regeneration, more recent evidence suggests that the process of reactive astrogliosis may actually be an attempt to promote CNS repair by removing excitotoxins and secreting several neurotrophic factors.

The apparent neuronotrophic role of astrocytes seems to be particularly important in the young brain, when astrocytes are assumed to be more growth supporting than in the adult (Nieto-Sampedro, 1988; Hatten et al., 1991; Smith and Miller, 1991), and microglial response follows a rapid timecourse. Although damage to the immature rat brain is often characterized by considerable anatomical reorganization and sprouting of target-depleted axons (Kolb et al., 1994; Kolb et al., 1996), the injury associated glial reaction playing a potential role in plasticity, has not been well studied. In this regard, in our laboratory we are working on an experimental model based on thalamocortical connectivity in the young brain in order to study the glial response after different grades of neuronal affectation (Acarin et al., 1996; Acarin et al., 1997, 1998a, 1998b). By injecting *N*-methyl-D-aspartate (NMDA), a glutamate analog, into the nine-day old neocortex we induce an excitotoxic cortical lesion, which causes sublethal affectation of target depleted thalamic nuclei. Thus, we are able to study and compare the glial response in the lesion site, where BBB breakdown and neuronal death occurs (anisomorphic gliosis), and in mildly affected thalamic nuclei, where tissue structure is maintained (isomorphic gliosis). Throughout the following chapter we will present and discuss some of these observations.

MICROGLIAL RESPONSE

Activation signals for microglial cells

The specific signals triggering the microglial response are not yet well understood, and while activation factors for microglial cells in the immature and adult brain may be similar, the source, effects and timecourse may differ. Due to the difficulty of working with in vivo models, most putative factors for microglial cells activation, proliferation or upregulation of macrophage antigens have been demonstrated in vitro. Factors that activate microglial cells in culture include interleukin-1 (IL-1) (Merrill, 1991; Ganter et al., 1992), specific microglial mitogens (Giulian et al., 1991) and a variety of colony-stimulating factors (CSF) such as macrophage colony stimulating factor (M-CSF), granulocyte-macrophage colony stimulating factor (GM-CSF) and interleukin-3 (IL-3) (Giulian et al., 1991; Brosnan et al., 1993; Raivich et al., 1993; Streit, 1993). Whether these factors activate microglial cells in situ is supported by observations that microglia are able to express receptors for CSF after injury in vivo (Raivich et al., 1991; Streit, 1993). Thus, it seems likely that CSF may constitute important activating factors for microglial cells after injury, but the precise role of these factors as well as their specific sources in the injured brain remain unknown.

Moreover, in related studies microglial cells were found to be extremely sensitive to changes in potassium concentrations of the extracellular medium, being susceptible to small depolarizing events (Kettenmann et al., 1993; Kettenmann and Ilschner, 1993). Such a mechanism is especially important in relation to the extracellular ionic concentrations that are disrupted by neuronal overexcitation following excitotoxicity and therefore may contribute to microglial activation in response to excitotoxic neuronal death. This possibility is supported by recent work reporting that extracellular ATP (and ADP to a lesser extent), which can accumulate by neuronal leakage after injury, can promote depolarization of microglial cells in culture via a specific membrane diphosphate purinoreceptor (Walz et al., 1993). This mechanism could act as an activation signal for microglial cells and may correspond to the increased nucleoside diphosphatase enzymatic activity seen in the plasma membrane of activated microglial cells under different types of lesions (Jorgensen et al., 1993; Jensen et al., 1994).

Potentially, all of these events are involved in the early activation of microglia, which occurs during the first 10 hours after developmental excitotoxin-induced injury (Acarin et al., 1996; Acarin et al., 1998a) as well as after aspiration lesions (Milligan et al., 1991), stab wounds (Kaur et al., 1987), and ischemic-hypoxic insults (Towfighi et al., 1995; Ivacko et al., 1996). In addition to these early activating signals, mainly leading to changes in microglial cell morphology and metabolism, later influences may depend on the extent of the lesion exerted. The presence of neuronal and/or axonal debris, massive tissue disruption and BBB breakdown induces microglial cells to become further activated to phagocytize, to show major histocompatibility complexes (MHC) I and

II expression and to upregulate other macrophage markers. In contrast, areas showing no neuronal or myelin degeneration, such as the sublethally damaged thalamus in our excitotoxic lesion model, usually have a less intense microglial response. In such areas, microglial cells usually do not show macrophage morphology but adopt a reactive ramified shape, the so called 'bushy-like cells'. These cells are not phagocytic and usually upregulate MHC class I expression but do not show MHC class II on their membrane (Acarin et al., 1998a) (fig. 15-1).

Contribution of peripheral monocytes/macrophages and microglia

One of the most debated issues in microglial biology concerns the relative contribution of blood-derived monocytes/macrophages and endogenous microglial cells to the brain's reaction to injury. Commonly used microglial markers such as OX42 antibody (Kaur et al., 1992), nucleoside diphosphatase (NDPase) demonstration (Jorgensen et al., 1993) or lectin binding (Streit and Kreutzberg, 1987; Acarin et al., 1996) show increased labeling of reactive microglial cells, but they unavoidably also stain monocyte-derived macrophages. Thus, the putative contribution of both cell types will remain unsolved until a marker is able to differentiate between these populations of cells. Obviously, the integrity or disruption of the BBB is crucial in the contribution of blood-borne cells.

The microglial response after CNS injury in the newborn characteristically involves a massive increase in the number of microglia/macrophages occurring around 3-5 days after the lesion (Acarin et al., 1996; Acarin et al., 1997) (fig. 15-1, 15-3). The dramatic natural increase in the number of microglial cells occurring in the normal rat neocortex between 5 and 20 days of age, which is attributed to the proliferation of resident microglia (Kaur et al., 1989), indicates that resident microglial cells can proliferate in response to developmental injury. Moreover, animal age may influence the microglial response to injury. Ameboid microglial cells have been demonstrated in white matter tracts like the corpus callosum of the normal postnatal rat brain until 15 days after birth (Ling et al., 1990; Boya et al., 1991; Kaur and Ling, 1991; Wu et al., 1992). These cells exhibit features indicative of phagocytosis and share common characteristics with their monocytic precursors (Yee et al., 1990; Ling et al., 1992). Accordingly, ameboid microglial cells are likely candidates to become activated and proliferate after developmental injury. However, several studies using irradiation and monocyte labelling techniques support the hypothesis that the majority of brain macrophages seen after an injury involving BBB disruption originate from blood monocytes (Kaur et al., 1987; Coffey et al., 1990; Milligan et al., 1991) and that ameboid microglial cells do not migrate to the site of a stab wound cortical injury (Leong and Ling, 1992). In light of these results, it is assumed that a considerable entry of blood-derived cells through the disrupted BBB at the site of injury provides an important population of phagocytic cells. Nevertheless, the three day delay post-injury before massive monocyte infiltration takes place suggests that an early activation of resident microglial cells may be important to trigger the recruitment of blood-borne monocytes.

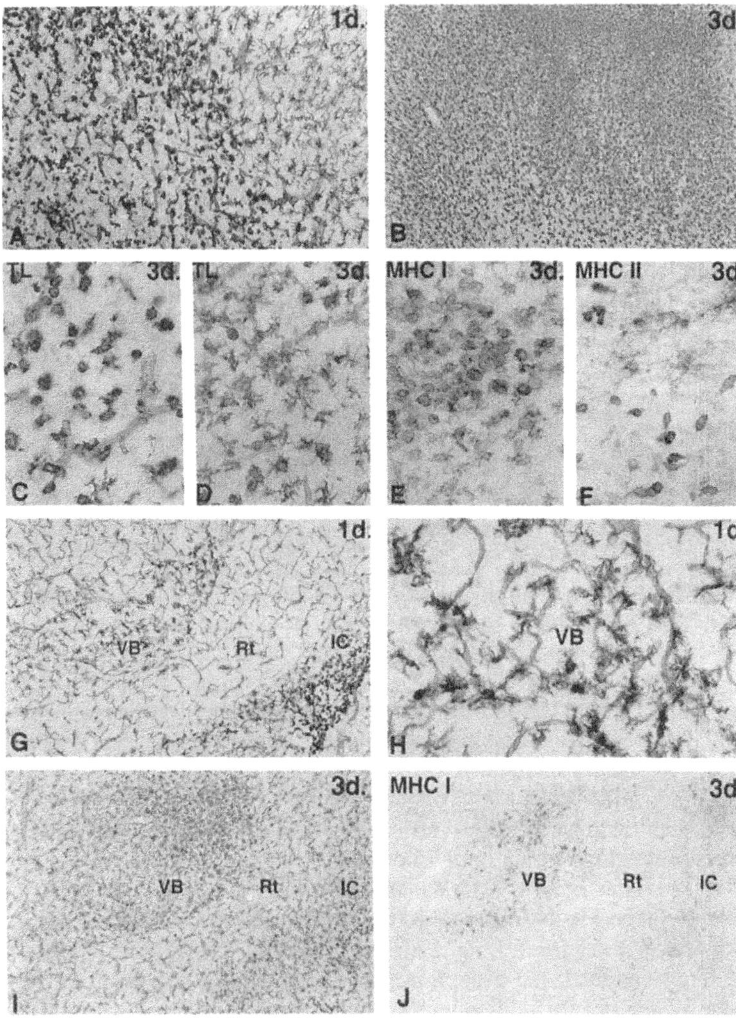

Figure 15-1. Microglial response after an excitotoxic injury in the postnatal day nine sensorimotor cortex. A-F are taken form the cortical degenerating area, and G-J from the sublethally affected thalamus at the more representative survival times. A-D and G-I, tomato lectin histochemistry for the visualization of microglial cells. E and J, MHC class I immunocytochemistry; F, MHC class II immunocytochemistry. Note the marked microglial response in the cortical areas, characterized by increased tomato lectin binding (A-D) and expression of MHC class I (E) and II (F), as well as changes in morphology towards ameboid and pseudopodic forms (C-F). In the thalamus and the internal capsule microglial cells change towards a reactive ramified shape (H), show increased lectin binding and only express MHC class I (J). A x49; B, G, I, J x31; C-F x1240; H x123.

Regardless of their origin, brain macrophages are responsible for the phagocytosis of axonal and/or neuronal debris in the degenerating areas, but the fate of these brain macrophages after repair is another interesting and controversial topic. Although no conclusions can be drawn from existing data, three possible mechanisms of elimination have been suggested: (i) observations that two-thirds of ameboid microglial cells present in the corpus callosum degenerate during the course of development and are engulfed by local companion cells (Kaur and Ling, 1991; Ling and Wong, 1993), lead the group of Ling and coworkers to the hypothesis that phagocytic cells may degenerate in situ (Kaur et al., 1987); (ii) subsequent transformation into ramified microglial-like cells, as it has been observed (Angelov et al., 1995) and demonstrated using bloodstream injection of carbon particles (Kaur et al., 1987); and (iii) re-entrance into the blood circulation, as suggested by our earlier observations in studies involving laser beam lesions (Gonzalez et al., 1985) and by the accumulation around blood vessels of round tomato lectin and MHC I positive cells as seen at 5 days after our neocortical excitotoxic lesions (Acarin et al., 1998a).

A different situation occurs in lesions where the BBB is not disrupted and peripheral cells are never observed. We have been able to study this event after excitotoxic cortical lesions, where no extravasation of plasma proteins can be detected in the sublethally affected thalamus (Acarin et al., 1998a). The microglial cell response seems to occur in the absence of peripheral cells and their secreted products. Noteworthy, as stated before, in this area microglial response is less intense and does not involve an increase in the number of cells (Acarin et al., 1998a) (fig. 15-1).

Interactions with the immune system

One of the most studied aspects of the microglial response to injury is the upregulation or de novo expression of immunomolecules such as MHC. These protein complexes have been classically considered as tools for antigen presentation to T lymphocytes in different tissues. Class I MHC proteins present antigen that has been synthesized within the antigen presenting cell itself, and class II MHC proteins present ingested antigens such as secreted products from infected, transformed or damaged cells. The observation that several stimuli could induce MHC expression on neural cells has lead to extensive study of MHC class I and MHC class II expression under distinct types of axonal, terminal or cell body degeneration in the rodent brain (Akiyama et al., 1988; Yee et al., 1990; Morioka and Streit, 1991; Finsen et al., 1993a; Finsen et al., 1993b; Cannella et al., 1995).

In the normal adult brain, MHC class I expression is restricted to endothelial cells and areas devoid of BBB such as the pineal gland (Pedersen et al., 1993). Upregulation generally occurs after any kind of degeneration or after an invasive lesion, but microglial cells also express MHC I in response to milder stimuli, such as changes caused by motoneuron regeneration (Streit et al., 1989), proximity to a degenerating area (Finsen et al., 1993a), or transient cerebral overpressure

(Kaur et al., 1995). As no phagocytosis is involved in any of these latter situations, these findings lead to the hypothesis that microglial expression of MHC class I antigen is a sensitive marker for variations in the brain microenvironment.

As the early postnatal brain is often considered to be immunologically immature because of developmental plasticity, incomplete myelination and the lack of a fully mature BBB, the study of the microglial expression of MHC class I and class II in the developing brain following diffferent types of injury may be relevant towards a better understanding of the role and significance of the expression of these molecules by microglial cells.

Normal developmental events such as tissue remodeling in the early postnatal corpus callosum provide the signals for MHC class I induction on ameboid microglial cells (Ling et al., 1991). Moreover, we have shown that upregulation of this immunomolecule on microglial cells of the young brain occurs both in response to excitotoxic cell death and BBB breakdown and in response to sublethal damage of thalamic neurons in the absence of BBB disruption (Acarin et al., 1998a) (fig. 15-1). Such findings imply that changes occurring in thalamic neurons and their extracellular milieu following cortical target depletion, like small variations in acid/base status, or the increased extracellular concentrations of glutamate or potassium ions, may provide a sufficient stimulus for MHC class I expression. In this sense it seems likely that MHC class I antigen expression more likely indicates a general state of microglial metabolic activation rather than a specific involvement in antigen presenting function.

In contrast, the presence of MHC class II in the normal immature brain is very low. *De novo* expression of this immunomolecule has been observed in phagocytic cells after motoneuron degeneration or ischemic-hypoxic injury in neonatal rats (Morioka and Streit, 1991; Greensmith and Navarrete, 1994) and in the degenerating optic tract when it achieves myelination (Yee et al., 1990). Moreover, in our lesion model we have observed MHC class II expression on round or pseudopodic cells observed in the excitotoxic cortical lesion site (Acarin et al., 1998a) (fig. 15-1), but none of these cells could be detected in the retrogradelly damaged thalamus where no phagocytosis is seen. Thus, it seems likely that MHC class II expression on microglia/macrophages occurs only when a phagocytic process is involved. However, recent studies have shown that MHC class II can be induced in the normal developing brain by the peripheral injection of interferon γ (IFNγ) or lipopolysaccharide (LPS) (Xu and Ling, 1994a; Xu and Ling, 1994c; Xu and Ling, 1994b) or the intraventricular injection of a neurotrophic factor such as the ciliary neurotrophic factor (CNTF) (Kahn et al., 1995). Additionally, and in an attempt to elucidate the putative MHC inducing signals, different groups have used in vitro models to show the ability of several other molecules such as tumor necrosis factor α (TNFα), IL-1ß, IL-3 and GM-CSF to induce MHC expression on microglia (Lassmann et al., 1991; Imamura et al., 1994; Fischer et al., 1995).

In summary, it seems likely that, in addition to phagocytosis, several inflammatory molecules and growth factors may induce MHC class II expression on microglial cells after brain damage. Although the presence of these factors has not yet been demonstrated in the injured developing brain, these putative inducing molecules may be produced locally by reactive astrocytes and microglial cells, may extravasate through the leaking BBB or may be secreted by infiltrating blood cells.

Moreover, it should be mentioned that expression of MHC II on microglia is selective and only involves a small population of cells. Thus, it is likely that this molecule may be related to the functional state of some activated microglia. As damaged tissue offers several potential sources of antigen that can be recognized as 'foreign', a putative function of MHC II expression could include an elevated state of immune alertness or processes other than antigen presentation, like non-immunological roles such as cell interaction and adhesion (Streit et al., 1989; Lampson et al., 1994).

Role of microglial response: detrimental or beneficial?

In spite of the great interest that the study of the microglial response to injury has engendered in the last decades, the functional significance of microglial cell activation in response to injury remains speculative. As various studies have indicated, microglial reactivity may produce both beneficial and/or detrimental effects.

The beneficial effects associated with microglial cells activation include the production of several molecules that are though to play a role in neuroprotection or are involved in tissue repair after injury. In this sense, the microglial secretion of basic fibroblast growth factor (bFGF) has been demonstrated both in vitro (Nakajima and Kohsaka, 1993a) and in vivo after injury (Frautschy et al., 1991). The beneficial effects of bFGF include the increase of neuronal resistance to excitotoxic injury (Mattson and Rychlik, 1990; Frim et al., 1993; Nozaki et al., 1993) and the induction of astrocytic proliferation and angiogenesis (Gomez-Pinilla et al., 1992; Rothwell and Relton, 1993). Neuroprotective effects have also been associated with the production of transforming growth factor beta (TGF-ß) which is expressed by microglial cells after excitotoxic injury and ischemia (Morgan et al., 1993; Wiessner et al., 1993), and has been involved in the regulation of tissue damage: promoting repair and down-regulating microglial cytotoxicity (Rothwell and Relton, 1993).

In addition, an important component in mediating inflammation and recruitment after CNS injury is the microglial production of several cytokines. The specific role of cytokines in the injury response is still unclear, but it has been suggested that perturbations leading to the oversecretion of these cytokines could be more harmful than the specific actions they exert (Merrill, 1992). Cytokines expressed by activated microglial cells include IL-1 (Woodroofe et al., 1991; Van Dam et al., 1992; Yao et al., 1992; Yabuuchi et al., 1993), IL-3 (Gebicke-Haerter et al., 1994), IL-6 (Woodroofe et al., 1991; Sawada et al., 1992) and TNFα (Hartung et al., 1992).

IL-1 promotes leukocyte adhesion to endothelial cells (Benveniste, 1993) as well as astrocytic hypertrophy and proliferation (Giulian et al., 1988a; Giulian et al., 1988b; Merrill, 1991; Balasingam et al., 1994). Moreover, IL-1 induces microglial secretion of toxic arachidonic acid and nitric oxide (Hartung et al., 1992), and its putative role in mediating excitotoxic and ischemic cell damage has recently been reported (Relton and Rothwell, 1992). IL-3 can induce the proliferation of microglial cells and trigger an increased expression of MHC class II (Imamura et al., 1994). Moreover, studies with transgenic mice have shown that IL-3 is a potent inflammation-promoting cytokine, and may thus be involved in neuronal damage and demyelination (Chiang et al., 1996). IL-6 is able to attenuate the excitotoxicity caused by glutamate analogs (Toulmond et al., 1992; Yamada and Hatanaka, 1994), increase the production of growth factors (Frei et al., 1989) and promote astrogliosis (Fattori et al., 1995). TNFα is thought to mimic some IL-1 effects such as leukocyte adhesion, astrogliosis, or microglial proliferation (Merrill, 1991; Benveniste, 1993; Rothwell and Relton, 1993; Balasingam et al., 1994), and has demonstrated protective actions against NMDA toxicity in hippocampal slices (Cheng et al., 1994). Thus, it seems evident that these cytokines have multiple and redundant effects on microglial cells themselves as well as on the other surrounding glial and neuronal cells.

In addition to growth factor secretion and cytokine release, the main detrimental effects associated with the microglial response are derived primarily from in vitro studies demonstrating the production of harmful molecules by stimulated microglial cells. Several laboratories have demonstrated the cytotoxic effect of microglial derived products such as glutamate (Piani et al., 1991; Streit, 1993), nitric oxide (Chao et al., 1992; Murphy et al., 1993), arachidonic acid (Hartung et al., 1992), proteases (Nakajima and Kohsaka, 1993a), reactive oxygen species (Colton and Gilbert, 1993; Giulian and Corpuz, 1993; Giulian and Vaca, 1993; Giulian et al., 1993), and non-identified factors acting via a NMDA receptor-mediated mechanism and implicated in delayed neuronal death after ischemic injury (Piani et al., 1991; Piani et al., 1992; Giulian et al., 1993; Giulian and Vaca, 1993; Giulian et al., 1993).

ASTROGLIAL RESPONSE

Activation signals for astrocytes

The cellular changes leading to astrocytic gliosis are common features of CNS damage and involve raises in GFAP levels, although the putative signals triggering this response are incompletely understood.

The first astroglial change observed after developmental excitotoxic injury is a mild increase in GFAP immunoreactivity occurring around 10 hours post-lesion (Acarin et al., 1998b) followed by higher immunostaining for GFAP and astroglial hypertrophy at later times. Noteworthy, for this early timepoint, when astrocytes are first activated, evident neuronal degeneration, BBB disruption and microglial cell activation (Acarin et al., 1998a) implies a contribution of these factors in triggering the astroglial activation.

Several factors secreted by early activated microglial cells and other inflammatory cells have been implicated in the astroglial response to injury. These factors include IL-1ß, a potent astrocytic mitogen (Giulian et al., 1988b), and TNFα, both of them specifically demonstrated in the excitotoxically injured developing brain at 4-6 hours post-lesion (Szaflarski et al., 1995; Hagan et al., 1996) (also see chapter 11). Neurotrophic factors such as glial-derived neurotrophic factor (GDNF), CNTF and several other growth factors may be released from damaged neurons or reactive glial cells and therefore may also contribute to astrocytic activation after injury (Frautschy et al., 1991; Logan and Berry, 1993; Levison et al., 1996).

Although is well known that astrocytes become reactive after different kinds of neuronal or axonal damage, including the excitotoxic-induced neuronal death as caused by high concentrations of excitatory amino acid analogs such as glutamate, the possible direct effect of these excitotoxins on astrocytes should be considered.

Since excitatory neurotransmitters are likely to provide signals regulating glial gene expression, proliferation and differentiation during development, glutamate analogs when injected into the brain conceivably could directly affect astrocytes, proabably through a receptor mediated mechanism. The presence of glutamate metabotropic receptors as well as functional ionotropic glutamate receptors such as AMPA and kainate on astrocytes of different brain regions supports this possibility (see Steinhauser and Gallo, 1996 for review), although the presence of functional NMDA receptors on astroglial cells is still debated. However, in vivo studies demonstrated the immunocytochemical detection of the NMDA-R1 subunit in astrocytes of the locus coeruleus (Van Bockstaele and Colago, 1996) and in astrocytes of the rat visual cortex from postnatal day 14 (Aoki et al., 1994). Another study showed astrocytic NMDA induced currents in cultured hippocampal slices, but a possible indirect effect of NMDA acting via neuronal receptor activation could not be excluded (Porter and McCarthy, 1995).

In spite of evidence that functional receptors for several excitatory amino acid analogs may be present on astroglial membranes, the question whether glial NMDA receptors represent functional channels is still unclear (Steinhauser and Gallo, 1996). Even if astrocytes can directly respond to excitatory amino acids, it is quite evident that the astroglial response to excitotoxic neuronal damage is largely due to the degeneration process itself, but the question of a direct effect exerted by glutamate analogs remains controversial.

Astroglial hypertrophy and hyperplasia

The increase in GFAP protein in the astrocyte cytoskeleton leads to reactive astrogliosis, which plays an important physical role in attempting to preserve or reestablish the structural integrity of brain tissue following damage. This physical purpose can be accomplished by hypertrophy of pre-existing cells and/or an increase in the number of astrocytes, both as a result of hyperplasia and migration from distant sites. Of these events, astrocytic hypertrophy seems to occur earliest (Miyake et al., 1992; Balasingam et al., 1994) and may or may not be followed by astrocytic hyperplasia, mainly depending on the severity of the injury.

The molecular signals triggering this increase in astrocyte GFAP levels after injury are not well understood. The higher immunoreactivity for GFAP observed after damage could occur by an increased gene transcription or translation of preexisting mRNAs. After hypoxia-ischemia or NMDA excitotoxicity in seven day old rats, a significant increase in GFAP mRNA is not observed until one day after the insult, and it is maintained for two weeks (Burtrum and Silverstein, 1993; Burtrum and Silverstein, 1994). Related studies also reported increased GFAP mRNA after other types of injury both in the early postnatal (Oblinger and Singh, 1993; Balasingam et al., 1996) and the adult brain (Hozumi et al., 1990). Thus it seems clear that an increase GFAP gene transduction and/or higher stability of the transcripts, both events reflected as an increase of mRNA, occurs in response to injury. However, initial injury-induced rises in GFAP protein appeared to be of greater magnitude than GFAP mRNA increases although both increases may even occur simultaneously (Oblinger and Singh, 1993; Burtrum and Silverstein, 1994). These findings suggest that in addition to increased GFAP mRNA levels an early dissociation of GFAP intermediate filaments leading to exposure of previously hidden epitopes may explain the early and dramatic GFAP immunostaining elicited by tissue injury (Balasingam et al., 1996) (fig.15-2).

Interestingly, when tissue structure is only marginally compromised and the integrity of the BBB is maintained (isomorphic lesions), as with the retrogradely affected thalamus of our experimental lesion model, no increase in the number of reactive cells is detected, but astroglial hypetrophy and increased GFAP immunostaining is observed (Acarin et al., 1998b) (fig. 15-2, 15-3). These changes are believed to represent a homeostatic reaction to restore the normal integrity of the affected neuronal network. In contrast, astroglial proliferation has been described after different types of anisomorphic brain injury (where BBB disruption occurs) such as stab wounds (Takamiya et al., 1988; Miyake et al., 1992; Janeczko, 1993; Schiffer et al., 1993), and ischemic infarction (Petito and Halaby, 1993), but the proliferation pattern and timecourse of the astrocytic response depends on animal age and type of injury. Moreover, observations that astrocytes transplanted into hosts younger than postnatal day 11 are able to migrate in a radial pattern (Hatton et al., 1993) suggests that both astrocytic proliferation and migration may contribute to the astroglial response seen after excitotoxic lesions (as well as other anisomorphic lesions) in the young brain.

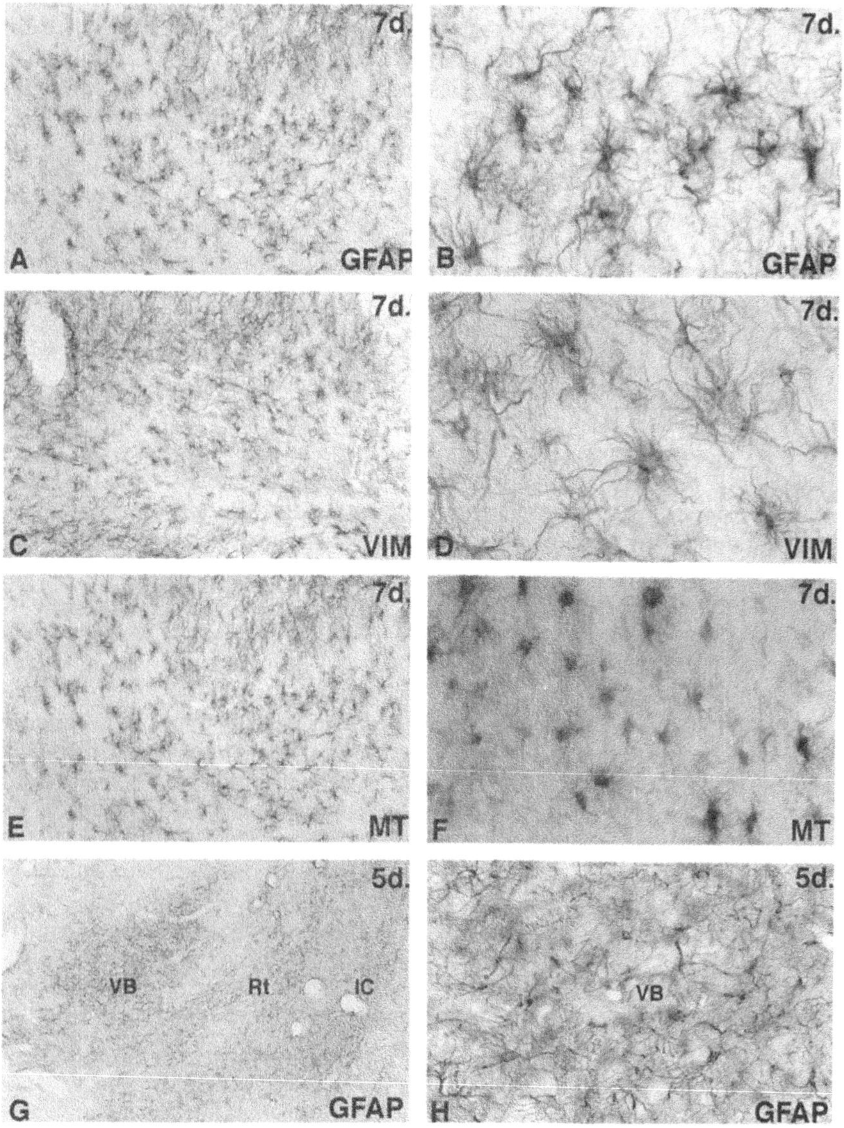

Figure 15-2. Astroglial response following an excitotoxic lesion in the postnatal day nine neocortex. Reactive astrocytes are present in the degenerating cortical site (A-F) as well as in the sublethally affected thalamus (G-H). In the cortical area maximal hypertrophy is achieved at 7 days post-lesion (A-F), reactive astrocytes increase GFAP expression (A-B), show de novo expression of vimentin (VIM) (C-D) and express metallothionein (MT) (E-F). In the thalamus, where no neuronal degeneration occurs, reactive astrocytes increase GFAP expression and become hypertrophic (G-H) but no expression of vimentin or metallothionein is observed. A, C, E x49; G x31; B, D, F, H x1542.

Changes in the astrocytic proliferative response to injury seem to follow a developmental timecourse. Dividing astrocytes already appear at 1 day after early postnatal brain damage, showing a maximal proliferation rate with lesions occurring in the one week old rat brain (Janeczko, 1988; Janeczko, 1994). As young astrocytes are able to react readily to the presence of injury-induced mitogenic factors, it is conceivable that both hypertrophy and hyperplasia occur after developmental excitotoxic lesions.

Astroglial vimentin expression

Several authors have suggested that astrocytic motility and proliferation may be linked to vimentin expression (Takamiya et al., 1988; Janeczko, 1993; Schiffer et al., 1993). An intermediate filament protein mainly found in cells of mesenchymal origin, vimentin is found in astrocytes during brain development but is lost in rats as their brain approaches maturity, i.e., around the second postnatal week, when GFAP becomes the major intermediate filament protein (Dahl, 1981). Astroglial de novo expression of vimentin in response to brain damage in adult and younger animals has been found mainly in the injury area (Takamiya et al., 1988; Schiffer et al., 1993; Schulz et al., 1993) (fig. 15-2) where a proliferative response is expected. In contrast, this expression of vimentin is absent after milder injures areas that do not provoke cell proliferation, BBB disruption or massive tissue destruction (Acarin et al., 1998b) (fig. 15-3).

In addition to a role in the remodeling of cytoskeletal components required for proliferation and migration, vimentin expression has also been related to other functions. Its presence during development and the reappearance after injury suggests that it could be related to cells in contact with relatively large volumes of extracellular fluid, being involved in providing communication between the fluid compartment and the adjacent tissue via intracellular transport mechanisms involving vimentin containing intermediate filaments (Pixley and De Vellis, 1984). Vimentin could also provide a structural link between the cytoskeleton and the cell plasma membrane (Hutchins and Casagrande, 1989) and may serve as a support for kinases such as Yes, contributing to the regulation of phosphorylation dependent cell mechanisms (Ciesielski-Treska et al., 1995). These findings clearly suggest that vimentin does not only play a structural role but is also implicated in the metabolic changes occuring in reactive astrocytes.

Metabolic changes in reactive astrocytes

The cytoskeletal modifications reflected by increased GFAP or vimentin expression are also accompanied by metabolic changes of reactive astrocytes. In addition to their structural function, reactive astrocytes are also thought to play a role in reestablishing the extracellular environment after injury. This is accomplished by increased astroglial buffering capacity for potassium ions and putative neurotoxic neurotransmitters, increased synthesis of oxidoreductive enzymes, proteases, protease inhibitors, glutamine synthase and glycogen synthesis (see Eddleston and Mucke, 1993 for review). In general, these metabolic changes would help remodel the extracellular matrix, reestablish the disrupted concentration of proteins and neurotransmitters in the extracellular milieu and degrade cellular debris from dying neurons.

At the same time, astrocytes are able to protect the microenvironment by restoring ion homeostasis and clearing putative toxic molecules generated during the injury response. In this regard, several studies have provided evidence of the formation of reactive oxygen species after seizures (Armstead et al., 1989), excitotoxicity (Schulz et al., 1995) and ischemia (Oliver et al., 1990). In all these processes hyperactivity of neuronal cells after excitatory amino acid receptor activation causes an influx of calcium ions that culminate in damage partially mediated by production of reactive oxygen species via nitric oxide synthase or phospholipase A2 activation (Pazdernik et al., 1992). Thus, an important aspect of restoring a physiologic extracellular milieu after excitotoxicity relies on the activation of antioxidant systems capable of neutralizing free radicals.

Metallothionein expression

Metallothionein is a small metal-binding protein that is over expressed in different tissues in response to stress and inflammation, where it is thought to play a key role in the detoxification and storage of heavy metals, regulation of copper and zinc homeostasis and scavenging of free radicals (Hidalgo et al., 1994b; Borghesi and Lynes, 1996). Although highest levels of metallothionein are present in the liver and kidney it is also found in other tissues like the CNS.

Brain metallothioneins I and II are mainly expressed in glial cells. Constitutive levels of the protein are expressed in astrocytes of the adult rodent brain (Young et al., 1991; Nishimura et al., 1992), but astroglial metallothionein expression is very low during development (Young et al., 1991; Ebadi et al., 1995; Penkowa and Moos, 1995). An increase in brain metallothionein expression is triggered by several factors which were first described to raise peripheral levels. In this sense, astroglial metallothionein levels can be upregulated in vitro by the well known inducers such as zinc, copper, bismuth, cadmium or glucocorticoids (Hidalgo et al., 1994a; Cowan et al., 1996). Although these factors induce metallothionein expression in cultures, in vivo studies are scarce.

Astroglial metallothionein induction in vivo was first described in patients with amyotrophic lateral sclerosis (Sillevis Smitt et al., 1992) and more recently in reactive astrocytes after neonatal cryoinjury (Penkowa and Moos, 1995). Lately, in our laboratory, metallothionein expression has been demonstrated both in reactive astrocytes and microglial cells in the spinal cord of the hypomyelinated jimpy mutant mouse (Vela et al., 1997) and in the lesion site of our excitotoxic lesioned immature brains (Acarin et al., 1998b) (fig. 15-2). Interestingly, no metallothionein expression in either glial cell type was observed in the sublethally damaged thalamus (fig. 15-3). Accordingly, several factors, such as the extravasation of plasma proteins (which can bind metals like zinc and copper) and blood cells after BBB disruption, as well as free radical formation due to excitotoxic neuronal damage and inflammation, which have been shown to increase metallothionein gene transcription (Dalton et al., 1994), may be responsible for glial metallothionein expression exclusively in the injury site. Moreover, cytokines like IFNγ, TNFα, IL-1 or IL-6, released to the site of injury by inflammatory cells and reactive microglia (Woodroofe et al., 1991; Benveniste, 1992) could also contribute to upregulate metallothionein expression (Sato et al., 1994; Sawada et al., 1994).

The precise role of metallothionein induction after injury is still unclear; it could be part of a protective response to minimize damage caused by oxidative stress (Hidalgo et al., 1988; Sato and Bremner, 1993) or increased metal levels after the excitotoxic process (Masters et al., 1994). Additionally, metallothionein can contribute to macrophage functions by inducing monocyte activation and increasing the superoxide burst in macrophages (Borghesi and Lynes, 1996), and participate in cellular adaptation mechanisms by supplying metals for nucleic acid metabolism, protein synthesis and other metabolic processes required for highly metabolic reactive glial cells. The question whether metallothionein is induced solely to protect the astrocyte or the microglial cell themselves or are secreted to influence the environment of other cells remains unknown.

Glial scar formation in the young brain

The formation of a glial scar after injury is the long term result of the glial reaction. The glial scar is mainly composed of hypertrophied astrocytic processes, but it also contains microglial projections and extracellular matrix. Although glial scars do not develop after embryonic or newborn brain injury (Moore et al., 1987), they form around 1-2 weeks after traumatic injury to rats of one week of age and older (Firkins et al., 1993; Janeczko, 1994; Acarin et al., 1998b).

In contrast to the absence of scars after lesions of the newborn brain, scar formation after injury to the one week old brain, has been related to a much larger invasion of macrophages and a slower clearing of degenerating debris (Firkins et al., 1993). However, it should be noted that the astrocytic response in sublethally damaged areas of the young brain distant from the lesion focus, where no massive tissue destruction or BBB breakdown exists, follows a transient time course and diminishes to basal levels without forming a glial scar by 7-14 days post-lesion (Acarin et al., 1998b) (fig. 15-3).

The functional consequences of glial scar formation have been strongly debated in the last decades. Glial scars may form a new glia limitans serving to isolate damaged tissue and thus prevents extension of the lesion process, but they may also represent potent inhibitors of axonal regeneration by mechanisms still not well understood (Bovolenta et al., 1992).

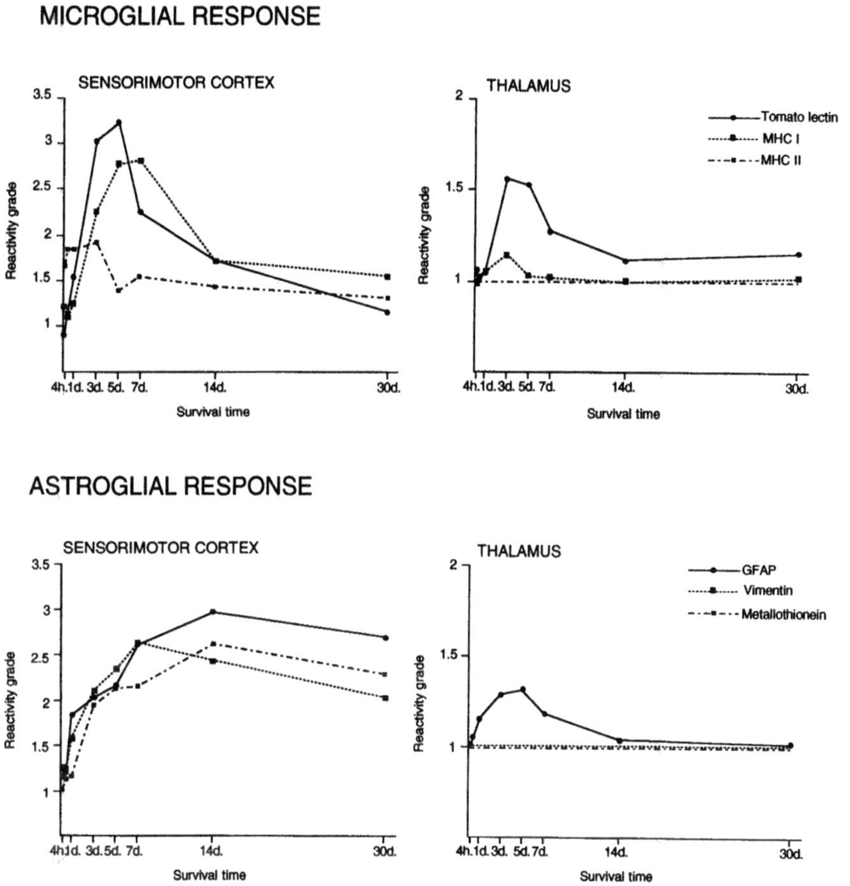

Figure 15-3. Graphs showing quantification of the microglial and astroglial response in an excitotoxic lesion site and in the sublethally affected thalamus at different survival times using a quantitative method previously described (Acarin et al., 1997). Note that 'reactivity grade' of 1 implies equivalence to the control hemisphere. Microglial response was evaluated using tomato lectin histochemistry and MHC class I and II immunocytochemistry. The astroglial response was studied with GFAP, vimentin and metallothionein immunocytochemistry. Glial response at the injury site is very fast and reaches high values; the microglial response diminished by 14 days post-injection but the glial scar (reflected by GFAP, vimentin and metallothionein immunostaining) remains until the last survival time. In comparison, thalamic glial response is milder and clearly transitory and devoid of MHC class II, vimentin or metallothionein expression.

Recent in vitro studies indicate that astrocytes, the major elements of glial scars, possess neuronotrophic properties related to the secretion of growth factors, adhesion molecules, and extracellular matrix proteins (Hatten et al., 1991). Moreover, young and embryonic astrocytes are generally thought to be more growth promoting. This characteristic can be attributed to an increased ability to secrete trophic molecules like platelet-derived growth factor (PDGF), nerve growth factor (NGF), brain-derived neurotrophic factor (BDNF), CNTF, TGF-ß, or adhesion molecules such as laminin or N-CAM (Eddleston and Mucke, 1993; Schwartz and Nishiyama, 1994). In addition, the growth promoting action of young astrocytes has been linked to the lack of expression of putative growth inhibitory molecules such as condroitin-6-sulfate proteoglycan or cytotactin/tenascin that are present in adult astrocytes (McKeon et al., 1991). Accordingly, an in vitro study showed that glial tissue removed from the lesion site of neonatal animals supported neurite outgrowth, while scars removed from adult animals did not (McKeon et al., 1991).

In view of these findings, a critical issue concerns whether the chronological age of the glial cell determines its capacity to support neurite growth, i.e. whether astrocytes possess a scheduled, age dependent expression of properties that promote axon and neurite growth. This hypothesis becomes evident when immature astrocytes are transplanted into the lesioned adult brain; young astrocytes promote functional recovery (Nieto-Sampedro, 1988; chapter 14) and induce suppression of glial scar formation by rapid migration from the lesion site and association with host blood vessels, reconstituting the BBB and restricting the immune response (Smith and Miller, 1991).

The apparent growth promoting attributes of immature astrocytes, as well as the regeneration observed after peripheral nerve lesions or after CNS lesions of some adult invertebrates, imply that in some instances the functional primitivity and/or lack of differentiation of the glial population correlates with their ability to support growth. In this context, the order of events after injury has been postulated to represent one of the main issues of mammalian CNS regeneration failure (Nieto-Sampedro, 1988). By giving priority to glial cell proliferation, the organism ensures the restitution of the blood supply and the glia limitans, but this may compromise functional regeneration because the later increase of axon-sprouting and neurotrophic factors occurs when axon sprouts and their putative targets are already separated by the compact glial scar (Nieto-Sampedro, 1988).

ACKNOWLEDGMENTS

The authors thank Miguel A. Martil for his outstanding technical help. This work was supported by DGICYT PB95-0662 of the Spanish Government, and fellowships to L.A. from CIRIT, Caixa de Sabadell Foundation, 'la Caixa' Fellowship Program and F.P.I. program of the Universitat Autònoma Barcelona.

REFERENCES

Acarin, L., B. Gonzalez, B. Castellano and A.J. Castro (1996) Microglial response to N-methyl-D-aspartate-mediated excitotoxicity in the immature rat brain. J. Comp. Neurol. *367:* 361-374.

Acarin, L., B. González, B. Castellano and A.J. Castro (1997) Quantitative analysis of microglial reaction to a cortical excitotoxic lesion in the early postnatal brain. Exp. Neurol. *147:* 410-417.

Acarin, L., B. Gonzalez, A.J. Castro and B. Castellano (1998a) Primary cortical glial reaction versus secondary thalamic glial response in the excitotoxically injured young brain. Microglial response and MHC class I and II expression. *Submitted.*

Acarin, L., B. Gonzalez, J. Hidalgo, A.J. Castro and B. Castellano (1998b) Primary cortical glial reaction versus secondary thalamic glial response in the excitotoxically injured young brain. Astroglial response and metallothionein expression. *Submitted.*

Akiyama, H., S. Itagaki and P.L. McGeer (1988) Major histocompatibility complex antigen expression on rat microglia following epidural kainic acid lesions. J. Neurosci. Res. *20:* 147-57.

Angelov, D.N., A. Gunkel, E. Stennert and W.F. Neiss (1995) Phagocytic microglia during delayed neuronal loss in the facial nucleus of the rat: Time course of the neuronofugal migration of brain macrophages. Glia *13:* 113-129.

Aoki, C., C. Venkatesan, C.G. Go, J.A. Mong and T.M. Dawson (1994) Cellular and subcellular localization of NMDA-R1 subunit immunoreactivity in the visual cortex of adult and neonatal rats. J. Neurosci. *14:* 5202-5222.

Armstead, W.M., R. Mirro, C.W. Leffler and D.W. Busija (1989) Cerebral superoxide anion generation during seizures in newborn pigs. J. Cereb. Blood Flow Metab. *9:* 175-179.

Ashwell, K. (1991) The distribution of microglia and cell death in the fetal rat forebrain. Dev. Brain Res. *58:* 1-12.

Balasingam, V., K. Dickson, A. Brade and V.W. Yong (1996) Astrocyte reactivity in neonatal mice: apparent dependence on the presence of reactive microglia/macrophages. Glia *18:* 11-26.

Balasingam, V., T. Tejadaberges, E. Wright, R. Bouckova and V.W. Yong (1994) Reactive astrogliosis in the neonatal mouse brain and its modulation by cytokines. J. Neurosci. *14:* 846-856.

Benveniste, E.N. (1992) Inflammatory cytokines within the central nervous system: sources, function, and mechanism of action. Am. J. Physiol. *263:* C11-C16.

Benveniste, E.N. (1993) Astrocyte-microglia interactions. In: Astrocytes: Pharmacology and Function. Murphy, S. ed. San Diego: Academic Press, Inc., 355-382.

Borghesi, L.A. and M.A. Lynes (1996) Stress proteins as agents of immunological change: some lessons from metallothionein. Cell Stress Chaper. *1:* 99-108.

Bovolenta, P., F. Wandosell and M. Nieto Sampedro (1992) CNS glial scar tissue: a source of molecules which inhibit central neurite outgrowth. Prog. Brain Res.. *94:* 367-379.

Boya, J., J.L. Calvo, A.L. Carbonell and A. Borregon (1991) A lectin histochemistry study on the development of rat microglial cells. J. Anat. *175:* 229-36.

Boya, J., A.L. Carbonell, J. Calvo and A. Borregon (1987) Ultrastructural study on the origin of rat microglia cells. Acta Anat *130:* 329-35.

Brosnan, C.F., B. Shafit-Zagardo, D.A. Aquino and J.W. Berman (1993) Expression of monocyte/macrophage growth factors and receptors in the central nervous system. Adv. Neurol. *59:* 349-361.

Burtrum, D. and F.S. Silverstein (1993) Excitotoxic injury stimulates glial fibrillary acidic protein mRNA expression in perinatal rat brain. Exp. Neurol. *121:* 127-132.

Burtrum, D. and F.S. Silverstein (1994) Hypoxic-Ischemic Brain Injury Stimulates Glial Fibrillary Acidic Protein mRNA and Protein Expression in Neonatal Rats. Exp. Neurol. *126:* 112-118.

Cannella, B., D.A. Aquino and C.S. Raine (1995) MHC II expression in the CNS after long-term demyelination. J. Neuropathol. Exp. Neurol. *54:* 521-530.

Castellano, B., B. Gonzalez, M.B. Jensen, E.B. Pedersen, B.R. Finsen and J. Zimmer (1991) A double staining technique for simultaneous demonstration of astrocytes and microglia in brain sections and astroglial cell cultures. J. Histochem. Cytochem. *39:* 561-8.

Chao, C.C., S. Hu, T.W. Molitor, E.G. Shaskan and P.K. Peterson (1992) Activated microglia mediate neuronal cell injury via a nitric oxide mechanism. J. Immunol. *149:* 2736-41.

Cheng, B., S. Christakos and M.P. Mattson (1994) Tumor necrosis factors protect neurons against metabolic excitotoxic insults and promote maintenance of calcium homeostasis. Neuron *12:* 139-153.

Chiang, C.S., H.C. Powell, L.H. Gold, A. Samimi and I.L. Campbell (1996) Macrophage/microglial-mediated primary demyelination and motor disease induced by the central nervous system production of interleukin-3 in transgenic mice. J. Clin. Invest. *97:* 1512-1524.

Ciesielski-Treska, J., G. Ulrich, S. Chasserotgolaz and D. Aunis (1995) Immunocytochemical localization of protein kinases Yes and Src in amoeboid microglia in culture: Association of Yes kinase with vimentin intermediate filaments. Eur. J. Cell Biol. *68:* 369-376.

Coffey, P.J., V.H. Perry and J.N.P. Rawlins (1990) An investigation into the early stages of the inflammatory response following ibotenic acid-induced neuronal degeneration. Neuroscience *35:* 121-132.

Colton, C.A. and D.L. Gilbert (1993) Microglia, an in vivo source of reactive oxygen species in the brain. Adv. Neurol. *59:* 321-6.

Cowan, K., D. Conklin and M. Aschner (1996) Metallothionein induction by bismuth in neonatal rat primary astrocyte cultures. Brain Res. *732:* 106-112.

Dahl, D. (1981) The vimentin-GFA protein transition in rat neuroglia cytoskeleton occurs at the time of myelination. J. Neurosci. Res. *6:* 741-748.

Dalmau, I., B.R. Finsen, N. Tonder, B. Gonzalez and B. Castellano (1997) Development of microglia in the prenatal rat hippocampus. J. Comp. Neurol. *377:* 70-84.

Dalton, T., R.D. Palmiter and G.K. Andrews (1994) Transcriptional induction of the mouse metallothionein-I gene in hydrogen peroxide-treated Hepa cells involves a composite major late transcription factor/antioxidant response element and metal response promoter elements. Nucleic Acids Res. *22:* 5016-5023.

Del Río-Hortega, P. (1920a) El 'tercer elemento' de los centros nerviosos. II. Intervención de la microglía en los procesos patológicos. Bol. Soc. Esp. Biol. *VIII:* 91-103.

Del Río-Hortega, P. (1920b) El 'tercer elemento' de los centros nerviosos. Poder fagocitario y movilidad de la microglía. Bol. Soc. Esp. Biol. *VIII:* 154-166.

Ebadi, M., P.L. Iversen, R. Hao, D.R. Cerutis, P. Rojas, H.K. Happe, L.C. Murrin and R.F. Pfeiffer (1995) Expression and regulation of brain metallothionein. Neurochem. Int. *27:* 1-22.

Eddleston, M. and L. Mucke (1993) Molecular profile of reactive astrocytes-implications for their role neurologic disease. Neuroscience *54:* 15-36.

Fattori, E., D. Lazzaro, P. Musiani, A. Modesti, T. Alonzi and G. Ciliberto (1995) IL-6 expression in neurons of transgenic mice causes reactive astrocytosis and increase in ramified microglial cells but no neuronal damage. Eur. J. Neurosci. *7:* 2441-2449.

Finsen, B.R., M.B. Jorgensen, N.H. Diemer and J. Zimmer (1993a) Microglial MHC antigen expression after ischemic and kainic acid lesions of the adult rat hippocampus. Glia *7:* 41-49.

Finsen, B.R., N. Tonder, G.F. Xavier, J.C. Sorensen and J. Zimmer (1993b) Induction of microglial immunomolecules by anterogradely degenerating mossy fibres in the rat hippocampal formation. J. Chem. Neuroanat. *6:* 267-275.

Firkins, S.S., C.A. Bates and D.J. Stelzner (1993) Corticospinal tract plasticity and astroglial reactivity after cervical spinal injury in the postnatal rat. Exp. Neurol. *120:* 1-15.

Fischer, H.G., C. Eder, U. Hadding and U. Heinemann (1995) Cytokine-dependent K+ channel profile of microglia at immunologically defined functional states. Neuroscience *64:* 183-191.

Frautschy, S.A., P.A. Walicke and A. Baird (1991) Localization of basic fibroblast growth factor and its mRNA after CNS injury. Brain Res. *553:* 291-9.

Frei, K., S. Bodmer, C. Schwerdel and A. Fontana (1986) Astrocyte-derived interleukin 3 as a growth factor for microglia cells and peritoneal macrophages. J. Immunol. *137:* 3521-3527.

Frei, K., U.V. Malipiero, T.P. Leist, R.M. Zinkernagel, M.E. Schwab and A. Fontana (1989) On the cellular source and function of interleukin 6 produced in the central nervous system in viral diseases. Eur. J. Immunol. *19:* 689-94.

Frim, D.M., T.A. Uhler, M.P. Short, Z.D. Ezzedine, M. Klagsbrun, X.O. Breakefield and O. Isacson (1993) Effects of biologically delivered NGF, BDNF and bFGF on striatal excitotoxic lesions. Neuroreport *4:* 367-70.

Ganter, S., H. Northoff, D. Mannel and P.J. Gebicke-Harter (1992) Growth control of cultured microglia. J. Neurosci. Res. *33:* 218-30.

Gebicke-Haerter, P.J., K. Appel, G.D. Taylor, A. Schobert, I.N. Rich, H. Northoff and M. Berger (1994) Rat microglial interleukin-3. J. Neuroimmunol. *50:* 203-214.

Giulian, D. and M. Corpuz (1993) Microglial secretion products and their impact on the nervous system. Adv. Neurol. *59:* 315-20.

Giulian, D., M. Corpuz, S. Chapman, M. Mansouri and C. Robertson (1993) Reactive mononuclear phagocytes release neurotoxins after ischemic and traumatic injury to the central nervous system. J. Neurosci. Res. *36:* 681-693.

Giulian, D., B. Johnson, J.F. Krebs, J.K. George and M. Tapscott (1991) Microglial mitogens are produced in the developing and injured mammalian brain. J. Cell Biol. *112:* 323-333.

Giulian, D. and K. Vaca (1993) Inflammatory glia mediate delayed neuronal damage after ischemia in the central nervous system. Stroke *24 (suppl):* I84-90.

Giulian, D., K. Vaca and M. Corpuz (1993) Brain glia release factors with opposing actions upon neuronal survival. J. Neurosci. *13:* 29-37.

Giulian, D., J. Woodward, D.G. Young, J.F. Krebs and L.B. Lachman (1988a) Interleukin-1 injected into mammalian brain stimulates astrogliosis and neovascularization. J. Neurosci. *8:* 2485-2490.

Giulian, D., D.G. Young, J. Woodward, D.C. Brown and L.B. Lachman (1988b) Interleukin-1 is an astroglial growth factor in the developing brain. J. Neurosci. *8:* 709-714.

Glenn, J.A., P.L. Booth and W.E. Thomas (1991) Pinocytotic activity in ramified microglia. Neurosci. Lett. *123:* 27-31.

Gomez-Pinilla, F., J.W.-K. Lee and C. Cotman (1992) Basic FGF in adult brain: cellular distribution and response to entorhinal lesion anf fimbria-fornix transection. J. Neurosci. *12:* 345-355.

Gonzalez, B., B. Castellano, G. Palacios and J. Garcia (1985) Migration and elimination of brain macrophages during recovery. An electron microscopy study. Neurosci. Lett. (Suppl.) *22:* S525.

Graeber, M.B. and W.J. Streit (1990) Microglia: immune network in the CNS. Brain Pathol. *1:* 2-5.

Greensmith, L. and R. Navarrete (1994) Effect of neonatal nerve injury on the expression of major histocompatibility complex antigens in the rat spinal cord. Neurodegeneration *3:* 235-242.

Hagan, P., S. Poole, A.F. Bristow, F. Tilders and F.S. Silverstein (1996) Intracerebral NMDA injection stimulates production of interleukin-1 beta in perinatal rat brain. J. Neurochem. *67:* 2215-2218.

Hansson, E. and L. Ronnback (1994) Astroglial modulation of synaptic transmission. Perspect. Develop. Neurobiol. *2:* 217-223.

Hartung, H.P., S. Jung, G. Stoll, J. Zielasek, B. Schmidt, J.J. Archelos and K.V. Toyka (1992) Inflammatory mediators in demyelinating disorders of the CNS and PNS. J. Neuroimmunol. *40:* 197-210.

Hatten, M.E., R.K. Liem, M.L. Shelanski and C.A. Mason (1991) Astroglia in CNS injury. Glia *4:* 233-243.

Hatton, J.D., M.H. Nguyen and H.S. U (1993) Differential migration of astrocytes grafted into the developing rat brain. Glia *9:* 113-119.

Hickey, W.F. and H. Kimura (1988) Perivascular microglial cells of the CNS are bone marrow-derived and present antigen in vivo. Science *239:* 290-2.

Hidalgo, J., L. Campmany, M. Borras, J.S. Garvey and A. Armario (1988) Metallothionein response to stress in rats: role in free radical scavenging. Am. J. Physiol. *255:* E518-E524.

Hidalgo, J., A. Garcia, A.M. Oliva, M. Giralt, T. Gasull, B. Gonzalez, H. Milnerowicz, A. Wood and I. Bremner (1994a) Effect of zinc, copper and glucocorticoids on metallothionein levels of cultured neurons and astrocytes from rat brain. Chem. Biol. Interact. *93:* 197-219.

Hidalgo, J., T. Gasull, M. Giralt and A. Armario (1994b) Brain metallothionein in stress. Biol. Signals *3:* 198-210.

Hozumi, I., D.A. Aquino and W.T. Norton (1990) GFAP mRNA levels following stab wounds in rat brain. Brain Res. *534:* 291-294.

Hutchins, J.B. and V.A. Casagrande (1989) Vimentin: changes in distribution during brain development. Glia *2:* 55-66.

Imamura, K., A. Suzumura, M. Sawada, C. Mabuchi and T. Marunouchi (1994) Induction of MHC class II antigen expression on murine microglia by interleukin-3. J. Neuroimmunol. *55:* 119-125.

Ivacko, J.A., R. Sun and F.S. Silverstein (1996) Hypoxic-ischemic brain injury induces an acute microglial reaction in perinatal rats. Pediatr. Res. *39:* 39-47.

Janeczko, K. (1988) The proliferative response of astrocytes to injury in the neonatal brain. A combined immunocytochemical and autoradiographic study. Brain Res. *456:* 280-285.

Janeczko, K. (1993) Co-expression of GFAP and vimentin in astrocytes proliferating in response to injury in the mouse cerebral hemisphere- a combined autoradiographic and double immunocytochemical study. Int. J. Devl. Neurosci. *11:* 139-147.

Janeczko, K. (1994) Age-dependent changes in the proliferative response of S-100 protein-positive glial cells to injury in the rat brain. Int. J. Dev. Neurosci. *12:* 431-440.

Jensen, M.B., B. Gonzalez, B. Castellano and J. Zimmer (1994) Microglial and astroglial reactions to anterograde axonal degeneration: A histochemical and immunocytochemical study of the adult rat fascia dentata after entorhinal perforant path lesions. Exp. Brain Res. *98:* 245-260.

Jorgensen, M.B., B.R. Finsen, M.B. Jensen, B. Castellano, N.H. Diemer and J. Zimmer (1993) Microglial and astroglial reactions to ischemic and kainic acid-induced lesions of the adult rat hippocampus. Exp. Neurol. *120:* 70-88.

Kahn, M.A., J.A. Ellison, G.J. Speight and J. Devellis (1995) CNTF regulation of astrogliosis and the activation of microglia in the developing rat central nervous system. Brain Res. *685:* 55-67.

Kaur, C., Y.G. Chan and E.A. Ling (1992) Ultrastructural and immunocytochemical studies of macrophages in an excitotoxin induced lesion in the rat brain. J. Hirnforsch. *33:* 645-652.

Kaur, C. and E.A. Ling (1991) Study of the transformation of amoeboid microglial cells into microglia labelled with the isolectin Griffonia simplicifolia in postnatal rats. Acta Anat. *142:* 118-25.

Kaur, C., E.A. Ling and W.C. Wong (1987) Origin and fate of neural macrophages in a stab wound of the brain of the young rat. J. Anat. *154:* 215-27.

Kaur, C., E.A. Ling and W.C. Wong (1989) Development of the various glial cell types in the cerebral cortex of postnatal rats. Acta Anat. *136:* 204-210.

Kaur, C., J. Singh, M.K. Lim, B.L. Ng, E.P.H. Yap and E.A. Ling (1995) The response of neurons and microglia to blast injury in the rat brain. Neuropathol. Appl. Neurobiol. *21:* 369-377.

Kettenmann, H., R. Banati and W. Walz (1993) Electrophysiological behavior of microglia. Glia *7:* 93-101.

Kettenmann, H. and S. Ilschner (1993) Physiological properties of microglia. Clin. Neuropathol. *12:* 306-308.

Kolb, B., R. Gibb and D. Vanderkooy (1994) Neonatal frontal cortical lesions in rats alter cortical structure and connectivity. Brain Res. *645:* 85-97.

Kolb, B., B. Petrie and J. Cioe (1996) Recovery from early cortical damage in rats .7. comparison of the behavioural and anatomical effects of medial prefrontal lesions at different ages of neural maturation. Behav. Brain Res. *79:* 1-13.

Kreutzberg, G.W. (1996) Microglia: a sensor for pathological events in the CNS. Trends Neurosci. *19:* 312-318.

Lampson, L.A., A. Grabowska and J.P. Whelan (1994) Class I and II MHC expression and its implications for regeneration in the nervous system. Prog. Brain Res. *103*: 308-317.

Lassmann, H., F. Zimprich, K. Rössler and K. Vass (1991) Inflammation in the nervous system. Rev. Neurol. *147:* 763-781.

Leong, S.K. and E.A. Ling (1992) Amoeboid and ramified microglia: their interrelationship and response to brain injury. Glia *6:* 39-47.

Levison, S.W., M.H. Ducceschi, G.M. Young and T.L. Wood (1996) Acute exposure to CNTF in vivo induces multiple components of reactive gliosis. Exp. Neurol. *141:* 256-268.

Ling, E.A. (1994) Monocytic Origin of Ramified Microglia in the Corpus Callosum in Postnatal Rat. Neuropathol. Appl. Neurobiol. *20:* 182-183.

Ling, E.A., C. Kaur and W.C. Wong (1991) Expression of major histocompatibility complex and leukocyte common antigens in amoeboid microglia in postnatal rats. J. Anat. *177:* 117-126.

Ling, E.A., C. Kaur and W.C. Wong (1992) Expression of major histocompatibility complex antigens and CR3 complement receptors in activated microglia following an injection of ricin into the sciatic nerve in rats. Histol. Histopathol. *7:* 93-100.

Ling, E.A., L.C. Kaur, T.Y. Yick and W.C. Wong (1990) Immunocytochemical localization of CR3 complement receptors with OX-42 in amoeboid microglia in postnatal rats. Anat. Embryol. *182:* 481-486.

Ling, E.A. and W.C. Wong (1993) The origin and nature of ramified and amoeboid microglia: a historical review and current concepts. Glia *7:* 9-18.

Logan, A. and M. Berry (1993) Transforming growth factor-ß1 and basic fibroblast growth factor in the injured CNS. Trends Pharmacol. Sci. *14:* 337-343.

Masters, B.A., E.J. Kelly, C.J. Quaife, R.L. Brinster and R.D. Palmiter (1994) Targeted disruption of metallothionein I and II genes increases sensitivity to cadmium. Proc. Natl. Acad. Sci. USA *91:* 584-588.

Mattson, M.P. and B. Rychlik (1990) Glia protect hippocampal neurons against excitatory amino acid-induced degeneration: involvement of fibroblast growth factor. Int. J. Devl. Neurosci. *8:* 399-415.

McKeon, R.J., R.C. Schreiber, J.S. Rudge and J. Silver (1991) Reduction of neurite outgrowth in a model of glial scarring following CNS injury is correlated with the expression of inhibitory molecules on reactive astrocytes. J. Neurosci. *11:* 3398-3411.

Merrill, J.E. (1991) Effects of interleukin-1 and tumor necrosis factor-alpha on astrocytes, microglia, oligodendrocytes, and glial precursors in vitro. Dev. Neurosci. *13:* 130-7.

Merrill, J.E. (1992) Tumor necrosis factor alpha, interleukin 1 and related cytokines in brain development: normal and pathological. Dev. Neurosci. *14:* 1-10.

Milligan, C.E., P. Levitt and T.J. Cunningham (1991) Brain macrophages and microglia respond differently to lesions of the developing and adult visual system. J. Comp. Neurol. *314:* 136-146.

Miyake, T., M. Okada and T. Kitamura (1992) Reactive proliferation of astrocytes studied by immunohistochemistry for proliferating cell nuclear antigen. Brain Res. *590:* 300-302.

Moonen, G., B. Rogister, P. Leprince, J.M. Rigo, P. Delree, P.P. Lefebvre and J. Schoenen (1990) Neurono-glial interactions and neural plasticity. Prog. Brain Res. *86:* 63-73.

Moore, I.E., J.M. Buontempo and R.O. Weller (1987) Response of fetal and neonatal rat brain to injury. Neuropathol. Appl. Neurobiol. *13:* 219-228.

Morgan, T.E., N.R. Nichols, G.M. Pasinetti and C.E. Finch (1993) TGF-beta 1 mRNA increases in macrophage/microglial cells of the hippocampus in response to deafferentation and kainic acid-induced neurodegeneration. Exp. Neurol. *120:* 291-301.

Morioka, T. and W.J. Streit (1991) Expression of immunomolecules on microglial cells following neonatal sciatic nerve axotomy. J. Neuroimmunol. *35:* 21-30.

Murphy, S., M.L. Simmons, L. Agullo, A. Garcia, D.L. Feinstein, E. Galea, D.J. Reis, D. Mincgolomb and J.P. Schwartz (1993) Synthesis of nitric oxide in CNS glial cells. Trends Neurosci. *16:* 323-328.

Nakajima, K. and S. Kohsaka (1993a) Characterization of brain microglia and the biological significance in the central nervous system. Adv. Neurol. *60:* 734-43.

Nakajima, K. and S. Kohsaka (1993b) Functional roles of microglia in the brain. Neurosci. Res. *17:* 187-203.

Nieto-Sampedro, M. (1988) Growth factor induction and order of events in CNS repair. In: Pharmacological approaches to the treatment of brain and spinal cord injury. Stein, D.G. and Sabel, A.B. eds. New York: Plenum Press, 301-337.

Nishimura, N., H. Nishimura, A. Ghaffar and C. Tohyama (1992) Localization of metallothionein in the brain of rat and mouse. J. Histochem. Cytochem. *40:* 309-315.

Norenberg, M.D. (1994) Astrocyte responses to CNS injury. J. Neuropathol. Exp. Neurol. *53:* 213-220.

Nozaki, K., S.P. Finklestein and M.F. Beal (1993) Basic fibroblast growth factor protects against hypoxia-ischemia and NMDA neurotoxicity in neonatal rats. J. Cereb. Blood Flow Metab. *13:* 221-8.

Oblinger, M.M. and L.D. Singh (1993) Reactive astrocytes in neonate brain upregulate intermediate filament gene expression in response to axonal injury. Int. J. Devl. Neurosci. *11:* 149-156.

Oliver, C.N., P.E. Starke-Reed, E.R. Stadtman, G.J. Liu, J.M. Carney and R.A. Floyd (1990) Oxidative damage to brain proteins, loss of glutamine synthetase activity, and production of free radicals during ischemia/reperfusion-induced injury to gerbil brain. Proc. Natl. Acad. Sci. U.S.A. *87:* 5144-5147.

Pazdernik, T.L., M. Layton, S.R. Nelson and F.E. Samson (1992) The osmotic/calcium stress theory of brain damage: are free radicals involved? Neurochem. Res. *17:* 11-21.

Pedersen, E.B., L.M. Fox, A.J. Castro and J.A. McNulty (1993) Immunocytochemical and electron-microscopic characterization of macrophage/microglia cells and expression of class II major histocompatibility complex in the pineal gland of the rat. Cell Tissue Res. *272:* 257-265.

Penkowa, M. and T. Moos (1995) Disruption of the blood-brain interface in neonatal rat neocortex induces a transient expression of metallothionein in reactive astrocytes. Glia *13:* 217-227.

Petito, C.K. and I.A. Halaby (1993) Relationship between ischemia and ischemic neuronal necrosis to astrocyte expression of glial fibrillary acidic protein. Int. J. Devl. Neurosci. *11:* 239-247.

Piani, D., K. Frei, K.Q. Do, M. Cuénod and A. Fontana (1991) Murine brain macrophages induce NMDA receptor mediated neurotoxicity in vitro by secreting glutamate. Neurosci. Lett. *133:* 159-162.

Piani, D., M. Spranger, K. Frei, A. Schaffner and A. Fontana (1992) Macrophage-induced cytotoxicity of N-methyl-D-aspartate receptor positive neurons involves excitatory amino acids rather than reactive oxygen intermediates and cytokines. Eur. J. Immunol. *22:* 2429-36.

Pixley, S.K.R. and J. De Vellis (1984) Transition between immature radial glia and mature astrocytes studied with a monoclonal antibody to vimentin. Dev. Brain Res. *15:* 201-209.

Porter, J.T. and K.D. McCarthy (1995) GFAP-positive hippocampal astrocytes in situ respond to glutamatergic neuroligands with increased in [Ca2+] i. Glia *13:* 101-112.

Raivich, G., J. Gehrmann and G.W. Kreutzberg (1991) Increase of macrophage colony-stimulating factor and granulocyte-macrophage colony-stimulating factor receptors in the regenerating rat facial nucleus. J. Neurosci. Res. *30:* 682-686.

Raivich, G., J. Gehrmann, M. Moreno-Floros and G.W. Kreutzberg (1993) Microglia: growth factor and mitogen receptors. Clin. Neuropathol. *12:* 293-295.

Relton, J.K. and N.J. Rothwell (1992) Interleukin-1 receptor antagonist inhibits ischaemic and excitotoxic neuronal damage in the rat. Brain Res. Bull. *29:* 243-6.

Rothwell, N.J. and J.K. Relton (1993) Involvement of cytokines in acute neurodegeneration in the CNS. Neurosci. Biobehav. Rev. *17:* 217-27.

Sato, M. and I. Bremner (1993) Oxygen free radicals and metallothionein. Free Radic. Biol. Med. *14:* 325-337.

Sato, M., M. Sasaki and H. Hojo (1994) Differential induction of metallothionein synthesis by interleukin-6 and tumor necrosis factor-alpha in rat tissues. Int. J. Immunopharmacol. *16:* 187-195.

Sawada, J., Y. Kikuchi, M. Shibutani, K. Mitsumori, K. Inoue and T. Kasahara (1994) Induction of metallothionein in astrocytes by cytokines and heavy metals. Biol. Signals *3:* 157-168.

Sawada, M., A. Suzumura and T. Marunouchi (1992) TNF alpha induces IL-6 production by astrocytes but not by microglia. Brain Res. *583:* 296-9.

Schiffer, D., M.T. Giordana, P. Cavalla, M.C. Vigliani and A. Attanasio (1993) Immunohistochemistry of glial reaction after injury in the rat - double stainings and markers of cell proliferation. Int. J. Devl. Neurosci. *11:* 269-280.

Schulz, J.B., D.R. Henshaw, D. Siwek, B.G. Jenkins, R.J. Ferrante, P. Bencipolloni, N.W. Kowall, B.R. Rosen and M.F. Beal (1995) Involvement of free radicals in excitotoxicity in vivo. J. Neurochem. *64:* 2239-2247.

Schulz, M.K., L. Acarin, G.L. Tillotson and A.J. Castro (1993) Glial response to neocortical injection of the excitotoxin N-methyl-D-aspartic acid (NMDA). Soc. Neurosci. Abstr. *19:* 671.

Schwartz, J.P. and N. Nishiyama (1994) Neurotrophic factor gene expression in astrocytes during development and following injury. Brain Res. Bull. *35:* 403-407.

Sillevis Smitt, P.A., H.G. Blaauwgeers, D. Troost and J.M. De Jong (1992) Metallothionein immunoreactivity is increased in the spinal cord of patients with amyotrophic lateral sclerosis. Neurosci. Lett. *144:* 107-110.

Smith, G.M. and R.H. Miller (1991) Immature type-1 astrocytes suppress glial scar formation, are motile and interact with blood vessels. Brain Res. *543:* 111-122.

Steinhauser, C. and V. Gallo (1996) News on glutamate receptors in glial cells. Trends Neurosci *19:* 339-345.

Streit, W.J. (1993) Microglial-neuronal interactions. J. Chem. Neuroanat. *6:* 261-266.

Streit, W.J., M.B. Graeber and G.W. Kreutzberg (1989) Expression of Ia antigen on perivascular and microglial cells after sublethal and lethal motor neuron injury. Exp. Neurol. *105:* 115-26.

Streit, W.J. and G.W. Kreutzberg (1987) Lectin binding by resting and reactive microglia. J. Neurocytol. *16:* 249-260.

Szaflarski, J., D. Burtrum and F.S. Silverstein (1995) Cerebral hypoxia-ischemia stimulates cytokine gene expression in perinatal rats. Stroke *26:* 1093-1100.

Takamiya, Y., S. Kohsaka, S. Toya, M. Otani and Y. Tsukada (1988) Immunohistochemical studies on the proliferation of reactive astrocytes and the expression of cytoskeletal proteins following brain injury in rats. Brain Res. *466:* 201-210.

Thomas, W.E. (1990) Characterization of the dynamic nature of microglial cells. Brain Res. Bull. *25:* 351-4.

Thomas, W.E. (1992) Brain macrophages: evaluation of microglia and their functions. Brain Res. Rev. *17:* 61-74.

Toulmond, S., X. Vige, D. Fage and J. Benavides (1992) Local infusion of interleukin-6 attenuates the neurotoxic effect of NMDA on rat striatal cholinergic neurons. Neurosci. Lett. *144:* 49-52.

Towfighi, J., N. Zec, J. Yager, C. Housman and R.C. Vannucci (1995) Temporal evolution of neuropathologic changes in an immature rat model of cerebral hypoxia: A light microscopic study. Acta Neuropathol. *90:* 375-386.

Van Bockstaele, E.J. and E.E.O. Colago (1996) Selective distribution of the NMDA-R1 glutamate receptor in astrocytes and presynaptic axon terminals in the nucleus locus coeruleus of the rat brain: An immunoelectron microscopic study. J. Comp. Neurol. *369:* 483-496.

Van Dam, A.M., M. Brouns, S. Louisse and F. Berkenbosch (1992) Appearance of interleukin-1 in macrophages and in ramified microglia in the brain of endotoxin-treated rats: a pathway for the induction of non-specific symptoms of sickness? Brain Res. *588:* 291-296.

Vela, J.M., I. Dalmau, B. Gonzalez and B. Castellano (1995) Morphology and distribution of microglial cells in the young and adult mouse cerebellum. J. Comp. Neurol. *361:* 602-616.

Vela, J.M., J. Hidalgo, B. González and B. Castellano (1997) Induction of metallothionein in astrocytes and microglia in the spinal cord from the myelin-deficient jimpy mouse. Brain Res. *767:* 345-355.

Walz, W., S. Ilschner, C. Ohlemeyer, R. Banati and H. Kettenmann (1993) Extracellular ATP activates a cation conductance and a K+ conductance in cultured microglial cells from mouse brain. J. Neurosci. *13:* 4403-4411.

Wiessner, C., J. Gehrmann, D. Lindholm, R. Topper, G.W. Kreutzberg and K.A. Hossmann (1993) Expression of transforming growth factor-beta-1 and interleukin-1-beta messenger RNA in rat brain following transient forebrain ischemia. Acta Neuropathol. *86:* 439-446.

Woodroofe, M.N., G.S. Sarna, M. Wadhwa, G.M. Hayes, A.J. Loughlin, A. Tinker and M.L. Cuzner (1991) Detection of interleukin-1 and interleukin-6 in adult rat brain, following mechanical injury, by in vivo microdialysis: evidence of a role for microglia in cytokine production. J. Neuroimmunol. *33:* 227-236.

Wu, C.H., C.Y. Wen, J.Y. Shieh and E.A. Ling (1992) A quantitative and morphometric study of the transformation of amoeboid microglia into ramified microglia in the developing corpus callosum in rats. J. Anat. *181:* 423-430.

Wu, C.H., C.Y. Wen, J.Y. Shieh and E.A. Ling (1993) A quantitative study of the differentiation of microglial cells in the developing cerebral cortex in rats. J. Anat. *182:* 403-413.

Xu, J. and E.A. Ling (1994a) Expression of major histocompatibility complex class II antigen on amoeboid microglial cells in early postnatal rat brain following intraperitoneal injections of lipopolysaccharide. Exp. Brain Res. *100:* 287-292.

Xu, J. and E.A. Ling (1994b) Upregulation and induction of major histocompatibility complex class I and II antigens on microglial cells in early postnatal rat brain following intraperitoneal injections of recombinant interferon-γ. Neuroscience *60:* 959-967.

Xu, J. and E.A. Ling (1994c) Upregulation and Induction of Surface Antigens with Special Reference to MHC Class II Expression in Microglia in Postnatal Rat Brain Following Intravenous or Intraperitoneal Injections of Lipopolysaccharide. J. Anat. *184:* 285-296.

Yabuuchi, K., M. Minami, S. Katsumata and M. Satoh (1993) In situ hybridization study of interleukin-1 beta mRNA induced by kainic acid in the rat brain. Mol. Brain Res. *20:* 153-161.

Yamada, M. and H. Hatanaka (1994) Interleukin-6 protects cultured rat hippocampal neurons against glutamate-induced cell death. Brain Res. *643:* 173-180.

Yao, J., J.E. Keri, R.E. Taffs and C.A. Colton (1992) Characterization of interleukin-1 production by microglia in culture. Brain Res. *591:* 88-93.

Yee, K.T., A.M. Smetanka, R.D. Lund and K. Rao (1990) Differential expression of class I and class II major histocompatibility complex antigen in early postnatal rats. Brain Res. *530:* 121-125.

Young, J.K., J.S. Garvey and P.C. Huang (1991) Glial immunoreactivity for metallothionein in the rat brain. Glia *4:* 602-610.

16

EFFECTS OF CONTRALATERAL LESIONS AND AGING ON THE NEURONAL AND GLIAL POPULATION OF THE CEREBRAL CORTEX OF THE RAT

María Angeles Peinado, María Jesús Ramírez, Juan Angel Pedrosa, Manuel Martínez, Adoración Quesada, María Luisa del Moral, Francisco José Esteban, José Rodrigo and José María Peinado

Glial cells and pericytes are strongly involved in the neurodegenerative processes of the central nervous system. In the present study, we analyzed the cytomorphological effects as well as the reactive changes of glial cells and pericytes of two different neurodegenerative processes: normal aging and changes induced by contralateral implantation of a microdialysis probe. Quantifications were made in the frontal and parietal cortex of the rat. The results show that the microdialysis induced lesions affected all contralateral cortical layers, the microglia being the most proliferating type of cells. The aging process mainly affected the layer V where density of astrocytes was increased. Pericytes in frontal cortex did not show significant changes under our experimental conditions. We detected a significant increase in the number of these cells only in the parietal cortex of aging rats. In addition, pericytes and perivascular microglial cells accumulated a great number of dense and foamy inclusions during aging. Finally, both neurodegenerative processes, lesion and aging, seems to produce glial hyperplasia and hypertrophia through a different mechanism which also differentially affects the neuronal sprouting response.

NEURODEGENERATION AND REACTIVE GLIOSIS

Current data have suggested several mechanisms underlying the degenerative processes responsible for cell damaging and neuronal death, including deficiencies in energy, release of excitatory amino acids, particularly glutamic acid or hyperproduction of free radicals such as nitric oxide. All these mechanisms may occur simultaneously, finally producing an increase in the cytosolic Ca^{++} concentration (For review see Lees, 1993). Glial cells and pericytes seem to be involved in the modulation and development of these processes as a consequence of disturbances of their regular functions, including maintenance of the blood brain barrier and brain homeostasis for astrocytes (Kimelberg et al., 1993), immuno network for microglial cells (Graeber and Streit, 1990; Streit et al., 1990; Davis et al., 1994), and synthesis of endothelial growth factor (EGF) for pericytes (Nomura et al.,1995).

Under unbalanced conditions, different chemical signals, from paracrine and autocrine origin, induce glial and pericyte proliferation. This process is known as reactive gliosis in the literature, and it has been extensively studied in several neurodegenerative processes (Banati et al., 1993; Norenberg, 1994) as well as during normal aging (Finch et al., 1993). However, numerous features of glial cells and pericyte reactivity are still unknown under physiological or pathological conditions, particulary those concerning the dynamic variations of the neuronal and glial population.

LESIONS AND GLIOSIS

The first reports about the reactivity of the glial cells during induced lesions were carried out at the beginning of the century. Firstly, Held (1909) demonstrated the presence of reactive changes in cells after brain punctures. After that, studies from Cajal (1913) and Del Río Hortega (1920; 1921) revealed that microglial cells may play a key role in the development of these lesions. Afterward, these changes were deeply analyzed by different histological methods (Penfield, 1927; Rand and Courville, 1932; Rand, 1952; Schultz and Pease, 1959; Konigsmark and Sidman, 1963; Murray and Walker, 1973; Stensaas and Stensaas, 1976; Castellano et al., 1991).

The reactivity of glial cells, mainly astrocytes and microglia, in response to different type of lesions is well documented (Hajos et al., 1990; Landis, 1994; Nieto-Sampedro et al., 1994). Thus, in the central nervous system (CNS), reactive astrocytes participate in the formation of epileptogenic scar tissue in response to injury (Reier, 1986; Linsay, 1986). In addition, these reactive cells are able to undergo hyperplasia (Jacneczko, 1989), although limited to the side of the injury (Mathewson and Berry, 1985; Hatten et al., 1991). Lesions of the CNS are also able to induce microglia proliferation (Andersson et al., 1992) as well as transformation of microglial ramified resting cells to the reactive forms (Giulian and Ingelman, 1993; McGeer et al., 1993; Giordana et al., 1994; Vela et al., 1995). Reactive microglia operate like active macrophages, showing a noteworthy increase in CR3 and MHC-I receptors (Graeber et al., 1988; Streit et al., 1988). In addition, these cells are able to release cytokines as interleukin-1, interleukin-6 and tumor necrosis factor alpha (TNF-alpha) (Griffin et al., 1989; Dickson et al., 1993). Interleukin-1 plays an important role in developing glial scar through the stimulation of the astrogliosis (Giulian and Lanchman, 1985; Giulian and Barker, 1986; Giulian et al., 1989).

Recent observations (Giordana et al., 1994) using the model of punctures induced cortex lesions and an antiserum to fibrillary acidic protein (GFAP) have shown that glial cells play an important role in the development of these lesions. After two days, an increase of GFAP-positive astrocytes was found in the white matter and molecular layer of the cortex near the damaged areas. After seven days, GFAP-positive astrocytes were more extensively distributed and appeared not only around the lesion but also in the white matter of the contralateral hemisphere. In this same model, and using specific markers for microglial cells (lectin and bromodioxiuridin), Giordana et al. (1994) have shown that microglia was the most proliferating type of glial cells, not only by mitosis of the microglial cells located near to the damaged area, but also themselves from blood precursors (Giordana et al., 1994). The punctures and lesions induced by chronic implant of microdialysis probes may share several neuromorphological and neurochemical aspects, as mentioned above (Van den Pol and Gallyas, 1990; Gallyas et al., 1992). However, analyses of the literature have indicated that the role of glial cells and pericytes in this type of traumatic lesion, have not been extensively studied yet, including quantitatives morphological studies.

AGING AND GLIOSIS

Aging is a normal physiological process, although it may cause functional disturbances mainly in aged populations. In addition to the numerous neurochemical findings, it is well documented that the aging brain suffers morphological alterations in neural and glial populations, which may reflect the neurodegeneration underlying this process (Goldman et al., 1994; Amaducci and Tesco, 1994; Schipper, 1996). The intensity of these changes is related to the area studied (Sturrock, 1980; 1990; Haug et al., 1983), animal used (Morgan et al., 1987) and methods applied (Haug et al., 1984). Therefore, age-related changes in the glial cell population of the cerebral cortex have yielded controversial results. In humans, either increases in the number of glial cells (Terry et al., 1987; Hansen et al., 1987) or no changes (Haug et al., 1984) had been described in the cerebral cortex of aged subjects. In monkeys (Brizzee, 1975; Peters et al., 1994) and mice (Heumann and Leuba, 1983) increased numbers of glial cells have also been found. In the rat cerebral cortex, the population of astrocytes seems to increase during aging (Amenta et al., 1994). However, Diamond et al., (1977) did not find changes in the number of astrocytes or oligodendrocytes. In addition, astrocytes in the rat cortex were characterized by an increase of the cytoplasmic surface, which plays a key role in the development of thick cell processes (Bjorklund et al., 1985). Although all aged glial cells seem to accumulate lipofucsin in their cytoplasm (Peters and Vaughan, 1981), it has been recently reported in striatum, hippocampus and periventricular brain region a senescent astrocyte phenotype which accumulate cytoplasmic inclusions called "gliosomes" (Schipper, 1996). These inclusions are histochemically and morphologically different from lipofucsin, showing affinity to Gomori stains, and highly intense nonenzymatic peroxidase activity. It was hypothesized that peroxidase activity could have promoted lipid peroxidation of adjacent neuropil constituents by oxidizing estrogens and catecholamines to free radicals, which are potentially neurotoxic (Schipper, 1996) and may contribute to age induced neurodegenerative process (Chao et al., 1996).

THE NEURODEGENERATION PRODUCED BY CONTRALATERAL LESIONS AND AGING COULD INFLUENCE THE REACTIVITY OF THE GLIAL POPULATION

The above information on glial changes after traumatic induced lesions and aging, may help the understanding of the neurodegeneration process underlying reactive gliosis. Recent studies carried out in our laboratory have demonstrated that microdialysis probes, implanted in the frontal cortex of young and aged wistar rats for seven days (fig. 16-1), produce, in the contralateral hemisphere, reactive changes in glial population which are particulary relevant in aged animals. The study was performed in the same area of the contralateral hemisphere where probes were implanted. Histological, immunocytochemical, ultrastructural, and morphometric quantitative analyses were used.

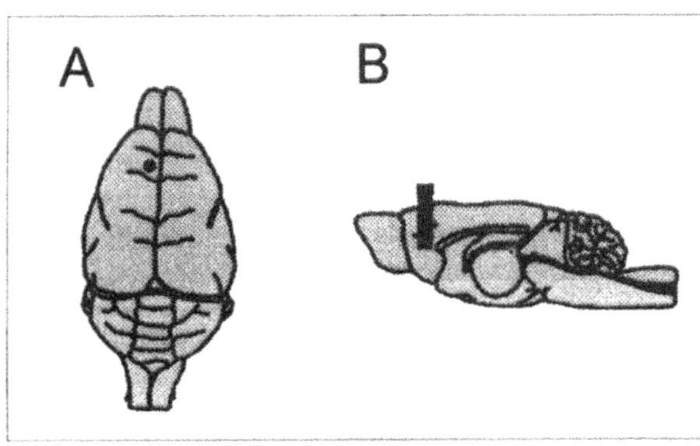

Figure 16-1. Schematic representation of dor-sal (A) and medial (B) view of the implanted micro-dialysis probes in the rat frontal cortex.

In order to correlate the reactive gliosis and the degree of neuronal damage, the neuronal population was also evaluated. Thus, it may be possible to know whether the neurodegeneration, underlying contralateral lesions and aging, could also involves neuronal death. Our recent data in the rat cortex have demonstrated the absence of neuronal loses neither in the contralateral area to the lesion nor in the aged cortex. However, some degree of neuronal degeneration as well as glial cell proliferation was observed (Peinado et al., 1993, Peinado et al., 1997). In fact, the cortex of both lesioned and aged animals, presented a certain number of neurons with an unusually high affinity for several histological stains (fig. 16-2).

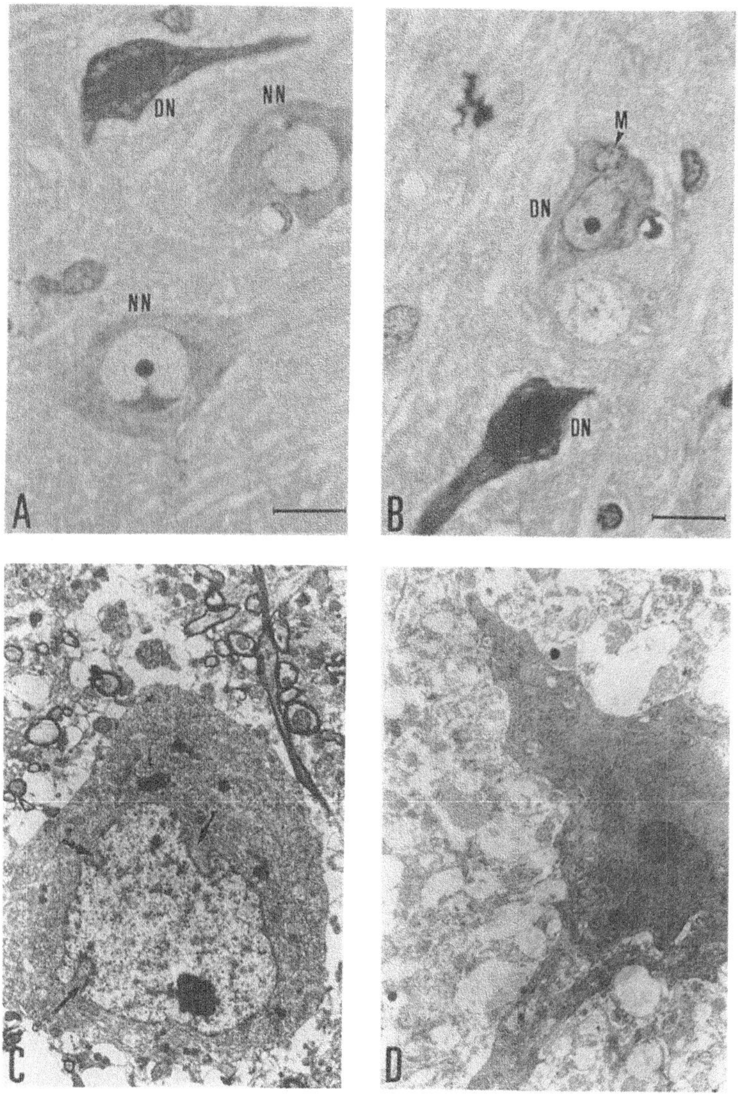

Figure 16-2. Light and electron microscopic characteristics of frontal and parietal cortex neurons in aged and aged-lesioned rats. (A) Toluidine blue stained semithin section of the frontal cortex of an aged rat with a dark neuron (DN) beside other non-damaged neurons (NN). (B) Toluidine blue stained semithin section of frontal cortex of an aged-lesioned rat with two dark neurons (DN) in different neurodegenerative stages; evident is a microglial perineuronal cell (M). (C) Ultrathin section of parietal cortex from an aged rat with a relative dark neuron containing lipofuscin inclusions (L) in their cytoplasm and a nucleus with some invaginations (arrows). (D) Ultrathin section of frontal cortex from an aged-lesioned rat with a dark neuron that shows a strongly disorganized structure. (Bar in A and B = 10μm; Magnifications: C = 8500x; D = 10000x).

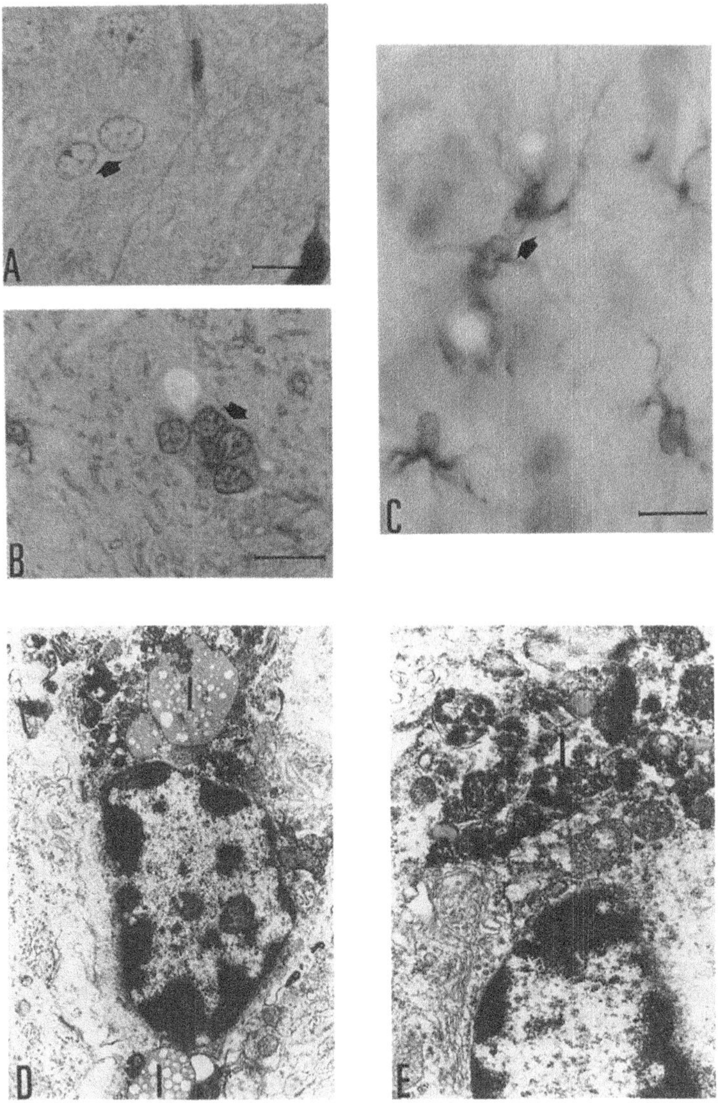

Figure 16-3. Light and electron microscopic characteristics of the glial cells from frontal and parietal cortex of aged and aged-lesioned rats. (A and B) Toluidine blue stained sections of parietal cortex from aged rats with paired (A) or clustered (B) astrocytes (arrow). (C) Anti-GFAP immunostained section of frontal cortex from an aged rat showing a pair of GFAP-positive astrocytes (arrow). (D and E) Ultrathin sections of the parietal cortex from aged rats showing a microglial cell (D) and an oligodendrocyte (E), respectively. Both glial cells showed numerous inclusions (I) with osmiophilic and foamy material. (Bar in A, B and C = 10μm; Magnifications: D = 9000x; D = 17700x).

These hyperchromatic neurons had been previously reported in several pathologies (Gallyas et al., 1990; 1992) as dark neurons (DN). Morphologically, they exhibited shrunken outlines, corkscrew-like dendrites and chromatin condensation. Early studies considered the presence of DN as possibly due to postmorten autolysis of unfixed nervous tissue (Cammermeyer, 1961). However, later on, Gallyas et al. (1990) revealed that DN are generated in vivo as an acute or delayed consequence of several pathological situations such as ischemia (Gallyas et al., 1990).

In addition to the changes described for neurons, glial cell populations also showed some degree of glial hyperplasia as demonstrated by the presence of some paired or clustered glial cells (fig. 16-3A, B and C).

We have also evaluated the degree of neurodegeneration and gliosis by quantification of the neuronal and glial cell populations in both contralateral lesioned and aged rats. Figures 16-4 and 16-5 summarize the dynamic changes found for both cells populations at the frontal cortex. The total number of neurons (TN) was unchanged in contralateral lesion and aging (fig. 16-4). However, a decreased number of non-damaged neurons (NN) and an increased number of DN were seen in the contralateral cortex to the lesion and in the normal aging (non lesioned) groups. Particularly, the degenerative effect was more evident in the contralateral area to the lesion than in aging (30 and 15% DN respectively).

Moreover, when the effects of contralateral lesion were studied in aged animals, the degree of degeneration was increased, being 43% of the total neurons DN.

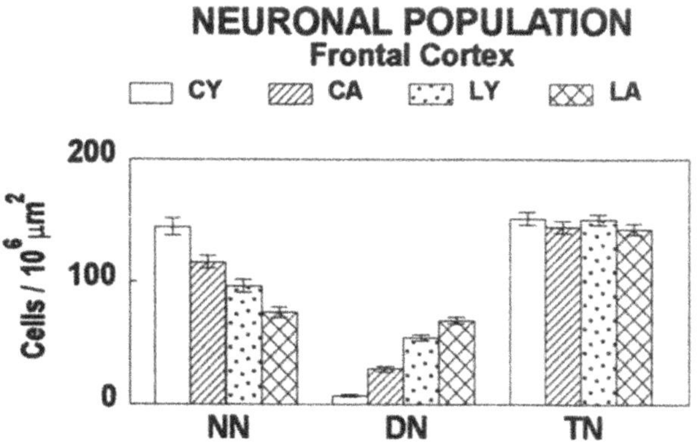

Figure 16-4. Neuronal density (MEAN±SEM in $10^6 \mu m^2$ of nervous tissue) of non-damaged neu-rons (NN), dark neurons (DN) and total neurons (TN) quantified in the frontal cortex of control young (CY), control aged (CA), lesioned young (LY) and lesioned aged (LA) rats.

Glial cells showed some degree of glial hyperplasia in both models although it was higher in the contralateral area to the lesion (22% increase in relation to non-lesioned animals) than in aging (20% increase in relation to young animals) (fig. 16-5). Particularly, lesioned aged animals displayed the highest increase in glial cells (40%) when compared to young unlesioned animals.

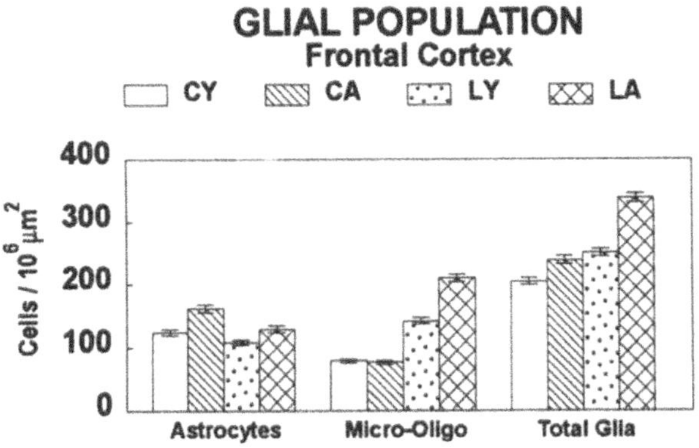

Figure 16-5. Glial density (MEAN±SEM in 10^6 µm^2 of nervous tissue) of astrocytes, microglia-oligodendrocytes and total glial cells quantified in the frontal cortex of control young (CY), control aged (CA), lesioned young (LY) and lesioned aged (LA) rats.

THE GLIAL REACTION DEPENDS ON THE TYPE OF NEURODEGENERATIVE MODEL

In the contralateral cortex to the induced lesions, our investigations demonstrated the absence of significant neuronal loss. However, it should be considered that we analyzed the neuronal population only seven days after stereotaxic surgery, as well as that neurons were counted in the contralateral hemisphere. In contrast, other authors detected an intense neurodegeneration in the lesioned area, where the formation of the scar make glial reaction more evident (Linsay et al., 1986; Reier, 1986; Giordana et al., 1994).

On the other hand, in aging, neuronal loss has been reported neuronal loss in cerebellar cortex (Jones et al., 1983), hippocampus (West et al., 1994), sustantia nigra (Pakkemberg et al., 1995), locus coerulus and basal nucleus of Meynert (McGeer, 1984). Nevertheless, other studies have shown neuronal stability in the cortex (Curcio and Coleman, 1982; Haug, 1984; Terry et al., 1987; Vincent et al., 1989; Peters et al., 1994).

Our results about dark neurons and glial hyperplasia in the contralateral area to the lesion and in aging, are compatible with the beginning of neurodegenerative mechanisms that initially only showed neuronal atrophy but finally would produce neuronal death. It has been postulated that neuronal death may occur by necrosis in traumatic lesions and by apoptosis in some neurodegenerative brain diseases with effective neuronal death as the Alzheimer disease (Cotman and Anderson, 1995). Currently, we ignore the mechanisms responsible for neurodegeneration in normal aging; however, according to previous reports (Landis, 1994), our results suggest that lesions and aging bear different glial reactions (see fig. 16-5). Perhaps these differential glial reactions allow us to speculate with different mechanisms of neural degeneration in both models.

In the contralateral cortex to the lesion, we have detected an increase in the number of oligodendrocytes-microglia (fig. 16-5). Although we have not direct evidences, we believe this increase could be mainly due to microglial cells. In fact studies using leu-7, a specific oligodendrocyte antibody marker, have revealed stability of these cells after cerebral injury (Schiffer et al., 1991). In addition, traumatic (Andersson et al., 1992) or ischemic (Morioka et al., 1993; Gehrmann et al., 1992; Lees, 1993) cerebral lesions produce microglial hyperplasia. On the other hand, it is generally accepted that brain injury induces activation and transformation of resting ramified microglial cells to reactive ameboide-phagocyte-like cells (Giulian and Ingelman, 1993; Giordana et al., 1994). Recently, it has been suggested (Couraud, 1994) that these cells could release interleukin-1, which may act on endothelial cells providing blood-macrophage entry to nervous tissue, and possible increase of the microglial phagocytic cell population. It was regarded that four days after injury, macrophage invasion could be completed (Marty et al., 1991). Therefore, in our study, after seven days of lesion, we hypothesize that the hyperplasia of microglial cells could be originated by resident microglia as well as by blood precursors.

In the contralateral hemisphere to the lesioned cortex, our results for young animals have shown a decrease in the number of astrocytes. It has been described that, in the lesioned hemisphere and close to the injured area, there is a significant increase in the number of glial cells, which mainly contribute to the genesis of the glial scar (Giordana et al., 1994). One of the glial cells that suffers hyperplasia as a consequence of the lesion are astrocytes (Fog and Pakkenberg, 1983; Janeczko, 1989). These cells can increase their number either through proliferation of activated astrocytes located in the damaged area, or from astrocytes migrated from other regions near to the damaged area (Dahl et al., 1991; Mucke and Eddleston, 1993). This additional source of astrocytes could be explained based on experimental evidence. In fact, astrocytes from primary cultures or from fetal central nervous tissue are able to produce an extensive migration after trasplantation to mature brain even up to 5cm (Goldberg and Bernstein, 1988a, 1988b). This last observation could explain the decrease in reactive astrocytes detected in the contralateral hemisphere to the lesion, based on their emigration towards the damaged areas.

In contrast to experimental lesions, aging causes astrocyte hyperplasia, while microglia and oligodendrocytes are unchanged. Moreover, in aged animals the presence of pair of astrocytes is observed frequently (fig. 16-3A, B and C). It has been described that aging is characterized by an increase in the number of astrocytes in mouse white matter (Bronson et al., 1993), rat hippocampus (Landfield et al., 1977; Lolova, 1991) and rat frontal cortex (Amenta et al., 1994). Aging studies have also detected astrocyte hypertrophy (Peters et al., 1991; Bonfanti et al., 1993; Berciano et al., 1995). These aging findings could be correlated with increases of specific astroglial enzymes such as glutamin-synthase (Cao et al., 1985; Peinado and Myers, 1988a; 1988b; Le Prince et al., 1995) and monoaminoxidase B (Wang et al., 1994).

The effects of aging on microglial population are heterogeneous (Vaughan and Peter, 1974; Peters et al., 1991; 1994). Studies using specific antibodies OX-42 and OX-6 against microglial cells showed no changes in the number of these cells, although their reactivity was increased (Lawson et al., 1990). This last observation was confirmed by Perry et al. (1993) who detected MHC-II in microglia from aging brains. In relation to oligodendrocytes, the few data found in the literature seem to indicate stability for this type of glial cells with aging (Vaughan and Peter, 1974; Peters et al., 1991).

Finally, our data show that the astrocyte nucleus increases its size, but only in contralateral lesioned animals (fig. 16-6). Microglia-oligodendrocytes increase their nucleus in aging but not in lesioned animals (fig. 16-6). Generally, with aging, all glial cell types are characterized by changing their ultrastructural features, including the presence of different inclusions (osmiphylic and foamy matherials) in their cytoplasm (fig. 16-3 D and E). These observations are similar to others for different species and brain areas (Vaughan and Peter, 1974; Peters et al., 1991; 1994).

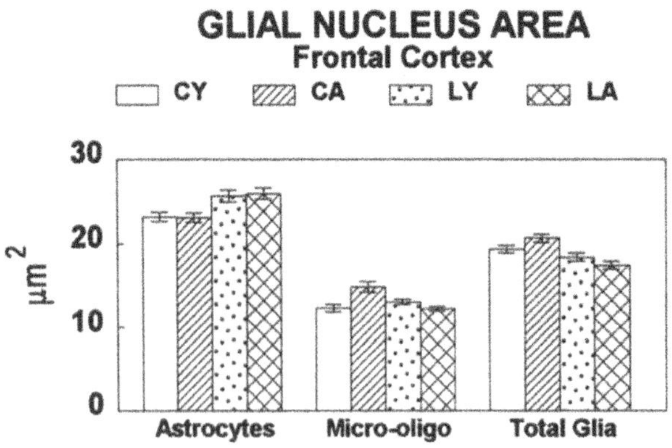

Figure 16-6. Nucleus area (MEAN AREA±SEM expressed in μm^2) of astrocytes, microglia-oligodendrocytes and total glial cells in the frontal cortex of control young (CY), control aged (CA), lesioned young (LY) and lesioned aged (LA) rats.

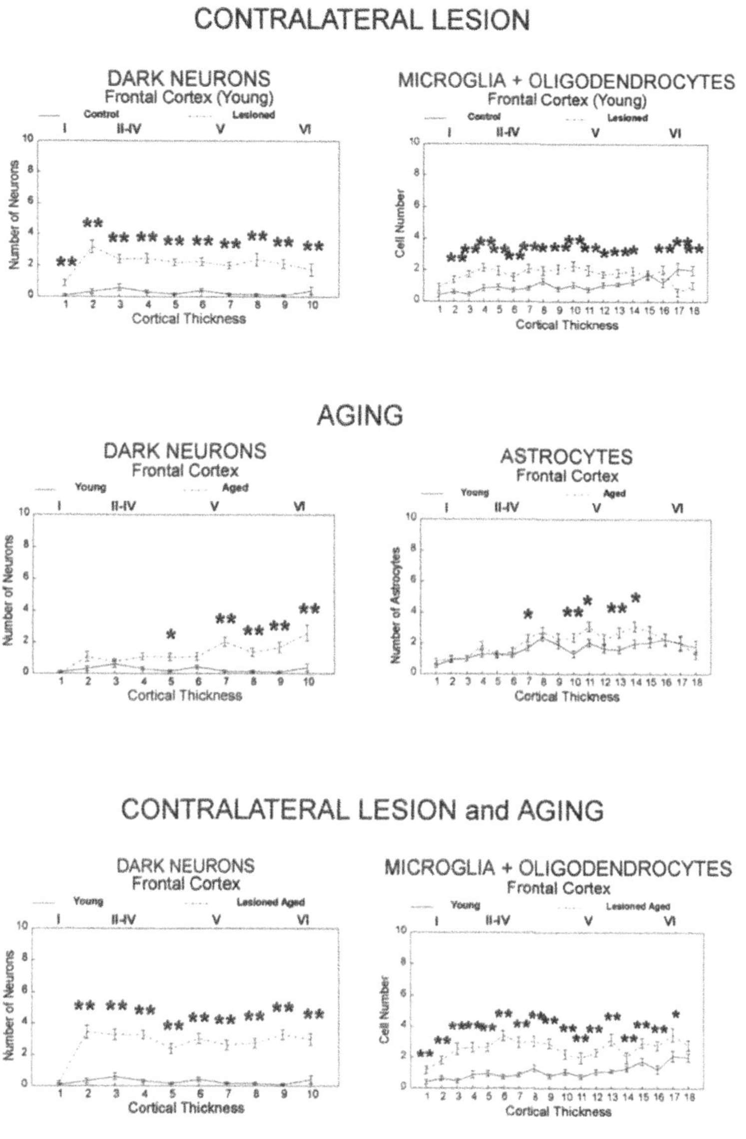

Figure 16-7. Distribution of dark neurons and glial cells across the cortical thickness of the rat frontal cortex in different experimental situations: "contralateral lesioned", "aged" and "aged-contralateral lesioned groups". The botton of each figure represents the different units of count (10 for dark neurons and 18 for glial cells) and its corresponding cortical layers (above). In the model of contralateral lesions, the graphs show the distribution of dark neurons and microglial-oligodendrocytes in young lesioned animals in relation to young unlesioned control animals. In the model of aging, the graphs show the distribution of dark neurons and astrocytes in aged animals in relation to young controls. In the model of contralateral lesion plus aging, the graphs show the distribution of dark neurons and microglia-oligodendrocytes in aged lesioned animals in relation to young unlesioned controls. (* $p<0.05$; ** $p<0.01$).

CORTICAL LAYERS WERE DIFFERENTLY AFFECTED BY LESION AND AGING

Another different feature between contralateral lesion and aging, was that neurodegeneration and glial proliferation were unequally distributed across the thickness of the cortex in both models (fig. 16-7). In fact, in the contralateral area to the lesion, the presence of dark neurons and the changes in microglia-oligodendrocytes were generally homogeneously distributed in the six cortical layers. However, with aging, the hyperplasia of astrocytes were particularly located in layer V. Moreover, this layer was characterized by the presence of a large number of DN in aged animals (fig. 16-7).

NEURONAL REGENERATION IS INFLUENCED BY GLIOSIS

The qualitative and quantitative morphological changes described for glial cells in the aging and lesion models seem to influence differently several neuromorphometric parameters (fig. 16-8). Thus, the area displayed by the neuronal pericaryon, nucleus and nucleolus of DN was decreased compared with NN. It has been reported that during normal aging and in neurodegenerative diseases, the mean size of the neurons appeared to be reduced (Finch, 1993); this feature could be due to the shrinkage of the dark neurons. Nevertheless, our data indicated that NN were able to sprout by increasing all these cytomorphometric parameters. This regenerative ability of the neurons was more significant in the contralateral area to the lesion than in aging. In fact, NN in young injured animals were able to increase significantly the size of its pericaryon, nucleus and nucleolus whereas in aged animals neurons were only able to increase their nucleolus (fig. 16-8). It could be suggested that these NN could supply the functions lost by DN.

Several findings have revealed that glial cells may have opposite functions-trophic or toxic on neurons. Thus, glycogen is predominantly stored in astrocytes, supplying neuronal energetic requirements (Magistretti et al., 1988). These glial cells are also able to release neurotrophic factors which are required for the sprouting response (Gage et al., 1988). However, astrocytes also may induce neurodegeneration by increasing lipid peroxidation and free radicals (Schipper, 1996). In relation to microglial cells, it has been suggested that they can modulate astrogliosis (Giulian and Lanchman, 1985; Giulian and Barker, 1986; Giulian et al., 1989). In addition, they may release several cytokines, and others factors that may also modulate the degenerative or regenerative status of neurons (Griffin et al., 1989).

EFFECTS OF LESIONS AND AGING ON PERICYTES

The brain pericytes are perivascular cells located outside of the endothelial cells (EC) and enclosed inside the basal lamina common to microvascular capillary endothelium (Sturrock, 1983; Peters, 1991). It has been suggested that pericytes are microglial precursors with phagocytic functions (Baron and Gallego, 1972; Sturrock, 1974). Today it is clear that pericytes belong to a different line of cells with specific functions different from those of microglial cells (Ling et al., 1980; Mato et al., 1996; Ookawara et al., 1996).

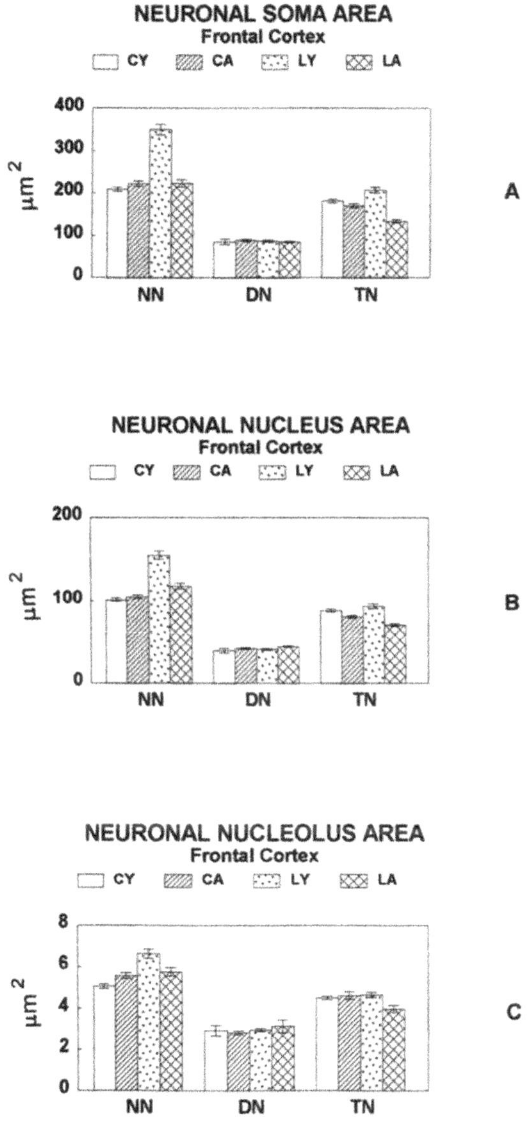

Figure 16-8. Neuronal soma (pericaryon) (A), nucleus (B) and nucleolus (C) areas (MEAN AREA± SEM expressed in μm^2) of non-damage neurons (NN), dark neurons (DN) and total neurons (TN) in the frontal cortex of control young (CY), control aged (CA), lesioned young (LY) and lesioned aged (LA) rats.

Different studies (Peters et al., 1991; Shepro and Morel, 1993; Ookawara et al., 1996) have demonstrated that pericytes are morphologically characterized by an elongated multibranched shape and oval nucleus with chromatin rather clamped beneath the nuclear envelope. Their cytoplasm is relatively dense and contains a small number of cytoplasmic organelles such as mitochondria and rough endoplasmic reticulum (Peters et al., 1991; Ookawara et al., 1996; Shepro and Morel, 1993). Pericytes are part of the blood-brain barrier and together with the vascular endothelial cells, astrocytes and perivascular microglial cells, play a major role in maintaining local and systemic homeostasis of the brain (Shepro and Morel, 1993; Mato et al., 1996).

Functionally, pericytes are characterized by their heterogeneity; these cells are able to regulate differentiation and proliferation of EC via the secrection of tumoral growth factor beta (TGF-ß) (Morel et al., 1993), and also synthesize and secrete a wide variety of vasoactive agents (Hudes et al., 1988). Earlier, an active role of pericytes in the formation of the basal membrane and extracelular matrix was reported (Schor et al., 1991). In addition, pericytes contain contractile proteins such as both smooth and non-smooth muscle isoforms of actin and myosim, tropomyosin, alpha-actin and desmin (Hermann and D'Amore,1985; Nehls and Drenckhahn, 1991). Moreover, the presence of smooth-muscle protein isoforms and cyclic guanosine-monophosphate (cGMP)-dependent protein kinase, supports a possible contractile function for pericytes (Joyce et al., 1984; Haeflinger et al., 1994). Therefore this function would be responsible for capillary diameter regulation and consequently the capillary perfusion (Tilton et al., 1979), inflammatory response (Sims, 1991) and opening of the interendothelial cell junctions (Braverman and Braverman, 1986). This responses seem to be modulated by different vasoactive amines (Murphy and Wagner, 1994).

Recent observations have suggested that pericyte proliferation may play a key role in the genesis of hypertension (Herman et al., 1987; Herman and Jacobson, 1988). Nomura et al. (1995) have reported increases in the number of endothelial cells and pericytes under hypoxic conditions. Our investigations have demonstrated that in lesioned or aged rats, the frontal cortex shows stability in the number of pericytes; however, in parietal cortex we found an increase in the number of the pericyte with aging. Peters et al. (1991), in the monkey visual cortex, reported stability of the pericyte population during aging. In addition, they found lipofuscin or body-like large globular inclusions and dark granular material inside the pericytes; these inclusions usually fill a substantial portion of their perikaryal cytoplasm (Peters et al., 1991). Our ultrastructural and cytomorphometric studies have demonstrated that contralateral lesion induced increases in nuclear size of pericytes that were even greater when aged animals were used (fig. 16-9). Moreover, during aging, the thickness of the vascular wall increased and pericytes and perivascular microglial cells were characterized by accumulation of flak-like precipitation and lipid inclusions (osmiphilic granules) in their cytoplasm and in the extracellular matrix. Similar observations were reported in the thoracic aortic wall of aged rats (Aliev et al., 1995).

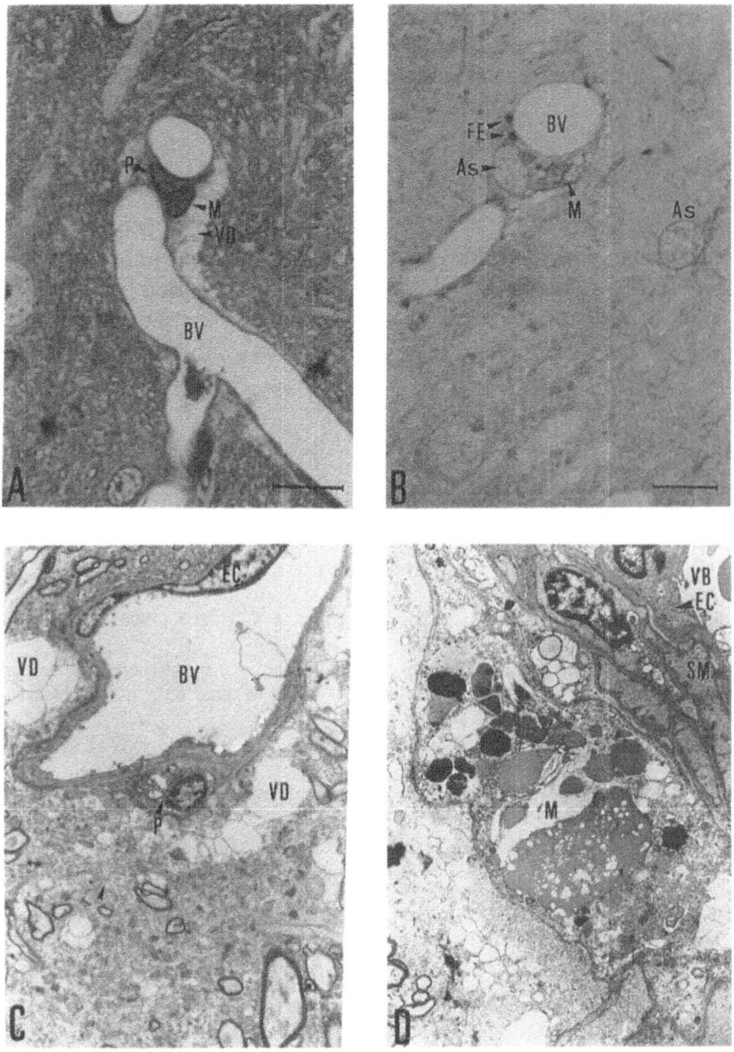

Figure 16-9. Light and electron microscopic characteristics of the perivascular cells in the frontal and parietal cortex of aged rats. (A) Toluidine blue stained section of the parietal cortex of an aged rat with a blood vessel (BV) sorrounded by a zone of vacuolar degeneration (VD); also can be seen a pericyte (P) and a perivascular microglial cell (M). (B) anti-GFAP/blue toluidine stained section of the frontal cortex of an aged rat showing a blood vessel (BV) sorrounded by different GFAP-positives astrocyte feed-end (FE) as well as a microglial cell in the perivascular space (M), whose cytoplasm containing many foamy inclusions; note the close association between the astrocyte (As) and the microglial perivascular cells (M). (C and D) Ultrathin sections of the parietal cortex of aged rats. In picture "C" the blood vessel (BV) is surrounded by several zones of vacuolar degeneration (VD) as well as a pericyte (P) with inclusions in its cytoplasm. In picture "D" it is shown the profile from a blood vesel (BV) with a proccess of a perivascular microglial cells (M) including numerous dark and foamy inclusions. (EC: endothelial cells; SM: smooth muscle cell).(Bar in A and B = 10μm; Magnifications: C = 3300x; D = 46600x).

Our investigations suggest that changes observed in the ultrastructure of the pericytes, particulary in the parietal cortex of aged lesioned animals, also may play an important role in the development of neurodegeneration and gliosis.

ACKNOWLEDGMENTS

The authors are indebted to Dr. G. Aliev for his invaluable help in the critical reading of the manuscript and M. Angeles Cruz and David Nesbitt for reviewing the English style of the manuscript. This work was supported by DGICYT PM 90-0146 and by Andalusian Regional Government Science and Education Council.

REFERENCES

Aliev, G., S. Miah, M. Turmaine, and G. Burnstock (1995) An ultrastructural and immunocytochemical study of thoracic aortic endothelium in aged Sprague-dawley rats. J. Submicr. Cytol. Path. *27*:477-490.

Amaducci, L., and G. Tesco (1994) Aging as a major risk for degenerative diseases of the CNS. Curr. Opin. Neurol. *7*: 283-286.

Amenta, F., S. Bongrani, S. Cadel, A. Ricci, B. Valsecchi, and Y.C. Zeng (1994) Neuroanatomy of aging brain. Ann. NY. Acad. Sci. *717*: 33-44.

Andersson, P.B., V.H. Perry, and S. Gordon (1992) The acute inflammatory response to lipopolysaccharide in the CNS parenchyma differs from that of other body tissues. Neuroscience *48*:169-186.

Banati, R.B., J. Gehrmann, P. Schubert, and G.W. Kreutzberg (1993) Cytotoxicity of microglia. Glia *7*: 111-118.

Baron, M., and A. Gallego (1972) The relation of the microglia with the pericytes in the cat cerebral cortex. Z. Zellforsch. *128*:42-57.

Berciano, M.T., M.A. Andrès, E. Calle, and M. Lafarga (1995) Age-induced hypertrophy of astrocytes in rat supraoptic nucleus: A cytological, morphometric, and immunocytochemical study. Anat. Rec., *243*:129-144.

Björklund, H., M. Eriaksdotter-Nilsson, D. Dahl, G. Rose, B. Hoffer, and L. Olsson (1985) Image analysis of GFA-positive astrocytes from adolescence to senescence. Exp. Brain Res. *58*: 163-170.

Bonfanti, L., D.A. Poulain, and D.T. Theodosis (1993) Radial glia-like cells in the supraoptic nucleus of the adult rat. J. Neuroendocrinol. *5*:1-5.

Braverman, M.S., and I.M. Braverman (1986) Three dimensional reconstruction of serial sections using a microcomputer graphics system. J. Invest. Dermatol. *86*: 290-294.

Brizzee, K.R., P. Klara, and J.E. Johnson (1975) Changes in microanatomy, neurocytology and fine structure with aging. In Ordy J.M. and K.R. Brizze (eds): Neurobiology of Aging. New York and London: Plenum Press. pp. 425-461.

Bronson, R.T., R.D. Lipman, and D.E. Harrison (1993) Age-related gliosis in the white matter of mice. Brain Res. *609*:124-128.

Cajal, S.R. (1913) Sobre un nuevo proceder de impregnación de la neuroglia y sus resultados en los centros nerviosos del hombre y animales. Trab. Lab. Inves. Biol. Madrid *14*: 155-162.

Cammermeyer, J. (1961) The importance of avoiding "dark" neurons in experimental neuropathology. Acta Neuropathol. *1*: 245-270.

Cao, D.H., M. Strolin, and P. Dostert (1985) Age-related changes in glutamine synthetase activity of rat brain liver and heart. Gerontology *31*: 95-100.

Castellano B., B. Gonzales, M.B. Jensen, E.B. Pedersen, B.R. Finsen, and J. Zimmer (1991) A double staining technique for simultaneous demonstration of astrocytes and microglia in brain sections and astroglial cell cultures. J. Histochem. Cytochem. *39*: 561-568.

Chao, C.C., S. Hu, and P.K. Peterson (1996) Glia: the not-so-innocent bystanders. J. Neurovirol. *2*:234-239.

Cotman, C.W., and K.J. Anderson (1995) A potential role for apoptosis in neurodegeneration and Alzheimer's disease. Mol. Neurobiol. *10*: 19-45.

Couraud, P.O. (1994) Interactions between lymphocytes, macrophages, and central nervous system. J. Leukocyte Biol. *56*:407-415.

Curcio, C.A., and P.D. Coleman (1982) Stability of neuron number in cortical barrels of ageing mice. J. Comp. Neurol. *212*: 158-172.

Dahl, D., H. Mansour, and A. Bignami (1991) Astrocytes colonize dorsal root ganglia transplanted into brain. Brain Res. Bull. *27*:169-173.

Davis E.J., T.D. Foster, and W.E. Thomas (1994) Cellular forms and functions of brain microglia. Brain Res. Bull. *34*:73-78.

Del Rio-Hortega, P. (1920) La microglía y su transformación en células en bastoncito y cuerpos gránulo-adiposos. Trab. Inst. Cajal Invest. Biol. *14*: 37-82.

Del Rio-Hortega, P. (1921) Estudios sobre la neuroglia. La glia de escasas radiaciones (oligodendroglia) Bol. Real. Soc. Esp. Hist. Nat. *14*: 5-122.

Diamond, M.C., R.E. Johnson, and M.W. Gold (1977) Changes in neuron number and size and glial number in the young, adult and ageing rat medial occipital cortex. Behavior. Biol. *20*: 409-418.

Finch, C.E. (1993) Neuron atrophy during aging: programmed or sporadic? Trends Neurosci. *16*: 104-110.

Fog, R., and H. Pakkenberg (1983) Mitotic activity in the mouse brain after cortical and striatal lesions. J. Hirnforsch. *24*: 341-343.

Gage, F.H., P. Olejniczak, and D.M. Armstrong (1988) Astrocytes are important for sprouting in the septohippocampal circuit. Exp. Neurol. *102*:2-13.

Gallyas, F., Z.H. Güldner, G. Zoltay, and J.R. Wolff (1990) Golgi-like demonstration of "dark" neurons with an argyrophil III method for experimental neuropathology. Acta Neuropathol. *79*: 620-628.

Gallyas, F., G. Zoltay, and W. Dames (1992) Formation of "dark" (argyrophilic) neurons of various origin proceeds with a common mechanism of biophysical nature (a novel hypothesis). Acta Neuropathol. *83*: 504-509.

Gehrmann, J., P. Bonnekoh, T. Miyazawa, K.A. Hossmann, and G.W. Kreutzberg (1992) Immunocytochemical study of an early microglial activation in ischemia. J. Cereb. Blood F. Met. *12*: 257-269.

Giordana, M.T., A. Attanasio, P. Cavalla, A. Migheli, M.C. Vigliani, and D. Schiffer (1994) Reactive cell proliferation and microglia following injury to the rat brain. Neuropathol. Appl. Neurobiol. *20*: 163-174.

Giulian, D., and L.B. Lachman (1985) Interleukin-1 stimulation of astroglial proliferation after brain injury. Science *228*: 497-499.

Giulian, D., and T.J. Baker (1986) Characterization of ameboid microglia isolated from developing mammalian brain. J. Neurosci. *6*: 2163-2178.

Giulian, D., J. Chen, J.E. Ingeman, J.K. George, and M. Naponen (1989) The role of mononuclear phagocytes in wound healing after traumatic injury to adult mammalian brain. J. Neurosci. *9*: 4416-4429.

Giulian, D., and J.E. Ingeman (1993) Colony-stimulating factors as promoters of ameboid microglia. J. Neurosci. *8*: 4707-4717.

Goldman, J.E., N.Y. Calingasan, and G.E. Gibson (1994) Aging and the brain. Curr. Opin. Neurol. *7*: 287-293.

Goldberg, W.J., and J.J. Berstein (1988a) Fetal cortical astrocytes migrate from cortical homografts throughout the host brain over the glial limitants. J. Neurosci. Res. *20*: 38-45.

Goldberg, W.J., and J.J. Bernstein (1988b) Migration of cultured fetal spinal cord astrocytes into adult host cervical cord and medulla following transplantation into thoracic spinal cord. J. Neurosci. Res. *19*: 34-42.

Graeber, M.B., W.J. Streit, and G.W. Kreutzberg (1988) Axotomy of the rat facial nerve leads to increased CR3 complement receptor expression by activated microglial cells. J. Neurosci. Res. *21*: 18-24.

Graeber, M.B. and W.J. Streit (1990) Microglia: Immune network in the CNS. Brain Pathol. *1*: 2-5.

Griffin, W.S.T., L.C. Stanley, C. Ling, L. White, V. Macleod, L.J. Perrot, C.L. White, and C. Araoz (1989) Brain interleukin 1 and S-100 immunoreactivity are elevated in Down syndrome and Alzheimer disease. Proc. Natl. Acad. Sci. USA. *86*: 7611-7615.

Haeflinger, I.O., A. Zzchauer, and D.R. Anderson (1994) Relaxation of retinal pericyte contractile tone through the nitric oxide-cyclic guanosine monophosphate pathway. Invest. Ophthalmol. Vis. Sci. *35*:991-997.

Hajos, F., N. Kalman, K. Zilles, A. Sleicher, and P. Sotony (1990) Remote astrocytic response as demonstrated by glial fibrillary acidic protein immunohistochemistry in the visual cortex of dorsal lateral geniculate nucleus lesioned rats. Glia *3*: 301-310.

Hansen, L.A., D.M. Armstromg, and R.D. Terry (1987) An immunohistochemical quantification of fibrous astrocytes in the aging human cerebral cortex. Neurobiol. Aging *8*: 1-6.

Hatten M.E., R.H.F. Liem, M.L. Shelanski, and C.A. Mason (1991) Astroglia in CNS injury. Glia *4*:233-243.

Haug, H., U. Barmwater, R. Eggers, D. Fisher, S. Kühl, and N.L. Sass (1983) Anatomical Changes in Ageing Brain: Morphometric Analysis of the Human Prosencephalo. In J. Cervós-Navarro and H.I. Sarkander (eds): Brain Ageing: Neuropathology and Neuropharmacology. New York: Raven Press. pp: 1-12.

Haug, H., S. Kühl, E. Mecke, N.L. Sass, and K. Wasner (1984) The significance of morphometric procedures in the investigation of age changes in cytoarchitectonic structures of human brain. J. Hirnforsh. *25*: 353-374.

Held, H. (1909) Über die Neuroglia marginales der menschlichen Großhirnrinde. Mschr. Psychiat. Neurol. *26*: 360-416.

Heumann, D., and G. Leuba (1983) Neuronal death in the development and aging of the cerebral cortex of the mouse. Neuropathol. Appl. Neurobiol. *5*: 433-456.

Herman, I., P.M. Newcomb, J.E. Coughlin, and S. Jacobson (1987) Characterization of microvascular cell cultures from normotensive and hypertensive rat brains: pericyte-endothelial cell interactions in vitro. Tissue Cell. *19*:197-206.

Herman, I., and S. Jacobson (1988) In situ analysis of microvascular pericytes in hypertensive rat brains. Tissue Cell. *20*:1-12.

Hermann, I., and P.A. D'Amore (1985) Microvascular pericytes contain muscle and nonmuscle actins. J. Cell Biol. *101*: 43-52.

Hudes, G.R., W. Li, J.H. Rockey, and P. White (1988) Prostacyclin is the major prostaglandin synthesized by bovine retinal capillary pericytes in culture. Invest. Opthalmol. Vis. Sci. *29*: 1511-152.

Janeczko, K. (1989) Spatiotemporal patterns of the astroglial proliferations in rat brain injured as the postmitotic stage of postnatal development: A combined inmunocytochemical and autoradiographic study. Brain Res. *485*:236-243.

Joyce, N.C., P. DeCamilli, and J. Boyles (1984) Pericytes, like vascular smooth muscle cells, are immunocytochemically positive for cyclic GMP-dependent protein kinase. Microvasc. Res. *28*: 206-219.

Kimelberg H.K., T. Jalonen, and W. Walz (1993) Regulation of the brain microenvironment: transmitters and ions. In S. Murphy (ed): Astrocytes: Pharmacology and Function. New York: Academic Press, Inc. pp.193-202.

Konigsmark, B.W., and R.L. Sidman (1963) Origin of brain macrophages in the mouse. J. Neuropath. Exp. Neurol. *22*: 643-676.

Landfield, P.W., G. Rose, L. Sandles, T.C. Wohstadter, and G. Lynch (1977) Patterns of astroglial hypertrophy and neuronal degeneration in the hippocampus of aged memory deficient rats. J. Gerontol. *32*:3-12.

Landis, D.M.D. (1994) The early reactions of non-neuronal cells to brain injury. Annu. Rev. Neurosci. *17*:133-151.

Lawson, L.J., V.H. Perry, P.Y. Dri, and S. Gordon (1990) Heterogeneity in the distribution and morphology of microglia in the normal adult mouse brain. Neuroscience *39*: 151-170 .

Lees, G.J. (1993) Contributory mechanisms in the causation of neurodegenerative disorders. Neuroscience *54*:287-322.

Le Prince, G., P. Delaere, C.C. Fages, T. Lefrançois, M. Touret, M. Salanon, and M. Tardy (1995) Glutamine Synthetase (GS) expression is reduced in senile dementia of the Alzheimer type. Neurochem. Res. *20*: 859-862.

Ling, E.A., D. Penney, and C.P. Leblond (1980) Use of carbon labelling to demonstrate the role of blood monocytes as precursors of the "amoeboid cells" present in the corpus callosum of postnatal rats. J. Comp. Neurol. *193*: 631-657.

Lindsay, R.M. (1986) Reactive gliosis. In: S. Federoff and A. Vernadakis (ed): Astrocytes. New York: Academic Press. pp. 231-262.

Lolova, I. (1991) Qualitative and quantitative glial changes in the hippocampus of aged rats. Anat. Anz. *172*:263-271.

Magistretti, P.J. (1988) Regulation of glycogenolysis by neurotransmitters in the central nervous system. Diabetes Metab. *14*:237-246.

Marty, S., I. Dusart, and M. Pechanski (1991) Glial changes following an excitotoxic lesion in the CNS. I Microglia/macrophages. Neuroscience *45*:529-539.

Mathewson, A.J., and M. Berry (1985) Observations on the astrocyte response to a cerebral stab wound in adult rats. Brain Res. *327*:61-69.

Mato, M., S. Ookawara, A. Sakamoto, E. Aikawa, T. Ogawa, U. Mitsuhashi, T. Masuzawa, H. Suzuki, M. Honda, Y. Yazaki, E. Watanabe, J. Luoma, S. Yla-Herttuala, I. Fraser, S. Gordon, and T. Kodama (1996) Involvement of specific macrophage-lineage cells arrounding arterioles in barrier and scavenger function in brain cortex. Proc. Natl. Acad. Sci. USA. *93*:3269-3274.

McGeer, P.L. (1984) The 12th J.A.F. Stevenson Memorial Lecture. Aging, Alzheimer's disease and the cholinergic system. Can. J. Physiol. Pharmacol. *62*: 741-54.

McGeer, P.l., T. Katawama, D.G. Walker, H. Akiyama, I. Tooyama, and E.G. McGeer (1993) Microglia in degenerative neurological diseases. Glia *7*: 119-123.

Morel, N.M.L., H.B. Hechtman, and D. Shepro (1993) Pericyte modulation of tubular formation and cytoskeletal actin distribution in pulmonary microvessels endothelial cells. J. Cell Physiol. Biochem. *3*:127-136.

Morgan, D.G., P.C. May, and C.E. Finch (1987) Dopamine and serotonine systems in human and rodent brain: Effects of ageing and neurodegenerative diseases. J. Am. Gerontol. Soc. *35*: 334-345.

Morioka, T., A.N. Kalehua, and W.J. Streit (1993) Characterization of microglial reaction after middle cerebral artery occlusion in rat brain. J. Comp. Neurol. *327*:123-132.

Mucke, L., and M. Eddleston (1993) Astrocytes in infections and immuno-mediated diseases of the central nervous system. FASEB J. *7*:1226-1231.

Murphy, D.D., and R.C. Wagner (1994) Differential contractile response of cultured microvascular pericytes to vasoactive agents. Microcirculation *1*:121-128.

Murray, H.M., and B.E. Walker (1973) Comparative study of astrocytes and mononuclear leukocytes reacting to brain trauma in mice. Exp. Neurol. *41*: 290-302.

Nehls, V., and D. Drenckhahn (1991) Heterogeneity of microvascular pericytes for smooth muscle type alpha-actin. J. Cell Biol. *113*:147-154.

Nieto-Sampedro, M., and F. Mora (1994) Active microglia, sick astroglia and Alzheimer type dementias. NeuroReport *5*: 375-380.

Nomura, M., S. Yamagishi, S. Harada, Y. Hayashi, T. Yamashima, J. Yamashita, and H. Yamamoto (1995) Possible participation of autocrine and paracrine vascular endothelial growth factors in hypoxia-induced proliferation of endothelial cells and pericytes. J. Biol. Chem. *270*: 28316-28324.

Norenberg, M.D. (1994) Astrocyte responses to CNS injury. J. Neuropath. Exper. Neurol. *53*: 213-220.

Ookawara, S., U. Mitsuhashi, Y. Suminaga, and M. Mato (1996) Study on distribution of pericyte and fluorescent granular perithelial (FGP) cell in the transitional region between arteriole and capillary in rat cerebral cortex. Anat. Rec. *244*:257-264.

Pakkenberg, H., B.B. Anderson, R.S. Burns, and B. Pakkenberg (1995) A stereological study of substantia nigra in young and old rhesus monkeys. Brain Res. *693*:201-206.

Peinado J.M., and R.D. Myers (1988a) In vivo analysis of cortical amino acid neurotransmitter collected in the rat by a new double lumen push-pull catheter system. Neuroscience *24*:1027-34.

Peinado J.M. and, R.D. Myers (1988b) Cortical aminoacid neurotransmitter release is altered by CCK perfused in frontal region of unrestrained aged rat. Peptides *9*:631-6.

Peinado, M.A., M. Martínez, J.A. Pedrosa, A. Quesada, and J.M. Peinado (1993) Quantitative morphological changes in neurons and glia in the frontal lobe of the aging rat. Anat. Rec. *237*:104-108.

Peinado, M.A., A. Quesada, J.A. Pedrosa, M. Martínez, F.J. Esteban, M.L. del Moral, and J.M. Peinado (1997) Light microscopic quantification of morphological changes during aging in neurons and glia of the rat parietal cortex. Anat. Rec. *247*:420-425.

Penfield, W. (1927) The mechanism of cicatricial contraction in the brain. Brain *50*:499-517.

Perry, V.H., M.K. Matyszak and S. Fearn (1993) Altered antigen expression of microglia in the aged rodent CNS. Glia *7*:60-67.

Peters, A., and D.W. Vaughan (1981) Central nervous system. In: J.E. Jhonson (ed): Aging and Cell Structure Vol.1. New York: Plenum Press. pp. 1-34.

Peters, A., S.L. Palay, and H. Webster (1991) The fine structure of the nervous system. Neurons and their supporting cells. 3rd ed. New York: Oxford University Press.

Peters, A., D. Leuha, M.B. Moss, and K.J. McNally (1994) The effects of aging on area 46 of the frontal cortex of the Rhesus monkey. Cereb. Cortex *6*:621-635.

Rand, C.W., and C.B. Courville (1932) Histologic studies in the brain in cases of fatal injury to the head. III. Reaction of microglia and oligodendroglia. Arch. Neurol. Psychiat. *27*:605-644.

Rand, C.W. (1952) The role of the astrocyte in the formation of cerebral scars. Los Angeles Neurol. Soc. *17*:57-70 (1952).

Reier, P.J. (1986) Gliosis following CNS injury: the anatomy of astrocytic scars and their influences on axonal elongation. In: S. Federoff and A. Vernadakis (eds): Astrocytes. New York: Academic Press. pp. 263-324.

Schiffer, D., M.T. Giordana, M.C. Vigliana, and P. Cavalla (1991) Relationship between a glial reaction to a stab wound and tumor development after receiving transplacental etylnitrosurea in the rat. Acta Neuropathol. *83*:30-38.

Schipper, H.M. (1996) Astrocytes, brain aging, and neurodegeneration. Neurobiol. Aging *17*:467-480.

Schor, A.M., A.E. Canfield, P. Sloan, and S.L. Schor (1991) Differentiation of pericytes in culture is accompanied by changes in the extracellular matrix. In vitro *27A*: 651-659.

Schultz, R.L., and D.C. Pease (1959) Cicatrix formation in rat cerebral cortex as revealed by electron microscopy. Amer. J. Path. *35*:1017-1041.

Sims, D.E. (1991) Recent advances in pericyte biology. Implications for health and disease. Can J. Cardiol. *7*:431-443.

Stensaas, S.S., and L.J. Stensaas (1976) The reaction of the cerebral cortex to chronically implanted plastic needles. Acta Neuropathol. *35*:187-203.

Streit, W.J., M.B. Graeber, and G.W. Kreutzberg (1988) Functional plasticity of microglia: A review. Glia *1*:301-307.

Streit, W.J. (1990) Microglia: Immune network in the CNS. Brain Pathol. *1*:2-5.

Sturrock, R.R. (1974) Histogenesis of the anterior limb of the anterior commissure of the mouse brain.III. An electron microscopic study of gliogenesis. J. Anat. *117*:37-53.

Sturrock, R.R. (1980) A comparative quantitative and morphological study of aging in the mouse neostriatum, indusium griseum and anterior comisure. Neuropathol. Appl. Neurobiol. *6*: 415-420.

Sturrock, R.R. (1983) Problems of glial identification and quantifications. In J. Cervós-Navarro and H.I. Sarkander (eds): The Aging Central Nervous System. Brain Aging. Neuropathology and Neuropharmacology. New York: Raven Press. pp.179-209.

Sturrock, N.N. (1990) A comparison of age-related changes in neuron number in the dorsal motor nucleus of the vagus and the nucleus ambiguous of the mouse. J. Anat. *173*:169-176.

Terry, R.D., R. de Teresa, A. Lawrence, and M.D. Hansen (1987) Neocortical cells counts in normal human adult ageing. Ann. Neurol. *21*:530-539.

Tilton, R.G., C. Kilo, and J.R. Williamson (1979) Pericyte-endothelial relationships in cardiac and skeletal muscle capillaries. Microvasc. Res. *18*: 325-335.

Van den Pol, A.N., and F. Gallyas (1990) Trauma induced Golgi-like staining of neurons a new approach to neuronal organization and response to injury. J. Comp. Neurol. *296*:654-673.

Vaughan, D.W., and A. Peters (1974) Neuroglial cells in the cerebral cortex of rats from young adulthood to old age: an electron microscopy study. J. Neurocytol. *3*:405-429.

Vela, J.M., I. Dalmau, B. Gonzalez, and B. Castellano (1995) Morphology and distribution of microglial cells in the young and adult mouse cerebellum. J. Comp. Neurol. *361*:602-616.

Vincent, S.L., A. Peters, and J. Tigges (1989) Effects of aging on the neurons within area 17 of Rhesus monkey cerebral cortex. Anat. Rec. *223*:329-341.

Wang J., D. Lieberman, H. Tabubo, J.P.M. Finberg, E.H. Oldfield, and K. Bankiewicz (1994) Effects of gliosis on dopamine metabolism in rat striatum. Brain Res. *663*:199-205.

17

THE ROLE OF SCHWANN CELL IN NERVE REGENERATION

Enrique Verdú and Xavier Navarro

Wallerian degeneration, axonal reaction and chromatolysis, axonal growth, target reinnervation and maturation of regenerated axons are sequential processes that occur after a peripheral nerve injury. In all these steps there is a close relationship between neurons and Schwann cells. In response to nerve injury, Schwann cells divide rapidly, migrate to the lesion zone, phagocyte myelin and axonal debris and create endoneurial bands of Büngner for regenerating axons to follow. Schwann cells secrete a number of neurotrophic factors, cell adhesion molecules and basement membrane components that can promote axonal growth. Early genes are expressed in both neurons and Schwann cells after nerve lesions, regulating the expression of later genes, that code for proteins responsible for phenotypic changes in each phase of nerve regeneration. The first steps that prepare surviving neurons for growth are mediated by cell-to-cell interactions and by soluble factors secreted by Schwann cells. Axonal regrowth is dependent on guiding substrates provided by extracellular matrix components and reactive Schwann cells in the distal degenerated nerve. Peripheral reinnervation is mediated by interactions between neuron and target cells, but Schwann cells also participate in the reformation and maintenance of synapses. Whilst during axonal growth neuronal gene expression is controlled by factors liberated by Schwann cells, in the final reinnervation and maturation steps Schwann cell and target cell gene expression is controlled by the regenerated neurons. The potential of Schwann cells to allow and promote regeneration may be applied in the development of cellular prostheses, being composed of a nerve chamber filled with an exogenous extracellular matrix and seeded with Schwann cells, which may provide an artificial substitute to the traditional autologous graft used to repair severe nerve lesions.

INTRODUCTION

In the peripheral nervous system (PNS), compression or section of nerves cause a loss of motor, sensory and autonomic neural control of target organs. Nerve fibers distal to the lesion are disconnected from the neuronal body and then degenerate. Histological studies show a progressive disintegration of the axonal cytoskeleton and the myelin sheaths. These changes are accompanied by metabolic changes in the soma of axotomized neurons, known as retrograde reaction and chromatolysis (Selzer, 1980; Fawcett and Keynes, 1990; Gordon, 1983). Wallerian degeneration serves to create a microenvironment distal to the injury site that is favorable for the axonal regrowth of surviving neurons, while retrograde reaction and chromatolysis represent metabolic changes necessary for regeneration and axonal elongation (Selzer, 1980; Fawcett and Keynes, 1990; Gordon, 1983). The functional significance of regeneration is to replace the distal nerve segment lost during Wallerian degeneration, allowing the reinnervation of target organs and the restitution of their corresponding functions.

During the regeneration process, important relationships are established between the regenerating neurons and their Schwann cells. These interactions have been included under the name of neural relation (Nieto-Sampedro, 1996). During development, Schwann cells derive from neural crest cells, that in the presence of several differentiation factors, such as neuregulins, generate immature Schwann cells (fig. 17-1). Further differentiation to form myelin-forming and non-myelin forming cells depends on axon-associated signals (Mirsky and Jessen, 1996; Topilko et al., 1996). The proliferating capacities of Schwann cells are lost in the mature peripheral nerve. However, after nerve injury Schwann cells became dedifferentiated and express capabilities for proliferation and secretion of different factors that support axonal regeneration.

The purpose of this chapter is to review the cellular and molecular aspects of nerve degeneration and regeneration, to explain the role played by Schwann cells and other non-neuronal cells, the neural relation and, finally, to analyze the rationale of using glial cells for the repair of injured nerves.

BIOLOGY OF WALLERIAN DEGENERATION

Axons have little or no capacity for synthesis of protein or lipids, containing few if any ribosomes. The cell body of the neurons is the main site of macromolecular synthesis and axons depend on their transport system for the supply of required materials synthesized in the soma. Following injury to a peripheral nerve, the nerve distal to the lesion undergoes Wallerian degeneration. Degeneration of isolated axons was first described in 1850 by Waller, who showed that the cell body was essential for the survival of axons and that the connection with end-organs cannot support axons in the absence of the cell body.

Axoplasmic changes

The first signs of Wallerian degeneration are observed within 24 hours of nerve injury, and they are prolonged for about two weeks. After 48 hours, all axons

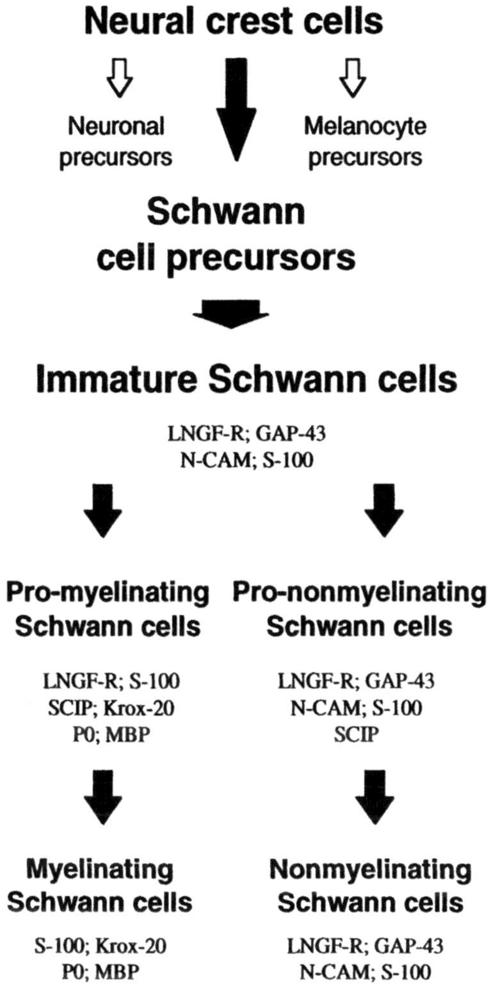

Figure 17-1. Origin and characteristic markers of Schwann cells during development.

show complete disruption of their internal structure with disappearance of neurofilaments and progressive disintegration of neurotubules. Some fibers become extremely swollen, the myelin sheaths acquire complex infoldings, and the cytoplasm of the Schwann cells show signs of increased activity, with the appearance of numerous mitochondria and considerable amounts of granular and agranular reticulum. The earliest changes in the elements of the axoplasm appear in the endoplasmic reticulum. The tubular elements become fragmented into rows of small vesicles. There is hypertrophy of mitochondria that occurs simultaneously with the destruction of the cytoskeleton. Thick particles of 160 to 300 Å become more abundant in the first stage of degeneration. Later on, it is impossible to distinguish them from the irregular axoplasmic clumps. Finally, myelin sheaths and axons are disintegrated and their debris found within the endoneural tubes (Vial, 1958; Lubinska and Jastreboff, 1977; Stoll et al., 1989).

Recruitment of macrophages and Schwann cell proliferation

From 2-3 days after injury there is an important infiltration of macrophages into the degenerating nerve (fig. 17-2), attracted by chemotactic factors, such as leukaemia inhibitory factor (LIF), synthesized and secreted by reactive Schwann cells (Curtis et al., 1994). This cytokine together with other factors secreted by infiltrated macrophages and Schwann cells may promote axonal growth (Donato et al., 1995). The recruitment of macrophages is the main pathway for phagociting myelin and axonal debris (Lunn et al., 1989; Stoll et al., 1989). On the other hand, during degeneration macrophages synthesize and secrete several factors, including interleukine-1 (IL-1), PDGF, FGF and TGF, that promote the mitosis of reactive Schwann cells (Heumann et al., 1987b; Nathan,1987). Mytogenic Schwann cells became phagocytes and help macrophages in the degradation of myelin debris (Holtz-man and Norvikoff, 1965; White et al., 1989).

Myelin degeneration

The first morphological change is the retraction of myelin at the nodes, a process that spreads peripherally and is followed by myelin and axon fragmentation. The main role of Schwann cells during Wallerian degeneration consists of degrading myelin. Histological studies have shown the presence of ovoid structures that are fragments of myelin sheaths and lipid droplets in the cytoplasm of Schwann cells (Stoll et al., 1989; Reichter et al., 1994; Liu et al., 1995), constituting the basic mechanism of reabsorption of myelin by reactive Schwann cells.

Treatment with silica produces a modest and transient reduction in the number of macrophages in degenerating nerves and a mild slowing of myelin degeneration, suggesting that the rate of myelin degeneration is related to the number of macrophages present in the nerve (Beuche and Friede, 1986; Müller and Minwegan, 1987). When macrophage infiltration in degenerating nerves was prevented by killing macrophages with whole body irradiation, myelin degradation was not affected at early time points, but was slowed at later times when macrophages would normally be present (Perry et al., 1995). These data support the view that Schwann cells can degrade myelin, but that macrophages accelerate and complete the degradation process (Tanaka et al., 1992).

Biochemical studies have reported that myelin lipid degradation is not evident until one week after the onset of Wallerian degeneration (Johnson et al., 1949; Mannell, 1952; Wood and Dawson, 1974). These biochemical data are consistent with the morphological results suggesting that Schwann cells can initiate myelin

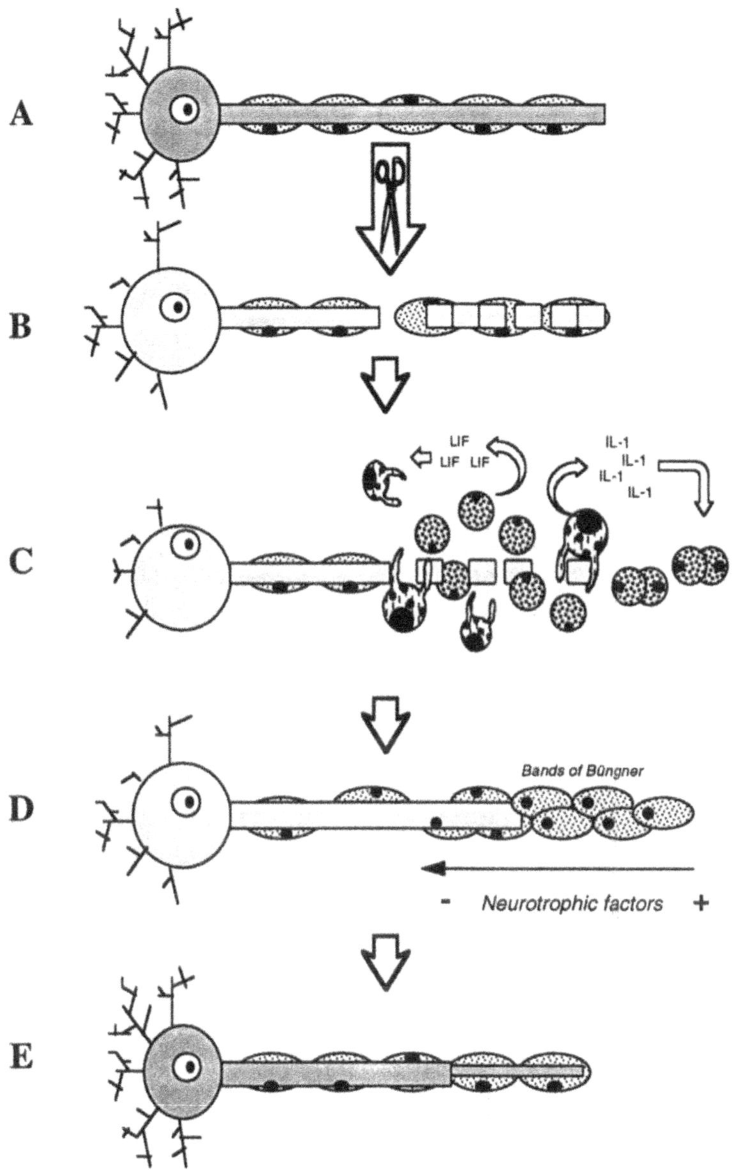

Figure 17-2. Schematic representation of the main features of degeneration and regeneration after peripheral nerve injury. (A) Normal appearance. (B) Transection of the fiber results in distal fragmentation of axon and myelin. (C) In the distal segment Schwann cells proliferate and macrophages migrate. Macrophages and Schwann cells phagocytose debris material. (D) The Schwann cells in the distal segment have lined up in bands of Büngner. Advancing axons are embedded in Schwann cell cytoplasm attracted by a gradient of neurotrophic factors. (E) Axonal connection with periphery, and maturation of the nerve fiber.

degradation, but that macrophages significantly increase and accelerate the degradative process. When nerves were labeled with tritiated acetate, oleate or glycerol before or after nerve injury, tritiated precursors were localized in cytoplasmic lipid droplets within Schwann cells, fibroblasts and extratubular macrophages 2 to 3 days after nerve lesion (Goodrum et al., 1994, 1995). Lipid droplets have been traditionally interpreted as representing breakdown products from the degenerating myelin. The presence of labeled lipid droplets in extratubular fibroblasts and macrophages, especially in the absence of extracellular labeled myelin debris, is evidence of the breakdown of myelin lipids, the extratubal secretion of breakdown products, and their uptake, incorporation, and storage by extratubal cells.

Within the first two weeks after the onset of Wallerian degeneration in nerves prelabeled with tritiated acetate, labeled myelin debris became localized into phagocytes within the basal lamina tubes. Free fatty acids, phospholipids, cholesterol, cholesteryl esters and glycerolipids were seen in lipid droplets from degenerating myelin. Some of these lipids may be reused during nerve regeneration and axonal remyelination (Goodrum et al., 1995; Rawlins et al., 1972). During Wallerian degeneration, infiltrating macrophages also secrete a 37 kD protein identified as apolipoprotein (apo) E (Goodrum and Novicki, 1988; Snipes et al., 1986; Gelman et al., 1987). Based on the apoE capacity to bind cholesterol in lipoprotein particles, it was hypothesized that locally synthesized apoE was responsible for the intercellular transport of cholesterol within the degenerating nerve and its uptake into remyelinating Schwann cells via low-density lipoprotein (LDL) receptors (Boyles et al., 1987) (fig. 17-3).

Degeneration of unmyelinated axons

The processes of degeneration and regeneration of unmyelinated nerve fibers have been less studied than those of myelinated fibers, despite the fact that unmyelinated fibers far outnumber myelinated fibers in mammalian peripheral nerves. The degenerative changes in the axons have been found to resemble those that take place in myelinated ones and all axons degenerate completely within 7 days. Axonal degeneration takes place external to the Schwann cells, but these cells tend to react by surrounding the degenerating axons with multiple flattened processes. The debris are removed both by Schwann cells and macrophages, and disappear by 15 days after transection. However, compact columns of Schwann cells analogous to bands of Büngner are not formed (Thomas and King, 1974).

Chromatolysis and metabolic reaction

The cell body of axotomized neurons typically undergoes a number of characteristic morphological alterations, which were first described by Nissl, and are collectively termed chromatolytic changes. Chromatolysis includes swelling and migration of the nucleus to an eccentric position, and an increase in nuclear and nucleolus size. Other changes observed are increased number of lysosomes, hypertrophy of the Golgi apparatus and the mitochondria and vacuolization of other organella. The disappearance of the prominent basophilic stained Nissl granules is particularly evident. These granules are ribosome clusters and ordered arrays of rough endoplasmic reticulum, which can no longer be visualized after axotomy, when they become disorganized, freeing polyribosomes and even ribonucleotides into the cytoplasm. Disorganization of the ribosome clusters indicates that morphological changes in the cell body are associated with increased protein synthesis, suggesting that chromatolysis represents an anabolic response of the injured neuron (Lieberman, 1971; Gordon, 1983). With a few exceptions, this interpretation is supported by evidence from studies of RNA and protein content in axotomized neurons, showing reduced DNA repression, increased RNA synthesis with transfer of RNA from the nucleus to the cytoplasm, and increased cellular protein content (Watson, 1974). The concentrations of enzymes of the oxidative pentose phosphate shunt, required for RNA synthesis, are also increased (Härkonen and Kauffman, 1974; Sinicropi and Kauffman, 1979), while the production of neurotransmitters and cytoskeleton proteins decreases (Boyle and Gillespie, 1970; Frizell and Sjöstrand, 1974a; O'Brien, 1978; Heiwell et al., 1979; Hoffman and Lasek, 1980).

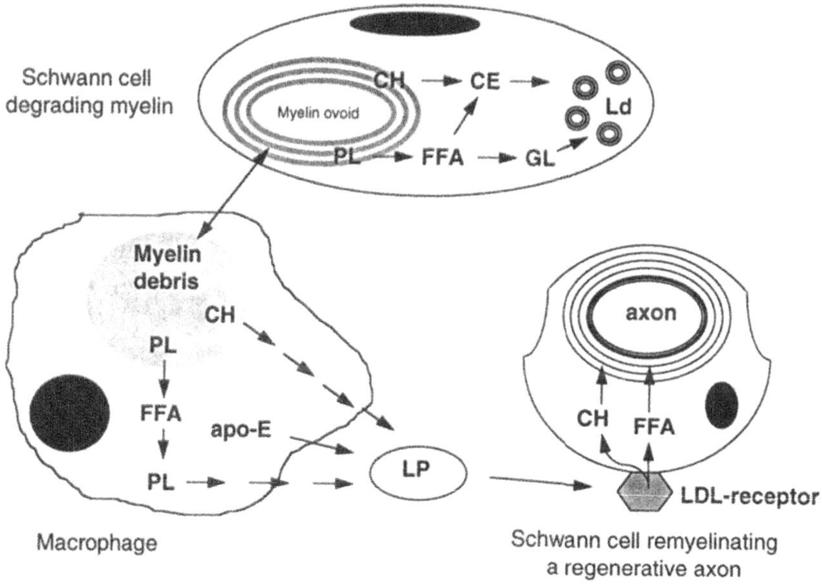

Figure 17-3. Myelin degradation and lipid product reutilization for remyelination. Myelin loses its periodicity and forms "myelin ovoids". Most of the hydrolysis of myelin phospholipids occurs within the macrophages and Schwann cells and yields free fatty acids (FFA). As much as half the FFA incorporated into phospholipids (PL) and all the cholesterol (CH) are complexed to apolipoprotein E (apo-E) to form lipoprotein particles (LP). The LP are subsequently taken up by myelinating Schwann cells via LDL receptors, and the cholesterol and fatty acyl moieties of the LP are reutilized in myelin synthesis.

Following axon interruption there is a depression of synaptic transmission, and the axotomized neuron becomes synaptically isolated by the interposition of astrocytes between presynaptic and postsynaptic membranes (Mathews and Nelson, 1975; Summer and Sunderland, 1973; Purves, 1975), although this condition is reversible if the axon regenerates and reinnervates target cells (Summer and Watson, 1971). Nerve conduction measurements during Wallerian degeneration showed a progressive loss of motor evoked responses between 3 to 7 days after injury, paralleling the disintegration of axons and myelin sheaths (Levenson and Rosenbluth, 1990).

AXONAL REGENERATION

Changes in axotomized neurons begin to reverse toward normal when the regenerating axons make peripheral connections. In as much as axotomy promotes the dedifferentiation of neurons to a growing state, the reformation of peripheral connections halts growth and promotes maturation, functional synaptic transmission and elaboration of peripheral processes.

Biology of axonal elongation: the growth cone

Ramón y Cajal (1913) first reported that a few hours after nerve section thin axonal sprouts emerged from the proximal nerve segment. The tips of these sprouts are called growth cones. Within 24 hours some growth cones arrive to the lesion site, and penetrate the fibrotic scar on about the second or third day. Some of them do not reach the distal nerve stump, because axonal growth is delayed and impeded into the scar. The growth cones that enter the distal nerve stump may continue elongation through endoneural tubes to eventually reinnervate peripheral target organs. Nerve regeneration is the result of an "ameboid" extension of growth cones. The required increase in axoplasm for this growth is the result of protein-synthetic activity in the nerve cell body. The protein thus formed passes distally to the fiber tip by axonal transport. The growing fiber tip initially lies on the surface of Schwann cells and other non-neuronal cells. The orientation of the advancing tip is subject to modifications depending on the physico-chemical characteristics of the adjacent distal milieu, neurotrophic and neurotropic factors produced by non-neuronal cells.

Structurally the growth cone shows two regions: a central domain (C-domain) and a peripheral domain (P-domain). Microtubule elongation takes place in the growth cone by addition of soluble tubulin onto the distal ends of microtubules that enter the C-domain (Baas et al., 1987; Gordon-Weeks and Mansfield, 1992; Gordon-Weeks, 1991). Growth cones may thus be considered as sites of assembly of the plasma membrane and the cytoskeleton (Bray, 1992). The P-domain shows several cytoplasmatic expansions called filopodia, filled with intermediate filaments and contractile proteins such as actin and myosin (Bridgman and Dailey, 1989; Letourneau and Shattuck, 1989; Bamburg et al., 1992). The mechanical activities of these proteins gives support to the consideration that filopodia act as "tentacles" of the growth cone, that pull or push exploring the microenvironment. The elongation of the growth cone is mediated by N-CAM and integrins that allow the adhesion of filopodia to trophic molecules of the extracellular matrix (Letourneau and Shattuck, 1989). Thus the contraction of cytoskeleton proteins promotes elongation along the Büngner bands (Bridgman, 1992). Simultaneously to the advance of growth cones, new microtubules are assembled in their C-domain.

Factors that may be involved in the control of microtubule assembly in growth cones include calcium, GTP, post-translational modification of tubulin and microtubule-associated proteins (MAPs) (Gordon-Weeks, 1991; Serrano and Avila, 1990). When the growth cone has moved on and the microtubules have become incorporated into the neurite cytoskeleton, tubulin is de-tyrosinated at its carboxy-terminal end and acetylated. These post-translational modifications correlate with an increase in the stability of the microtubules to depolymerization by cold shock and depolymerizing agents (Schulze et al., 1987). However, MAPs are still necessary for stabilizing the microtubules in a permanent cytoskeleton. The MAPs are a set of structural proteins that bind to microtubules in vitro. Two classes have been distinguished, MAP (MAP1-MAP5) and tau proteins; both classes induce polymerization of tubulin and remain bound to the newly formed microtubules. Tau proteins bind to several tubulin molecules simultaneously, thereby enhancing tubulin polymerization. MAPs act similarly, but they differ in being composed of two domains: while one binds to the microtubule, the other extends outward and may cross-link the microtubule to other cell components. All known MAPs are highly phosphorylated proteins, and thus substrates for phosphorylating enzymes, including cyclic AMP-dependent protein kinase and calcium-calmodulin type II kinase (Mathus, 1992), and growth associated proteins (GAPs), particularly GAP-43. GAP-43 expression is closely related to periods of active growth cone function. The levels of GAP-43 dramatically increase within regenerating nerves after axonal injury (Benowitz and Lewis, 1983; Skene and Willard, 1981), mainly concentrated at the cytosolic face of the plasma membrane in the growth cone (Meiri et al., 1986; Skene et al., 1986).

Growth cones also respond to neurotrophic factors secreted by Schwann cells and target organs. It has been reported that NGF induces rapid reactivation of growth cone motility and advancement. In fact, growth cone membrane shows receptors for these neurotrophic factors. Their responses can be mediated by G0 protein, a major component of the growth cone membrane that modulates several protein kinases and intracellular calcium levels (Phelan et al., 1992; Strittmatter et al., 1991).

Behavior of Schwann cells during regeneration

During Wallerian degeneration there is a marked proliferation of cellular elements in the peripheral nerve. Schwann cell multiplication starts 1-5 days after nerve severance, the peak activity occurring at approximately day 3 and thereafter slowly declining during the following weeks (Bradley and Asbury, 1970). A second wave of Schwann cell mitosis occurs during the regenerative phase. Mitosis takes place in Schwann cells adjacent to regenerating axons and precedes the formation of myelin (Pellegrino and Spencer, 1985). As they proliferate within the basement membrane of the endoneural connective tissue sheath, Schwann cells form distinct rows called bands of Büngner (fig. 17-2). The bands of Büngner are thought to be important pathways for the regenerating axons, both as guidelines and as a trophic source. The loss of axolemma induces the Schwann cell to express on its plasma membrane several receptors to neurotrophic factors and to produce and release neurotrophic molecules. These molecules bind to receptors and thereby become concentrated upon the Schwann cell surface. As regenerating axons enter the distal portion of the nerve, they are guided haptotactically along the Schwann cell substratum by the binding of neurotrophic factors to their own neurotrophic receptors. The receptors on the axolemma, by virtue of their slow-dissociation binding characteristics, can bind and retain neurotrophic molecules. The close apposition of regenerating axolemma and Schwann cell plasmalemma would facilitate the receptor-mediated, intercellular transfer of neurotrophic factors. Concurrently, axolemmal contact acts to suppress Schwann cell expression of both neurotrophic factors and their receptors. With continued elongation of the regenerating axons and concomitant Schwann cell suppression, a dynamic gradient is established, with the most highly activated Schwann cells located distally. This gradient of substrate-adsorbed neurotrophic factors maintains the proper directionality of axonal regeneration. When the neuronal fibers reinnervate target organs, Schwann cells return to a quiescent state (Taniuchi et al., 1988).

In determining the rate of axonal outgrowth several factors seem to play a role, such as the nature of the lesion, the species and the method of assessment. Values obtained after transection and suture of a nerve are always worse than those after a crush. Regeneration rates in rodent nerves fall within the range of 2-3.5 mm/day after transection and repair and 3-4 mm/day after crush (table 17-1), although in humans it seems to be slower (Lundborg, 1988). The regeneration rate appears to be similar for different types of nerve fibers after the same lesion (Verdú et al., 1995; Verdú and Navarro, 1997).

Within the first few days after transection, myelinated axons in the proximal stump produce a great number of collateral and terminal sprouts which advance distally. At an early stage "regenerating units", clusters of nonmyelinated axons, are a characteristic feature of the regenerating nerve. A single Schwann cell and its basal lamina may completely surround a great number of sprouts originating from the same myelinated fiber. During the first weeks various types of regenerating units can be observed, some of them containing several Schwann cells surrounded by a single basal lamina. With time, the average number of axons per group continuously diminishes. Each regenerating unit may be considered as the anatomical expression of the regenerative activity from a single myelinated axon. The fate of regeneration units after the first few months has been impossible to study, because of the further reorganization of the nerve trunk into a large number of miniature compartments, each surrounded by a new perineurium. Each regenerating fascicle contains a great number of axons, thus including several regenerating units at an earlier stage. The stimulus to compartmentation is probably a disturbance of the endoneurial environment resulting from damage to the perineurium. The formation of numerous miniature fascicles expresses the need for the restitution of the normal endoneurial environment around nerve fibers as quickly as possible by restoring the perineurial barrier (Lundborg, 1988).

Table 17-1. *Rates of axonal regeneration obtained in different experimental studies*

Nerve (axons)	Assessment technique	Rate (mm/day)	Lesion	Reference
Rabbit sciatic (sensory)	Pinch test	4.4	Crush	Guttmann et al. 1942
Rabbit hypoglossal and vagus (motor)	Radiotracer	3.0-4.0	Crush	Frizell and Sjöstrand, 1974
Rat sciatic (motor)	Radiotracer	3.7	Crush	Black and Lasek, 1976
Rat sciatic (motor)	Pinch test	4.5	Crush	Berenberg et al., 1977
Rat sciatic (sensory)	Pinch test	4.3	Crush	McQuarrie et al., 1977
Rat sciatic (motor)	Radiotracer	4.4	Crush	Forman and Berenberg, 1978
Rat sciatic (adrenergic)	NE uptake	3.9	Crush	McQuarrie et al., 1978
Rat sciatic (sensory)	Radiotracer	2.6	Crush	Pestronk et al., 1980
Rabbit hypoglossal (motor)	Radiotracer	3.5	Crush	Danielsen et al., 1986
Rat sciatic (motor)	Nerve conduc.	2.6	Crush	Cuppini, 1993
Mouse sciatic (sudomotor)	Silicone mold	2.3	Crush	Verdú et al., 1995
Mouse sciatic (motor)	Nerve conduc.	2.0	Crush	Verdú et al., 1995
Rabbit sciatic (sensory)	Pinch test	3.5	Section and suture	Guttmann et al., 1942
Dog peroneal (sensory)	Pinch test	2.8	Section and suture	Tsuge et al., 1975
Rat sciatic (sensory)	Pinch test	3.2	Section and suture	Forman et al., 1979
Rabbit hypoglossal (motor)	Radiotracer	2.0	Section and suture	Danielsen et al., 1986

Typical growth cones have not been seen in regenerating unmyelinated fibers in situ (Aguayo et al., 1973), although growth cones are a prominent feature of axonal growth in unmyelinated neurons in culture (Bunge, 1973). Multiple sprouts from a single injured unmyelinated fiber not only invade the distal stump but regenerate for long distances within it. This sprouting, combined with the paucity of Schwann cell proliferation, results in a marked increase in the number of unmyelinated axons per Schwann cell in electron micrographs (Aguayo et al., 1973). These axons are about half the diameter of normal unmyelinated axons.

With time the number of surviving fibers in the distal stump decrease and the proportion of fibers with diameters in the normal range increases (Bray and Aguayo, 1974).

After transection, axonal sprouts from the proximal nerve segment cross the interstump gap and reach the distal segment. Sprouts that do not take an extraneural course may either approach a Schwann cell column or may grow at random into connective tissue layers of the nerve. The Schwann cell columns are invaded by numerous sprouts, coming from parent axons in the proximal segment. Sprouts reaching a band of Büngner initially lie on the surface of the Schwann cells, but later become embedded in the Schwann cell cytoplasm. The basal lamina originally surrounding the Schwann cell column may rupture and disappear as the diameter of the regenerating fibers increases. Since an excess number of sprouts invade the distal Schwann cell columns, the total number of axons in the distal segment may initially considerably exceed the number in the same nerve proximal to the lesion. With time the number of distal axons decreases, as sprouts which do not make peripheral connections atrophy and disappear. The axons linking up the periphery mature and increase in diameter (Lundborg, 1988).

Changes in phenotypic expression

Peripheral nerve regeneration after experimental lesions is a well-characterized paradigm that closely mimics development in terms of the reciprocal regulation of early versus late Schwann cell genes. In particular, the expression of myelin markers decreases dramatically as a consequence of the axonal degeneration that occurs distal to the injury site, whereas markers of immature and nonmyelinating Schwann cells are readquired (Scherer and Salzer, 1996; Scherer et al., 1994). Schwann cells that undergo division as a result of nerve injury possess a multipotentiality similar to that expressed during normal development. In crushed nerves, axonal degeneration and Schwann cell dedifferentiation are followed by axonal regeneration and Schwann cell redifferentiation, which occur in a proximo-to-distal direction. Myelinating and unmyelinating Schwann cells are distinguished by a number of morphological and molecular properties (Lemke, 1992; Mirsky and Jessen, 1996). The former associate with axons in a 1:1 relationship, elaborate the myelin sheath, and express myelin-specific genes such as those encoding myelin protein zero (P0) and myelin basic protein (MBP). In contrast, nonmyelinating Schwann cells are associated with several small diameter axons and express markers such as p75 low-affinity nerve growth factor receptor (LNGF-R) and glial fibrillary acidic protein (GFAP).

During degeneration and regeneration, Schwann cells show a number of changes on the surface and within the cytoplasm. Intracellularly, intermediate filament protein expression changes, as would be anticipated considering the modification in cell shape during proliferative and reparative phases. Two intermediate filament proteins, vimentin and GFAP, are normally expressed in myelinating and nonmyelinating cells respectively. Upon loss of axonal contact, vimentin expression increases in the distal segment and becomes predominant, while GFAP decreases. This reverses with axonal regrowth and reestablishment of Schwann cell-axonal contact (Gilmore and Sims, 1995). Schwann cells are instructed by signals from the advancing axons to develop into myelinated or unmyelinated fibers. Schwann cells of the Büngner bands acquire a new basal lamina replacing the old disrupted one. Schwann cells instructed to form myelinated fibers surround the corresponding axons individually, developing short lengths of myelin (internodes) along the regenerated nerve. The regenerating axon of an originally myelinated neuron instructs the Schwann cells of the distal segment to form myelin, while unmyelinated fibers remain unmyelinated even when regenerating into a distal nerve originally containing myelinated fibers. Apparently each neuron is coded to be myelinated or nonmyelinated, the instruction to the Schwann cell to form myelin requiring a direct axon-Schwann cell contact (Politis and Spencer, 1981). However, denervated Schwann cells do not persist if the regenerating neurites are prevented from entering the distal segment of an injured nerve (Weinberg and Spencer, 1978).

Neurotrophic factors

In physiological conditions, peripheral tissues and non-neuronal cells synthesize and secrete several neurotrophic factors, needed for maintenance of the homeostatic state of intact neurons. Schwann cells and several peripheral tissues (muscle, skin, etc.) have a basal expression of neurotrophic factors, that play biological functions in promoting survival and maintenance of sensory, motor and autonomic peripheral neurons (Raivich and Kreutzberg, 1994; Lewin and Barde, 1996). After nerve injury the expression of neurotrophins increases, in an attempt to support the survival and growth of regenerating axons and reinnervation of denervated target organs. There are several families of neurotrophic factors secreted by non-neuronal cells:

-*Neurotrophins*: this family includes nerve growth factor (NGF), brain-derived neurotrophic factor (BDNF), and several neurotrophins (NT-3, NT-4 and NT-5). NGF is required for the survival of nociceptive and sympathetic neurons, while BDNF and NT-3 are required for the survival of neurons innervating mechanoreceptors and propioceptors (Lewin and Barde, 1996). There are two different receptor types for neurotrophins. The first type, named tyrosine kinase receptor (trk) is a high-affinity receptor specific for one neurotrophic factor. In mammals, three different trk receptors have been identified: trk-A which binds NGF, trk-B which binds BDNF and trk-C which preferentially binds NT-3 (Rodríguez-Tebar and Barde, 1988; Sutter et al., 1979). All three trk receptors are single-chain glycoproteins of 140-150 kD molecular weight, with an intracellular tyrosine kinase domain. Binding of the neurotrophin to its specific receptor induces receptor dimerization, activation of tyrosine kinase and cross-phosphorylation of the juxtaposed receptor molecule, followed by an intracellular signal cascade (Jing et al., 1992). The second type of neurotrophin receptor is a single 75 kD glycoprotein (p75), which exhibits low-affinity neurotrophin binding, fast to very fast dissociation characteristics, and a lack of specificity on the basis of affinity for different neurotrophins (Rodríguez-Tebar et al., 1990, 1992). The binding of neurotrophins to p75 receptor induces an increase of second messengers mediated by G-proteins (Feinstein and Larhammer, 1990). Following axotomy, neurotrophin receptors are rapidly upregulated on the distal portion of the nerve, preferentially on denervated Schwann cells and growth cones of regenerating axons (Raivich and Kreutzberg, 1994).

-*Epidermal growth factors* (EGFs): the main factors of this family in the PNS are epidermal growth factor (EGF) and schwannoma-derived growth factor (SDGF). Both are synthesized and secreted by regenerative neurons and required for the proliferation of Schwann cells and fibroblasts (Raivich and Kreutzberg, 1994). The receptor for EGF is a 170 kD single-chain glycoprotein, which became a classical model of tyrosine kinase receptors (Ullrich and Schlessinger, 1990; Yarden and Schlessinger, 1987). It contains an extracellular, glycosylated portion involved in ligand binding, a transmembrane domain and a cytoplasmic tyrosine kinase. Appropriate ligand binding induces dimerization, cross-phospho-rylation of the adjacent receptor, followed by phosphorylation of a number of cytoplasmic signal transduction molecules including phospholipase-C, MAP-kinase and GTPase-activating protein (GAP) (Prigent and Lemoine, 1992). The receptor is located in Schwann cells, fibroblasts and regenerating neurons (Raivich and Kreutzberg, 1994).

-*Neuregulins*: are a large number of glycoproteins similar to glial growth factor (GGF), neu-differentiating factor (NDF) and heregulins. These factors promote Schwann cell proliferation and maturation when regenerative axons have reinnervated target organs (Falls et al., 1993; Raivich and Kreutzberg, 1994). The neuregulin receptor is similar to the EGF receptor (Raivich and Kreutzberg, 1994).

-*Insulin-like growth factors* (IGFs): IGF-1 and IGF-2 are two relatively small polypeptides structurally related to proinsulin. Both factors are synthesized by target organs, particularly by muscle cells, and induce neuronal regeneration and Schwann cell proliferation (Kanje et al., 1989; Near et al., 1992; Raivich and Kreutzberg, 1994). Most of the IGF actions are mediated by the high-affinity IGF receptor type I, where IGF-1 binds with high affinity. IGF-1 also binds with low affinity to the insulin receptor. Both receptors are closely related glycoproteins consisting of two alpha and two beta subunits, cross-linked by disulfide bridges. The alpha subunits are extracellular glycoproteins mediating specific ligand binding, while the beta subunits consist of a short extracellular stretch and transmembrane and intracellular domains. The intracellular portion contains a tyrosine kinase domain. The type II IGF receptor presents only a short cytoplasmic portion, without tyrosine kinase activity, but IGF-2 binding to this site can lead to calcium influx (Nishimoto et al., 1987; Raivich and Kreutzberg, 1994).

-Platelet-derived growth factors (PDGFs): are potent fibroblast and Schwann cell mitogens. PDGF-AA, PDGF-BB and PDGF-AB are synthesized and secreted by infiltrated macrophages, and may induce the rapid increase of endoneural fibroblasts, corresponding to the spatial and temporal pattern of injury-induced breakdown of the blood-nerve barrier, in the axotomized nerve (Raivich and Kreutzberg, 1987). There are two types of PDGF receptor, but only the beta receptor is expressed in the PNS, localized in Schwann cells and fibroblasts. These receptors are closely related single-chain glycoproteins with intrinsic tyrosine kinase and ligand binding leading to the dimerization and activation of tyrosine kinase (Raivich and Kreutzberg, 1994).

Extracellular matrix components

Nerve regeneration is mediated by cell-to-cell and cell-to-extracellular matrix interactions. The majority of extracellular matrix (ECM) molecules serve as guiding cues to regenerating axons for elongation until their target organs. These molecules may be considered as neurotropic factors. The main constituents identified in the ECM are fibronectin (types I, II and III), laminin, vitronectin, collagen (types I, II, III and IV), proteoglycans and C-tenascin. The ECM has also been shown to bind to and regulate the activity and stability of several growth factors, most notably basic-fibroblast growth factor (FGF) and transforming growth factor (TGF-beta) (Reichardt and Tomaselli, 1991, Letourneau et al., 1994).

Integrin receptors are heterodimeric glycoprotein complexes expressed on the surfaces of neurons and non-neuronal cells, that appear to be the primary cellular receptors for many ECM constituents. These receptors are composed of one alpha chain noncovalently associated with a beta subunit. In addition, neuronal cells also express on their surface other receptors for ECM constituents, as laminin, tenascin, hyaluronic acid and proteoglycan receptors (Reichardt and Tomaselli, 1991). After a nerve lesion, the expression of integrins increases in neurons and Schwann cells. However, specific integrins are expressed in particular subpopulations of neurons, suggesting that different ECM constituents have biological effects on selective neurons (Letourneau et al., 1994). Nerve injury also increases the expression of laminin, tenascin, collagen, hyaluronic acid, chondroitin-sulfate and fibronectin receptors. Immunocytochemical studies showed that all these receptors are located in Schwann cells and some (collagen type I and III) in endoneural fibroblasts (Siironen et al., 1992; Battisti et al., 1992; Tona et al., 1993; Martini, 1994; Fruttiger et al., 1995; Braunewell et al., 1995). Knowledge of the mechanisms by which ECM components exert their effects is still sketchy; however, in the case of integrin-mediated interactions with the ECM, a picture has begun to emerge. Upon binding of ligands, integrins appear to transduce signals across the plasma membrane in two ways: by interacting directly with cytoskeletal proteins, and by regulating the levels of second messengers (Reichardt and Tomaselli, 1991; Strittmatter et al., 1991).

Finally, it is well known that ECM acts as a reservoir of FGF and mitogenic agents for Schwann cells. Within a few days after a nerve lesion, ECM components contribute to the survival and maintenance of axotomized neurons; ECM molecules are probably the first survival factors acting on neurons, before neurotrophic factors (Reichardt and Tomaselli, 1991). In addition, the basal lamina components are necessary for Schwann cell migration and re-myelination of regenerated axons (Eldridge et al., 1989).

GENETIC CONTROL OF INJURED NERVES

The morphological, biochemical and physiological changes observed in injured nerves are the results of changes in the gene expression of neurons and Schwann cells (fig. 17-4). In neuronal cells, these changes promote the reorganization of cell priorities, in terms of the enzymes and cofactors required to achieve an effective and continued growth of the axonal appendage, while in Schwann cells they allow for the proliferation and secretion of several factors that induce axonal regeneration.

Neuronal control

There are two types of genes in neuronal cells: genes whose transcription is activated rapidly and transiently within minutes of stimulation, termed as cellular immediate early genes (IEGs), and late response genes (LRGs), whose expression is induced (or repressed) more slowly, over a time frame of hours, through a mechanism that is generally dependent on new protein synthesis. IEGs encode regulatory proteins that control the expression of LRGs, while proteins with specific functions in neuronal responses are encoded by LRGs (Sheng and Greeberg, 1990).

In general, IEGs share the following characteristics: i) their expression is low or undetectable in quiescent cells, but is rapidly induced at the transcription level within minutes of stimulation; ii) this transcriptional induction is transient and independent of new protein synthesis, and iii) the mRNAs transcribed from these genes often have a very short half-time (Sheng and Greeberg, 1990).

Axotomy induces changes in neuronal IEGs, although with several differences in their dynamic expression depending on the type of nerve and the type of injury (Leah et al., 1991; Herdegen et al., 1992; Haas et al., 1993). IEGs may be activated by an influx of Ca^{2+} in neuronal cells (Greenberg et al., 1986; Morgan and Curran, 1986). During a few seconds after a nerve lesion, the axoplasm of axotomized axons is in continuity with the extracellular medium (enriched with Ca^{2+}), before the plasmatic membrane is sealed. This influx of Ca^{2+} ions causes, together with the suppression of retrograde transport of neurotrophic factors, the axonal reaction and the activation of IEGs (Lunn et al., 1990; Haas et al., 1993).

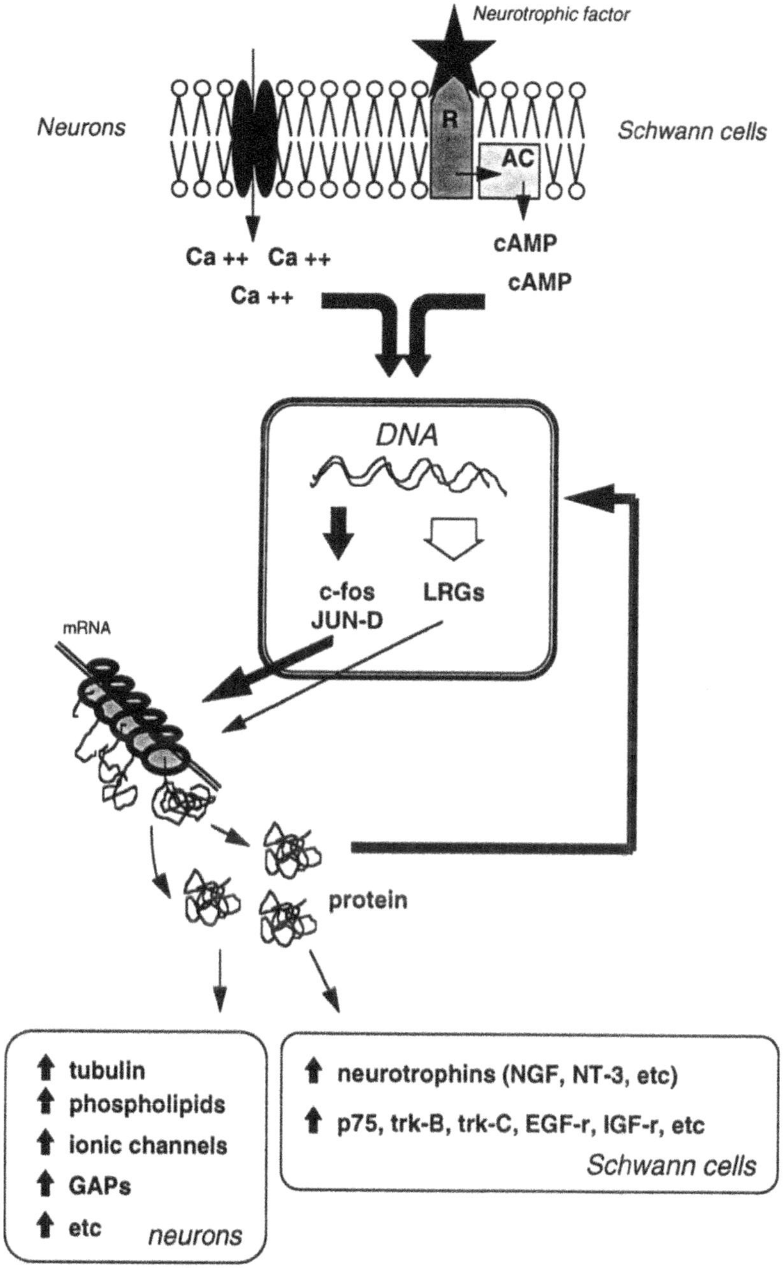

Figure 17-4. Changes in the gene response and phenotypic expression induced by nerve injury in neurons and Schwann cells. Neuronal immediate early genes (IEGs) are activated by an influx of Ca^{2+}, whereas Schwann cell IEGs are activated by mitogenic factors that raise intracellular cAMP levels. IEGs encode regulatory proteins that control the expression of late response genes (LRGs), while proteins with specific functions in neuronal and Schwann cell responses are encoded by LRGs.

Schwann cell control

Nerve injury also induces the activation of IEGs in Schwann cells. It has been reported that from 15 to 60 minutes after injury there is an increase of mRNA for c-fos gene in Schwann cells of the lesioned nerve and its levels remain elevated for some days after injury (Pyykonen and Koistinaho, 1991; Plan-tinga et al., 1994). In this case, IEGs are activated by mitogenic factors liberated by infiltrating macrophages, that raise intra-cellular cAMP levels of activated Schwann cells (Sheng and Greenberg, 1990). The IEGs cause an upregulation of several LRGs that encode for several proteins, such as NGF, p75 receptor, BDNF, glial maturation factor (GMF), N-CAM proteins, glial derived nexin (GDN), laminin, T3-receptors and GAPs (Bolin and Shooter, 1993; Barakat, 1993; Raffroni and Bradshaw, 1993), and a downregulation of other LRGs that encode for myelin, CNTF and GFAP (Sendtner et al., 1992). The coordinated changes of several LRGs induce Schwann cell activation during nerve degeneration and regeneration.

PERIPHERAL REINNERVATION AND NEURAL MATURATION

The final purpose of nerve regeneration is to restore appropriate axon-target cell reconnections and thus to recover the functions lost after injury. The interactions of axonal terminals with receptive target cells and new changes in Schwann cell behavior account for the processes underlying functional recovery.

Re-establishment of synaptogenesis

Peripheral reinnervation is mediated by biological factors or signals between regenerative axons and peripheral tissues. Muscle reinnervation is a typical example; after axotomy, there is an increase in the synthesis and secretion of IGF by denervated muscle cells, and IGF has attractive effects upon motoneuron axons. When these axons reinnervate the muscle, the establishment of new neuron-myocyte contacts causes a decrease of IGF synthesis (Glazner and Ishii, 1995). On another hand, agrin also plays an important role in the synaptogenesis of muscle cells. Agrin is a polypeptide produced by motoneurons, that causes the reorganization and aggregation of acetyl choline-receptors in the sarcoplasm where neuromuscular junction will be reformed (Bowe and Fallon, 1995).

In addition, muscle cells also secrete agrin at the synaptic site. Muscle reinnervation is thus induced by both trophic factors and membrane proteins.

Peripheral reinnervation determines important changes in the regenerated neurons, as they have to change their growth cones to presynaptic buttons. This transformation is coincidental with GAP-43 phosphorylation (Skene, 1989), that causes inhibition of several biochemical reactions leading to inactivation of protein-kinases of the growth cone and inhibition of GAP-43 binding to calmodulin. Both mechanisms obstruct polymerization of new microtubules and stop the advance of the regenerative axon (Strittmatter et al., 1991). During this transformation, there is an important increase in axonal transport of vesicles that fuse with growth cone membrane vesicles filled with neurotransmitters and mitochondria (Kelly and Grote, 1993).

The expression and repression of neuronal and Schwann cell genes lie under these structural and metabolical changes. When peripheral fibers reinnervate target organs, synaptogenesis causes a rapid up-regulation of the levels of mRNAs encoding myelin-specific proteins and lipids (Johnson et al., 1988) and neurotrophins and CNTF (Sendtner et al., 1992), but also a down-regulation of the expression of p75NGF receptor and NGF genes (Johnson et al., 1988). The reestablishment of Schwann cell-axonal contact in the distal segment of the injured nerve results in a rapid increase of cAMP levels in Schwann cells, which may induce these genic changes (LeBlanc and Podulso, 1990; LeBlanc et al., 1992). SCIP (suppressed cAMP-inducible POU) is a transcription factor activated by variations in intracellular levels of cAMP, that mediates phenotypic changes of Schwann cells that produce axonal myelination (Scherer et al., 1994).

Role of Schwann cells

Schwann cells also play a role in the reinnervation of target organs. Schwann cells of the non-myelinating variety, or terminal Schwann cells, cover the neuromuscular junctions and seem to participate in the maintenance and repair of neuromuscular synapses. Terminal Schwann cells are able to sense synaptic transmission occurring at the junction. During stimulation of the motor nerve there are transmembrane Ca^{2+} fluxes in these Schwann cells (Jahromi et al., 1992; Reist and Smith, 1992) through ligand- and voltage-gated channels (Robitaille, 1995). Terminal Schwann cells also change their gene expression in response to perturbations of synaptic transmission. During transmission block with tetrodotoxine, Schwann cells rapidly upregulate the expression of GFAP; while the application of muscarinic agonists prevents this upregulation (Georgiou et al., 1994). In response to muscle denervation by transection of its nerve, terminal Schwann cells grow to surround fragments of the degenerating nerve endings, and begin to synthesize and release acetylcholine, detectable as miniature endplate potentials in the muscle fibers (Brockes, 1984). In addition, terminal Schwann cells extend processes that grow through the muscle, and may act as an excellent guide for regenerating axons during reinnervation (fig. 17-5) (Reynolds and Woolf, 1992). Some studies suggest that intramuscular nerve sprouts are invariably associated with these Schwann cell processes (Duvovy and Svizenska, 1994).

Finally, Schwann cells also appear to participate in the disruption of neuromuscular junctions during degenerative lesions; local injection of myotoxic drugs causing muscle degeneration promotes the appearance of Schwann cell processes that separate the nerve terminals from injured muscle fibers (Jirmanová, 1975).

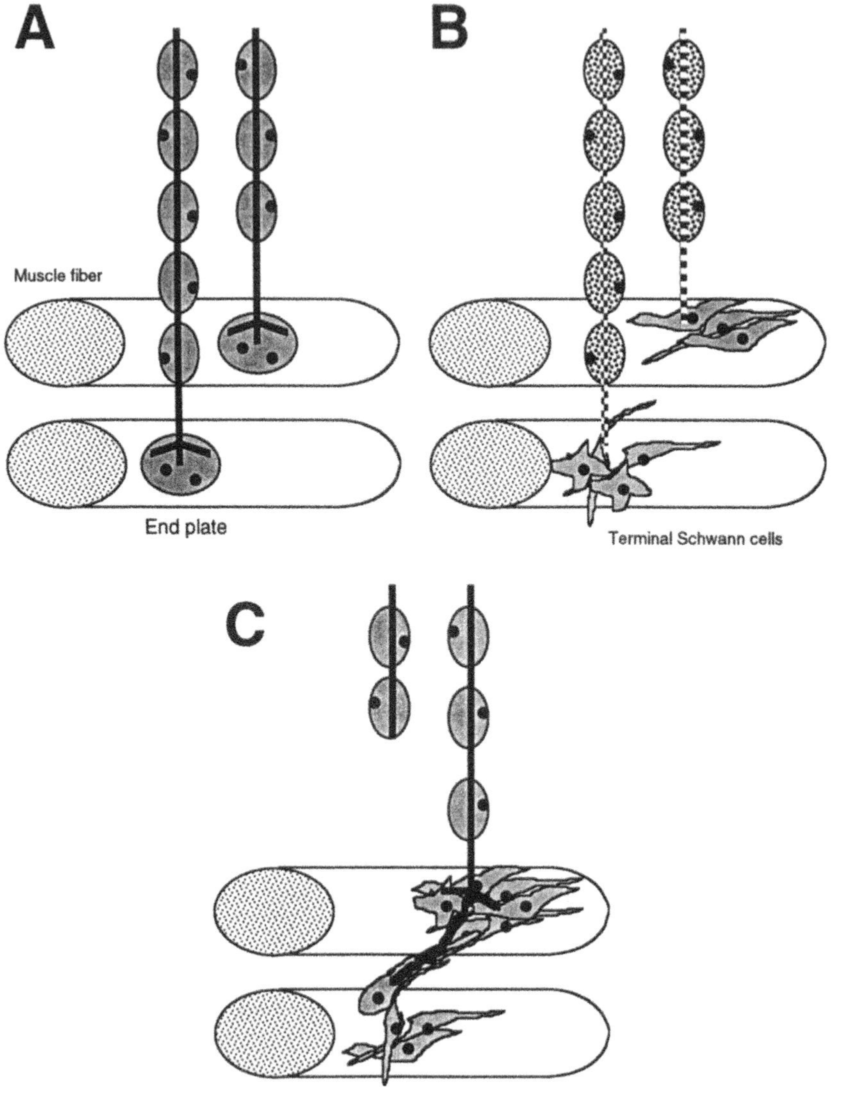

Figure 17-5. Schematic representation of the role of terminal Schwann cells in guiding muscle reinnervation by regenerating axons. (A) The normal innervation of two muscle fibers by two myelinated motor axons each forming a neuromuscular junction. Terminal Schwann cells cover the neuromuscular junction. (B) After nerve lesion, the axons degenerate distally to the lesion site. The terminal Schwann cells begin to extend processes over the muscle fibers. (C) The Schwann cells form a bridge across the vacant junctions and one axon regenerates across this site. Processes of the terminal Schwann cells have extended from each endplate and interconnect the two endplates. The regenerating axon has reinnervated the first fiber and extends a terminal branch by growing onto the Schwann cell processes.

Maturation of regenerated nerves

The last step involved in nerve regeneration consists of the maturation of the regenerated axon and its terminal connections. Maturation includes progressive changes leading to the restoration of normal fiber diameter and normal electrical properties of the nerve. During this step myelin sheath thickness and axonal diameter increase, supported by metabolic changes that sustain an increased synthesis of myelin lipids and proteins and cytoskeletal proteins in Schwann cells and axons (Guth, 1956; Sunderland, 1991). However, complete maturation of the regenerated fibers is rarely achieved. Several studies have shown that axons regenerating across a gap or through suture or graft repairs never achieve their normal size (Fields and Ellisman, 1986a; LeBeau et al., 1988; Giannini et al., 1989; Archibald et al., 1995; Gómez et al., 1996). The mean fiber and axon diameters were only about 75% of those in the normal sciatic nerve, even at long term when the functional recovery achieved was stabilized (Butí et al., 1996; Gómez et al., 1996). The reduction in size of mature regenerated fibers may reflect the failure of axons to appropriately reinnervate target tissue and re-establish functional connections. Regenerated fibers that remain disconnected from the periphery remain small in diameter and eventually atrophy and die-back (Sanders and Young, 1944; Weiss et al., 1945; Giannini et al., 1998). With more prolonged time of regeneration, axonal diameter, g ratio and internodal length may achieve closer to normal values (Fields and Ellisman, 1986a; Archibald et al., 1995).

Morphological maturation is accompanied by improving functional restitution, with a progressive recovery of nerve conduction velocity and synaptic transmission efficiency. However, the conduction velocity of the regenerated nerve also remains below normal values for a long time after lesion (Sanders and Whitterbridge, 1946; Fields and Ellisman, 1986b). From clinical and experimental studies, it is known that motor, sensory and autonomic functions do not improve to the point of perfect restitution, except may be after lesions occurring in childhood, and that considerable variability may appear between different types of function (Guth, 1956; Buchthal and Kühl, 1979; Dyck et al., 1988; Navarro et al., 1994). There are interesting qualitative differences between the functions conveyed by different types of nerve fibers; large motor and sensory axons have comparatively lower reinnervating capabilities than sudomotor or nociceptive axons, mainly when the nerve lesion is severe. After peripheral nerve injuries, small nociceptive and sympathetic nerve fibers achieve a faster and higher degree of reinnervation than large myelinated motor and sensory fibers, the differences being more marked with the severity of the lesion (Navarro et al., 1994; Verdú et al., 1995). This is despite the findings that unmyelinated and myelinated fibers have a similar regeneration rate and are seen by immunohistochemical techniques to arrive by the same time to the denervated skin (Navarro et al., 1997). There are also important differences in sprouting capabilities between nerve fibers; small axons are able to sprout and reinnervate larger territories than large myelinated ones. After partial denervation of the hindpaw, unmyelinated sudomotor axons of intact nerves collaterally branch to reinnervate sweat glands that are outside the skin area supplied by their nerve of origin (Kennedy and Sakuta, 1984), whereas sprouts from alphamotor axons do not reinnervate adjacent denervated muscles and branch only within a restricted area in their own muscle (Brown et al., 1980; Torigoe, 1985). The same differences have been sustained for large and small sensory nerve fibers (Devor et al., 1979; Jackson and Diamond, 1984). These differences in reinnervating ability seem closely related to the functions conveyed by the axon types. Presumably the constraints on collateral and terminal sprouting of large nerve fibers to low-threshold mechanoreceptors and to muscle exist to preserve the precise spatial organization of innervation necessary to retain the correct discrimination and coordination of motion. On the other hand, the warning aspects of nociceptive and thermal sensitivity and the thermoregulatory sweating can be well accomplished in a more diffuse manner, without precise topographical innervation.

STRATEGIES TO IMPROVE PERIPHERAL NERVE REGENERATION

After peripheral nerve injuries, the capability of severed axons to regenerate and recover functional connections is dependent on the site and type of lesion and the distance over which axons must regrow to span the injury area. After nerve crush, regeneration is usually successful because the continuity of the endoneurial tubes, through which axons regrow, is preserved (Sunderland, 1991). This is in contrast to the limited growth across a gap imposed after complete severance or resection of a nerve, when endoneurial continuity is completely disrupted and the neural environment must be reconstituted by regenerating elements. Various methods of surgical repair have been used in attempts to improve regeneration after peripheral nerve lesion.

Surgical repair

The classic and still preferred method involves the mobilization of the nerve to overcome any gap created between the sectioned nerve stumps, followed by epineurial suturing of the proximal and distal stumps (Kline, 1990). The advantage of the epineurial suture technique is its simplicity and the minimal surgical trauma required. However, the technique does not ensure an absolutely correct matching of the fascicular structures over the nerve trunk. It was demonstrated that the epineurial suture may cause misalignment and considerable displacement of fascicles in spite of a perfect superficial appearance of the epineurial adaptation (Edshage, 1964). Fascicular suture repair may be an alternative method. The objective of fascicular repair is to achieve an optimal orientation by approximating and adapting fascicles or groups of fascicles individually. The repair does not resist much tension, and can therefore usually be carried out only as a primary procedure when no resection is required. Fascicular repair requires optical magnification and can be performed only by a skilled and experimented microsurgeon (Lundborg, 1988, Kline, 1990). Moreover, fascicular repair is difficult in lesions involving distal nerves of the limbs, where fasciculation of the nerve is extensive and intermingled (Sunderland, 1991).

When the length of the gap created by tissue destruction and nerve retraction is too long to allow apposition and suture without tension, nerve grafts are usually employed. If left unrepaired, the regenerative sprouting of fibers in the proximal stump forms a neuroma, as the growing axons do not find a favorable terrain to elongate through the extraneural ambiance to reach the distal stump. The purpose of introducing a graft between the two ends of a transected nerve is to offer a mechanical guideline as well as an optimal environment for the advancing axonal sprouts. In this respect the Schwann cells of the graft and their basal lamina play an essential role in promoting neurite growth. If a thin nerve graft is placed in a healthy well-vascularized bed, it will survive and will be able to fulfill these purposes. It has been demonstrated by isotope techniques that most transplanted Schwann cells in such a situation survive, multiply, form Büngner bands and remain confined to the grafted segments (Aguayo, 1981). During the first days the graft survives by diffusion from the surrounding tissues; it is then rapidly revascularized, starting on the third postoperative day (Almgren, 1974). Thick and long grafts have more difficulties in surviving because of longer diffusion distances and delayed revascularization (Lundborg, 1988). The nerve graft may come from the same individual (autograft), from another individual of the same species (homograft) or from a different species (heterograft) (Sunderland, 1978). Nerve homografts have been used in the fresh state and after storage, and their usefulness for bridging gaps has been tested both experimentally and clinically. Experimentally, the reaction to the implant is subject to considerable variation. The homograft may retain its specific features for only a few days, but thereafter it is removed by a combination of necrosis and macrophage activity. Myelin fragmentation, invasion of the graft by lymphocytes and macrophages and patchy necrosis have been well documented. Vascular breakdown, destruction of the elements of the graft and massive invasion of lymphocytes are all evidence that the grafted foreign tissue provokes an active immune reaction from the host, and immunosuppressive therapy is required. Clinically the use of homografts has not been promising or encouraging. The histological appearance of such grafts examined a considerable time after implantation is one of fibrosis, in which all traces of nerve structure have been lost and there is no evidence of innervation. The poor results are clearly due to the specific tissue immunity reaction (Sunderland, 1978). Nerve tissue heterografting has been also tried, both experimentally and clinically, with results that have been consistently disappointing. Histological studies showed that the graft is rapidly invaded by macrophages and soon becomes the site of a marked cellular reaction. At later times it is not revascularized, undergoes a patchy necrosis, and becomes encased in fibrous tissue. On both homo- and heterografts, functional recovery is low or absent (Sunderland, 1978). Immunosuppressive therapy is needed for reducing the graft rejection, but its secondary complications usually overcome the benefits (Evans et al., 1994).

It is generally agreed that gaps in injured nerves are most successfully bridged by the use of autologous nerve grafts. There is a considerable experimental evidence that, after implantation, nerve autografts behave in much the same way as does the distal segment of a severed nerve. Wallerian degeneration occurs somewhat more slowly in the graft than in the distal stump, but autografts retain their essential structure for considerable periods and do not show immunoreactivity. Experimental evidence also indicates satisfactory reinnervation of such grafts and the rapid growth of axons through them (Sunderland, 1978; Pover and Lisney, 1989; Butí et al., 1992; Gómez et al., 1996). However, clinical results are usually poor, with only 25% of cases achieving significant motor recovery after graft repair of nerves injured at the distal forearm (Young et al., 1980). Moreover, autograft repair implies some problems such as the need for a second surgical step, elimination of the donor nerve function, a limited supply of donor nerves, and the mismatch between nerve and graft dimensions.

Tubulization repair

Tubulization, the implantation of nerve guides as sleeves or tubes to bridge gaps, is an alternative to the repair of transected nerves. This experimental paradigm has provided an useful model for studying the cellular and biochemical events during peripheral nerve regeneration (Williams et al., 1983; LeBeau et al., 1988; Liu, 1996), but it has been also applied as a method for repairing lesioned nerves in experimental (Ducker and Hayes, 1968; Fields et al., 1989; Butí et al., 1992, 1996) and clinical settings (Lundborg et al., 1997). The spatial-temporal progress of peripheral nerve regeneration within these neural chambers is well known (Williams et al., 1983, Liu, 1996). Histological studies have shown that during the first week after tube implantation in the transected sciatic nerve, a loosely organized matrix which displayed numerous fibrin strands appeared along the chamber. Red blood cells were common both within and around the matrix with irregular distribution. Fibroblasts and Schwann cells were consistently observed in the most proximal and distal sites, indicating that migration of these cells into the matrix proceeded from both nerve stumps. Few nonmyelinated axons were observed in the proximal site of some chambers. By 2 weeks, migrating fibroblasts and Schwann cells had advanced toward the midpoint, with Schwann cells organized in columns. At the proximal site axons were already found, showing the typical appearance of regenerating units. The majority of these axons were nonmyelinated an very few myelinated. By 3 weeks, nonneuronal cells had completely bridged the chamber, neoformed vessels were also observed along the tube, and nonmyelinated axons were seen distally, while myelinated axons arrived at the midpoint of the tube. Finally, by the fourth week, axons had regenerated across the whole chamber. Myelinated axons advanced up to the end of the chamber length, and their myelin sheath thickness was noticeably greater in proximal sites. With longer time, the regenerated nerve increases in caliber (fig. 17-6). The new nerve cable is usually centered in the tube lumen, composed of small nerve fascicles with myelinated and unmyelinated axons, and surrounded by a thin perineurium (fig. 17-7).

Furthermore, biochemical and biological studies showed that the fluid collected from synthetic chambers one week after implantation contains neurotrophic activities for several neuronal types, including sensory, motor and sympathetic neurons. These trophic activities are in association with proteins present in the fluid (Longo et al., 1983; LeBeau et al., 1988), secreted by infiltrated Schwann cells and other non-neuronal cells. In addition, immunohistological studies have demonstrated that the initial fibrin matrix, formed during the first week following tube implantation, contains fibronectin, collagen and laminin (Longo et al., 1984), ECM components that promote axonal regeneration and Schwann cell migration.

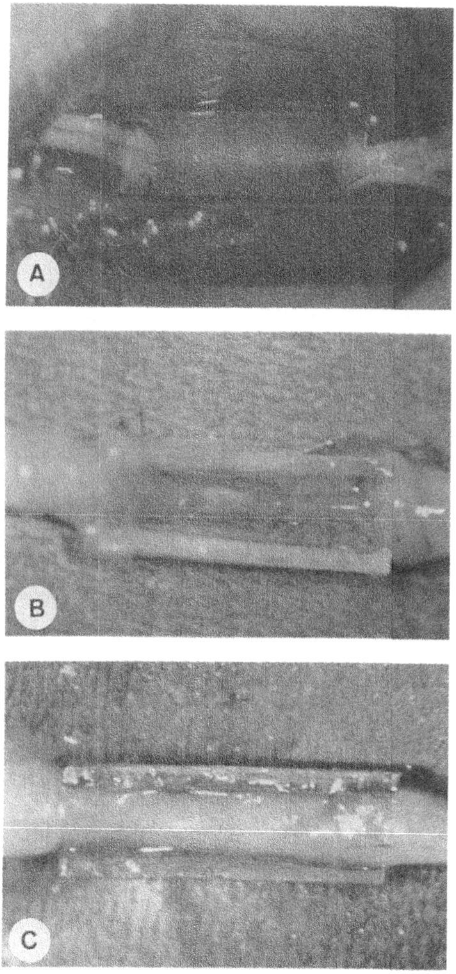

Figure 17-6. Photographs of (A) a silicone tube implanted to repair a 4 mm gap in the mouse sciatic nerve, (B) the thin nerve cable reconstituted 14 days after implantation, and (C) the regenerated nerve 6 months after implantation. The proximal nerve stump is at the left side.

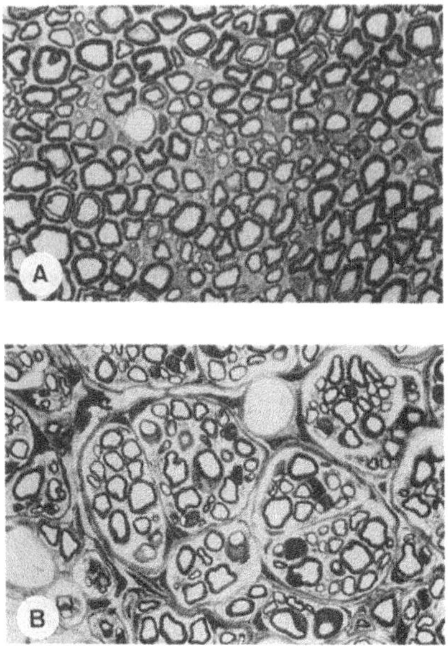

Figure 17-7. Light micrographs of cross-sections of the mouse sciatic nerve regenerated through (A) an autologous nerve graft and (B) through a silicone tube, 3 months after re-section and repair. Final magnification x590.

The ECM molecules contain an adhesive tripeptide (Arg-Gly-Asp) sequence which is recognized by integrin receptors on responsive cells (Ruoslahti and Pierschbacher, 1987), such as endothelial and Schwann cells. After a few days postlesion, when the fibrin cable is already populated by ingrowing cells, fibrin is lysed and in its place collagen fibrils appear amidst a reconstituted ECM that condenses around Schwann cells to form a basal lamina (Liu, 1992). Regenerating axons are enclosed within the Schwann cell cytoplasm or adhere to the Schwann cell surface enveloped by its basal lamina.

There are several theoretical advantages of tubulization versus other nerve repair methods. Tube repair limits the ingrowth of fibroblasts into the repair site and thus avoids excessive collagen and scar formation, provides directional guidance for the regenerating axons and prevents axonal escape into the surrounding tissues. In addition, trophic factors from the injured nerve may become concentrated within the lumen of the tube and serve to facilitate axonal growth (Lundborg et al., 1982). Accordingly, tubulization allows successful regeneration over longer gaps than in unrepaired nerves. However, a limit to regeneration also exists within nerve chambers depending upon the length of the gap. Regenerating axons are able to bridge empty chambers implanted in the rat sciatic nerve if the gap is 10 mm or less, but fails in most cases with 15 mm gaps (Lundborg et al., 1982; Williams et al., 1983); in the mouse, regeneration is successful in mostinstances with a 4 mm gap, but does not occur with gaps 6 mm or longer within synthetic tubes (Henry et al., 1985; Madison et al., 1987; Butí et al., 1996; Gómez et al., 1996). On the other hand, the number of axons regenerating through nerve guides bridging gaps below the limit length for regeneration has been shown to be comparable to that obtained after neurorraphy or nerve graft (Archibald et al., 1995; Lundborg et al., 1982; Rosen et al., 1989).

Functional reinnervation of distal target organs may be similar with short gaps (Archibald et al., 1995; Dellon and Mackinnon, 1988; Navarro et al., 1994), but it is usually lower with tube repair than with autografts when applied to bridge medium or long gaps (fig. 17-8) (Fields and Ellisman, 1986a; Rosen et al., 1989; Navarro and Kennedy, 1991).

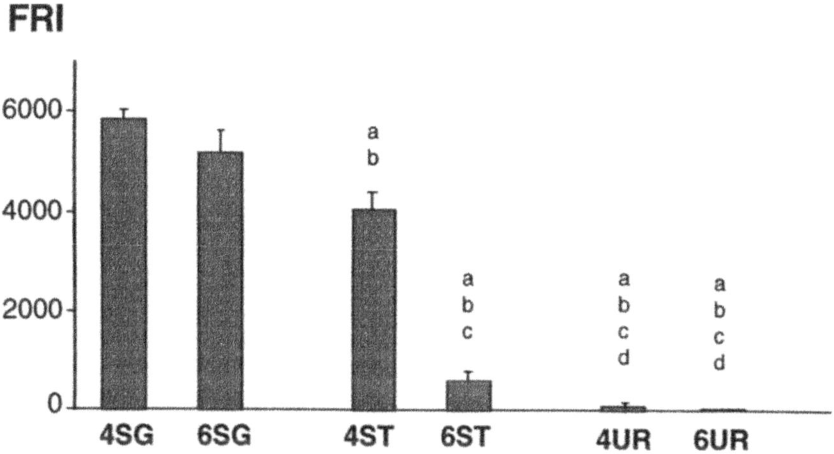

Figure 17-8. Histograms representing the functional recovery index (FRI), that averages the degree of motor, sensory and sudomotor functional reinnervation, achieved in groups of mice after resection of the sciatic nerve leaving 4 or 6 mm gaps, repaired by an autologous graft (SG) or a silicone tube (ST) or left unrepaired (UR). $p<0.05$ a vs 4SG, b vs 6SG, c vs 4ST, d vs 6ST.

Moreover, the physicochemical characteristics of the tube wall, mainly in terms of permeability, durability and cell adhesiveness, also influence the chances of regeneration through long gaps (Madison et al., 1987; Aebischer et al., 1988; Navarro et al., 1996), and further advances in the production of tubular biomaterials may improve the effectiveness of tubulization.

Introduction of exogenous factors in nerve chambers

The manipulation of the local environment within the tube lumen has been used in attempts to improve and accelerate the growth of axons regenerating through the guide. Effective regeneration in tubular chambers is dependent on the formation of an initial matrix across the gap (Williams et al., 1983, 1987). Therefore, the possibility of securing this initial step by placing a preformed matrix inside the nerve chamber at the time of implantation has been investigated. Prefilling the lumen with components of the naturally formed intratubular matrix, such as fibrin, collagen and laminin-containing gels, has been reported to enhance peripheral nerve regeneration in vivo (Williams et al., 1987; Madison et al., 1987; Kljavin and Madison, 1991), promoting the generation of a larger-diameter matrix that supports axonal growth across longer gaps than when the chamber is empty. However, compared with saline solution prefilled chambers, gel substrates containing neurite-promoting agents impair the regeneration process, probably by physically impeding the migration of regenerating elements (Valentini et al., 1987; Labrador et al., 1995). Among several matrices tested, those composed of collagen or laminin, at the maximal dilution that still allows gelification, were the most successful in promoting regeneration (Labrador and Navarro, unpublished observations). Thus, the density and the longitudinal alignment of the exogenous intratubular matrix seem to be relevant factors to allow regenerating axons to grow over tubuled long gaps.

The effects of growth factors on neurite outgrowth and their ability to prevent neuronal death after mechanical injury suggest that they may be useful in promoting regeneration after injury. The in vivo use of exogenous neurotrophic factors, applied locally or systemically, to enhance peripheral regeneration is under active study. Several studies have reported the beneficial effects of several neurotrophic factors in different experimental models. The addition of NGF inside a tubular guide partially prevents neuronal loss in dorsal root ganglia following nerve transection and tubulization repair (DaSilva and Langone, 1989) and facilitates the regeneration of rodent peripheral nerves (Rich et al., 1989; Otto et al., 1987; Derby et al., 1993; Hollowell et al., 1990; Chen et al., 1989; Eppley et al., 1991). A silicone chamber filled with bFGF also prevents the loss of sensory neurons after sciatic nerve transection (Otto et al., 1987). GMF-b (glial maturation factor) applied in polyethylene tubes stimulates axonal regeneration of the transected rat sciatic nerve (Harman et al., 1991). IGF-I and IGF-II, locally administered by means of a miniosmotic pump, increase the rate of sciatic nerve regeneration in rats (Kanje et al., 1990; Glazner et al., 1993; Svenningsen et al., 1994) and enhance motor nerve regeneration (Near et al., 1992). CNTF potentiates peripheral nerve regeneration when applied either by osmotic pumps (Newman et al., 1996) or by intraperitoneal injection (Sahenk et al., 1994), prevents degeneration of motor neurons after facial nerve lesion in newborn rats (Sendtner et al., 1990) and improves muscle reinnervation (Ulenkate et al., 1994). The local application of GDNF to the transected facial nerve has been shown to prevent the massive motor neuron death and atrophy that normally follows axotomy in the neonatal period (Yan et al., 1995). Finally, BDNF applied by osmotic pumps also enhances functional recovery after peripheral nerve transection (Utley et al., 1996).

In general, these strategies to promote nerve regeneration based on tubulization repair have tested the introduction of only one tropic or trophic factor, within a limited range of concentrations and time windows, and have been performed in different animal models and after a variety of injuries. Some early reports suggested that an appropriate mixture of different neurotrophic factors, extracellular matrix components and hormones may potentiate peripheral nerve regeneration better than any one individual component (Müller et al., 1987).

Cellular prostheses: towards an artificial nerve graft

When peripheral axons regenerate into the distal stump of a transected nerve, they establish functional connections with the persisting Schwann cell population, which result in the expression of neurite-promoted changes of Schwann cell behavior, such as ensheathing and myelinogenesis. If the distal nerve stump is rendered acellular before coaptation with the proximal stump, axons still grow within the autograft, but only in the presence of Schwann cells (Hall, 1986a). The first wave of Schwann cells repopulating these acellular grafts is supposed to come from cells that proliferate and migrate from the proximal stump through the suture scar. When the transected proximal stump is injected with antimitotic agents in order to arrest Schwann cell division after injury, axons grow well into distal cellular grafts, where they become associated with the resident population of Schwann cells. However, when the autograft is acellular, very few neurites regrow (Hall et al., 1986b).

Regeneration across synthetic tubes is dependent on the presence of a nerve stump as a distal insert, that provides both cellular and humoral factors for regeneration (Williams et al., 1984; Le Beau et al., 1988). If the distal nerve stump is excluded from the chamber, regeneration through tubular guides is significantly reduced or even absent (Lundborg et al., 1982a; Weis and Schröder, 1989). Similarly, the inability of axons to regenerate along tubes with a long interstump gap can be explained because the intratubular volume is too large to allow the formation of an initial fibrin-containing matrix and migration of non-neuronal cells into it. Using a flat glass chamber, Torigoe et al. (1996) have shown that transected axons regenerate following Schwann cells that migrate from the proximal nerve stump.

In order to overcome the gap length limits, a paradigm was developed by placing a small piece of nerve inside the nerve chamber used to repair a resected peripheral nerve. This piece of nerve transplant would act by dividing the long gap into two smaller gaps that axons are able to regenerate, by providing neurotrophic factors and supportive elements to the regenerating axons of the proximal stump. With this paradigm, it has been found that a nerve transplant, irrespective of its size, when placed inside a tube resulted in higher numbers of myelinated and unmyelinated fibers growing through the gap (Jenq and Coggeshall, 1986), increased the maximal gap length that the nerve is able to bridge (Smahel and Jentsch, 1986), and improved functional recovery of target organs (Navarro and Kennedy, 1991). These benefits, however, are not achieved when the nerve transplant is not autologous or is rendered acellular by repeated freezing-thawing cycles (unpublished observations).

From all the information available, it seems appropriate to propose that there are two key points determining the success of tubulization repair. First, the initial formation of an adequate acellular matrix bridging the proximal and distal stumps, and second, the repopulation of this matrix by migrating Schwann cells, that will stimulate and guide the regenerating axonal sprouts. The hypothesis to construct cellular prostheses, composed of a biocompatible nerve chamber filled with components of the ECM (collagen, laminin, etc.) and seeded with competent Schwann cells, seems at this stage to be a reasonable option to enhance nerve regeneration and to allow for an artificial substitute to the classical autologous nerve graft used to repair severe lesions of the PNS.

Over the past few years several of the techniques which would be required to develop an autotransplantation program using Schwann cells prepared in culture have been achieved. These include the abilities to isolate adult-derived Schwann cells from peripheral nerves (Morrisey et al., 1991) and to expand these cells in culture with mitogenic agents (Levi, 1996). It is substantially more difficult to extract Schwann cells from adult than from embryonic or perinatal nerve tissue because the adult tissue contains more developed epineurium and perineurium composed of fibroblast-derived cells. Dissociation of perinatal nerves yields a high percentage of Schwann cell population in the culture, and subsequent treatment with antimitotic agents can remove the few remaining fibroblasts. Conversely, the dissociation of adult nerves results in relatively poor yields and a higher percentage of fibroblasts that can not be easily eliminated by antimitotic agents. Within a few days in vitro, the fibroblasts often outgrow and overtake the Schwann cell population. Considerable progress has been made in the development of methods for the isolation of Schwann cells from adult nerves and for their expansion in vitro (Askanas et al., 1980; Morrisey et al., 1991; Rutkowski et al., 1992; Levi and Bunge, 1994; Levi, 1996; Casella et al., 1996).

The essential features of these methods included: i) the maintenance of nerve fragments as explants for a period of time in vitro, which improved the starting cell yield; ii) the sequential transferal of the explants to new dishes, which gradually depleted them of fibroblasts; iii) gentle enzymatic digestion of the explants with collagenase and dispase, and iv) the expansion of the Schwann cells in vitro using media containing several mitogenic agents. Applying these and other advances of Schwann cell culture has already allowed the development of the so-called cellular prostheses.

This experimental paradigm has been used experimentally by several authors for repairing lesioned nerves. Keeley et al. (1993) reported that nerve regeneration through autograft repair was significantly better than with artificial grafts (composed of a tube filled with collagen matrix) including or not cellular elements (fibroblasts or Schwann cells), and that among these, the artificial graft composed of live Schwann cells in collagen matrix was superior to other repair options. Walking track and electrophysiological analysis showed a similar pattern. In contrast, Kim et al.. (1994) showed that Schwann cell implants induced the return of function comparable to that obtained with sural nerve grafts and higher than with only collagen implants.

The morphological assessment suggested a tendency towards higher numbers of myelinated axons in Schwann cell implants than in sural nerve grafts. Anatomical analyses using fluorescent dye showed both high viability of prelabeled Schwann cells 120 days after transplantation and migration as far as 30 mm away from the implant site. However, the origin and density of the seeded Schwann cells and the type of matrix used to retain them in the nerve chamber are factors that may influence the outcome. Guénard et al.. (1992) have observed that the density with which Schwann cells were seeded in tubes influenced the results of regeneration. The number of myelinated axons in cables regenerated through channels seeded with a low Schwann cell density (40 x 10^6 cells/ml) was close to that found in empty channels, whereas myelinated axon counts increased significantly when the seeding density was doubled or tripled. These results suggest that by increasing both the Schwann cell seeding density and the luminal volume chamber, the regenerative potential of artificial nerve grafts may approach that of nerve autografts. In addition, heterologous Schwann cells seeded in artificial chambers induced an immune reaction that impeded regeneration. Myelinated axons did not extend farther than 2 mm within tubes seeded with heterologous Schwann cells, whereas myelinated axons bridged the whole gap in tubes seeded with syngenic Schwann cells at a similar density. When human Schwann cells seeded in tubes were transplanted in athymic nude rats to repair a sciatic nerve gap, the number of myelinated axons and the cross-sectional area of the regenerated cable were significantly greater than with chambers containing only a diluted laminin-containing gel (Levi et al., 1994). The human Schwann cells are capable of surviving in the injured peripheral nerve of the immune-deficient rat at least 35 days postimplantation. During this time some of the transplanted human Schwann cells were capable of myelinating regenerating rat axons, but were less successful than the host Schwann cells migrating into the chamber (Levi et al., 1994).

Although still partial and incomplete, all these results suggest ways in which Schwann cells could be used clinically for repair injuries to the PNS in a near future.

ACKNOWLEDGMENTS

This work was supported by grants from the FIS (95/0033-02), the CICYT (SAF97-0147) and the Societat Catalana de Neurologia, Spain.

REFERENCES

Aebisher, P., V. Guenard, S.R. Winn, R.F. Valentini, and P.M. Galletti (1988) Blind-ended semipermeable guidance channels support peripheral nerve regeneration in the absence of distal nerve stump. Brain Res. *454*: 179-187.

Aguayo, A.J. (1981) Schwann cell survival in auto-, allo- and xenografts. In A. Gorio, H. Millesi (eds): Post-traumatic peripheral nerve regeneration: experimental basis and clinical implications. New York: Raven Press, pp 319-321.

Aguayo, A.J., J.M. Peyronnard, and G.M. Bray (1973) A quantitative ultrastructural study of regeneration from insolated proximal stumps of transected unmyelinated nerves. J Neuropathol Exp Neurol. *32*: 256-270.

Almgren, K.G. (1974) Revascularisation of free peripheral nerve grafts. An experimental study in the rabbit. Acta Orthopaedica Scand. *suppl 154*: 1-104.

Archibald, S.J., J. Shefner, C. Krarup, and R.D. Madison (1995) Monkey median nerve repaired by nerve graft or collagen nerve guide tube. J Neurosci. *15*: 4109-4123.

Askanas, V., W.K. Engel, M.C. Dalakas, J.V. Lawrence, and L.S. Carter (1980) Human Schwann cells in tissue culture. Histochemical and structural studies. Neurology. *37*: 329-337.

Baas, P.W., L.A. White, and S.R. Heidermann (1987) Microtubule polarity reversal accompanies regrowth of amputated neurites. Proc Natl Acad Sci USA. *84*: 5272-5276.

Bamburg, J.R., R.O. Lockerbie, B.W. Bernstein, T.E. Morgan, M. Whitney, and M.E. Adams (1992) Actin dynamics in neuronal growth cones. In P.C. Letourneau, S.B. Kater and E.R. Macagno (eds): The nerve growth cone. New York: Raven Press, pp. 27-38.

Barakat, I. (1993) Nuclear triiodothyronine receptor expression is regulated by axon-Schwann cell contact. Neuroreport. *5*: 137-140.

Battisti, W.P., Y. Shinar, M. Schwartz, P. Levitt, and M. Murray (1992) Temporal and spatial patterns of expression of laminin, chondroitin sulphate proteoglycan and HNK-1 immunoreactivity during regeneration in the goldfish optic nerve. J Neurocytol. *21*: 557-573.

Benowitz, L.I., and E.R. Lewis (1983) Increased transport of 44000-49000 dalton acidic proteins during regeneration of the goldfish optic nerve: a two dimensional gel analysis. J Neurosci. *3*: 2153-2163.

Berenberg, R.A., D.S. Forman, D.K. Wood, A. Da Silva, and J. Demarre (1977) Recovery of peripheral nerve function after axotomy: effect of triiodothyronine. Exp Neurol. *57*: 349-363.

Beuche,W., and R.L. Friede (1986) Myelin phagocytosis in Wallerian degeneration of peripheral nerves depends on silica-sensitive, bg/bg-negative and Fc-positive monocytes. Brain Res. *378*: 97-106.

Black, M.M., and R.J. Lasek (1979) Slowing of the rate of axonal regeneration during growth and maturation. Exp Neurol. *63*: 108-119.

Bolin, L.M., and E.M. Shooter (1993) Neurons regulate Schwann cells genes by diffusible molecules. J Cell Biol. *123*: 237-243.

Bowe, M.A., and J.R. Fallon (1995) The role of agrin in synapse formation. Ann Rev Neurosci. *18*: 443-462.

Boyle, F.C., and J.S. Gillespie (1970) Accumulation and loss of noradrenaline central to a constriction on adrenergic nerves. Eur J Pharmacol. *12*: 77-84.

Boyles, J.K., D.Y. Hui, K.H. Weisgraber, R.E. Pitas, and R.W. Mahley (1987) Expression of apolipoprotein B, E (LDL) receptors and uptake of apolipoprotein E-containing lipoproteins by the regenerating rat sciatic nerve. Abstr Soc Neurosci. *13*: 294.

Bradley, W.G., and A.K. Asbury (1970) Duration of synthesis phase in neurilemma cells in mouse sciatic nerve regeneration. Exp Neurol. *26*: 275-282.

Braunewell, K.H., R. Martini, R. LeBaron, H. Kresse, A. Faissner, B. Schmitz, and M. Schachner (1995) Up-regulation of a chondroitin sulphate epitope during regeneration of mouse sciatic nerve: evidence that the immunoreactive molecules are related to the chondroitin sulphate proteoglycans decorin and versican. Eur J Neurosci. *7*: 792-804.

Bray, D. (1992) Cytoskeletal basis of nerve axon growth. In P.C. Letourneau, S.B. Kater and E.R. Macagno (eds): The nerve growth cone. New York: Raven Press, pp 7-17.

Bray, G.M., and A.J. Aguayo (1974) Regeneration of peripheral unmyelinated nerves. Fate of the axonal sprouts which develop after injury. J Anat. 117: 517-530.

Bridgman, P.C., and M.E. Dailey (1989) The organization of myosin and actin in rapid frozen nerve growth cones. J Cell Biol. *108*: 95-109.

Bridgman, P.C. (1992) Functional anatomy of the growth cone in relation to its role in locomotion and neurite assembly. In P.C. Letourneau, S.B. Kater and E.R. Macagno (eds): The nerve growth cone. New York: Raven Press, pp 39-53.

Brockes, J.P. (1984) Assays for cholinergic properties in cultured rat Schwann cells. Proc R Soc London Ser B. *222*: 121-134.

Brown, M.C., R.L. Holland, W.G. Hopkins, and R.J. Keynes (1980) An assessment of the spread of the signal for terminal sprouting within and between muscles. Brain Res. *210*: 145-151.

Buchthal, F., and V. Kühl (1979) Nerve conduction, tactile sensibility, and the electromyogram after suture or compression of peripheral nerve: a longitudinal study in man. J Neurol Neurosurg Psychiat. *42*: 436-451.

Bunge, M.B. (1973) Fine structure of nerve fibers and growth cones of isolated sympathetic neurons in culture. J Cell Biol. *56*: 713-725.

Butí, M., E. Verdú, R. Labrador, J.J. Vilches, J. Forés, and X. Navarro (1996) Influence of physical parameters of nerve chambers on peripheral nerve regeneration and reinnervation. Exp Neurol. *137*: 26-33.

Butí, M., E. Verdú, and X. Navarro (1992) Evaluación de un modelo experimental de reparación por tubulización en lesiones de nervios periféricos. Rev Neurol (Barc) *20*: 259-265.

Casella, G.T.B., R. Bunge, and P.M. Wood (1996) Improved method for harvesting human Schwann cells from mature peripheral nerve and expansion in vitro. Glia *17*: 327-338.

Cuppini, R. (1993) Age-related changes in maturation of regenerated motor innervation. J Neurol Sci. 114: 99-103.

Curtis, R., S.S. Scherer, R. Somogyi, K.M. Adryan, N.Y. Ip, Y. Zhu, R.M. Lindsay, and P.S. DiStefano (1994) Retrograde axonal transport of LIF is increased by peripheral nerve injury: correlation with increased LIF expression in distal nerve. Neuron *12*:, 191-204.

Chen, Y., L.T. Wang-Bennett, and N.J. Coker (1989) Facial nerve regeneration in the silicone chamber: the influence of nerve growth factor. Exp Neurol. *103*: 52-60.

Da Silva, C.F., and F. Langone (1989) Addition of nerve growth factor to the interior of a tubular prothesis increases sensory neuron regeneration in vivo. Brazilian J Med Biol Res. *22*: 691-694.

Danielsen, N., G. Lundborg, and M. Frizell (1986) Nerve repair and axonal transport: outgrowth delay and regeneration rate after transection and repair of rabbit hypoglossal nerve. Brain Res *376*: 125-132.

Dellon, A.L., and S.E. Mackinnon (1988) An alternative to the classical nerve graft for the management of the short nerve gap. Plast Reconstr Surg. *82*: 849-856.

Derby, A., V.W. Engleman, G.E. Frierdich, G. Neises, S.R. Rapp, and D.G. Roufa (1993) Nerve growth factor facilitates regeneration across nerve gaps: morphological and behavioral studies in rat sciatic nerve. Exp Neurol. *119*: 176-191.

Devor, M., D. Schonfeld, Z. Seltzer, and P.D. Wall (1979) Two modes of cutaneous reinnervation following peripheral nerve injury. J. Comp. Neurol. *185*: 211-220.

Donato, R., S. Cheema, D. Finkelstein, P. Bartlett, and W. Morrison (1995) Role of leukaemia inhibitory factor (LIF) in rat peripheral nerve regeneration. Ann Acad Med Singapore *24*: 94-100.

Dubovy, P., and I. Svizenska (1994) Denervated skeletal muscle stimulates migration of Schwann cells from the distal stump of transected peripheral nerve: an in vivo study. Glia *12*: 99-107.

Ducker, T.B., and G.J. Hayes (1968) Experimental improvements in the use of a silastic cuff for peripheral nerve repair. J Neurosurg. *28*: 582-587.

Dyck, P.J., E.H. Lambert, M.B. Wood, and R.L. Linscheid (1988) Assessment of nerve regeneration and adaptation after median nerve reconnection and digital neurovascular flap transfer. Neurology *38*: 1586-1591.

Edshage, S. (1964) Peripheral nerve suture. A technique for improved intraneural topography. Acta Chirurg Scand. *suppl 331*: 1-104.

Eldridge, C.F., M.B. Bunge, and R.P. Bunge (1989) Differentiation of axon-related Schwann cells in vitro. II. Control of myelin formation by basal lamina. J Neurosci. *9*: 625-638.

Eppley, B.L., R.V. Snyder, T.M. Winkelmann, and D.G. Roufa (1991) Efficacy of nerve growth factor in regeneration of the mandibular nerve. A preliminary report. J Oral Maxillofac Surg. *49*: 61-68.

Evans, P.J., R. Midha, and S.E. Mackinnon (1994) The peripheral nerve allograft: a comprehensive review of regeneration and neuroimmunology. Progr Neurobiol. *43*: 187-233.
Falls, D.L., K.M. Rosen, G. Corfas, W.S. Lane, and G.D. Fischbach (1993) ARIA, a protein that stimulates acetylcholine receptor synthesis, is a member of the neu ligand family. Cell *72*: 801-815.
Fawcett, J.W., and R.J. Keynes (1990) Peripheral nerve regeneration. Annu Rev Neurosci. *13*: 43-60.
Feinstein, D., and D. Larhammer (1990) Identification of a conserved protein motif in a group of growth factor receptors. FEBS Lett. *272*: 7-11.
Fields, R.D., and M.H. Ellisman (1986a) Axons regenerated through silicone tube splices. I. Conduction properties. Exp Neurol. *92*: 48-60.
Fields, R.D., and M.H. Ellisman (1986b) Axons regenerated through silicone tube splices. II. Functional morphology. Exp Neurol. *92*: 61-74.
Fields, R.D., J.M. LeBeau, F.M. Longo, and M.H. Ellisman (1989) Nerve regeneration though artificial tubular implants. Progr Neurobiol. *33*: 87-134.
Forman, D.S., and R.A. Berenberg (1978) Regeneration of motor axons in the rat sciatic nerve studied by labelling with axonally transported radioactive proteins. Brain Res. *156*: 213-225.
Forman, D.S., D.K. Wood, and S. Da Silva (1979) Rate of regeneration of sensory axons in transected rat sciatic nerve repair with epineural sutures. J Neurol Sci. *44*: 55-59.
Frizell, M., and J. Sjöstrand (1974a) The axonal transport of [3H] fucose labelled glyoproteins in normal and regenerating peripheral nerves. Brain Res. *78*: 109-123.
Frizell, M., and J. Sjöstrand J. (1974b) Transport of proteins, glycoproteins, and cholinergic enzymes in regenerating hypoglossal neurons. J Neurochem. *22*: 845-850.
Fruttiger, M., M. Schachner, and R. Martini (1995) Tenascin-C expression during Wallerian degeneration in C57BL/Wlds mice: possible implications for axonal regeneration. J Neurocytol. *24*: 1-14.
Gelman, B.B., N. Rifia, J.F. Goodrum, T.W. Boulin, and M.R. Krigman (1987) Apolipoprotein E is released by rat sciatic nerve during segmental demyelination and remyelination. J Neuropathol Exp Neurol. *46*: 644-652.
Georgiou, J., R. Robitaille, W.S. Trimble, and M.P. Charlton (1994) Synaptic regulation of glial protein expression in vivo. Neuron *12*: 443-455.
Giannini, C., A.C. Lais, and P.J. Dyck (1989) Number, size, and class of peripheral nerve fibers regenerating after crush, multiple crush, and graft. Brain Res. *500*: 131-138.
Gilmore, S.A., and T.J. Sims (1995) Glial cells and regeneration in the peripheral nervous system. In H. Kettenmann, and B.R. Ranson BR (eds): Neuroglia. New York: Oxford University Press, pp 829-842.
Glazner, G.W., and D.N. Ishii (1995) Insulin-like growth factor gene expression in rat muscle during reinnervation. Muscle Nerve *18*: 1433-1442.
Glazner, G.W., S. Lupien, J.A. Miller, and D.N. Ishii (1993) Insulin-like growth factor-II increases the rate of sciatic nerve regeneration in rats. Neuroscience *54*: 791-797.
Gómez, N., J. Cuadras, and X. Navarro (1996) Histologic assessment of sciatic nerve regeneration following resection and graft or tube repair in the mouse. Restor Neurol Neurosci. *10*:187-196.
Goodrum, J.F., T. Earnhardt, N. Goines, and T.W. Bouldin (1994) Fate of myelin lipids during degeneration and regeneration of peripheral nerve. An autoradiographic study. J Neurosci. *14*: 357-367.
Goodrum, J.F., and D.L. Novicki (1988) Macrophage-like cells from explant cultures of rat sciatic nerve produce apolipoprotein E. J Neurosci Res. *20*: 457-462.
Goodrum, J.F., J.E. Weaver, N.D. Goines, and T.W. Bouldin (1995) Fatty acids from degenerating myelin lipids are conserved and reutilized for myelin synthesis during regeneration in peripheral nerve. J Neurochem. *65*: 1752-1759.
Gordon, T. (1983) Dependence of peripheral nerves on their target organs. In G. Burnstock, R. O'Brien, and G. Vrbova G (eds): Somatic and autonomic nerve-muscle interactions. New York: Elsevier, pp. 289-325.
Gordon-Weeks, P.R., and S.G. Mansfield (1992) Assembly of microtubules in growth cones. The role of microtubule-associated proteins. In P.C. Letourneau, S.B. Kater, and E.R. Macagno (eds): The nerve growth cone. New York: Raven Press, pp 55-64.

Gordon-Weeks, P.R. (1991) Control of microtubule assembly in growth cones. J Cell Sci. *suppl 15*: 45-49.
Greenberg, M.E., E.B. Ziff, and L.A. Greene (1986) Stimulation of neuronal acetylcholine receptors induces rapid gene transcription. Science *234*: 80-83.
Guénard, V., N. Kleitman, T.K. Morrisey, R.P. Bunge, and P. Aebischer (1992) Syngeneic Schwann cells derived from adult nerves seeded in semipermeable guidance channels enhance peripheral nerve regeneration. J Neurosci. *12*: 3310-3320.
Guth, L. (1956) Regeneration in the mammalian peripheral nervous system. Physiol. Rev. *36*: 441-478.
Gutman, E., L. Guttman, P.B. Medawar, and J.Z. Young (1942) The rate of regeneration of the nerve. J Exp Biol. *19*: 14-44.
Haas, C.A., C. Donath, and G.W. Kreutzberg (1993) Differential expression of immediate early genes after transection of the facial nerve. Neuroscience *53*: 91-99.
Hall, S.M. (1986a) Regeneration in cellular and acellular autografts in the peripheral nervous system. Neuropathol Appl Neurobiol. *12*: 27-46.
Hall, S.M. (1986b) The effect of inhibiting Schwann cell mitosis on the re-innervation of acellular autografts in the peripheral nervous system of the mouse. Neuropathol Appl Neurobiol. *12*: 401-414.
Härkönen, M.H.A., and F.C. Kauffman (1974) Metabolic alterations in the axotomized superior cervical ganglion of the rat. II. The pentose phosphate pathway. Brain Res. *65*: 141-157.
Harman, K., J. Katnick, R. Lim, A. Zaheer, and J.C. de la Torre (1991) Glia maturation factor b stimulates axon regeneration in transected sciatic nerve. Brain Res. *564*: 332-335.
Heiwell, P.O., A. Dahlström, P.A. Larsson, and S. Booj (1979) The intra-axonal transport of acetylcholine and cholinergic enzimes in rat sciatic nerve during regeneration after various types of axonal trauma. J Neurobiol. *10*: 119-136.
Henry, E.W., T.H. Chiu, E. Nyilas, T.M. Brushart, P. Dikkes, and R.L. Sidman (1985) Nerve regeneration through biodegradable polyesther tubes. Exp Neurol. *90*: 652-676.
Herdegen, T., C.E. Fiallos-Estrada, W. Schmid, R. Bravo, and M. Zimmermann (1992) The transcription factors c-JUN, JUN D and CREB, but not FOS and KNOX-24, are differentially regulated in axotomized neurons following transcetion of rat sciatic nerve. Mol Brain Res. *14*: 155-165.
Heumann, R., D. Lindholm, C. Bandtlow, M. Meyer, M.J. Radeke, T.P. Misko, E. Shooter, and H. Thoenen (1987) Differential regulation of mRNA encoding nerve growth factor and its receptor in rat sciatic nerve during development, degeneration and regeneration: role of macrophages. Proc Natl Acad Sci USA. *84*: 8735-8739.
Hoffman, P.N., and R.J. Lasek (1980) Axonal transport of the cytoskeleton in regenerating motor neurons: constancy and change. Brain Res. *202*: 317-333.
Holtzman, E., and A.B. Novikoff (1965) Lysosomes in the rat sciatic nerve following crush. J Cell Biol. *27*: 651-669.
Hollowell, J.P., A. Villadiego, and K.M. Rich (1990) Sciatic nerve regeneration across gaps within silicone chambers: long-term effects of NGF and consideration of axonal branching. Exp Neurol. *110*: 45-51.
Jackson, P.C., and J. Diamond (1984) Temporal and spatial constraints of the collateral sprouting of low-threshold mechanosensory nerves in the skin of rats. J. Comp. Neurol. *226*: 336-345.
Jahromi, B.S., R. Robitaille, and M.P. Charlton MP (1992) Transmitter release increases intracellular calcium in perisynaptic Schwann cells in situ. Neuron *8*: 1069-1077.
Jenq, C.B., and R.E. Coggeshall (1986) The effects of an autologous transplant on patterns of regeneration in rat sciatic nerve. Brain Res. *364*: 45-56.
Jing, S., P. Tapley, and M. Barbacid (1992) Nerve growth factor mediates signal transduction though trk homodimer receptors. Neuron *9*: 1067-1079.
Jirmanová, I. (1975) Ultrastructure of motor end-plates during pharmacologically-induced degeneration and subsequent regeneration of skeletal muscle. J Neurocytol. *4*: 141-155.
Johnson, A.C., A.R. McNabb AR, and R.J. Rossiter (1949) Chemical studies of peripheral nerve during Wallerian degeneration. I. Lipids. Biochem J. *45*: 500-508.
Johnson, E.M., M. Taniuchi, and P.S. DiStephano (1988) Expression and possible function of nerve growth factor receptors on Schwann cells. Trends Neurosci. *11*: 299-304.

Kanje, M., A. Skottner, J. Sjöberg, and G. Lundborg (1989) Insulin-like growth factor I (IGF-I) stimulates regeneration of the sciatic nerve. Brain Res. *486*: 396-398.

Kanje, M., A. Skottner, and J. Sjöberg (1990) Insulin-like growth factor I (IGF-I) and regeneration of the sciatic nerve of the rat. Restor Neurol Neurosci. *1*: 211-215.

Keeley, R., T. Atagi, E. Sabelman, J. Padilla, P. Kalcik, J. Agras, L. Eng, T.W. Wiedman, K. Nguyen, A. Sudekum, and J. Rosen (1993) Synthetic nerve graft containing collagen and synthetic Schwann cells improves functional, electrophysiological, and histological parameters of peripheral nerve regeneration. Restor Neurol Neurosci. *5*: 353-366.

Kelly, R.B., and E. Grote (1993) Protein targeting in the neuron. Ann Rev Neurosci. *16*: 95-127.

Kennedy, W.R., and M. Sakuta (1984) Collateral reinnervation of sweat glands. Ann. Neurol. *15*: 73-78.

Kim, D.H., S.E. Connolly, D.G. Kline, R.M. Voorhies, A. Smith, M. Powell, T. Yoes, and J.K. Daniloff (1994) Labeled Schwann cells transplants versus sural nerve grafts in nerve repair. J Neurosurg. *80*: 254-260.

Kline, D.G. (1990) Surgical repair of peripheral nerve injury. Muscle Nerve *13*: 843-852.

Kljavin, I.J., and R.D. Madison (1991) Peripheral nerve regeneration within tubular prostheses: effects of laminin and collagen matrices on cellular ingrowth. Cell Materials *1*: 17-28.

Labrador, R.O., M. Butí, and X. Navarro (1995) Peripheral nerve repair: role of agarose matrix density on functional recovery. Neuroreport *6*: 2022-2026.

Leah, J.D., T. Herdegen, and R. Bravo (1991) Selective expression of Jun proteins following axotomy and axonal transport block in peripheral nerves in the rat: evidence for a role in the regeneration process. Brain Res. *566*: 198-207.

LeBeau, J.M., M.H. Ellisman, and H.C. Powell (1988) Ultrastructural and morphometric analysis of long-term peripheral nerve regeneration though silicone tubes. J Neurocytol. *17*: 161-172.

LeBlanc, A.C., and J.F. Podulso (1990) Axonal modulation of myelin gene expression in the peripheral nerve J Neurosci Res. *26*: 317-326.

LeBlanc, A.C., A.J. Windebank, and J.F. Podulso JF (1992) P0 gene expression in Schwann cells is modulated by increase of cAMP which is dependent on the presence of axons. Mol Brain Res. *12*: 31-38.

Lemke, G. (1992) Myelin and myelination. In Z.W. Hall (ed): An introduction to molecular neurobiology. Sunderland: Sinauer, pp 281-312.

Letourneau, P.C., and T.A. Shattuck (1989) Distribution and possible interactions of actin-associated proteins and cell adhesion molecules of nerve growth cones. Development *105*: 505-519.

Letourneau, P.C., M.L. Condic, and D.M. Snow (1994) Interactions of developing neurons with the extracellular matrix. J Neurosci. *14*: 915-928.

Levenson, D., and J. Rosenbluth (1990) Electrophysiologic changes accompanying Wallerian degeneration in frog sciatic nerve. Brain Res. *523*: 230-236.

Levi, A.D.O., and R.P. Bunge (1994) Studies of myelin formation after transplantation of human Schwann cells into the severe combined immunodeficient mouse. Exp Neurol. *130*: 41-52.

Levi, A.D.O., V. Guenard, P. Aebicher, and B.P. Bunge (1994) The functional characteristics of Schwann cells cultured from human peripheral nerve after transplantation into a gap within the rat sciatic nerve. J Neurosci. *14*: 1309-1319.

Levi, A.D.O. (1996) Characterization of the technique involved in isolating Schwann cells from adult human peripheral nerve. J Neurosci Meth. *68*: 21-26.

Lewin, G.R., and Y.A. Barde YA (1996) Physiology of the neurotrophins. Ann Rev Neurosci. *19*: 289-317.

Lieberman, A.R. (1971) The axon reaction: a review of the principal features of perikarial responses to axon injury. Int Rev Neurobiol. *14*: 49-124.

Liu, H.M., L.H. Yang, and Y.J. Yang (1995) Schwann cell properties. III. C-fos expression, bFGF production, phagocytosis and proliferation during Wallerian degeneration. J Neuropathol Exp Neurol. *54*: 487-496.

Liu, H.M. (1996) Growth factors and extracellular matrix in peripheral nerve regeneration, studied with a nerve chamber. J Periph Nerv System. *1*: 97-110.

Liu, H.M. (1992) The role of extracellular matrix in peripheral nerve regeneration: a wound chamber study. Acta Neuropathol. *83*: 469-474.

Longo, F.M., E.G. Hayman, G.E. Davis, E. Rouoslahti, E. Engvall, M. Manthorpe, and S. Varon (1984) Neurite-promoting factors and extracellular matrix components accumulating in vivo within nerve regeneration chambers. Brain Res. *309*: 105-117.

Longo, F.M., M. Manthorpe, S.D. Skaper, G. Lundborg, and S. Varon (1983) Neurotrophic activities accumulate in vivo within silicone nerve regeneration chambers. Brain Res. *261*: 109-117.

Lubinska, L., and P. Jastreboff (1977) Early course of Wallerian degeneration in myelinated fibers of the rat phrenic nerve. Brain Res. *130*: 47-63.

Lundborg, G., L.B. Dahlin, N. Danielsen, R.H. Gelberman, F.M. Longo, H.C. Powell, and S. Varon (1982) Nerve regeneration in silicone chambers: influence of gap length and of distal stump components. Exp Neurol. *76*: 361-375.

Lundborg, G., B. Rosén, L. Dahlin, N. Danielsen, and J. Holmberg (1997) Tubular versus conventional repair of median and ulnar nerves in the human forearm: early results from a prospective, randomized, clinical study. J Hand Surg. *22*: 99-106

Lundborg, G. (1988) Nerve injury and repair. London: Curchill Livingstone, pp 149-195.

Lunn, E.R., M.C. Brown, and V.H. Perry (1990) The pattern of axonal degeneration in the peripheral nervous system varies with different types of lesion. Neuroscience *35*: 157-165.

Lunn, E.R., V.H. Perry, M.C. Brown, H. Rosen, and S. Gordon (1989) Absence of Wallerian degeneration does not hinder regeneration in peripheral nerve. Eur J Neurosci. *1*: 27-33.

Madison, R.D., C.F. Da Silva, P. Dikkes, R.L. Sidman, and T.H. Chiu (1987) Peripheral nerve regeneration with entubulation repair: Comparison of biodegradable nerve guide versus polyethylene tubes and the effects of a laminin-containing gel. Exp Neurol. *95*: 378-390.

Mannell, W.A. (1952) Wallerian degeneration in the rat. A chemical study. Can J Med Sci. *30*: 173-179.

Martini, R.(1994) Expression and functional roles of neural cell surface molecules and extracellular matrix components during development and regeneration of peripheral nerves. J Neurocytol. *23*: 1-28.

Mathews, M.R., and V.H. Nelson (1975) Detachment of structurally intact nerve endings from chromatolytic neurones of rat superior cervical ganglion during the depression of synaptic transmission induced by post-ganglionic axotomy. J Physiol. *245*: 91-135.

Matus, A. (1991) Microtubule-associated proteins and neuronal morphogenesis. J Cell Sci. *suppl 15*: 61-67.

McQuarrie, I.G., B. Grafstein, C.F. Dreyfus, and M.D. Gershon (1978) Regeneration of adrenergic axons in rat sciatic nerve: effect of a conditioning lesion. Brain Res. *141*: 21-34.

McQuarrie, I.G., B. Grafstein, and M.D. Gershon (1977) Axonal regeneration in rat sciatic nerve: effect of a conditioning lesion and of dcb-AMP. Brain Res. *132*: 443-453.

Meiri, K.F., K.H. Pfenninger, and M.B. Willard (1986) Growth-associated protein, GAP-43, a polypeptide that is induced when neurons extend axons, is a component of growth cones and corresponds to pp46, a major polypeptide of a subcellular fraction enriched in growth cones. Proc Natl Acad Sci. *83*: 3537-3541.

Mirsky, R., and K.R. Jessen (1996) Schwann cell development, differentiation and myelination. Curr Opin Neurobiol. *6*: 89-96.

Morgan, J.I., and T. Curran (1986) Role of ion flux in the control of c-fos expression. Nature *322*: 552-555.

Morrisey, T.K., N. Kleitman, and R.P. Bunge (1991) Isolation and functional characterization of Schwann cells derived from adult peripheral nerve. J Neurosci. *11*: 2433-2442.

Müller, H., L.R. Williams, and S. Varon (1987) Nerve regeneration chamber: evaluation of exogenous agents applied by multiple injections. Brain Res. *413*: 320-326.

Müller, H.W., and P. Minwegan (1987) Nonresident macrophages in peripheral nerve rat. Effect of silica on migration, myelin phagocytosis, and apolipoprotein E expression during Wallerian degeneration. J Neurosci Res. *18*: 222-229.

Nathan, C.F. (1987) Secretory products of macrophages. J Clin Invest. *79*: 319-326.

Navaro, X., E. Verdú, G. Wendelschafer-Crabb, and W.R. Kennedy WR (1997) Immunohistochemical study of skin reinnervation by regenerative axons. J Comp Neurol. *380*: 164-174.

Navarro, X., and W.R. Kennedy (1991) The effects of autologous nerve transplants on motor and sudomotor reinnervation by regenerative axons. Brain Res. *565*: 181-187.

Navarro, X., F.J. Rodríguez, R.O. Labrador, M. Butí, D. Ceballos, N. Gómez, J. Cuadras, and G. Perego (1996) Peripheral nerve regeneration through bioresorbable and durable nerve guides. J Periph Nerv System. *1*: 57-64.

Navarro, X., E. Verdú, and M. Butí (1994) Comparison of regenerative and reinnervating capabilities of different functional types of nerve fibers. Exp Neurol. *129*: 217-224.

Near, S.L., R. Whalen, J.A. Miller, and D.N. Ishii (1992) Insulin-like growth factor II stimulates motor nerve regeneration. Proc Natl Acad Sci. *89*: 11716-11720.

Newman, J.P., A.N. Verity, S. Hawatmeh, W.E. Fee, and D.J. Terris (1996) Ciliary neurotrophic factor enhances peripheral nerve regeneration. Arch Otolaryngol Head Neck Surg. *122*: 399-403.

Nieto-Sampedro, M. (1996) Plasticidad neural: una propiedad básica que subyace desde el aprendizaje a la reparación de lesiones. In F. Mora (ed): El cerebro íntimo. Ensayos sobre neurociencia. Barcelona: Ariel, pp 66-96.

Nishimoto, I., Y. Hata, E. Ogata, and I. Kojima (1987) Insulin-like growth factor II stimulates calcium influx in competent BALB/c 3T3 cells primed with epidermal growth factor. J Biol Chem. *262*: 12120-12126.

O'Brien, R.A.D. (1978) Axonal transport of acetylcholine, choline acetyltransferase and cholinesterase in regenerating peripheral nerve. J Physiol. *282*: 91-103.

Otto, D., K. Unsicker, and C. Grothe (1987) Pharmacological effects of nerve growth factor and fibroblast growth factor applied to the transectional sciatic nerve on neuron death in adult rat dorsal root ganglia. Neurosci Lett. *83*: 156-160.

Pellegrino, R.G., and P.S. Spencer (1985) Schwann cell mitosis in response to regenerating peripheral axons in vivo. Brain Res. *341*: 16-25.

Perry, V.H., J.W. Tsao, S. Fearn, and M.C. Brown (1995) Radiation-induced reductions in macrophage recruitment have only slight effects on myelin degeneration in sectioned peripheral nerves of mice. Eur J Neurosci. *7*: 271-280.

Pestronk, A., D.B. Drachman, and J.W. Grifin (1980) Effects of aging on nerve sprouting and regeneration. Exp Neurol. *70*: 65-82.

Phelan, K.A., E.H. Sherr, J.M. Aletta, and L.A. Greene (1992) Regulation of growth cone formation and function by neurotrophic factors. In P.C. Letourneau, S.B. Kater, and E.R. Macagno (eds): The nerve growth cone. New York: Raven Press, pp 151-166.

Plantinga, L.C., J. Verhaagen, S.L. Wong, P.M. Edwards, P.R. Bär, and W.H. Gispen (1994) The neurotrophic peptide Org 2766 does not influence the expression of the immediate early gene c-fos following sciatic nerve crush in the rat. Int J Dev Neurosci. *12*: 117-125.

Politis, M.J., and P.S. Spencer (1981) A method to separate spatially the temporal sequence of axons. Schwann cell interaction during nerve regeneration. J Neurocytol. *10*: 221-232.

Pover, C.M., and S.J.W. Lisney (1989) Influence of graft size on peripheral nerve regeneration in cats. J Neurol Sci. *90*: 179-185.

Prigent, S.A., and N.R. Lemoine (1992) The type 1 (EGFR-related) family of growth factor receptors and their ligands. Progr Growth Factor Res. *4*: 1-25.

Purvès, D. (1975) Functional and structural changes in mammalian sympathetic neurones following interruption of their axons. J Physiol. *252*: 429-463.

Pyykonen, I., and J. Koistinaho (1991) C-fos protein like immunoreactivity in non-neuronal cells of rat peripheral nerve after transection. Acta Neuropathol. *82*: 66-71.

Raffroni, S., and R.A. Bradshaw (1993) The receptors for nerve growth factor and other neurotrophins. Ann Rev Biochem. *62*: 823-850.

Raivich, G., and G.W. Kreutzberg (1987) Expression of growth factor receptors in injured nervous tissue. II. Induction of specific platelet-derived growth factor binding in the injured PNS is associated with a breakdown in the blood-nerve barrier and endoneural interstitial oedema. J Neurocytol. *16*: 701-711.

Raivich, G., and G.W. Kreutzberg (1994) Pathophysiology of glial growth factor receptors. Glia *11*: 129-146.

Ramón Cajal, S. (1913) Estudios sobre la degeneración y regeneración del sistema nervioso. Tomo I. Degeneración y regeneración de los nervios. Madrid.

Rawlins, F.A., G. Villegas, E.T. Hedley-White, and B.G. Uzman (1972) Fine structural localization of cholesterol-1,2-3H in degenerating and regenerating mouse sciatic nerve. J Cell Biol. *52*: 615-625.

Reichardt, L.F., and K.J. Tomaselli (1991) Extracellular matrix molecules and their receptors: functions in neural development. Ann Rev Neurosci. *14*: 531-570.
Reichert, F., A. Saada, and S. Rotshenker (1994) Peripheral nerve injury induced Schwann cells to expresss two macrophage phenotypes: phagocytosis and galactose-specific lectin MAC-2. J Neurosci. *4*: 3231-3245.
Reist, N.E., and S.J. Smith (1992) Neurally evoked calcium transients in terminal Schwann cells at the neuromuscular junction. Proc Natl Acad Sci. *89*: 7625-7629.
Reynolds, M.L., and C.J. Woolf (1992) Terminal Schwann cells elaborate extensive processes following denervation of the motor endplate. J Neurocytol. *21*: 50-66.
Rich, K.M., T.D. Alexander, J.C. Pryor, and J.P. Hollowell (1989) Nerve growth factor enhances regeneration though silicone chambers. Exp Neurol. *105*: 162-170.
Robitaille, R. (1995) Purinergic receptors and their activation by endogenous purines at perisynaptic glial cells of the frog neuromuscular junction. J Neurosci. *15*: 7121-7131.
Rodríguez-Tebar, A., and Y. Barde (1988) Binding characteristics of brain-derived neurotrophic factor to its receptors on neurons from the chick embryo. J Neurosci. *8*: 3337-3342.
Rodríguez-Tebar, A., G. Dechant, Y. Barde (1990) Binding of brain-derived neurotrophic factor to nerve growth factor receptors. Neuron *4*: 487-492.
Rosen, J.M., H.N. Pham, G. Abraham, L. Harold, and V.R. Hentz (1989) Artificial nerve graft compared to autograft in a rat model. J Rehab Res Dev. *26*: 1-14.
Ruoslahti, E., and M.D. Pierschbacher (1987) New perspectives in cell adhesion: RGD and integrins. Science *238*: 491-497.
Rutkowski, J.L., G.I. Tennekoon, and J.E. McGuillicuddy (1992) Selective culture of mitotically active human schwann cells from adult sural nerves. Ann Neurol. *31*: 580-586.
Sahenk, Z., J. Seharaseyon, and J.R. Mendell (1994) CNTF potentiates peripheral nerve regeneration. Brain Res. *655*: 246-250.
Sanders, F.K., and D. Whitterbridge (1946) Conduction velocity and myelin thickness in regenerating nerve fibres. J Physiol. *105*: 152-174.
Sanders, F.K., and J.Z. Young (1944) The role of the peripheral stump in the control of fibre diameter in regenerating nerves. J Physiol *103*: 119-131.
Scherer, S.S., and J.L. Salzer (1996) Axon-Schwann cell interactions in peripheral nerve regeneration. In K.R. Jessen, and W.D. Richardson (eds): Glial cell development. Oxford: Bios Scientific, pp 165-196.
Scherer, S.S., D.Y. Wang, R. Kuhn, G. Lemke, L. Wrabetz, and J. Kamholz (1994) Axons regulate Schwann cell expression of the POU transcription factor SCIP. J Neurosci. *14*: 1930-1942.
Schulze, E., D.J. Asai, J.C. Bulinski, and M. Kirschner (1987) Post-translational modification and microtubule stability. J Cell Biol. *105*: 2167-2177.
Selzer, M.E. (1980) Regeneration of peripheral nerve. In Summer, A.J. (ed): The physiology of peripheral nerve disease. Philadelphia: Saunders, pp 358-431.
Sendtner, M., G.W. Kreutzberg, and H. Thoenen (1990) Ciliary neurotrophic factor prevents the degeneration of motor neurons after axotomy. Nature *345*: 440-441.
Sendtner, M., A. Stöckli, and H. Thoenen (1992) Synthesis and localization of ciliary neurotrophic factor in the sciatic nerve of the adult rat after lesion and during regeneration. J Cell Biol. *118*: 139-148.
Serrano, L., and J. Avila (1990) Structure and function of tubulin regions. In Avila, J. (ed): Microtubule proteins. Florida: CRC Press, pp 67-88.
Sheng, M., and M.E. Greenberg (1990) The regulation and function of c-fos and other immediate early genes in the nervous system. Neuron *4*: 477-485.
Siironen, J., M. Sandberg, V. Vuorinen, and M. Roytta (1992) Expression of type I and III collagens and fibronectin after transection of rat sciatic nerve. Reinnervation compared with denervation. Lab Invest. *67*: 80-87.
Sinicropi, D.V., and F.C. Kauffman (1979) Retrograde alteration of 6-phosphogluconate dehydrogenase in axotomized cervical ganglia of the rat. J Biol Chem. *254*: 3011-3017.
Skene, J.H.P., R.D. Jacobson, G.J. Snipes, C.B. McGuire, J.J. Norden, and J.A. Freeman (1986) A protein induced during nerve growth (GAP-43) is a major componen of growth-cone membranes. Science *233*: 783-785.
Skene, J.H.P., and M. Willard (1981) Axonally transported proteins associated with axon growth in rabbit central and peripheral nervous system. J Cell Biol. *89*: 96-103.

Skene, J.H.P. (1989) Axonal growth-associated proteins. Ann Rev Neurosci. *12*: 127-156.
Smahel, J., and B. Jentsch (1986) Stimulation of peripheral nerve regeneration by an isolated nerve segment. Ann Plast Surg. *16*: 494-501.
Snipes, G.J., C.B. McGuire, J.J. Norden, and J.A. Freeman (1986) Nerve injury stimulates the secretion of apolipoprotein E by non-neuronal cells. Proc Natl Acad Sci. *83*: 1130-1134.
Stoll, G., J.W. Griffin, C.Y. Li, and B.D. Trapp (1989) Wallerian degeneration in the peripheral nervous system: participation of both Schwann cells and macrophages in myelin degradation. J Neurocytol. *18*: 671-683.
Strittmatter,S.M., D. Valenzuela, T. Vartanian, Y. Sudo, M.X. Zuber, and M.C. Fishman (1991) Growth cone transduction: G0 and GAP-43. J Cell Sci. *suppl 15*: 27-33.
Sumner, B.E.H., and S. Sunderland (1973) Quantitative electron microscopy on the injured hypoglossal nucleus in the rat. J Neurocytol. *2*: 315-328.
Sumner, B.E.H., and W.E. Watson (1971) Retraction and expansion of the dendritic tree of motor neurones of adult rats induced in vivo. Nature *233*: 273-275.
Sunderland, S. (1991) Nerve injuries and their repair. A critical appraisal. Edinburgh: Churchill Livingstone.
Sunderland, S. (1978) Nerves and nerve injuries. Edinburgh: Churchill Livingstone.
Sutter, A., R.J. Riopelle, R.M. Harris-Warrick, and E.M. Shooter (1979) NGF receptors: characterization of two distinct classes of binding sites on chick embryo sensory ganglia cells. J Biol Chem. *254*: 5972-5982.
Svenningsen, A.F., M. Kanje, and J. Sjöberg (1994) Insulin-like growth factor-II (IGF-II) stimulates regeneration and proliferation in the rat sciatic nerve. Neurosci Res Comm. *15*: 111-118.
Tanaka, K., Q.L. Zhang, and H.D.F. Webster (1992) Myelinated fiber regeneration after sciatic nerve crush. Morphometric observations in young adult and aging mice and the effects of macrophage suppression and conditioning lesions. Exp Neurol. *18*: 53-61.
Taniuchi, M., H.B. Cark, J.B. Schweitzer, and E.M. Johnson (1988) Expression of nerve growth factor receptors by Schwann cells of axotomized peripheral nerves: ultrastructural location, suppression by axonal contact, and binding properties. J Neurosci. *8*: 664-681.
Thomas, P.K., and R.H.M. King (1974) The degeneration of unmyelinated axons following nerve section. An ultrastructural study. J Neurocytol. *3*: 497-512.
Tona, A., G. Perides, F. Rahemtulla, and D. Dahl (1993) Extracellular matrix in regenerating rat sciatic nerve: a comparative study on the localization of laminin, hyaluronic acid, and chondroitin sulfate proteoglycans, including versican. J Histochem Cytochem. *41*: 593-599.
Topilko, P., P. Murphy, and P. Charnay (1996) Embryonic development of Schwann cells. Multiple roles for neuregulins along the pathway. Mol Cel Neurosci. *8*: 71-75.
Torigoe, K., H.F. Tanaka, A. Takahashi, A. Awaya, and K. Hashimoto (1996) Basic behavior of migratory Schwann cells in peripheral nerve regeneration. Exp Neurol. *137*: 301-308.
Torigoe, K. (1985) Distribution of motor nerve sproutings in the mouse gastrocnemius muscle after partial denervation. Brain Res. *330*: 273-282.
Tsuge, K., Y. Ikuta, and M. Sakaue (1975) A new technique for nerve suture. The anchoring funicular suture. Plast Reconstruct Surg. *56*: 496-500.
Ulenkate, H.J.L.M., E.C.A. Kaal, W.H. Gispen, F.G.I. Jannekens (1994) Ciliary neurotrophic factor improves muscle fibre reinnervation after facial nerve crush in young rats. Acta Neuropathol. *88*: 558-564.
Ullrich, A., and J. Schlessinger (1990) Signal transduction by receptors with tyrosine kinase activity. Cell *61*: 203-212.
Utley, D.S., S.L. Lewin, E.T. Cheng, A.N. Verity, D. Sierra, and D.J. Terris (1996) Brain-derived neurotrophic factor and collagen tubulization enhance functional recovery after peripheral nerve transection and repair. Arch Otolaryngol Head Neck Surg. *122*: 407-413.
Valentini, R.F., P. Aebischer, S.R. Winn, and P.M. Galletti (1987) Collagen- and laminin-containing gels impede peripheral nerve regeneration though semipermeable nerve guidance channels. Exp Neurol. *98*: 350-356.
Verdú, E., M. Butí, and X. Navarro (1995) The effect of aging on efferent nerve fibers regeneration in mice. Brain Res. *696*: 76-82.
Verdú, E., and X. Navarro (1997) Comparison of immunohistochemical and functional reinnervation of skin and muscle after peripheral nerve injury. Exp Neurol. *146*: 187-198.

Vial, J.D. (1958) The early changes in the axoplasm during wallerian degeneration. J Biophys Biochem Cytol. *4*: 551-567.

Waller, A.V. (1850) Experiments on the section of the glossopharyngeal and hypoglossal nerves of the frog, and observation on the alterations produced thereby in the structure of their primitive-fibres. Philos Trans R Soc Lond (Biol). *140*: 423-473.

Watson, W.E. (1974) The binding of actinomycin D to the nuclei of axotomized neurones. Brain Res. *65*: 317-322.

Weinberg, H.J., and P.S. Spencer (1978) The fate of Schwann cells insolated from axonal contacts. J Neurocytol. *7*: 555-569.

Weis, J., and J.M. Schröder (1989) Differential effects of nerve, muscle, and fat tissue on regenerating nerve fibers in vivo. Muscle Nerve *12*: 723-734.

White, F.V., A.D. Toews, and J.F. Goodrum (1989) Lipid metabolism during early stages of Wallerian degeneration in the rat sciatic nerve. J Neurochem. *52*: 1085-1092.

Williams, L.R., N. Danielsen, H. Müller, and S. Varon (1987) Exogenous matrix precursors promote functional nerve regeneration across a 15 mm gap within a silicone chamber in the rat. J Comp Neurol. *264*: 284-290.

Williams, L.R., F.M. Longo, H.C. Powell, G. Lundborg, and S. Varon (1983) Spatial-temporal progress of peripheral nerve regeneration within a silicone chamber. Parameters for a bioassay. J Comp Neurol. *218*: 160-170.

Williams, L.R., H.C. Powell, G. Lundborg, and S. Varon (1984) Competence of nerve tissue as distal insert promoting nerve regeneration in a silicone chamber. Brain Res. *293*: 201-211.

Wood, J.G., and R.M.C. Dawson (1974) Lipid and protein changes in sciatic nerve during Wallerian degeneration. J Neurochem. *22*: 631-635.

Yan, Q., C. Matheson, and O.T. Lopez (1995) In vivo neurotrophic effects of GDNF on neonatal and adult facial motor neurons. Nature *373*: 341-344.

Yarden, Y., and J. Schlessinger (1987) Epidermal growth factor induces rapid, reversible aggregation of the purified epidermal growth factor receptor. Biochemistry *26*: 1443-1451.

Young, V.L., C.R. Wray, and P.M. Weeks (1980) The results of nerve grafting in the wrist and hand. Ann Plast Surg. *5*: 212-215.

IV. GLIA IN PATHOLOGY

18

GLIAL ABNORMALITIES IN GENETICALLY DETERMINED DISORDERS OF MYELIN

José Miguel Vela, Ishar Dalmau,
Berta González and Bernardo Castellano

A deficiency of myelin can result either from failure to produce the normal amount of myelin during development (hypomyelination) or from myelin breakdown after it is formed (demyelination). Diseases involving myelin include acquired disorders having an allergic or infectious basis, toxic and nutritional disorders, and genetically determined disorders. A growing number of mutations in protein genes that lead to CNS hypomyelination in mice and other species have been identified and are models of particular value to investigate the structural roles and interactions of proteins in myelin. However, in recent years, most studies in mutant animals are focused on the development and relations of glial cells, since hypomyelination is associated with diverse and complex glial abnormalities affecting differentiation, cell cycle and numbers of glial cells. Oligodendroglial cells show the most severe abnormalities, including oligodendrocyte death. A wide range of reactive alterations in the astrocyte line and in microglial cells have also been described.

INTRODUCTION

A deficiency of myelin can result either from myelin breakdown after it is formed (demyelination) or from failure to produce the normal amount of myelin during development (hypomyelination). Demyelination can be subdivided into primary and secondary categories. Primary demyelination involves the early destruction of myelin with relative preservation of axons and other neighboring elements, although, subsequently, these structures may be affected (Hauw et al., 1992; Quarles et al., 1994; Raine, 1984; Traugott and Raine, 1984). Examples of primary demyelination are multiple sclerosis (MS), which is the most common demyelinating disease of the CNS in man, Guillain-Barré syndrome (GBS), a demyelinating disease of the human PNS, and experimental allergic encephalomyelitis (EAE) and experimental allergic neuritis (EAN), which are considered to be good animal models of MS and GBS, respectively. On the other hand, secondary demyelination includes those disorders in which myelin loss is involved after damage to neurons, axons, blood vessels or glia. Consequently, secondary demyelination is a common sequela of a multitude of conditions initially affecting components other than myelin. Myelin may be secondarily damaged by a number of non-specific alterations affecting the normal integrity of the nervous tissue, such as trauma, infarct necrosis or hemorrhage, and may also be altered after lethal or non-lethal neuronal injuries primarily affecting the viability of neurons and/or the axonal integrity, e.g., secondary demyelination subsequent to neuronal loss in Alzheimer's disease or in the case of Wallerian degeneration. However, it might be noted that, strictly speaking, these disorders primarily affecting neurons/axons and causing secondary myelin destruction are not demyelinating diseases because the axons are destroyed as well, degeneration thus involving the whole nerve fiber (Quarles et al., 1994; Raine, 1984; Traugott and Raine, 1984). We will use the term demyelination to specifically refer to those diseases in which myelin is primarily and selectively affected (primary demyelination).

Diseases involving myelin may also be classified on the basis of their etiology. Diseases causing hypomyelination and demyelination are both included in two categories: acquired diseases and genetically determined disorders. Likewise, acquired diseases can be subdivided into allergic/infectious diseases and toxic/nutritional disorders. It must be noted that the most pronounced changes are similar regardless of etiology: a marked decrease of myelin proteins and lipids, accumulation of myelin breackdrown products within phagocytes, replacement of myelin by hypertrophied astrocytes, and oligodendreglial disfunctions. However, the magnitudes of these changes vary considerably from disease to disease and each disease show some differential features (Duncan, 1995; Hogan and Greenfield, 1984; Quarles et al., 1994; Raine, 1984; Traugott and Raine, 1984). This chapter mainly deals with those genetically determined disorders in which the formation or the maintenance of myelin is affected and only some general features of acquired diseases are discussed.

ACQUIRED DISEASES

Allergic (Inflammatory) and infectious diseases

Allergic and infectious diseases involving myelin are demyelinating diseases that are characterized by the presence of perivascular lesions devoid of myelin and containing inflammatory cells (Raine, 1984; Traugott and Raine, 1984). The sclerotic plaques in MS, the most common example of this family, are grossly visible areas of myelin loss that are the end product of the fusion of small perivascular lesions in which local demyelination had occurred (fig. 18-1A, B). Early lesions contain many lymphocytes, plasma cells and macrophages, axons are preservated, and oligodendrocytes appear to be viable under the microscope. In chronic MS, demyelinated plaques show astrocytic reaction, a marked reduction in the number of inflammatory cells (fig. 18-1C, D) and it is not uncommon to find old plaques showing axonal destruction and loss of oligodendrocytes (Boccaccio and Steinman, 1996; Sobel, 1995). It is interesting to note that inflammatory demyelinating diseases are not restricted to the CNS and also involve the PNS. The GBS is a common PNS disease included within this family. Lesions in GBS are not visible grossly, but microscope examination reveals a multifocal intense inflammation associated with demyelination (Said, 1996).

The particular cause of most acquired demyelinating diseases, including MS, is unknown. It has been hypothesized that these diseases may result from an autoimmune process related to the different proteins of CNS or PNS myelin since immunological mechanisms are clearly implicated in the pathogenesis of these inflammatory diseases (Gulcher et al., 1994; Hartung, 1993; Merrill et al., 1992; Stinissen et al., 1997). The possibility to produce experimental inflammatory demyelination in the EAE and EAN animal models through a single injection of nervous tissue or isolated myelin encephalitogenic antigens give support to this hypothesis (Lisak and Behan, 1975; Swanborg, 1995; Wekerle et al., 1994). The role of cellular and humoral immunological responses in causing the demyelination has also not been fully defined. Whereas T-cell infiltrates seem to be predominantly implicated in MS lesions (Raine, 1991; Utz and McFarland, 1994), a role for humoral immunity is strongly suggested in GBS by findings that sera from GBS patiens cause demyelination in appropiate text systems and that plasma exchange is an effective therapy in many of these patiens (Hartung et al., 1995a; Thornton and Griggs, 1994). A viral etiology has also been proposed and debated for decades (Cook and Dowling, 1980; Kurtzke, 1993; Sarchielli et al., 1993).

In the case of MS, elevated titers of antibody to measles, rubella and herpes zoster virus have been found in the sera of a significant number of MS patients. A precedent infection has also been proposed for GBS (Hartung et al., 1995b; Weiner and Fleming, 1984). Prions have also been implicated as a possible cause in MS (Wojtowicz, 1993). However, it is not always clear whether the immunological activity associated with these diseases is related primarily to an antecedent infection or is autoimmune in nature. Even in the cases in which the viral participation has been accepted, such as the acute disseminated encephalomyelitis in man and the demyelinating diseases caused by canine distemper virus in dogs, visna retrovirus in sheeps and Theiler's murine encephalomyelitis virus in mice, it is unclear what is the amount of damage directly caused by the virus and what by immune mechanisms (Lampert, 1978; Quarles et al., 1994; Raine, 1984). Not the virus per se, but the viral persistence and repeated viral reactivations that the defense system is not able to stop, is postulated as a key factor in MS (Simon and Neubert, 1996). Finally, age, nutritional status, endocrine status, genetic predisposition and environmental factors are also believed to contribute to the pathogenesis of these acquired demyelinating diseases (Bernard and Kerlero de Rosbo, 1992; Duquette and Girard, 1993: Ebers and Sadovnick, 1994; Hutter, 1993; Sayetta, 1986; Tienari, 1994; Oksenberg et al., 1996).

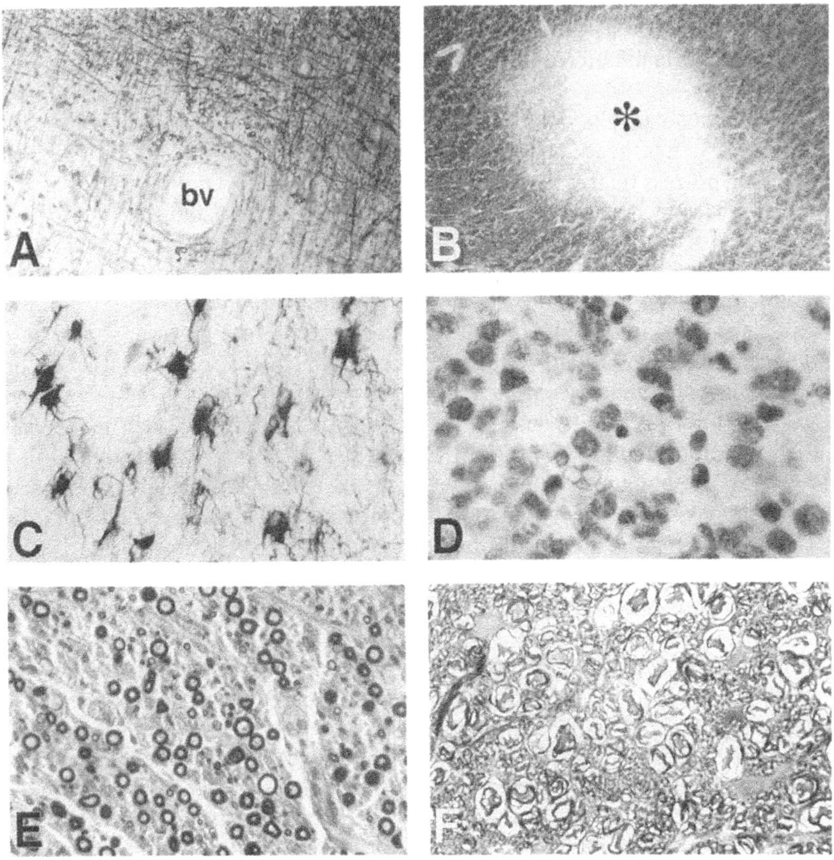

Figure 18-1. General features of myelin diseases. A, B: MS plaques revealed by Klüber-Barrera staining. The myelin loss is evidenced as a reduction or absence in the myelin staining. Photomicrograph A shows a typical MS lesion adjacent to a blood vessel (bv). Photomicrograph B shows a chronic plaque (asterisk) characterized by a pronounced demyelination and a poor celularity, C, D: Glial reactivity associated with myelin loss in MS. Astrocytes, visualized by GFAP immunocytochemistry, are hypertrophied (C) and the number of reactive microglia/macrophages is severely increased in early MS lesions, as demonstrated by CD68 immunocytochemistry (D). E, F: Myelin deficiency secondary to dietary deficiency of thiamine (vitamin B1). Photomicrograph E shows a peripheral nerve section of a patient affected by nutritional deficiency of thiamine (beriberi). Note that an important percentage of nerve fibers is poorly myelinated and is devoid of apparent osmium tetroxide myelin staining. Photomicrograph F shows a Nissl stained semithin section of the spinal cord of a rat nourished by a diet completely deficient in thiamine. Note that not only myelin reduction but also myelin uncompactation and desestructuration is evidenced in white matter. Photomicrographs A, B, C, D and E are cortesy from Dra. Arantxa Ortega (Dept. of Pathology, Hospital Vall d'Hebrón, Barcelona). (magnification A: x24; B: x60; C-D: x510; E: x320; F: x800).

Acquired toxic and nutritional disorders

These diseases involving myelin are characterized by the presence of non-inflammatory lesions in which, if myelin is broken down, the phagocytosis is usually accomplished by cells of local origin. This group of diseases is represented by a collection of alterations all secondary to the action of exogenous myelinotoxic compounds or caused by undernourishment or specific metabolic deficiences. In some cases there may be complete recovery of myelin after the toxic is removed from the diet or the metabolic deficiency is compensated (Blakemore, 1984; Quarles et al., 1994; Raine, 1984; Traugott and Raine, 1984). Both human and animal examples of CNS and PNS affectation have been described.

Hexachlorophene causes a reversible CNS demyelination accompanied by intramyelinic vacuolation and degenerative changes in oligodendroglia. This agent is of clinical significance because it was used in hospitals as an antiseptic agent for humans, particularly for the control of bacterial colonization in the skin of newborns (Rose et al., 1975; Tripier et al., 1981; Yajima and Suzuki, 1979). Intoxication with isonicotinic acid hydrazide and cuprizone, active chelaters of copper, also produce intramyelinic edema, oligodendroglial degeneration and reversible CNS demyelination (Johnson and Ludwin, 1981; Love, 1988; Ludwin, 1978). Acrylamide intoxication is associated with Schwann cell damage and PNS demyelination (LoPachin et al., 1992). Ethanol consumption during gestation and throughout lactation induces glial abnormalities, retardation in glial development, aberrant myelination and hypomyelination (Lancaster et al., 1984; Phillips and Krueger, 1992; Pinazo-Duran et al., 1993) whereas the chronic ethanol consumption in the adult mediates CNS demyelination (Tarnowska-Dziduszko et al., 1995). Some metals, such as lead and mercury, and some organic compounds of these heavy metals, are common environmental pollutants that have been employed as chemicals in several industries, in agriculture and as gasoline additives (Chang, 1990). Lead and other related compounds cause both CNS and PNS hypomyelination by inhibiting myelinogenesis when administered to developing rats and produce severe demyelination in adult animals (Brashear et al., 1978; Tennekoon et al., 1979; Toews et al., 1980; Powell et al., 1982; Coria and Monton, 1988; Winder et al., 1983).

Tellurium intoxication causes degeneration of myelinogenic cells and hypomyelination in the neonatal CNS and PNS (Hammang et al., 1986; Jackson et al., 1989), and it causes a demyelinating neuropathy accompanied by astrocytic reaction in the adult CNS (Smialek et al., 1994). Administration of organometals, such as triethyltin, results in a severe myelin loss which is accompanied by vacuolar edema in the myelin intraperiod line (Chang, 1990; O'Shaughnessy and Losos, 1986; Konat, 1984). The injection of ethidium bromide into the CNS causes demyelination and oligodendrocyte and astrocyte destruction (Blakemore, 1982; Yajima and Suzuki, 1979). Last, biological toxins (e.g., diphtheria toxin) cause marked PNS demyelinative changes (Hildebrand, 1989; Thomas, 1984) and cytotoxines such as tumor necrosis factor could lead to myelin destruction and may be toxic to myelinogenic cells (Selmaj and Raine, 1988; Jenkins and Ikeda, 1992).

Studies on undernourished animals have demonstrated that food deprivation during the first etapes of myelinogenesis, in which there is a burst of oligodendroglial cell proliferation and active synthesis of metabolites, produces a reduction in myelin formation (hypomyelination). Nutritional deprivation during this vulnerable period of development causes a generalized reduction in brain size, but the depression of myelin-specific lipids and proteins is greater than that of other brain components (Royland et al., 1992; Yeh, 1988). The hypomyelination is accompanied by a failure of oligodendroglia to mature and to initiate myelin formation (Lai et al., 1980; Wiggins, 1982). Dietary deficiences of specific substances can also cause myelin deficits. Deficiences of proteins, essential fatty acids, and several vitamins, including thiamine (B_1), cobalamin (B_{12}), and B_6 are associated with a retardation of myelination and/or failure to myelinate properly (hypomyelination) and demyelination (fig. 18-1E, F) (Claus et al., 1985; Morre et al., 1978; van der Westhuyzen et al., 1983). Copper deficiency has also been shown to cause hypomyelination during development and demyelination in the adult CNS (Dake and Amemiya, 1991; Zimmerman et al., 1976). Finally, there are some human demyelinating diseases of obscure etiology with show specific tropism to particular CNS areas, the central pontine myelinolysis and the Marchiafava-Bignami disease, in which alcoholism, hyponatremia/hypernatremia and undernourishment seem to be implicated (Charness, 1993; Messert et al., 1979).

GENETICALLY DETERMINED (HEREDITARY) DISORDERS

The group of hereditary diseases of myelin covers a large number of conditions and variants affecting CNS, PNS, and both CNS and PNS, that are ussually determined by the nature and severity of the genetic affectation (Duncan, 1995; Hogan and Greenfield, 1984; Nave, 1995). There are, however, distinctive features that unify the various diseases in this group. Clinically, mutants exhibit generalized body tremors and seizures that become manifest early in the life, usually prior to 10 years of age in humans and during the first postnatal weeks in rodents. Biochemically, the myelin deficit affects all myelin proteins and lipids, although the deficit in the myelin constituent which is specifically affected by the mutation is usually higher than that of other myelin constituents. Morphologically, mutants show a reduction in the number and thickness of myelin sheaths and abnormalities in the ultrastructure and periodicity of myelin. The myelin deficiency is generalized through the white matter and no plaques or perivascular lesions containing inflammatory cells are seen. The myelin deficit is primarily due to a deficiency in the formation of myelin (hypomyelination) and is related to diverse abnormalities in oligodendrocytes. Astroglial hypertrophy and reactive alterations in microglial cells are also common. In some cases, although the defective oligodendrocytes are able to form some myelin lamellae around the axon, they are unable to correctly mantain the myelin sheath or they degenerate. In these cases, hypomyelination is accompanied by myelin destruction (demyelination) and the phagocytosis of myelin and cell debris is accomplished by local microglial cells (Duncan, 1995; Hogan and Greenfield, 1984; Hudson and Nadon, 1992; Jones, 1972; Quarles et al., 1994; Nave, 1995; Nave, 1994; Raine, 1984). The term "dysmyelination" has also been used to refer to these hypomyelination plus demyelination observed in some mutants. According to the definition of Poser (1968), dysmyelination is a genetically determined disorder of myelinogenesis in which "myelin initially formed was abnormally constituted, thus inherently unstable, vulnerable, and liable to degeneration".

A growing number of inherited CNS myelin disorders have been identified in different species and are models of particular value to investigate abnormalities in glial cells. The pattern and degree of both the myelin deficiency and the associated glial abnormalities are highly dependent on the nature and severity of the mutation. Thus, when possible, we have grouped the different CNS disorders on the basis of their genetic affectation.

PLP Mutants

The myelin proteolipid proteins (PLPs) have been shown to exist in at least two isoforms, the PLP and the less abundant DM-20 protein, and represent 50% of the total protein in CNS myelin (Campagnoni, 1988). The gene encoding PLP is located on the X chromosome and is made up of seven exons spanning 15 Kb. The alternate usage of a cryptic splice site located within the third exon of the gene is involved in the formation of PLP and DM-20 (Mikoshiba et al., 1991; Ikenaka et al., 1991).

Mutants having a mutation in the PLP gene have been of special interest for the study of CNS disorders, since the PLP protein is absent in PNS myelin. PLP mutants exhibit sex-linked inheritance and show myelin deficiency in the CNS whereas the PNS develops normally. Mutations involving PLP have been identified in several species: mice, rats, dogs, rabbits and humans.

Jimpy mouse

The jimpy mutation was the first recognized (Phillips, 1954) and is now the best characterized. Jimpy is a sex-linked recessive mutation of the mouse which produces a drastic myelin deficiency in affected hemizygous males (fig. 18-2A, B) (Privat et al., 1972; Robain and Mandel, 1974; Sidman et al., 1964). Jimpy hemizygous males (jp/Y) develop severe motor tremors beginning at approximately 10 days postpartum, subsequently tonic seizures, and die as a result of respiratory paralysis by 3-4 weeks. In contrast, the heterozygous female carrier of the jimpy gene (jp/+) have normal longevity and there is only an early and partial hypomyelination that is followed by a progressive compensation and complete myelin recovery after the fifth month (Bartlett and Skoff, 1986).

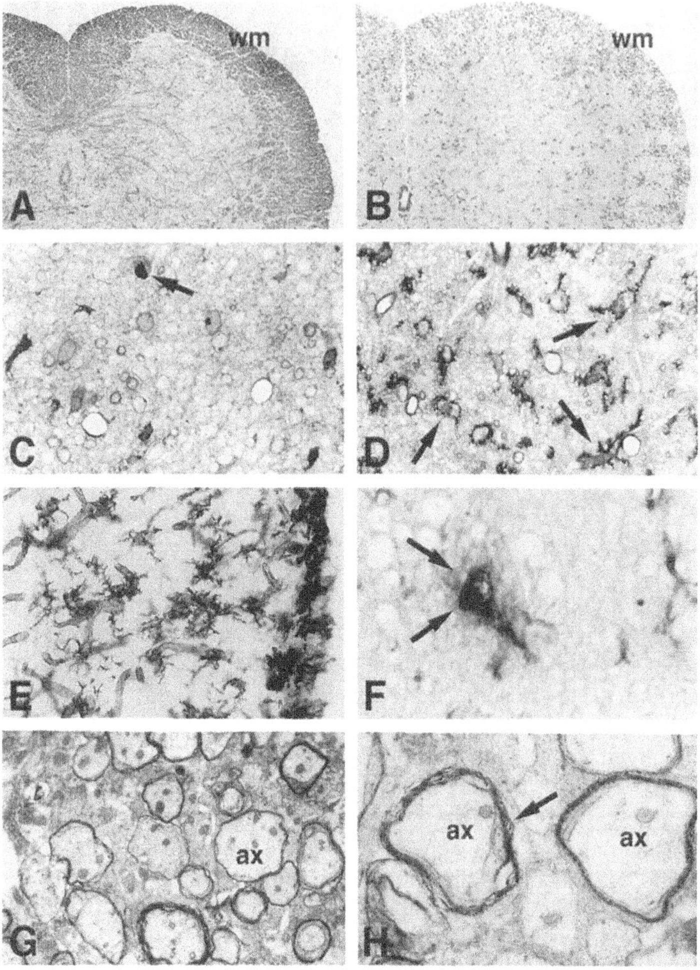

Figure 18-2. A, B: Paraffin sections from 20 day-old normal (A) and (B) jimpy spinal cords. Myelin in the white matter (wm) are visualized by immunocytochemistry for myelin basic protein. Note the absence or scarcity of immunoreactivity in the jimpy white matter. C: Semithin section from the white matter of a 20 day-old jimpy spinal cord. The myelin content is severely reduced and the scarce myelin sheaths present in the jimpy white matter have a reduced diameter. Note the presence of a pyknotic cell (arrow) in the jimpy white matter parenchyma. D, E: Microglial cells in the spinal cord of 20 day-old jimpy animals visualized by nucleoside diphosphatase staining. Semithin section in photomicrograph D shows the presence of numerous microglial cells in the mutant white matter (arrows). The vibratome section in photomicrograph E shows that jimpy microglial cells display reactive morphological features. Reactive microglia are characterized by a shortening and coarsening of their cell processes and swelling of their cell body. F: Paraffin section from the spinal cord of a 12 day-old jimpy animal. The photomicrograph shows a tomato lectin-positive microglial cell surrounding or engulfing a TUNEL-positive apoptotic oligodendrocyte (arrows). G, H: Electron photomicrographs from the white matter of quaking animals at 3 months of age. Note that myelin sheaths surrounding axons (ax) have a reduced thickness and myelin is poorly compacted (arrows). (magnification A: x18; B, x45; C-D: x585; E: x220; F: x750; G: x4160; H: x8300).

The genetic defect in jimpy consists in a point mutation in the PLP gene that produces an A to G transition at the splice acceptor signal for exon 5, which induces an aberrant splicing and subsequent loss of exon 5 and formation of a truncated form of PLP and DM-20 (Dautigny et al., 1986; Macklin et al., 1987; Nave et al., 1986). According to biochemical studies, jimpy male mice show a 95% reduction in the amount of CNS myelin (Yanagisawa and Quarles, 1986). The histology is that of an almost total lack of myelin in the entire CNS at all ages studied (Sidman et al., 1964). The myelin deficiency is evidenced as a reduction in both the number and thickness of myelinated fibers (fig. 18-2C) (Privat et al., 1972; Robain and Mandel, 1974). In addition, scattered myelinated fibers in the white matter display an abnormal compaction and are devoid of a normal intraperiod line (Duncan et al., 1989).

In recent years most studies in jimpy are focused on abnormalities in glial cells. Abnormalities in jimpy oligodendrocytes include deficiencies in myelin components and in metabolic enzymatic activities (Castellano, 1987; Hogan and Greenfield, 1984) as well as severe alterations affecting their development and differentiation. There is evidence for the protracted proliferation of oligodendroglial cells in the jimpy CNS (Privat et al., 1982; Skoff, 1982; Vela et al., 1997a). In addition, proliferative oligodendroglial cells exhibit a cell cycle defect as revealed by comparing between the number of cells which incorporate tritiated thymidine and the number of mitotic cells (Knapp and Skoff, 1987). These authors observed that many jimpy oligodendroglial cells incorporate thymidine without undergoing timely cell division and suggested that the cell cycle of jimpy oligodendroglial cells is lengthened. These abnormalities affecting the cell cycle are further supported by us in a recent paper reporting that proliferative oligodendroglial cells in jimpy show an abnormal expression of the proliferating cell nuclear antigen (Vela et al., 1997a). Surprisingly, although oligodendroblasts actively proliferate in jimpy males, mature oligodendrocytes are sparsely distributed and their number is severely reduced (Castellano, 1987; Farkas-Bargeton et al., 1972; Ghandour and Skoff, 1988; Skoff, 1976). This is explained by the premature death of jimpy oligodendrocytes (fig. 18-2C, F) (Knapp et al., 1986; Vela et al., 1996; Vermeesch et al., 1990). Glial cell death in jimpy is restricted to oligodendrocytes and occurs through an apoptotic mechanism, as demonstrates in situ labeling of nuclear DNA fragmentation and specific glial cell markers (Vela et al., 1996).

Alterations in the astrocyte line have also been largely described in jimpy. Astrocytes exhibit a pronounced gliosis (Privat et al., 1972; Skoff, 1976) and hypertrophied astrocytes showed an increased GFAP immunoreactivity (Best et al., 1988; Dupouey et al., 1980) and GFAP mRNA expression (Li and Bartlett, 1991). The astrocytic hypertrophy, however, is not accompanied by hyperplasia. Most quantitative studies indicate that there is not an increase in the number of astrocytes in jimpy (Skoff, 1976; Best et al., 1988). Accordingly, autoradiographic studies at the electron microscope level indicate that jimpy astrocytes do not show increased thymidine incorporation (Skoff, 1982). The cell cycle of jimpy astrocytes also show cell cycle phase irregularities, as revealed in vivo by analyzing the expression of the proliferating cell nuclear antigen (Vela et al.,

1997a) and in vitro by using the 5-bromodeoxyuridine incorporation method (Knapp and Skoff, 1993). In particular, Knapp and Skoff (1993) found that the total cell cycle of cultured jimpy astrocytes is lengthened by 5-6 hours. There is also a number of complex abnormalities affecting the metabolism of jimpy astrocytes. Astrocytes in jimpy show increased intracellular pH (Knapp et al., 1993) and abnormal metabolic response to K^+-induced stimulation, both in brain slices (Keen et al., 1976) and in vitro (Hertz et al., 1980). Finally, there is an induction of the expression of metallothionein in jimpy and astrocytes account for the majority of cells synthesizing this metal binding and antioxidant protein (Vela et al., 1997b).

Concerning microglia, we have examined the morphology and distribution of microglial cells in the jimpy spinal cord (Vela et al., 1995). The spinal cords of jimpy mice show an intense microglial reaction, as demonstrates by both severe alterations in the morphology and drastic increase in the number of microglial cells in jimpy. Morphologically, microglial cells show a reactive morphology, characterized by a shortening and coarsening of their cell processes, swelling of their cell body and accumulation of heterogeneous bodies and lipid inclusions in their cytoplasm (fig. 18-2D, E). In addition, reactive microglial cells are found in close association with nerve fibers and surrounding or engulfing cells showing nuclear pyknosis. Further study of the association between degenerative cells and microglia reveal that degenerative cells are oligodendrocytes and we found a remarkable association between apoptotic oligodendroglial cells and microglia (fig. 18-2F) (Vela et al., 1996). Jimpy microglial cells were closely attached to and even surrounding apoptotic cells by means of their cell processes. The mean microglial densities in the jimpy white matter were about threefold (10-12 days) and fivefold (20-22 days) higher than in the normal, whereas in the gray matter, the microglial density in jimpy was about 60% higher than in normal at both ages (Vela et al., 1995). This marked increase in the numbers of microglial cells populating the jimpy spinal cord correlates well with the observation that jimpy microglial cells actively proliferate (Vela et al., 1997a). Finally, microglial cells in jimpy, but not in normal animals, express metallothionein (Vela et al., 1997b). This protein is upregulated in injured tissues and plays a protective antioxidant role by acting as a free radical scavenger (Sato and Bremner, 1993).

Jimpy myelin synthesis deficient mouse

Myelin synthesis deficient (jimpy^{msd}; jp^{msd}) is another sex-linked mutation affecting mice, has the same position on the linkage map as jimpy, and is through to be allelic with jimpy. The genetic defect consist in a point mutation in exon 6 of the PLP gene which converts an alanine to a valine (Gencic and Hudson, 1990). As in the case of jimpy, the jp^{msd} mice have reduced levels of mRNA encoding PLP, although in contrast to PLP mRNA in jimpy mice, the PLP mRNA is of normal size, indicating no splicing defect (Gardinier and Macklin, 1988).

This mutation in the mouse reduce the amount of CNS myelin and produce glial abnormalities. In fact, the disease can be distinguished morphologically from jimpy only by their severity. Jimpy and jp^{msd} seem identical in every way except that jpmsd CNS has about twice as much myelin as jimpy CNS. Another difference between normal and jp^{msd} mice is an altered ratio of PLP to DM-20. Over 10 days of age in normal mice, the PLP protein is present at a significantly higher level than DM-20 protein, whereas in jp^{msd} the two proteins are present at approximately equivalent levels, albeit at significantly reduced levels (Gardinier and Macklin, 1988). Thus, the defect in the jp^{msd} mice include an inability of oligodendrocytes to shift from the early stage, with low PLP expression and DM-20 predominance, to the later stage, with high expression of PLP and low expression of DM-20. Surprisingly, when the mRNA ratio was analyzed, the PLP mRNA levels were higher than DM-20 mRNA levels, suggesting that in jp^{msd} oligodendrocytes either the translation of PLP mRNA is less efficient than DM-20 translation, or PLP protein is turned over more rapidly than DM-20 protein (Macklin et al., 1991).

Rumpshaker mouse

Rumpshaker (rsh; jp^{rsh}) is another PLP mutantion in the mouse resulting in CNS hypomyelination (Griffths et al., 1990). The mutation in these mice produces a single nucleotide difference in the fourth exon of the PLP gene: the codon 186 was changed from ATT (Ile) to ACT (Thr). Unlike jimpy and similar to jp^{msd}, the mutation do not results in a splicing defect since PLP mRNAs appeared normal in size (Schneider et al., 1992). Hypomyelination in the rumpshaker mouse is characterized by the presence of unmyelinated axons with intervening astrocytic processes, showing similarities to the jimpy mouse. However, unlike jimpy, these mice have normal longevity, can breed, and have a full complement of morphologically normal oligodendrocytes. The hypomyelination is less severe than in jimpy and it is not caused by a lack of oligodendrocytes since their number is not reduced and they do not degenerate (Fanarraga et al., 1992; Schneider et al., 1992). Additionally, the thin myelin sheaths present in rsh spinal cord show a marked deficiency in PLP, but DM-20 levels are relatively normal (Fanarraga et al., 1992). Analysis of the mRNA PLP/DM-20 ratio also revealed a reduction of PLP mRNA and an increase in the relative abundance of the DM-20 isoform (Schneider et al., 1992). Astrocyte GFAP levels are also increased in this murine mutant (Fanarraga et al., 1992).

Myelin deficient rat

The myelin-deficient (md) rats have a sex-linked recessive mutation resulting in severe hypomyelination of the CNS, whereas myelination of the PNS is normal (Csiza and de Lahunta, 1979). The genetic defect consists in a point mutation in exon III of the PLP gene that produces an A---C transversion and creates an additional restriction site (Boison and Stoffel, 1989). The disorder causes tremor and ataxia, noticeable at 10-12 days of age, and seizures beginning 16-21 days after birth. Most animals die 24-28 days postnatally. The morphological and biochemical studies on the myelin-deficient rat also demonstrate many similarities with the jimpy mutant mice. There are scattered myelinated fibers that are abnormally compacted and lack a normal intraperiod line (Duncan et al., 1987). The md contain abnormal immature oligodendroglia and the number of mature oligodendrocytes is reduced. The virtual absence of normal mature oligodendrocytes is accompanied by active proliferation of immature cells and, as in jimpy, the oligodendroglia die prematurely (Boison and Stoffel, 1989; Dentinger et al., 1985; Jackson and Duncan, 1988). Oligodendrocytes show a prominent dilatation of their nuclear envelope and rought endoplasmic reticulum (Dentinger et al., 1982). Astrocytes are hypertrophied and the number of reactive microglia, containing large cytoplasmic lipid droplets, is also greater than in normal animals in the oldest mutants (Dentinger et al., 1985; Jackson and Duncan, 1988).

Shaking pup

The shaking pup is another sex-linked mutation affecting the spaniel dog. The mutation is located in the PLP gene and leads to severe hypomyelination of the CNS with associated tremor and seizures (Griffiths et al., 1981a; Griffiths et al., 1981b; Nadon et al., 1990). PNS myelination is unaffected (Nadon and Duncan, 1996). As md rats and jimpy mice, this canine hypomyelinating mutant have low levels or absence of PLP and the number of undifferentiated glial cells is higher than in control canine brains. There are a marked reduction in the number of mature oligodendrocytes with many of these cells having distended rough endoplasmic reticulum. Oligodendrocyte death is not noticeable or is not a marked feature in shaking pups (Duncan et al., 1983). There is no available studies focused on astrocytes and microglia in this PLP mutant.

Paralytic tremor rabbit

Paralytic tremor is a neurological mutation of Chinchilla rabbits that produces strong CNS hypomyelination (Sypecka et al., 1993). The genetic defect consist in a single nucleotide change (T---A) in exon 2 which results in the substitution of a histidine by a glutamine at position 36 (Tosic et al., 1994a; Tosic et al., 1994b). Morphological studies indicate that number of oligodendrocytes is not reduced, but myelination is slower, irregular and defective. At 4 weeks, the amount of myelin in the mutant represents 25% of the normal level (Tosic et al., 1993).

Pelizaeus-Merzbacher (human disease)

The Pelizaeus-Merzbacher disease is an hereditary human disease that involves the X chromosome-linked gene for PLP. More than 30 different mutations in this gene have been defined and associated with Pelizaeus-Merzbacher disease (Seitelberger, 1995). Different genetic alterations in the PLP gene, including point mutations (Hodes et al., 1997), deletions (Kleindorfer et al., 1995) and duplications (Inonue et al., 1996), have been found to lead to this disease.

Most patients with Pelizaeus-Merzbacher disease, like jimpy and other PLP mutants, share a profound deficit of PLP in the CNS whereas the PNS is normal. The myelin deficiency is associated with a lack of normal mature oligodendrocytes (Koeppen et al., 1988).

MBP mutants

The myelin basic proteins (MBPs) are a class of myelin proteins consisting of multiple polypeptide chains varying in molecular mass from 14 to 21.5 KDa. At least six isoforms have been identified in mice. MBPs represent 30% of the protein content of CNS myelin. The mouse MBP is encoded by a single gene with seven exons (Campagnoni, 1988; Ikenaka et al., 1991)

Shiverer mouse

Shiverer is an autosomal recessive mutation in the mouse. Both sexes are affected and is characterized by generalized tremors, seizures, and a shortened life span (Chernoff, 1981). At about 12 days of age, mutant mice develop a generalized tremor, resembling a shiver, that accompanies movement by the animal. When animals are at rest, no tremors can be seen. From about 30 days onward, the mice have tonic seizures that increase in frequency and severety. Mutants die between 50 and 100 days of age. The genetic defect is located on the chromosome 18 and affects the MBP structural gene. The normal MBP gene is composed of seven exons, whereas the gene encoding MBP in shiverer contains a deletion with only the first two exons remaining (Roach et al., 1985).

The CNS is hypomyelinated but the amount of PNS myelin appears normal and only subtle abnormalities occur in PNS. Although MBP is present in both CNS and PNS, it is less abundant and apparently plays a less prominent role in the PNS. The CNS myelin, wherever present, is not well compacted and lacks the major dense line (Inoue et al., 1981; Privat et al., 1979). Unlike jimpy and other PLP mutants, hypomyelination does not result from oligodendrocyte death. In fact, there appear to be elevations in the oligodendrocyte density and in process elaboration by oligodendrocytes. Oligodendrocytes, however, have an abnormal shape, display numerous microprocesses which may be associated with either myelinated or unmyelinated axons, contain abnormal vacuoles derived from the Golgi apparatus, increased numbers of microtubules and do not normally assemble the cytoskeleton (Cammer et al., 1985; Dyer, 1995; Inoue et al., 1983; Nagaike et al., 1982; Privat et al., 1979; Rosenbluth, 1980). In relation to astrocytes, it has been described a considerable astrogliosis in the white matter

(Cammer et al., 1985). There is no evidence of microgliosis, sudanophilia or macrophage infiltration.

Shiverer myelin deficient mouse

The shiverer myelin deficient (shiverer^{mld}; shi^{mld}) mutant mouse (Doolittle and Schweikart, 1977) has also a mutation in the MBP structural gene. The MBP gene in shimld/shimld mice has undergone a duplication and an inversion affecting the order of the exons. Shiverer and shi^{mld} were shown to be alleles, but the myelin deficient mutant is less severely affected and low levels of MBP may be detected (Readhead and Hood, 1990). Defects of shi^{mld} myelination are comparable to those in shi with respect to the percentage of axons myelinated, sheath thickness, errors in the wrapping and targeting of myelin and abnormal oligodendrocyte shape. However the major dense line is present in a few shi^{mld} myelin sheaths and a few sheaths show myelin basic protein by immunocytochemistry, while neither phenomenon is seen in shiverer at this age in the same CNS regions (Mikoshiba et al., 1987; Shen et al., 1985).

Quaking

Quaking is an autosomal recessive mutant mouse that has a single mutation on the chromosome 17 which affects mouse CNS myelination. The nature of the primary gene defect in quaking is, however, unknown. The location of quaking mutation on chromosome 17 is remote from the known genes for myelin-specific proteins (MBP on 18; MAG on 7; PLP on X).

These mice exhibit frequent seizures that appear when animals are 10 to 12 days old and reaches full expression by 3 weeks. Morphologically, mutants exhibit a severe hypomyelination in the CNS (Sidman et al., 1964). The peripheral myelination is grossly normal (Sidman et al., 1964), although ultrastructural observations reveal the occurrence of disproportions in axon caliber and myelin sheath tickness (Samorajski et al., 1970). The quaking CNS include reduction in the numbers of myelin sheaths (less severe than in jimpy) and myelin sheaths are abnormal or distorted, uncompacted and thinner than wild type (fig. 18-2G, H) (Billings-Gagliardi et al., 1980; Nagara and Suzuki, 1981; Samorajski et al., 1970). Quaking mice show increased rather than decreased numbers of oligodendrocytes, but the volume of myelin per oligodendrocyte varies between 2-11% of normal (Friedrich, 1975). Degenerating cells are rare or absent (Billings-Gagliardi et al., 1980). The nuclei and perikarya of quaking oligodendrocytes are normal in size or smaller than normal oligodendrocytes (Friedrich, 1975). Many oligodendroglia appear as abnormal cells filled with membraneous inclusions and show a prominent dilation of smooth walled vesicles and cisterns (Nagara and Suzuki, 1981). Both, aberrant myelin figures and oligodendroglial abnormalities are very prominent by 5-15 postnatal days, but such changes gradually subside. The density of astroglia observed by GFAP immunostaining is higher in the white matter of quaking mice than in the controls (Mikoshiba et al., 1979). In addition, it has been described the presence of edematous or vacuolated astrocytes in quaking (Zhabotinskii et al., 1983). Microglial cells show a normal shape, there is no apparent increase in the density of microglia and phagocytic cells are not present (Billings-Gagliardi, 1980; Castellano, 1987; Samorajski et al., 1970).

FURTHER RESEARCH

Disorders involving myelin include a wide spectrum of neurological diseases that have a great significance for human health. Despite their heterogeneity, these diseases are characterized by a myelin deficiency that is accompanied by a variety of abnormalities in glial cells. Actually, the myelin deficiency must be considered as a secondary phenomenon that results from the primary affectation of glial cells. In particular, abnormalities affecting directly or indirectly the function, differentiation pattern and/or the viability of oligodendrocytes play a determinant role in CNS myelin diseases. Disturbances affecting astrocytes and microglia may also have a special significance. In fact, several findings point to the existence of a regulatory network involving a variety of soluble factors and cell to cell interactions that influence the growth and differentiation of all glial cell types during development and myelinogenesis. Animals displaying either hereditary or experimentally-induced myelin diseases may be models of particular value to analyze disturbances in extracellular signals implicated in the crosstalk between glial cells. Future investigations focused on the understanding of glial cell relations may be of vital importance in order to design new therapeutic strategies for myelin diseases.

ACKNOWLEDGMENTS

We would like to thank Miguel A. Martil for his excellent technical assistance. The work of our group presented in this review was supported by DGICYT project PB95-0662.

REFERENCES

Bartlett, W.P. and Skoff, R.P. (1986) Expression of the jimpy gene in the spinal cords of heterozygous female mice. I. An early deficit followed by compensation, J. Neurosci., *6*: 2802-2812.

Bernard, C.C. and Kerlero de Rosbo N. (1992) Multiple sclerosis: an autoimmune disease of multifactorial etiology, Curr. Opin. Immunol., *4*:760-765.

Best, T.T., Skoff, R.P. and Bartlett, W.P. (1988) Astroglial plasticity in hemizygous and heterozygous jimpy mice, Int. J. Dev. Neurosci., *6*:39-57.

Billings-Gagliardi, S., Adcock, L.H. and Wolf, M.K. (1980) Hypomyelinated mutant mice: description of jpmsd and comparison with jp and qk on their present genetic backgrounds, Brain Res., *194*:325-338.

Blakemore, W.F. (1982) Ethidium bromide induced demyelination in the spinal cord of the cat, Neuropathol. Appl. Neurobiol., *8*:365-375.

Blakemore, W.F. (1984) The response of oligodendrocytes to chemical injury, Acta Neurol. Scand. Suppl., *100*:33-38.
Boccaccio, G.L. and Steinman, L. (1996) Multiple sclerosis: from a myelin point of view, J. Neurosci. Res., *45*:647-654.
Boison, D. and Stoffel, W. (1989) Myelin-deficient rat: a point mutation in exon III (A---C, Thr75---Pro) of the myelin proteolipid protein causes dysmyelination and oligodendrocyte death, EMBO J., *8*:3295-3302.
Brashear, C.W., Kopp, V.J. and Krigman, M.R. (1978) Effect of lead on the developing peripheral nervous system, J. Neuropathol. Exp. Neurol., *37*:414-425.
Cammer, W., Sacchi, R., Kahn, S. and Sapirstein, V. (1985) Oligodendroglial structures and distribution shown by carbonic anhidrase immunostaining in the spinal cords of developing normal and shiverer mice, J. Neurosci. Res., *14*:303-316.
Campagnoni, A.T. (1988) Molecular biology of myelin proteins from the central nervous system, J. Neurochem., *51*:1-14.
Castellano, B. (1987) Estudio histoquímico de la actividad TPPasa/NDPasa en neuronas y células gliales del sistema nervioso central. Doctoral Thesis, Autonomous University of Barcelona, 312 pp.
Chang, L.W. (1990) The neurotoxicology and pathology of organomercury, organolead, and organotin, J. Toxicol. Sci., *15*:125-151.
Charness, M.E. (1993) Brain lesions in alcoholics, Alcohol Clin. Exp. Res., *17*:2-11.
Chernoff, G.F. (1981) Shiverer: an autosomal recessive mutant mouse with myelin deficiency, J. Hered., *72*:128.
Claus, D., Eggers, R., Warecka, K. and Neundorfer, B. (1985) Thiamine deficiency and nervous system function disturbances, Eur. Arch. Psychiatry Neurol. Sci., *234*:390-394.
Cook, S.D. and Dowling, P.C. (1980) Multiple sclerosis and viruses: an overview, Neurology, *30*:80-91.
Coria, F. and Monton, F. (1988) Recovery of the early cellular changes induced by lead in rat peripheral nerves after withdrawal of the toxin, J. Neuropathol. Exp. Neurol., *47*:282-290.
Csiza C.K. and de Lahunta, A. (1979) Myelin deficiency (md): a neurologic mutant in the Wistar rat, Am. J. Pathol., *95*:215-223.
Dake, Y. and Amemiya, T. (1991) Electron microscopic study of the optic nerve in copper deficient rats, Exp. Eye Res., *52*:277-281.
Dautigny, A., Mattei, M.-G., Morello, D., Alliel, P.M., Pham-Dinh, D., Amar, L., Arnaud, D., Simon, D., Mattei, J.-F., Guenet, J.-L., Jollès, P. and Avner, P. (1986) The structural gene coding for myelin-associated proteolipid protein is mutated in jimpy mice, Nature, *321*:867-869.
Dentinger, M.P., Barron, K.D. and Csiza, C.K. (1985) Glial and axonal development in optic nerve of myelin deficient rat mutant, Brain Res., *344*:255-266.
Dentinger, M.P., Barron, K.D. and Csiza, C.K. (1982) Ultrastructure of the central nervous system in a myelin deficient rat, J. Neurocytol., *11*:671-691.
Doolittle, D.P. and Schweikart, K.M. (1977) Myelin deficient, a new neurological mutant in the mouse, J. Hered., *68*:331-332.
Duncan, I.D., Griffiths, I.R. and Munz, M. (1983) "Shaking pups": a disorder of central myelination in the spaniel dog. III. Quantitative aspects of glia and myelin in the spinal cord and optic nerve, Neuropathol. Appl. Neurobiol., *9*:355-368.
Duncan, I.D., Hammang, J.P. and Trapp, B.D. (1987) Abnormal compact myelin in the myelin-deficient rat: absence of proteolipid protein correlates with a defect in the intraperiod line, Proc. Natl. Acad. Sci. USA, *84*:6287-6291.
Duncan, I.D., Hammang, J.P., Goda, S. and Quarles, R.H. (1989) Myelination in the jimpy mouse in the absence of proteolipid protein, Glia, *2*:148-154.
Duncan, I.D. (1995) Inherited disorders of myelination of the central nervous system. In H. Kettenmann and B.R. Ransom (Eds.), Neuroglia, Oxford University Press, New York. pp. 990-1009.
Dupouey, P., Lucas, C.V., Gomes, D. and Jacque, C. (1980) Immunohistochemical localization of the myelin basic protein and of the glial fibrillary acidic protein: Comparative study in normal, quaking and jimpy mice, J. Neurosci. Res., *5*:387-398.
Duquette, P. and Girard, M. (1993) Hormonal factors in susceptibility to multiple sclerosis, Curr. Opin. Neurol. Neurosurg., *6*:195-201.

Dyer, C.A., Philibotte, T.M., Billings-Gagliardi, S. and Wolf, M.K. (1995) Cytoskeleton in myelin-basic- protein-deficient shiverer oligodendrocytes, Dev. Neurosci., *17*:53-62.

Ebers, G.C. and Sadovnick, A.D. (1994) The role of genetic factors in multiple sclerosis susceptibility, J. Neuroimmunol., *54*:1-17.

Fanarraga, M.L., Griffiths, I.R., McCulloch, M.C., Barrie, J.A., Kennedy, P.G. and Brophy, P.J. (1992) Rumpshaker: an X-linked mutation causing hypomyelination: developmental differences in myelination and glial cells between the optic nerve and spinal cord, Glia, *5*:161-170.

Farkas-Bargeton, E., Robain, O. and Mandel, P. (1972) Abnormal glial maturation in the white matter in jimpy mice. An optical study, Acta Neuropathol., *21*:272-281.

Friedrich, V.L.Jr. (1975) Hyperplasia of oligodendrocytes in quaking mice, Anat. Embriol. (Berl.), *147*:259-271.

Gardinier, M.V. and Macklin, W.B. (1988) Myelin proteolipid protein gene expression in jimpy and jimpy(msd) mice, J. Neurochem., *51*:360-369.

Gencic, S. and Hudson, L.D. (1990) Conservative amino acid substitution in the myelin proteolipid protein of jimpymsd mice, J. Neurosci., *10*:117-124.

Ghandour, M.S. and Skoff, R.P. (1988) Expression of galactocerebroside in developing normal and jimpy oligodendrocytes in situ, J. Neurocytol., *17*:485-498.

Griffiths, I.R., Duncan, I.D. and McCulloch, M. (1981b) Shaking pups: a disorder of central myelination in the spaniel dog. II. Ultrastructural observations on the white matter of the cervical spinal cord, J. Neurocytol., *10*:847-858.

Griffiths, I.R., Duncan, I.D., McCulloch, M. and Harvey, M.J. (1981a) Shaking pups: a disorder of central myelination in the spaniel dog. Part 1. Clinical, genetic and light microscopical observations, J. Neurol. Sci., *50*:423-433.

Griffiths, I.R., Scott, I., McCulloch, M.C., Barrie, J.A., McPhilemy, K. and Cattanach, B.M. (1990) Rumpshaker mouse: a new X-linked mutation affecting myelination: evidence for a defect in PLP expression, J. Neurocytol., *19*:273-283.

Gulcher, J.R., Vartanian, T. and Stefansson, K. (1994) Is multiple sclerosis an autoimmune disease?, Clin. Neurosci., *2*:246-252.

Hammang, J.P., Duncan, I.D. and Gilmore, S.A. (1986) Degenerative changes in rat intraspinal Schwann cells following tellurium intoxication, Neuropathol. Appl. Neurobiol., *12*:359-370.

Hartung, H.P. (1993) Immune-mediated demyelination, Ann. Neurol., *33*:563-567.

Hartung, H.P., Pollard, J.D., Harvey, G.K. and Toyka, K.V. (1995a) Immunopathogenesis and treatment of the Guillain-Barre syndrome--Part I, Muscle Nerve, *18*:137-153.

Hartung, H.P., Pollard, J.D., Harvey, G.K. and Toyka, K.V. (1995b) Immunopathogenesis and treatment of the Guillain-Barre syndrome--Part II, Muscle Nerve, *18*:154-164.

Hauw, J.J., Delaere, P., Seilhean, D. and Cornu, P. (1992) Morphology of demyelination in the human central nervous system, J. Neuroimmunol., *40*:139-152.

Hertz, L., Chaban, G. and Hertz, E. (1980) Abnormal metabolic response to excess potassium in astrocytes from the jimpy mouse, a convulsing neurological mutant, Brain Res., *181*:482-487.

Hildebrand, C. (1989) Myelin sheath remodelling in remyelinated rat sciatic nerve, J. Neurocytol., *18*:285-294.

Hodes, M.E., Blank, C.A., Pratt, V.M., Morales, J., Napier, J. and Dlouhy, S.R. (1997) Nonsense mutation in exon 3 of the proteolipid protein gene (PLP) in a family with an unusual form of Pelizaeus-Merzbacher disease, Am. J. Med. Genet., *69*:121-125.

Hogan, E.L. and Greenfield S. (1984) Animal models of genetic disorders of myelin. In P. Morell (Ed.), Mielin, 2nd Ed., Plenum Press, New York. pp. 489-534.

Hogan, E.L. and Greenfield S. (1984) Animal models of genetic disorders of myelin. In P. Morell (Ed.), Mielin, 2nd Ed., Plenum Press, New York. pp. 489-534.

Hudson, L.D. and Nadon, N.L. (1992) Amino acid substitutions in proteolipid protein that cause dysmyelination. In R.E. Martenson (Ed.), Myelin biology and chemistry, Telford Press, New Jersey. pp. 677-702.

Hutter, C. (1993) On the causes of multiple sclerosis, Med. Hypotheses, *41*:93-96.

Ikenaka, K., Okano, H., Tamura, T. and Mikoshiba, K. (1991) Recent advances in studies on genes for myelin proteins, Develop. Growth & Differ., *33*:181-192.

Inoue, K., Osaka, H., Sugiyama, N., Kawanishi, C., Onishi, H., Nezu, A., Kimura, K., Yamada, Y. and Kosaka, K. (1996) A duplicated PLP gene causing Pelizaeus-Merzbacher disease detected by comparative multiplex PCR, Am. J. Hum. Genet., *59*:32-39.

Inoue, Y., Inoue, K., Terashima, T., Mikoshiba, K. and Tsukada, Y. (1983) Developmental changes of oligodendroglia in the posterior funiculus of "shiverer" mutant mouse spinal cord, with special reference to myelin formation, Anat. Embryol. (Berl.), *168*:159-171.

Inoue, Y., Nakamura, R., Mikoshiba, K. and Tsukada, Y. (1981) Fine structure of the central myelin sheath in the myelin deficient shiverer mouse, with special reference to the pattern of myelin formation by oligodendroglia, Brain Res., *219*:85-94.

Jackson, K.F. and Duncan, I.D. (1988) Cell kinetics and cell death in the optic nerve of the myelin deficient rat, J. Neurocytol., *17*:657-670.

Jackson, K.F., Hammang, J.P., Worth, S.F. and Duncan, I.D. (1989) Hypomyelination in the neonatal rat central and peripheral nervous systems following tellurium intoxication, Acta Neuropathol. (Berl.), *78*:301-309.

Jenkins H.G. and Ikeda, H. (1992) Tumor necrosis factor causes an increase in axonal transport of protein and demyelination in the mouse optic nerve, J. Neurol. Sci., *108*:99-104.

Johnson, E.S. and Ludwin, S.K. (1981) The demonstration of recurrent demyelination and remyelination of axons in the central nervous system, Acta Neuropathol. (Berl.), *53*:93-98.

Jones, T.C. (1972) Hereditary disease. In K. Beneyrechke, F.M. Garner and T.C. Jones (Eds.), Pathology of laboratory animals, Vol. II, Springer-Verlag, New York. pp. 1982-2048.

Keen, P., Osborne, R.H. and Pehrson, U.M.M. (1976) Respiration and metabolic compartmentation in brain slices from a glia-deficient mutant, the jimpy mouse, J. Physiol., *254*:22-23.

Kleindorfer, D.O., Dlouhy, S.R., Pratt, V.M., Jones, M.C., Trofatter, J.A. and Hodes, M.E. (1995) In-frame deletion in the proteolipid protein gene of a family with Pelizaeus-Merzbacher disease, Am. J. Med. Genet., *55*:405-407.

Knapp, P.E. and Skoff, R.P. (1987) A defect in the cell cycle of neuroglia in the myelin deficient jimpy mouse, Dev. Brain Res., *35*:301-306.

Knapp, P.E. and Skoff, R.P. (1993) Jimpy mutation affects astrocytes: lengthening of the cell cycle in vitro, Dev. Neurosci., *15*:31-36.

Knapp, P.E., Skoff, R.P. and Redstone, D.W. (1986) Oligodendroglial cell death in jimpy mice: an explanation for the myelin deficit, J. Neurosci., *6*:2813-2822.

Koeppen, A.H., Barron, K.D., Csiza, C.K. and Greenfield, E.A. (1988) Comparative immunocytochemistry of Pelizaeus-Merzbacher disease, the jimpy mouse, and the myelin-deficient rat, J. Neurol. Sci., *84*:315-327.

Konat, G. (1984) Triethyllead and cerebral development: an overview, Neurotoxicology, *5*:87-96.

Kurtzke, J.F. (1993) Epidemiologic evidence for multiple sclerosis as an infection, Clin. Microbiol. Rev., *6*:382-427.

Lai, M., Lewis, P.D. and Palet, A.J. (1980) Effects of undernutrition on gliogenesis and glial maturation in rat corpus callosum, J. Comp. Neurol., *193*:965-972.

Lampert, P.W. (1978) Autoimmune and virus-induced demyelinating diseases. A review, Am. J. Pathol., *91*:176-208.

Lancaster, F.E., Phillips, S.M., Patsalos, P.N. and Wiggins, R.C. (1984) Brain myelination in the offspring of ethanol-treated rats: in utero versus lactational exposure by crossfostering offspring of control, pairfed and ethanol treated dams, Brain Res., *309*:209-216.

Li, X. and Bartlett, W.P. (1991) Developmental expression of glial fibrillary acidic protein and glutamine synthetase mRNAs in normal and jimpy mice, Mol. Brain Res., *9*:313-317.

Lisak, R.P. and Behan, P.O. (1975) Experimental autoimmune demyelinating diseases: experimental allergic encephalomyelitis and experimental allergic neuritis, Biomedicine, *22*:81-87.

LoPachin, R.M., Castiglia, C.M. and Saubermann, A.J. (1992) Acrylamide disrupts elemental composition and water content of rat tibial nerve. II. Schwann cells and myelin, Toxicol. Appl. Pharmacol., *115*:35-43.

Love, S. (1988) Cuprizone neurotoxicity in the rat: morphologic observations, J. Neurol. Sci., *84*:223-237.

Ludwin, S.K. (1978) Central nervous system demyelination and remyelination in the mouse: an ultrastructural study of cuprizone toxicity, Lab. Invest., *39*:597-612.

Macklin, W.B., Gardinier, M.V., King, K.D., and Kampf, K. (1987) An AG-GG transition at a splice site in the myelin proteolipid protein gene in jimpy mice results in the removal of an exon, FEBS Lett., *223*:417-421.

Macklin, W.B., Gardinier, M.V., Obeso, Z.O., King, K.D. and Wight, P.A. (1991) Mutations in the myelin proteolipid protein gene alter oligodendrocyte gene expression in jimpy and jimpymsd mice, J. Neurochem., *56*:163-171.

Merrill, J.E., Graves, M.C. and Mulder, D.G. (1992) Autoimmune disease and the nervous system. Biochemical, molecular, and clinical uptake, West J. Med., *156*:639-646.

Messert, B., Orrison, W.W., Hawkins, M.J. and Quaglieri, C.E. (1979) Central pontine myelinolisis. Considerations on etiology, diagnosis, and treatment, Neurology, *29*:147-160.

Mikoshiba, K., Nagaike, K., Aoki, E. and Tsukada, Y. (1979) Biochemical and immunohistochemical studies on dysmyelination of quaking mutant mice in vivo and in vitro, Brain Res., *177*:287-299.

Mikoshiba, K., Okano, H., Inoue, Y., Fujishiro, M., Takamatsu, K., Lachapelle, F., Baumann, N. and Tsukada, Y. (1987) Immunohistochemical, biochemical and electron microscopic analysis of myelin formation in the central nervous system of myelin deficient (mld) mutant mice, Brain Res., *432*:111-121.

Mikoshiba, K., Okano, H., Tamura, T. and Ikenaka, K. (1991) Structure and function of myelin protein genes, Annu. Rev. Neurosci., *14*:201-217.

Morre, D.M., Kirksey, A. and Das, G.D. (1978) Effects of vitamin B-6 deficiency on the developing central nervous system of the rat. Myelination, J. Nutr., *108*:1260-1265.

Nadon, N.L. and Duncan, I.D. (1996) Molecular analysis of glial cell development in the canine "shaking pup" mutant, Dev. Neurosci., *18*:174-184.

Nadon, N.L., Duncan, I.D., Hudson, L.D. (1990) A point mutation in the proteolipid protein gene of the "shaking pup" interrupts oligodendrocyte development, Development, *110*:529-537.

Nagaike, K., Mikoshiba, K., Tsukada, Y. (1982) Dysmyelination of shiverer mutant mice in vivo and in vitro, J. Neurochem., *39*:1235-1241.

Nagara, H. and Suzuki, K. (1981) Chronological study of oligodendroglial alterations and myelination in quaking mice, Neuropathol. Appl. Neurobiol., *7*:135-149.

Nave, K.-A., Lai, C., Bloom, F.E. and Milner, R.J. (1986) Jimpy mutant mouse: a 74-base deletion in the mRNA for myelin proteolipid protein and evidence for a primary defect in RNA splicing, Proc. Natl. Acad. Sci. USA, *83*:9264-9268.

Nave, K.-A. (1994) Neurological mouse mutants and the genes of myelin, J. Neurosci. Res., *38*:607-612.

Nave, K.-A. (1995) Neurological mouse mutants: a molecular-genetic analysis of myelin proteins. In H. Kettenmann and B.R. Ransom (Eds.), Neuroglia, Oxford University Press, New York. pp. 571-586.

O'Shaughnessy, D.J. and Losos, G.J. (1986) Peripheral and central nervous system lesions caused by triethyl- and trimethyltin salts in rats, Toxicol. Pathol., *14*:141-148.

Oksenberg, J.R., Seboun, E. and Hauser, S.L. (1996) Genetics of demyelinating diseases, Brain Pathol., *6*:289-302.

Phillips, D.E. and Krueger, S.K. (1992) Effects of combined pre- and postnatal ethanol exposure (three trimester equivalency) on glial cell development in rat optic nerve, Int. J. Dev. Neurosci., *10*:197-206.

Phillips, R.J.S. (1954) Jimpy, a new totally sex-linked gene in the house mouse, Z. Vererbungsl., *86*:322-326.

Pinazo-Duran, M.D., Renau-Piqueras, J. and Guerri, C. (1993) Developmental changes in the optic nerve related to ethanol consumption in pregnant rats: analysis of the ethanol-exposed optic nerve, Teratology, *48*:305-322.

Poser, C.M. (1968) Diseases of the myelin sheath. In J.F. Minckler (Ed.), Pathology of the nervous system, Vol. I, McGraw-Hill, New York. pp. 767-820.

Powell, H.C., Myers, R.R. and Lampert, P.W. (1982) Changes y Schwann cells and vessels in lead neuropathy, Am. J. Pathol., *109*:193-205.

Privat, A., Jacque, C., Bourre, J.M., Dupouey, P. and Baumann, N. (1979) Absence of the major dense line in myelin of the mutant mouse "shiverer", Neurosci. Lett., *12*:107-112.

Privat, A., Robain, O. and Mandel, P. (1972) Aspects ultrastructuraux du corps calleux chez la souris Jimpy, Acta Neuropathol., *21*:282-295.

Privat, A., Valat, J., Lachapelle, F., Baumann, N. and Fulcrand, J. (1982) Radioautographic evidence for the protracted proliferation of glial cells in the central nervous system of jimpy mice, Dev. Brain Res., *2*:411-416.

Quarles, R.H., Morell, P. and McFarlin, D.E. (1994) Diseases involving myelin. In G.J Siegel, B.W Agranoff, R.W. Albers and P.B. Molinoff (Eds.), Basic Neurochemistry: molecular, cellular and medical aspects, 5th Ed., Raven Press, Ltd., New York. pp. 771-792.
Raine, C.S. (1991) Multiple sclerosis: a pivotal role for the T cell in lesion development, Neuropathol. Appl. Neurobiol., *17*:265-274.
Raine, C.S. (1984) The neuropathology of myelin diseases. In P. Morell (Ed.), Mielin, 2nd Ed., Plenum Press, New York. pp. 259-310.
Readhead, C. and Hood, L. (1990) The dysmyelinating mouse mutations shiverer (shi) and myelin deficient (shimld), Behav. Genet., *20*:213-234.
Roach, A., Takahashi, N., Pravtcheva, D., Ruddle, F. and Hood, L. (1985) Chromosomal mapping of mouse myelin basic protein gene and structure and transcription of the partially deleted gene in shiverer mutant mice, Cell, *42*:149-155.
Robain, O. and Mandel, P. (1974) Etude quantitative de la myélinisation et de la croissance axonale dans le corps calleux et le cordon postérieur de la moëlle chez la souris jimpy, Acta Neuropathol., *29*:293-309.
Rose, A.L., Wisniewski, H.M. and Cammer, W. (1975) Neurotoxicity of hexachlorophene: new pathological and biochemical observations. J. Neurol. Sci., 24:425-435.
Rosenbluth, J. (1980) Central myelin in the mouse mutant shiverer. J. Comp. Neurol., 194:639-648.
Royland, J., Konat, G.W., Kanoh, M. and Wiggins, R.C. (1992) Down regulation of myelin-specific mRNAs in the mechanism of hypomyelination in the undernourished developing brain. Dev. Brain Res., 65:223-226.
Said, G. (1996) Guillain-Barre syndrome: morphological lesions and their relations with clinical manifestations. Rev. Neurol. (Paris), 152:344-348.
Samorajski, T., Friede, R.L. and Reimer, P.R. (1970) Hypomyelination in the quaking mouse. A model for the analysis of disturbed myelin formation. J. Neuropathol. Exp. Neurol., 29:507-523.
Sarchielli, P., Trequattrini, A., Usai, F., Murasecco, D. and Gallai, V. (1993) Role of viruses in the etiopathogenesis of multiple sclerosis. Acta Neurol. (Napoli), 15:363-381.
Sayetta, R.B. (1986) Theories of the etiology of multiple sclerosis: a critical review. J. Clin. Lab. Immunol., 21:55-70.
Schneider, A., Montague, P., Griffiths, I., Fanarraga, M., Kennedy, P., Brophy, P. and Nave, K.-A. (1992) Uncoupling of hypomyelination and glial cell death by a mutation in the proteolipid protein gene. Nature, 358:758-761.
Seitelberger, F. (1995) Neuropathology and genetics of Pelizaeus-Merzbacher disease. Brain Pathol., 5:267-273.
Selmaj, K.W. and Raine, C.S. (1988) Tumor necrosis factor mediates myelin and oligodendrocyte damage in vitro. Ann. Neurol., 23:339-346.
Shen, X.Y., Billings-Gagliardi, S., Sidman, R.L. and Wolf, M.K. (1985) Myelin deficient (shimld) mutant allele: morphological comparison with shiverer (shi) on a B6C3 mouse stock. Brain Res., 360:235-247.
Sidman, R.L., Dickie, M.M. and Appel, S.H. (1964) Mutant mice (quaking and jimpy) with deficient myelination in central nervous system. Science, 144:309-311.
Simon, J. and Neubert, W.J. (1996) The pathogenesis of multiple sclerosis: reconsideration of the role of viral agents and defence mechanism. Med. Hypotheses, 46:537-543.
Skoff, R.P. (1982) Increased proliferation of oligodendrocytes in the hypomyelinated mouse mutant-jimpy. Brain Res., 248:19-31.
Skoff, R.P. (1976) Myelin deficit in the jimpy mouse may be due to cellular abormalities in astroglia. Nature, 264:560-562.
Smialek, M., Gajkowska, B. and Otrebska, D. (1994) Electron microscopy studies on the neurotoxic effect of sodium tellurite in the central nervous system of the adult rat. J. Hirnforsch, 35:223-232.
Sobel, R.A. (1995) The pathology of multiple sclerosis. Neurol. Clin., 13:1-21.
Stinissen, P., Raus, J. and Zhang, J. (1997) Autoimmune pathogenesis of multiple sclerosis: role of autoreactive T lymphocytes and new immunotherapeutic strategies. Crit. Rev. Immunol., 17:33-75.
Swanborg, R.H. (1995) experimental autoimmune encephalomyelitis in rodents as a model for human demyelinating disease. Cli. Immunol. Immunopathol., 77:4-13.
Sypecka, J., Tosic, M., Dolivo, M., Domanska-Janik, K. and Matthieu, J.M. (1993) Developmental expression of major myelin proteins in the hypomyelinated pt mutant rabbit. Acta Neurobiol. Exp. (Warsz), 53:281-284.

Tarnowska-Dziduszko, E., Bertrand, E. and Szpak, G.M. (1995) Morphological changes in the corpus callosum in chronic alcoholism. Folia Neuropathol., 33:25-29.
Tennekoon, G., Aitchison, C.S., Frangia, J., Price, D.L. and Goldberg, A.M. (1979) Chronic lead intoxication: effects on developing optic nerve. Ann. Neurol., 5:558-564.
Thomas, P.K. (1984) The peripheral nervous system as a target for toxic substances. Acta Neurol. Scand. Suppl., 100:21-26.
Thornton, C.A. and Griggs, R.C. (1994) Plasma exchange and intravenous immunoglobulin treatment of neuromuscular disease. Ann. Neurol., 35:260-268.
Tienari, P.J. (1994) Multiple sclerosis: multiple etiologies, multiple genes?. Ann. Med., 26:259-269.
Toews, A.D., Krigman, M.R., Thomas, D.J. and Morell, P. (1980) Effect of inorganic lead exposure on myelination in the rat. Neurochem. Res., 5:605-616.
Tosic, M., Dolivo, M., Amiguet, P., Domanska-Janik, K. and Matthieu, J.M. (1993) Paralytic tremor (pt) rabbit: a sex-linked mutation affecting proteolipid protein-gene expression. Brain Res., 625:307-312.
Tosic, M., Dolivo, M., Domanska-Janik, K. and Matthieu, J.M. (1994a) Myelin proteolipid protein mutation in the rabbit: a new model of Pelizaeus-Merzbacher disease. Scheweiz Arch. Neurol. Psychiatr., 145:24-26.
Tosic, M., Dolivo, M., Domanska-Janik, K. and Matthieu, J.M. (1994b) Paralytic tremor (pt): a new allele of the proteolipid protein gene in rabbits. J. Neurochem., 63:2210-2216.
Traugott, U. and Raine, C.S. (1984) The neurology of myelin diseases. In P. Morell (Ed.), Mielin, 2nd Ed., Plenum Press, New York, pp. 311-335.
Tripier, M.F., Berard, M., Toga, M., Martin-Bouyer, G., Le Breton, R. and Garat, J. (1981) Hexaclorophene and the central nervous system. Toxic effects in mice and baboons. Acta Neuropathol. (Berl.), 53:65-74.
Utz, U. and McFarland, H.F. (1994) The role of T cells in multiple sclerosis: implications for therapies targeting the T cell receptor. J. Neuropathol. Exp. Neurol., 53:351-358.
van der Westhuyzen, J., Cantrill, R.C., Fernandes-Costa, F. and Metz, J. (1983) Effect of a vitamin B-12-deficient diet on lipid and fatty acid composition of spinal cord myelin in the fruit bat. J. Nutr. 113:531-537.
Vela, J.M., Dalmau, I., Acarín, L., González, B. and Castellano, B. (1995) Microglial cell reaction in the gray and white matter of the jimpy spinal cord. An enzyme histochemical study at the light and electron microscope level. Brain Res., 694:287-298.
Vela, J.M., Dalmau, I., González, B. and Castellano, B. (1997a) Abnormal expression of PCNA in the spinal cord of the hypomyelinated jimpy mutant mice. Brain Res., 747:130-139.
Vela, J.M., Dalmau, I., González, B. and Castellano, B. (1996) The microglial reaction in spinal cords of jimpy mice is related to apoptotic oligodendrocytes. Brain Res., 712:134-142.
Vela, J.M., Hidalgo, J., González, B. and Castellano, B. (1997b) Induction of metallothionein in astrocytes and microglia in the spinal cord from the myelin-deficient jimpy mutant mouse. Brain Res., 767:345-355.
Vermeesch, M.K., Knapp, P.E., Skoff, R.P., Studzinski, D.M. and Benjamins, J.A. (1990) Death of individual oligodendrocytes in jimpy precedes expression of proteolipid protein. Dev. Neurosci., 12:303-315.
Weiner, L.P. and Fleming, J.O. (1984) Viral infections of the nervous system. J. Neurosurg., 61:207-224.
Wekerle, H., Kojima, K., Lannes-Vieira, J., Lassmann, H. and Linington, C. (1994) Animal models. Ann. Neurol., 36 Suppl. S47-S53.
Wiggins, R.C. (1982) Myelin development and nutritional insufficiency. Brain Res., 257:151-175.
Winder, C., Garten, L.L. and Lewis, P.D. (1983) The morphological effects of lead on the developing central nervous system. Neuropathol. Appl. Neurobiol., 9:87-108.
Wojtowicz, S. (1993) Multiple sclerosis and prions. Med. Hypotheses, 40:48-54.
Yajima, K. and Suzuki, K. (1979) Ultrastructural changes of oligodendroglia and myelin sheaths induced by ethidium bromide. Neuropathol. Appl. Neurobiol., 5:49-62.
Yanagisawa, K. and Quarles, R.H. (1986) Jimpy mice: quantitation of myelin-associated glycoprotein and other proteins. J. Neurochem., 47:322-325.
Yeh, Y.Y. (1988) Maternal dietary restriction causes myelin and lipid deficits in the brain of offspring. J. Neurosci. Res., 19:357-363.

Zhabotinskii, Iu.M., Chumasov, E.I. and Shefer, V.F. (1983) Hereditary demyelination in mutant quaking mice (electron microscopy study). Arkh. Patol., 45:53-60.

Zimmerman, A.W., Matthieu, J.M., Quarles, R.H., Brady, R.O. and Hsu, J.M. (1976) Hypomyelination in copper-deficient rats. Prenatal and postnatal replacement. Arch. Neurol., 33:111-119.

19

GLIAL CHANGES IN AGING AND ALZHEIMER'S DISEASE

María Asunción Morán and Pilar Gómez-Ramos

Glial changes were studied in autopsied brains from 10 neurologically normal subjects (4 young to mature and 6 elderly individuals), and 7 Alzheimer's disease patients. Cortices from hippocampal and parahippocampal regions, as well as temporopolar, insular, and frontal areas, were selected. Single and double immunocytochemical techniques were used, with antibodies recognizing both activated astrocytes and microglia, as well as AβP amyloid and a protein present in neurofibrillary degeneration. Counterstaining with thioflavine-S was also employed to visualize the degree of maturation of senile plaques. Thionin counterstaining allowed to distinguish whether neurofibrillary tangles were extracellular ones. A semi-quantification of medium densities of activated and reactive glial cells was systematically done in all cases. While glial reactivity was found to increase in a moderate way as age advances, it appeared dramatically increased in all Alzheimer's disease patients, whatever their age. A non-demented individual with abundant senile plaques and neurofibrillary degeneration in hippocampus and parahippocampal cortices, presented an intermediate glial reactivity as compared with that of all of the other controls and the Alzheimer's disease brains. Activated astrocytes and microglial cells were seen to associate with senile plaques in a direct proportion with the degree of cytotoxicity of every plaque, as demonstrated for the number of neurites with neurofibrillary degeneration. A highest glial reactivity was also found around extracellular neurofibrillary tangles, as well as related with blood vessels. All of these neuropathological signs eliciting glial reactivity, represent inflammatory focuses, that might become targets for a future treatment of Alzheimer's disease.

MICROGLIA AND ASTROCYTES IN AGING

Activation stages of microglial cells

Microglia are cells showing a surprisingly wide functional plasticity. It means that they change in morphology, immunophenotype, and function, while being confronted with a changing micro-environment, as a part of normal development or resulting of a pathological insult (Streit et al. 1988). Thus, any study on microglia needs to consider several cell populations, representing progressive states of activation as an answer to increasing micro-environmental demands. Since del Rio Hortega pioneer and elegant statements on these dynamic cells, that supposed a major contribution to the recent microglia knowledge (del Rio Hortega,1932), it is commonly considered that *resting* or *ramified* microglia are those cells showing a small, either oval or irregularly shaped perikaryon, and several highly branched processes. In white matter, most of these processes branch from the two opposite poles of the elongated cell body, being oriented in the same direction as the axonic bundles, while microglia processes extend throughout a wider space in the gray matter. *Activated* microglia show larger, plump, irregularly shaped somata, with short, thickened processes which either become regressive or even disappear in the *reactive* brain macrophages, with round, heavily stained somata and few or no processes. During developmental stages, when massive selective elimination of neuronal processes occurs (Killackey, 1984), numerous ameboid macrophages, that might derive from invasion of monocytes in early development (Ling and Wong, 1993), undertake a phenotype very similar to that of reactive microglial cells found in the adult. Following sequential changes, some monocytes might also evolve into resting microglia, that once activated up to a reactive stage, are capable of recovering their phagocytic functions. Since all transitional stages between microglial types may be evenly seen, it is actually difficult to find a clear-cut classification for every concrete cell present in the adult. Furthermore, in some neurologically intact, aged individuals, as well as in patients with different neuropathological disorders, clusters of 2-5 either activated or reactive microglial cells may also be found, likely representing a highly evolved stage of activation, where several cells joint and collaborate in undertaking their phagocytic functions.

Not only conditions related to aging and neurological disorders, but also all types of tissue damage involving neuronal injury, seem capable to cause microglial activation (Review in Thomas, 1992). As a result of their wide range of responsiveness, a myriad of proteins and functions have been found to be related with activated microglia, mainly when acting like macrophages. So, as likely effectors of the immune system, reactive microglial cells may uptake foreign antigens by phagocytosis, presenting them on their cell surface, bound to major histocompatibility complex (MHC) classes I and II molecules. Besides, they secrete factors which are involved in both the acute phase reaction and the classical complement cascade (Frei et al., 1987). Moreover, microglia contribute to inflammatory reactions taking place in the central nervous system (Giulian, 1987). They have been found to form part of the blood-brain barrier (glia limitants) (Lassmann et al., 1991). Also, reactive microglia produce interleukin-1 (Giulian et al., 1986) that is selectively increased at the brain injured sites (Giulian et al., 1989), and which mitogenic action on astrocytes has been well documented (Giulian and Lachman.,1985). All of these abilities make microglia a complex, but specially interesting polymorphous cellular population, from which a lot of knowledge is yet waiting to be learned.

Reactivity of astrocytes

As new data on the wide variety of functions played by astrocytes are known (Kimelberg and Norenberg, 1989), the necessity of considering the likely existence of numerous types of astrocytes becomes commonly accepted. A major classification distinguishes into two main types of cells: fibrous and protoplasmic astrocytes. Fibrous astrocytes are localized throughout the white matter, being immunoreactive for the glial fibrillary acidic protein (GFAP) antibody, raised against intermediate astrocytic filaments (Bignami and Dahl, 1974). Instead, protoplasmic astrocytes are generally in the gray matter, and do not immunoreact with the GFAP antibody, supposedly because their intermediate filaments content is under a minimal threshold needed to demonstrate this immunoreactivity (Duffy et al., 1980). Except for astrocytes forming part of the subpial layer I and, to a lesser extent, for perivascular astrocytes, young and adult healthy individuals do not show GFAP positive cells in gray matter, thus making the presence of GFAP immunoreactive astrocytes in the neuropil indicative of a pathological state referred to as "gliosis", which prevalence increases as age advances (Beach et al., 1989). In fact, it has been observed that astrocytes confronted to a wide range of tissue injuries react to them by increasing their number (Nieto-Sampedro et al., 1985), as well as their cytoskeletal consistency, thus becoming strongly positive for the GFAP antibody. So, a general agreement exists that reactivity of astrocytes is directly proportional to their degree of GFAP immunolabeling. When measured in cerebrospinal fluid, GFAP concentrations showed a significant correlation with age, both in non-demented as in demented individuals, as recently corroborated (Wallin et al., 1996). Related to Alzheimer's disease (AD), determinations of GFAP in brain homogenates showed an amount of GFAP significantly higher in AD patients, as compared with controls and individuals with other neurodegenerative disorders (Delacourte, 1990).

Alterations of microglia and astrocytes in aging

As age advances, a number of functional and structural changes take place in the human brain (Kemper, 1984). A part of these changes are related with microglial cells, that while showing a low (if any) immunoreactivity for antibodies against class II MHC in young and healthy adults, express this immunoreactivity in normal elderly brains (McGeer et al., 1988; Mattiace et al., 1990). Most positive cells present in brains of old healthy people display an activated morphology, although some of them appear still to be fully ramified. A possible interpretation for this observation is that up-regulation of MHC class II is an early consequence of activation, so that the threshold of detection of MHC molecules expressed is reached prior to the onset of any visible morphologic change (McGeer et al., 1993). When most sensitive MHC class II antisera, such as LN3, are used, a uniform reticular array of resting microglial cells are seen throughout gray matter, as well as white matter of elderly brains (Perlmutter et al., 1992; Carpenter et al.,1993). Furthermore, it has also been shown that an increased number of activated and even reactive microglia are found in gray matter depending on a wide range of circumstances, some of them related with either more persistent clinical, or just agonal, states (Mattiace et al., 1990).

Most relevant for this review, it has been demonstrated that numerous senile plaques (SP), as well as a relatively abundant neurofibrillary degeneration, may accumulate in normal elderly brains (Tomlinson et al., 1968, Arriagada et al., 1992). And both SP and neurons undergoing neurofibrillary degeneration (NFT), when at a mature stage of evolution, appear closely related with activated glial cells, in a manner that we will try to summarize in the following paragraphs. Besides, an increased glial reactivity is seen around blood vessels, mainly in the vicinity of impaired blood vessel walls. Regarding to astrocytes, some hypertrophy is also found in many cases at the interface between gray and white matter. This fact has been interpreted as likely due to a special vulnerability to physical stress inherent at interfaces between dissimilar structures (Beach et al., 1989), that could be compensated by an up-regulation of homeostasis-related astrocytic functions.

Another beneficial role of astrocytes in aging may be that played by corpora amylacea, part of which seem to have an astrocytic origin, and accumulate rests of death neurons, myelin and oligodendroglia, thus preventing potential injuries in neuropil areas heavily laden of both organic and inorganic materials (Singhrao et al., 1995). Moreover, astrocytes are a major source of apolipoprotein E (Pitas et al., 1987), which allelic form E2 appears to protect from AD, as firmly supported by epidemiological data (Corder et al., 1994).

MICROGLIA AND ASTROCYTES IN ALZHEIMER'S DISEASE

Neuropathological signs of Alzheimer's disease

AD, the most common cause of dementia affecting mainly to elderly people, is a severe syndrome characterized by a gradual and irreversible loss of memory, accompanied by a dramatic impairment of the highest intellectual functions. Neuropathologically, it is diagnosed by the presence of both SP and NFT in different brain cortices (Khachaturian, 1985). Classical studies demonstrated a positive correlation between the degree of dementia of AD patients and the number of SP found in their brains (Blessed et al., 1968). However, the consistency of such a correlation is a subject of controversy, since it has been convincingly shown that neurologically intact aged individuals may also have abundant SP (Delaère et al., 1990; Katzman et al., 1988). Although AD type neurofibrillary changes are neither exclusive of this syndrome, the association between their presence in isocortices of AD patients and the degree of dementia, has been found to be stronger than that of SP (Crystal et al., 1988; Wilcock and Esiri, 1982; McKee et al., 1991; Arriagada et al., 1992). Furthermore, NFT may also be present, but to a lesser extent, in normal aging (Braak and Braak, 1991; Hirano and Zimmerman, 1962), as well as in pathologies different to AD (Iqbal et al., 1987).

The decrease of synaptic density within association cortices is another variable proposed to be related with dementia scores found in AD (Terry, 1992). It has been claimed that this synaptic loss have to contribute to the severe disconnection syndrome suffered by AD patients. Furthermore, it has been pointed out that in AD there is an accentuated neuronal death, as compared with that found in aging brains, which not only affects to areas showing age-related neuronal loss, but rather extends to other areas apparently preserved of neuronal death in normal aging (Kemper, 1984; West et al., 1994; Gómez-Isla et al,. 1996).

Senile plaques

A consistency in the extent of correlation of SP with the increase in neurological damage is still lacking, mainly due to insufficient concretion of SP types, part of them likely with a different etiopathogeny. At this respect, a major point seems to be the conformational state of the Aß protein (AßP) that, once processed from its much longer precursor molecule, may either accumulate in a non-fibrillar, preamyloid form, or acquire a ß pleated configuration, thus reaching an amyloid, thioflavine-S positive structure. Preamyloid (commonly named as diffuse) plaques, have been demonstrated not to produce degeneration in neuronal bodies nor cellular processes belonging to the nearby neuropil (Yamaguchi et al., 1988, 1989). They may be very abundant in neurologically intact aging (Mackenzie, 1996), and if quantified using techniques (such as most of the classical ones) which are unable of distinguishing between types of SP, their high number may cause a wrong diagnosis of AD (Crystal et al., 1988; Duykaerts et al., 1990; Morán et al., 1992; Gómez-Ramos et al., 1993).

All types of SP other than the diffuse one are formed by fibrillar (thioflavine-S positive) AßP, and contain degenerating neurons intermingled with dystrophic neurites, as well as activated and reactive glial cells. While it is controversial whether diffuse SP evolve into more mature, amyloid SP, a certain way of evolution is generally assumed within amyloid SP, in a manner that the more mature the SP, the more dystrophic neurites they have, at least some of them undergoing neurofibrillary degeneration at the most mature stages of SP evolution (Crystal et al., 1988). Thus, while in non-demented elderly individuals, when present, most SP are of diffuse type, in AD an increased number of SP are positive for thioflavine-S, accumulating abundant dystrophic neurites and neurofibrillary degeneration (Wang and Muñoz, 1995). This fact explains the reason why the distinction between types of SP is crucial, since only a subset of these structures contain neurofibrillary changes, which positive correlation with the degree of dementia has been repeatedly shown (Wilcock and Esiri, 1982; McKee et al., 1991; Arriagada et al., 1992).

According with morphological and structural criteria, amyloid SP are usually classified into primitive, neuritic, classical and "burned-out" subtypes. While primitive SP show a quite homogeneously distributed volume of fibrous AßP, with a moderate number of dystrophic neurites, that may or may not contain neurofibrillary degeneration (fig. 19-1A), a higher proportion of dystrophic neurites are present in neuritic SP (fig. 19-1B).

Classical SP show a compact amyloid core in their center, surrounded by a halo, with very similar characteristics as these described for primitive SP (fig. 19-1C). A subset of classical SP contain abundant neurites with mature neurofibrillary degeneration in their halo (fig. 19-1D). "Burned-out" SP are formed by focal deposits of compact amyloid, devoid of any halo around them.

Neurofibrillary degeneration

This degeneration consists on the accumulation of bundles of both paired helical and straight filaments (Kidd, 1963; Gibson et al., 1976), and may be present in three different types of structures, referred to as NFT, SP associated neurites, and neuropil threads. Some neuronal populations have been found to be particularly prone to develop neurofibrillary changes, the same in AD and other pathologies as, to a lesser extent, in normal aging (Hirano and Zimmerman, 1962). So, in isocortices only pyramidal neurons have been described to become NFT (Braak and Braak, 1986), those involved in long cortico-cortical connections seeming to be most vulnerable to this degeneration (Lewis et al., 1986). And NFT appear involved in such degeneration following a consistent pattern, thus allowing to classify AD related changes into neuropathological stageings (Braak and Braak, 1991). A progressive sequence of events in the cytoskeleton of vulnerable neurons could result in NFT and neuropil threads (Braak et al., 1994), with a partially known crucial process culminating in the conversion of hyperphosphorylated tau into insoluble molecules (Morishima-Kawashima et al., 1993). These long-lasting insoluble fibrillary conglomerates of molecules form tangled bundles that remain in the neuropil once the corresponding neurons die, being referred to as extracellular or ghost NFT (Bancher et al., 1989).

Neuronal death

Though some controversy exists on the extent, and even the existence, of a significant neuronal loss during normal aging (review in Kemper, 1984), actually there is a general consensus on the occurrence of a massive neuronal death in AD. So, quantifications of hippocampal cortices by classical methods, resulted in a 5 times faster neuronal dropout in AD than in elderly controls (Ball, 1977).

Recent quantifications using modern stereological techniques, found differences in the pattern of hippocampal neuronal loss between normal aging and AD, thus suggesting that AD is not an inevitable consequence of aging, but involves a

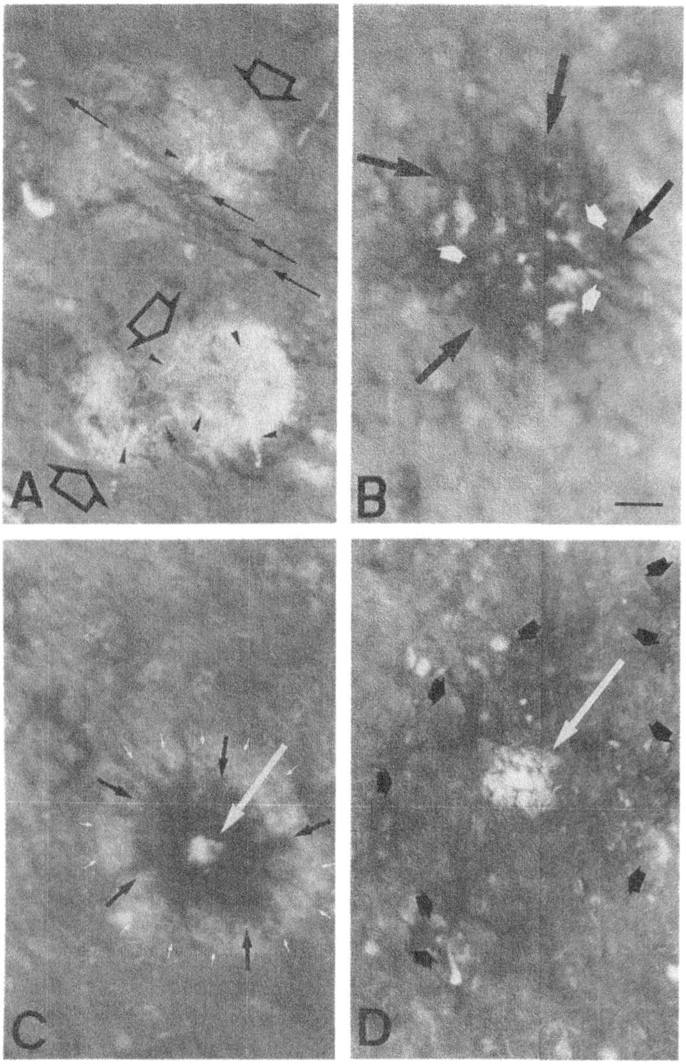

Figure 19-1. Alzheimer's disease sections microphotographs, immunoreacted with either LN-3 (for activated microglia) or GFAP (for reactive astrocytes), and counterstained with thioflavine-S. When observed with ultraviolet light, cells immunoreacted for both LN-3 and GFAP stain in black, while proteins with ß pleated configuration are seen in white. A-C: LN-3, hippocampal areas CA2 (A), CA3 (B) and subiculum (C). D: GFAP, para-hippocampal cortex. A: Primitive senile plaques (open arrows) with scarce neurofibrillary degeneration (arrowheads), and regularly oriented processes of microglia, with an elongated, resting-looking appearance (long arrows). B: Neuritic senile plaque, containing AßP and neurofibrillary degeneration (white arrows), in close apposition with clusters of reactive microglia (black arrows). C: Classic senile plaque, with a crown of reactive microglia (black arrows) immediately around its amyloid core (long white arrow), and an external homogeneous halo (small white arrows). D: Classic senile plaque with hypertrophied astrocytes (black arrows) surrounding its periphery (long white arrow). Bar in B = 30μm in all figures.

specific disease process (West et al., 1994). It has been pointed out that apoptosis might play a major role for the occurrence of neuronal death in AD (Cotman and Anderson, 1995). Whatever the complex etiological factors involved in this disease, a significant neuronal loss has also been found in other cortices of AD patients, such as the entorhinal cortex (Gómez-Isla et al., 1996). Any neuronal death in hippocampal and parahippocampal cortices may be most crucial for AD, since these cortices are considered to be specially involved in the memory impairment that characterizes AD from its initial stages (Hyman et al., 1984).

Microglia and astrocytes in senile plaques

As analyzing the relationship of SP with glial cells, a clear distinction needs to be done between diffuse (preamyloid), and amyloid (thioflavine-S positive) SP. While a most likely neuronal origin has been proposed by several authors for the AßP accumulated in diffuse SP (Allsop et al., 1989; Pappolla et al., 1991; Probst et al., 1991), both microglia (Wisniewski et al., 1989) and astrocytes (Siman et al., 1989) have been suggested as cells of origin of the compact core of AßP in the classical type of amyloid SP. Whatever diffuse SP may or may not precede the development of more mature SP (review in Selkoe, 1989), the degree of association of activated glial cells with diffuse SP seems to be much weaker than that seen with more mature primitive, classical and burned-out SP (Perlmutter et al., 1991; Mackenzie et al., 1995). Most authors agree that activated microglia appear only occasionally associated with diffuse SP, and describe that when found in the vicinity of preamyloid SP, microglial cells commonly show a resting phenotype. Instead, clusters of reactive microglial cells are tipically present within classical SP, just in close contact with their amyloid core (Wegiel and Wisniewsky, 1988; Perlmutter et al., 1992) (fig. 19-1C).

With regard to astrocytes, AßP has been described to induce a GFAP immunoreactive phenotype in vitro, in an aggregation-dependent manner (Pike et al., 1994). In keeping with this observation, in vivo, GFAP immunoreactive cells are seen associated with thioflavine-S positive SP, but not with diffuse SP. The degree of association of reactive astrocytes with mature SP has been reported to correlate with the abundance of neurofibrillary degeneration present in each SP type, as shown by using antibodies against hyperphosphorylated tau protein (Pike et al., 1995). Moreover, a crown of robust GFAP immunoreactive astrocytes surround the periphery of almost every classic SP, their processes being extended towards the compact amyloid core (Wisniewski and Wegiel, 1992) (fig. 19-1D).

Both microglia and astrocytes have been proposed to be involved in cascades contributing to plaques formation, and thus, to AD etiopathogeny. So, an experiment consistent in adding a synthetic homologue of 1-42 AßP to glial cells in vitro, found an increase in the glial secretion of both interleukin-1 and basic fibroblast growth factor, accompanied by the proliferation, and the morphological transformation into an activated phenotype, of microglia (Araujo and Cotman, 1992). Interleukin-1 has also been reported to be released from lipopolysaccharide-activated human monocytes, considered to be closely related in both lineage and functions to microglia, after incubation with AßP (Lorton et al., 1996). Since interleukin-1 and basic fibroblast growth factor are known to elevate the synthesis of AßP precursor protein, and interleukin-1 shows a strong mitogenic activity on astrocytes (Giulian and Lachman, 1985), which in turn are capable of producing more interleukin-1, cascades with a high potential of both AßP production and gliotic reaction, may arise in this way. Furthermore, activated glia appear to release other immunoinflammatory molecules, including complement proteins and acute phase reactants, whose activation products colocalize with SP and dystrophic neurites (Eikelenboom and Stam, 1982; McGeer et al., 1989; Eikelenboom and Veerhius, 1996; Rogers et al., 1996). Once more, AßP has been claimed to bind and activate C1q, thus being a possible candidate to initiate complement cascades, that could account for a significant amount of AD neurotoxicity, as recently reviewed by Chen et al (1996). However, AßP accumulation appears not to be a sufficient cause for AD neurodegeneration. In fact, diffuse SP contain early-stage complement components, but are weakly immunoreactive, if at all, for the crucial late-stage components (C5b-9), that actually lyse cells (Lue and Rogers, 1992). This observation is in keeping with the previously mentioned data that AßP deposition does not necessarily correlate with AD neurodegeneration, since abundant SP, most of them of the diffuse type, appear also in brains of non-demented elderly individuals (Delaere et al., 1990; Arriagada et al., 1992).

Microglia, astrocytes and neurofibrillary degeneration

A consistent and close association of extracellular neurofibrillary degeneration with both activated microglia and astrocytes has been repeatedly found in AD brains, at both light and electron microscope levels (Probst et al., 1982; Cras et al., 1991; Perlmuter et al., 1992). Such an association has not been demonstrated with intracellular neurofibrillary degeneration, and it is thought to be mainly due to neuronophagia of dying neurons (Peters et al., 1991). Moreover, activated glial cells have been proposed to participate in the gradual transformation of extracellular NFT, that may result in a secondary deposition of AßP on a subset of them (Yamaguchi et al., 1991), as well as in posttranslational changes of epitopes from paired and straight filaments (Schmidt et al., 1988). A specially interesting issue is that extracellular NFT constitute inflammatory foci with some potential for contributing to damage to neighbouring neurons.

Microglia, astrocytes and neuronal death

Immunoinflammatory processes mediated by neuroglia in AD, appear to involve not only to mature amyloid SP and extracellular neurofibrillary degeneration, but also to endothelial cells from the cerebral microvasculature (Pereira et al., 1996). Through these modified endothelial cells, inflammatory cascades could reach wide territories of the cerebral cortex, thus contributing to unidentified causes of the massive neuronal death -and loss of synapses- observed in AD. In keeping with this possibility, a recent study using immunocytochemistry for the membrane attack complex (C5b-9), and synaptophysin, showed that measures of inflammation accounted for significant percentages of the variance in synaptophysin immunoreactivity (Lue et al., 1996). In the same study, C5b-9 immunoreactivity was more highly correlated with synapse estimates than with NFT formation or AßP deposition, thus pointing out to the closest relationship of microglial activation with AD neuronal death, already suggested by other authors (Castellano and Gonzalez, 1991).

OBSERVATIONS IN A SERIES OF CASES

Cases and techniques

To study aging related glial changes, brains from 10 neurologically intact subjects (4 young to mature and 6 elderly individuals) were used. Moreover, to characterize glial modifications occurring in advanced age AD patients, brains from 7 elderly AD patients were included in the series (table 19-1). All brains were obtained at autopsy. The neuropathological criteria followed for AD diagnosis adhered to NIH/ADRDA criteria (Khachaturian, 1985). Cortices from hippocampal and parahippocampal regions as well as temporo-polar, insular, and frontal areas, were studied. Tissue fixation was performed by immersion, following conditions specified in table 19-1.

Antibodies used

i) For microglia, antisera against HLA-DR class II: Dako (1:32), Becton Dickinson (1:33), and LN3 from ICN, Biomedicals (1:12). ii) For fibrous astrocytes: GFAP from Sigma (1:400). iii) For AßP: 4G8 Mab against a synthetic peptide, corresponding to aminoacids 17-24 of AßP (1:1000) (Kim et al., 1988). iv) For neurofibrillary degeneration: 150 Mab (1:2) (Riederer et al., 1990; Ulloa et al., 1994), raised against the hyperphosphorylated cytoskeletal protein MAP-1B.

Single and double immunocytochemical protocols

For single immunocytochemical protocols, free floating 40μm thick sections were processed following a biotin-extravidin method, with diaminobenzidine (Sigma) as a chromogen, except for the 4G8 Mab, which followed the biotin extravidin-phosphatase alkaline method, after formic acid pre-treatment, and with fast red as a chromogen. In every case, complete series of sections were stained for either thionin or thiflavine-S, and additional sections were counterstained with either of these labelings. Sections immunoreacted for the 4G8 antibody could not be counterstained for thioflavine-S, due to modifications in the ß configuration during formic acid pretreatment. Selected serial sections were consecutively immunoreacted for either LN3 or GFAP and the 150 antisera.

When double immunocytochemical procedures were followed, one of the chromogens was diaminobenzidine, either with or without intensification with nickel ammonium sulfate (Kawamata et al., 1993), while the other chromogen was 4-chloro-naphthol (Sigma).

Semi-quantification of glial activation

Antibodies against HLA-DR class II antigens consistently labeled both microglia and pericytes. While the intensity of labeling obtained by antisera from Dako and Becton Dickinson has been described to depend on the degree of microglial activation, the labeling obtained after using LN3 was always neat, even as showing microglial cells with a resting appearance. Thus, the previously mentioned morphological criteria were basic, mainly for classifying cells immunoreacted with LN3. With regard to astrocytes, their degree of activation was considered to be directly proportional to the thickness of astrocytic processes stained for GFAP. Moreover, the GFAP antibody rendered a dotted labeling which extended through wide areas in the neuropil, and was interpreted to be due to thin diameter processes crossing the brain parenchyma in a plane perpendicular to the cutting plane. In every case, medium densities of activated/reactive glial cells, as well as mature (thioflavine-S positive) neurofibrillary degeneration and SP, were semi-quantified in series of sections immunoreacted for either microglia or astrocytes, and sections stained, or counterstained, for thioflavine-S. Medium densities of activated microglia and astrocytes, as well as of mature SP and NFT, were estimated by consistently analyzing the same cortical areas at 10X and 20X, under the light microscope. Five degrees of medium densities were considered: 0, absent; +, scarce; ++, moderate; +++, frequent, and ++++, very frequent (table 19-1).

Observations in aging

In young control individuals, resting microglial cells appeared distributed in an uniform reticular array throughout the white matter, filling also the outer part of layer I, just bellow the pial surface, and extending in a very moderate number towards layer II. A scarce number of microglial cells, most of them adjacent to (or sending processes towards) blood vessels (perivascular microglia), as well as perivascular cells (Graeber and Streit, 1990), were also immunolabeled in other cortical layers, mainly in the superficial ones. The association of cells stained for microglial markers with blood vessels seen in white matter was identical to that found in gray matter. Superimposed to the labeling found in young brains, middle-aged individuals showed a more intense immunoreactivity in white matter, as well as the presence of a higher number of immunoreactive microglial cells in gray matter, where more stained cells were distinguished in layer II and even in inner layers, the last ones showing a partially activated phenotype. Besides, the number of both perivascular microglia and perivascular cells positive for antisera against HLA-DR antigens, was higher. In elderly individuals, ramified microglia appeared distributed in a regular array, which extended throughout the white as well as the gray matter. Moreover, a higher intensity of labeling was seen after using both Dako's and Becton-Dickinson's antisera. Also, more activated microglia, and even some reactive cells, were eventually found. When present, as in case C-18 (table 19-1), frequent reactive microglia were located within mature SP, as well as in areas with exceptionally abundant extracellular NFT. We also observed that groups of 2-3 ramified cells surrounded a certain proportion of pyramidal neurons, mainly at the interface between layer VI and the white matter.

Table 19-1.

Code	Age (years)	Sex	Autolysis	Fixation (fix. and time)	Reactive astroglia	Reactive microglia	NFD	AßP
J-2	30	M	10h.	formol 36d.	+	+	0	0
J-1	37	M	3h.	formol 45d.	+	+	0	0
M-1	45	M	9h.	formol 90d.	+	+	0	0
M-2	62	F	8h.	formol ND	++	++	0	0
C-10	81	M	4,5h.	4% PF 25h.	++	++	++	+++
C-15	85	F	2h.	4% PF 47h.	++	++	0	+
C-18	85	M	8h.	4% PF 48h.	+++	+++	+++	+++
C-22	86	M	10,5h.	4% PF 20h.	++	++	+	0
C-25	92	F	5h.	4% PF 48h.	++	++	+	+
C-14	98	F	3,5h.	4% PF 29h.	++	++	+	0
A-43	75	M	5,5h.	4% PF 32h.	++++	ND	++++	+
A-21	78	F	9h.	4% PF 32h.	++++	ND	+++	+++
A-28	79	M	3h.	4% PF 52h.	++++	++++	++++	+++
A-46	81	F	3,5h.	4% PF 96h.	++++	++++	+++	+++
A-57	81	F	4,5h.	formol 14h.	++++	++++	+++	+++
A-38	90	F	5h.	4% PF 24h.	++++	ND	+++	+++
A-45	98	F	7h.	4% PF 24h.	++++	++++	+++	+++

J-, M-, C-: neurologically intact (control brains), young, middle-aged, and elderly individuals, respectively. A-: Alzheimer's disease patients. Sex: M (male), F (female). Autolysis time: delay between death and brain fixation. Fixative 4% PF: 4% paraformaldehyde in 0.1 phosphate buffer, pH 7.4, at 4°C. NFD: neurofibrillary degeneration. AßP: amyloid ß protein. ND: not determined. Semiquantification: 0, absent; +, scarce; ++, moderate; +++, frequent; ++++, very frequent.

With regard to astrocytes, the GFAP labeling in gray matter increased in parallel with age, the same as previously mentioned for the microglial reactivity. So, while in young cases only fibrous astrocytes of white matter and subpial layer

I were immunoreactive for GFAP, middle-aged cases showed more GFAP-positive astrocytes, that progressively extended into layers II, III and, finally, into inner layers. Most astrocytes showing immunoreactive processes in layers II-IV were found in the vicinity of blood vessels, to which walls they closely surrounded.

As seen in table 19-1, a positive relation was found in neurologically intact individuals between older age and higher glial activation. However, an interesting exception to this general tendency was present in case C-18 (table 19-1), which showed numerous NFT in limbic cortices (as much as to be classified at Braak's early stage IV) (Braak et al., 1991), and had quite abundant amyloid SP in all cortices studied. Accordingly, a high density of activated glial cells was seen in this case. Though C-15 (without NFT, and with almost no amyloid SP) had the same age as C-18, glial activation was lower in C-15 than in C-18. Likewise, C-18 showed a higher glial activation than C-22, C-25 and C-24 (all of them with very scarce extracellular NFT and mature SP, in spite of being older than C-18).

Observations in Alzheimer's disease

A dramatic activation of every type of microglial cells and astrocytes was systematically observed in gray matter in all of the AD brains studied. In fact, an abundant number of activated and reactive cells were consistently found, mainly located in layers II and III, as well as in layer V. Frequently, these activated cells appeared associated in groups, thus representing a highest degree of glial reactivity. Nevertheless, in white matter the microglial array presented a more homogeneous appearance, showing a moderate-to-low activation degree. Of course, glial reactivity was highest in layer I, as well as in the vicinity of (and within) the blood vessel walls. Related to the interface between gray and white matter, in sections immunoreacted for microglia and counterstained with thionin, we consistently found numerous either resting or lightly activated microglia cells around cell bodies of pyramidal neurons. The same interface region presented a dense array of hypertrophied astrocytes, with profuse, thick, processes. Differently from the general relation between glial activation and age, previously mentioned for non-demented individuals, glial activation in AD patients did not depend on their age (table 19-1). It was rather closely associated with the presence of both extracellular NFT and mature amyloid SP, which number (not only in the hippocampus, but also in isocortices) was high in all cases. Thus, case C-18, though still being neurologically normal, represented a transitional state between control and AD eldery cases, and as such, he showed more NFT and amyloid SP than usually seen at his age.

Glial reactivity in senile plaques

Thioflavine-S staining allowed us to distinguish diffuse (preamyloid) from amyloid SP. In 4G8 immunoreactive AßP accumulations located in regions without

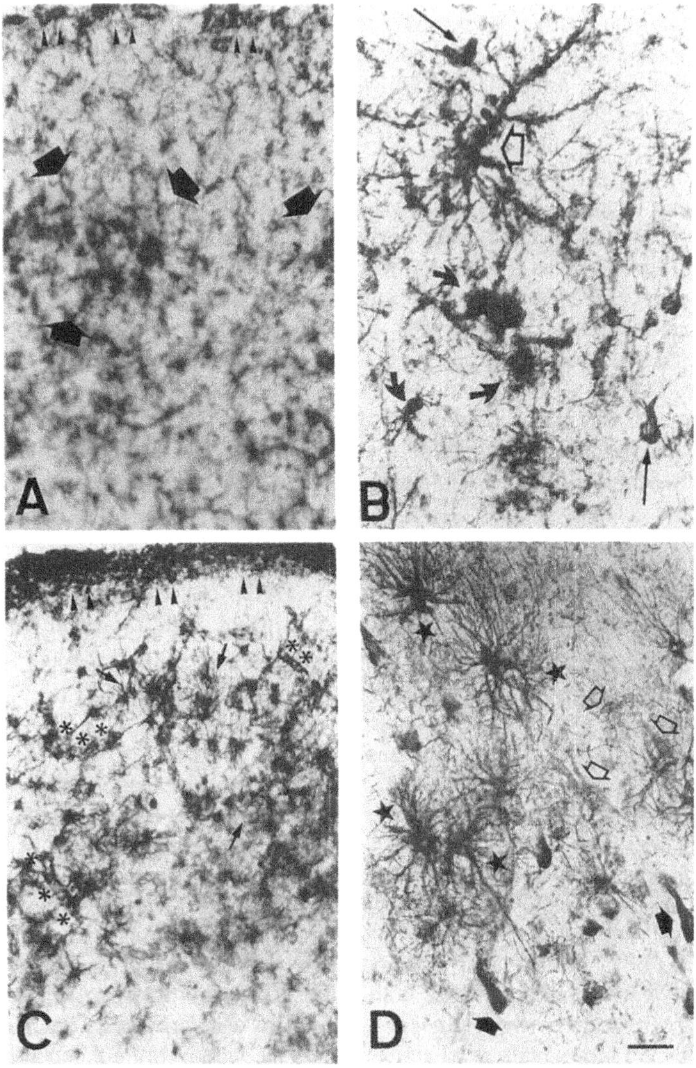

Figure 19-2. Microphotographs of Alzheimer's brain sections double-immunoreacted with A-150 (for neurofibrillary degeneration), and either LN-3 (for activated microglia) (A and B), or GFAP (for reactive astrocytes) (C and D). Both A and C show layers I-III from the entorhinal cortex, being oriented with glia limitants upwards (double arrowheads). A significant glial reactivity is evident, mainly in the subpial layer I and in layer II, where clusters of reactive microglia (thick arrows in A), as well as abundant hypertrophied astrocytes (thin arrows in C) are seen. Reactive astrocytes are also present around blood vessels (asterisks in C). B is from entorhinal layer II, showing activated (open arrows) and reactive (curved arrows) microglial cells. Some A-150 positive neurofibrillary tangles are also seen (long arrows). D is from the CA1 hippocampal cortex, with reactive astrocytes (stars), and both A-150 positive intracellular neurofibrillary tangles (small solid arrows) and weakly GFAP immunoreactive extracellular neurofibrillary tangles (small open arrows). Bar in D represents 80μm in A and C, and 37μm in B and D.

thioflavine-S positive SP (as seen in adjacent sections), neither activated microglia nor GFA positive astrocytes were ever present. Also, a weak or absent reactivity of local glial cells was found to be associated with primitive SP, which most of the times did not disturb the resting phenotype of microglia (fig. 19-1A), and did not appear surrounded by GFAP positive astrocytes. However, an increased activation of both microglia and astrocytes was observed either within (microglia), or around (astrocytes), mature SP accumulating dystrophic neurites (fig. 19-1B, D). This glial activation was highest in classical SP, that showed the amyloid core in close apposition with several reactive microglia cells (fig. 19-1C), and were surrounded by clusters of activated astrocytes (fig. 19-1D), which processes extended also up to reach the amyloid core, in a crown-like fashion.

When sections immunoreacted for microglia or for astrocytes were counterstained with thioflavine-S, most clusters of activated glial cells in gray matter demonstrated to be localized in close association with thioflavine-S positive SP. Besides, a consistent association of activated glia and blood vessels (either with or without AßP accumulations), was also observed in gray and white matter.

Glial activation related to structures with neurofibrillary degeneration

No association could be found between activated glia and mature (thioflavine-S positive) neuropil threads in cortical layers with scarce NFT. However, sections immunoreacted for the 150 Mab (which labels early neurofibrillary degeneration) showed parenchymal areas where an abundant number of activated microglial cells were coincidental with sites containing dense 150 positive, thioflavine-S negative, neuropil threads. While some of these parenchymal areas corresponded to regions where vascular pathology is frequently present, such as the dentate gyrus, in other cases they belonged to structures without any apparent vascular problem, nor mature SP, such as the presubiculum.

Regarding to mature NFT, they appeared closely associated with labeled microglia around their apical dendrites. Though this association was also observed with early NFT, microglial cells showed activated phenotypes mainly (if not exclusively) in sites where most mature NFT (the extracellular ones) were present (fig. 19-2B). With regard to GFAP positive astrocytes, they clustered around extracellular NFT, intermingling with NFT so closely as to make them immunoreactive for GFAP too (fig. 19-2D). In cortical areas where extracellular NFT were specially abundant, such as the entorhinal layer II, clusters of activated glial cells were consistently found, giving to these cortices a layered appearance in sections immunoreacted for either microglia or astrocytes (fig. 19-2A,C).

CONCLUDING REMARKS

As age advances, glial reactivity increases in a moderate manner. However, in AD patients there is a dramatic increase of glial reactivity, that seems not to be dependent on individual age. A non-demented (control) case with abundant SP and NFT (Braak's early stage IV), presented an intermediate glial reactivity as compared with that of all of the other controls and the AD brains, and actually higher than that usually found at his age.

This case may represent a transitional stage from normal to demented brains, the same as others previously reported in the literature, which careful study could be worthy pursuing in order to reach new therapeutic tools for slowing the progress, and even delaying the onset of the disease.

In this sense, a review of relevant publications on 15 clinical trials of conventional antiinflammatory drugs, suggests that some symptoms of AD were prevented or attenuated by these treatments (Breitner, 1996; Rogers et al., 1996). Whatever the key factors -most of them to be identified yet- involved in AD etiopathogeny, these promising epidemiological data point out the convenience of further studying glia related factors, since they might become targets for a future treatment of such a devastating disease.

ACKNOWLEDGMENTS

The skillful technical assistance of Vicenta Gómez and Milagros Guerra is very much acknowledged by the authors, who also appreciate the generosity of Dr H Wisniewski and Dr K S Kim for their anti AßP (4G8) antibody, as well as of Dr F Jiménez and Dr F Moya for their 150 antibody. This study was supported by Fondo de Investigaciones Sanitarias de la Seguridad Social 93/0198.

REFERENCES

Allsop, D., S.I. Haga, C. Haga, S.I. Ikeda, D.M.A. Mann and T. Ishii (1989) Early senile plaques in DS brains show a close relationship with cell bodies of neurons. Neuropathol. Appl. Neurobiol. *15*:531-542.

Araujo, D.M. and C.W. Cotman (1992) ß-Amyloid stimulates glial cells in vitro to produce growth factors that accumulate in senile plaques in Alzheimer's disease. Brain Res. *569*:141-145.

Arriagada, P.V., K. Marzloff and B.T. Hyman (1992) Distribution of Alzheimer-type pathologic changes in nondemented elderly individuals matches the pattern in Alzheimer's disease. Neurology *42*:1681-1688.

Ball, M.(1977) Neuronal loss, neurofibrillary tangles and granulovacuolar degeneration in the hippocampus with aging and dementia: A quantitative study. Acta Neuropathol. *37*:111-118.

Bancher, C., C. Brunner, H. Lassmann, H. Budka, K. Jellinger, G. Wiche, F. Seitelberger, I Grundke-Iqbal, K Iqbal and H. M. Wisniewski (1989) Accumulation of abnormally phosphorylated tau precedes the formation of neurofibrillary tangles in Alzheimer's disease. Brain Res. *477*:90-99.

Beach, T.G., R. Walker and E.G. McGeer (1989) Patterns of gliosis in Alzheimer's disease and aging cerebrum. Glia *2*:420-436.

Bignami, A. and D. Dahl (1974) Astrocyte-specific protein and neuroglial differentiation. An immunofluorescence study with antibodies to the glial acidic protein. J. Comp. Neurol., *153*:27-38.

Blessed, G., B.E. Tomlinson and M.Roth (1968) The association between quantitative measurements of dementia and senile changes in the cerebral gray matter of elderly subjects.Br.J.Psychiat.*114*:797-811.

Braak, E., H. Braak and E.-M. Mandelkow (1994) A sequence of cytoskeleton changes related to the formation of neurofibrillary tangles and neuropil threads. Acta Neuropathol. *87*:554-567.

Braak, H. and E. Braak (1986) Ratio of pyramidal cells versus non-pyramidal cells in the frontal isocortex and changes in ratio with ageing and Alzheimer's disease. Prog. Brain Res. *70*: 185-212.

Braak, H. and E. Braak (1991) Neuropathological stageing of Alzheimer-related changes. Acta Neuropathol. *82*:239-259.

Breitner, J.C.S. (1996) Inflammatory processes and antiinflammatory drugs in Alzheimer's disease: A current appraisal. Neurobiol. Aging *17*:789-794.

Carpenter, A.F., P.W. Carpenter and W.R. Markesbery (1993) Morphometric analysis of microglia in Alzheimer's disease. J. Neuropathol. Exp. Neurol. *52*:601-608.

Castellano, B. and B. González (1991) ¿Es la enfermedad de Alzheimer el resultado de una reacción inmunológica mediada por lascélulas gliales? En F. Bermejo and N. Acarin (eds): Demencias. Inicio de una Década ¿Los Años del Progreso? Barcelona, N.C.R. pp.3-15.

Chen, S., R.C.A. Frederickson and K.R. Brunden (1996) Neuroglial-mediated immunoinflammatory responses in Alzheimer's disease. Complement activation and therapeutic approaches. Neurobiol. Aging *5*:781-787.

Corder, E.H., A.M. Saunders, N.J. Risch, W.J. Strittmatter, D.E. Schmechel, P.C. Gaskell, J.B. Rimmler, P.A. Locke, P.M. Conneally, K.E. Schmader, G.W. Small, A.D. Roses, J.L. Haines and M.A. Pericak-Vance (1994) Apolipoprotein E type 2 allele decreases the risk of late-onset Alzheimer's disease. Nat. Genet. *7*:180-183.

Cotman C.W. and A.J. Anderson (1995) A potential role for apoptosis in neurodegeneration and Alzheimer's disease. Mol. Neurobiol. *10*:19-45.

Cras, P., M. Kawai, S. Siedlak and G. Perry (1991) Microglia are associated with the extracellular neurofibrillary tangles of Alzheimer disease. Brain Res. *558*: 312-314.

Crystal, H., D. Dickson, P. Fuld, D. Masur, R. Scott, M. Mehler, J. Masdeu, C. Kawas, M. Aronson and L. Wolfson (1988) Clinico-pathologic studies in dementia: Nondemented subjects with pathologically confirmed Alzheimer's disease. Neurology *38*: 1682-1687.

del Rio Hortega,P.(1932) Microglia. In W. Penfield (ed.): Cytology and Cellular Pathology of the Nervous System, vol.2, New York: Paul B. Hoeber, pp.481-534.

Delacourte, A. (1990) General and dramatic glial reaction in Alzheimer brains. Neurology *40*:33-37.

Delaère, P., C. Duyckaerts, C. Masters, K. Beyreuther, F. Piette and J.J. Hauw (1990) Large amounts of neocortical ßA4 deposits without neuritic plaques nor tangles in a psychometrically assessed, non demented person. Neurosci. Lett. *166*:87-93.

Duffy, P.E., M. Rapport and L. Graf (1980) Glial fibrillary acidic protein and Alzheimer-type senile dementia. Neurology, *30*:778-782.

Duyckaerts, C., P. Delaère, J.-J. Hauw. A.L. Abbamondi-Pinto, S. Sorbi, I. Allen, J.P. Brion, J. Flament-Durand, L.Duchen, J.Kauss, W.Schote, J. Lowe, A. Probst, R. Ravid, D.F. Swaab, K. Renkawek and B. Tomlinson (1990) Rating of lesions in senile dementia of the Alzheimer type: Concordance between laboratories.A European multicenter study under the auspices of EURAGE. J. Neurol. Sci.*97*:295-323.

Eikelenboom, P. and F.C. Stam (1982) Immunoglobulins and complement factors in senile plaques. Acta Neuropathol. (Berlin). *57*:239-242.

Eikelenboom, P. and Veerhuis (1996) The role of complement and activated microglia in the pathogenesis of Alzheimer's disease. Neurobiol. Aging *17*:673-680.

Frei, K., C. Siepl., P. Groscurth., S. Bodmer., C. Schwerdel and A. Fontana (1987) Antigen presentation and tumor cytotoxicity by interferon-gamma-treated microglial cells. Eur. J. Immunol. *17*:1271-1278.

Gibson, P.H., M. Stones and B.E. Tomlinson (1976) Senile changes in the human neocortex and hippocampus compared by the use of electron and light microscope. J. Neurol. Sci. *27*:389-405.

Giulian, D. (1987) Ameboid microglia as effectors of inflammation in the central nervous system. J. Neurosci. Res. *18*:155-171.

Giulian, D. and L. B. Lachman (1985) Interleukin-1 stimulation of astroglial proliferation after brain injury. Science *223*:497-499.

Giulian, D., J. Chen., J. E. Ingerman., J. K. George and M. Naponen (1989) The role of mononuclear phagocytes in wound healing after traumatic injury to adult mammalian brain. J. Neurosci. *9*:4416-4429.

Giulian, D., T. J. Baker., L.-C. N. Shih, and L. B. Lachman (1986) Interleukin-1 of the central nervous system is produced by ameboid microglia. J. Exp. Med.*164*:594-604.

Gómez-Isla, T., J.L. Price, D.W. Mckeel, J.C. Morris, J.H. Growdon and B.T. Hyman (1996) Profound loss of layer II entorhinal cortex neurons occurs in very mild Alzheimer's disease. J. Neurosci. *16*:4491-4500.

Gómez-Ramos, P., A. Jiménez., J.L. Cebrián, M.A. Morán (1993) Importancia de la degeneración neurofibrilar en el diagnóstico de la enfermedad de Alzheimer. Med. Clin. (Barc.) *101*:60-63.

Graeber, M.B. and W.J. Streit (1990) Perivascular microglia defined. Trends Neurosci. *13*:366.

Hirano, A. and H.M. Zimmerman (1962) Alzheimer's neurofibrillary changes. A topagraphic study. Arch. Neurol.*7*:227-242.

Hyman, B., G. Van Hoesen, L. Kromer, A. Damasio, C. Barns (1984) Alzheimer's disease: cell-specific pathology isolates the hippocampal formation. Science *255*:1168-70.

Iqbal, K., I. Grundke-Iqbal and H.M. Wisniewski (1987) Alterations in the neuronal cytosketeton in Alzheimer's disease and related conditions. In G. Perry (ed): Alterations in the neuronal cytoskeleton in Alzheimer disease, New York: Plenum Press, pp.109-136.

Katzman, R., R. Terry, R. De Teresa, T. Brown, P. Davies, P. Fuld, X. Renbing and A. Peck (1988) Clinical, pathological, and neurochemical changes in dementia: A subgroup with preserved mental status and numerous neocortical plaques. Ann. Neurol. *23*:138-144.

Kawamata, T. I. Tooyama, T. Yamata, D.G. Walker and P.L. McGeer (1993) Lactotransferrin immunocytochemistry in Alzheimer and normal brain. Am.J.Pathol.*142*:1574-1585.

Kemper, T. (1984) Neuroanatomical and neuropatho-logical changes in normal aging and in dementia. In M.L.Albert (ed): Clinical Neurology of Aging, New York: Oxford University Press. pp.9-52.

Khachaturian, Z. S. (1985) Diagnosis of Alzheimer's disease. Arch. Neurol. 42:1097-1105.

Kidd, M (1963) Paired helical filaments in electron microscopy of Alzheimer's disease. Nature, Lond. *197*:192-193.

Killackey, H. P.(1984) Glia and the elimination of transient cortical projections. Trends. Neurosci. 7:225-226.

Kim, K.S., D.L. Miller, V.J. Sapienza, C.J. Chen, C. Bai, I. Grundke-Iqbal, J. R. Currie and H.M. Wisniewski (1988) Production and characterization of monoclonal antibodies reactive to synthetic cerebrovascular amyloid peptide. Neurosci. Res. Commun. *2*:121-130.

Kimelberg, H.K. and M.D. Norenberg (1989) Astrocytes. Sci. Am. *260*:44-52.

Lassmann, H., F. Zimprich., K. Vass and W. F. Hickey (1991) Microglial cells are a component of the perivascular glia limitans. J. Neurosci. Res. *28*:236-243.

Lewis, D.A., M.J. Campbell, G.W. Huntley, D.L. Benson, R.D. Terry and J.H. Morrison (1986) Laminar and regional specificity of tangle and plaque distribution in Alzheimer's disease (AD): A quantitative study of visual and auditory cortices. Soc. Neurosci. Abst. *12*:943.

Ling, E.-A. and W.-C. Wong (1993) The origin and nature of ramified and amoeboid microglia: A historical review and current concepts. Glia *7*:9-18.

Lorton D., J.-M. Kocsis, L. King, K. Madden and K.R. Brunden (1996) ß-amyloid induces increased release of interleukin-1ß from lipopolysaccharide-activated human monocytes. J. Neuroimmunol. *67*:21-29.

Lue, L.-F, L. Brachova, W.H. Civin and J. Rogers (1996) Inflammation, Aß deposition, and neurofibrillary tangle formation as correlates of Alzheimer's disease neurodegeneration. J. Neuropathol. Exp. Neurol. *55*:1083-1088.

Lue, L.-F. and J. Rogers (1992) Full complement activation fails in diffuse plaques of Alzheimer's disease cerebellum. Dementia *3*:308-313.

Mackenzie, I.R.A., C.J. Hao and D.G. Muñoz (1995) Role of microglia in senile plaque formation. Neurobiol. Aging *16*:797-804.

Mackenzie, I.R.A., R.S. McLachlan, C.S. Kubu and L.A. Miller (1996) Prospective neuropsychological assessment of nondemented patients with biopsy proven senile plaques. Neurology *46*:425-429.

Mattiace, L.A., P. Davies and D.W. Dickson (1990) Detection of HLA-DR on microglia in the human brain as a function of both clinical and technical factors. Am. J. Pathol. *136*:1101-1114.

McGeer, P.L., H. Akiyama, S. Itagaki, and E.G. McGeer (1989) Immune system response in Alzheimer's disease. Can. J. Neurol. Sci. *16*:516-527.

McGeer, P.L., S. Itagaki and E.G. McGeer (1988) Expression of the histocompatibility glycoprotein HLA-DR in neurological disease. Acta Neuropathol. *76*:550-557.
McGeer, P.L., T.Kawamata, D.G. Walker, H.Akiyama, I.Tooyama and E.G. McGeer (1993) Microglia in degenerative neurological disease. Glia, *7*:84-92.
McKee, A.C., K.S. Kosik and N.W. Kowal (1991) Neuritic pathology and dementia in Alzheimer's disease. Ann. Neurol. *30*:156-165.
Morán, M.A., J.L. Cebrián, P. Gómez-Ramos, A. Cabello, S. Madero and E.J. Mufson (1992) Diagnóstico de la enfermedad de Alzheimer. Valoración de las placas seniles de tipo difuso. Med. Clin.(Barc.) *98*:19-23.
Morishima-Kawashima, M., M. Hasegawa, K. Takio, M. Suzuki, K. Tatani and Y. Ihara (1993) Ubiquitin conjugated with amino-terminally processed tau in paired helical filaments. Neuron *10*:1151-1160.
Nieto-Sampedro, M., R.P. Saneto, J. de Vellis and C.W. Cotman (1985) The control of glial populations in brain: Changes in astrocyte mitogenic and morphogenic factors in response to injury. Brain Res. *343*:320-328.
Pappolla, M.A., R.A. Omar and H.V. Vinters (1991) Image analysis microspectroscopy shows that neurons participate in the genesis of a subset of early primitive (diffuse) senile plaques. Am. J. Pathol. *139*:599-607.
Pereira, H.A., P. Kumar and P. Grammas (1996) Expression of CAP37, a novel inflammarory mediator, in Alzheimer disease. Neurobiol. Aging *17*:753-759.
Perlmutter, L.S., S.A. Scott and H.C. Chui (1991) The role of microglia in the cortical neuropathology of Alzheimer disease. Bull. Clin. Neurosci. *56*: 120-130.
Perlmutter, L.S., S.A. Scott, E. Barrón and H. C. Chui (1992) MHC class II-positive microglia in human brain: association with Alzheimer lesions. J. Neurosci. Res. *33*:549-558.
Peters, A., S.L. Palay and H. de F. Webster (1991). In: The Fine Structure of the Nervous System, New York: Oxford University Press. pp.273-311.
Pike, C.J., B.J. Cummings and C.W. Cotman (1995) Contributions of ß-amyloid to reactive astrocytosis in Alzheimer's disease. In K.Iqbal, J.A. Mortimer, B.Winblad and H.M. Wisniewski (eds): Research Advances in Alzheimer's Disease and Related Disorders. Sussex: John Wiley & Sons Ltd, pp.619-627.
Pike, C.J., B.J. Cummings, R. Monzavi and C.W. Cotman (1994) ß-Amyloid-induced changes in cultured astrocytes parallel reactive astrocytosis associated with senile plaques in Alzheimer's disease. Neuroscience *63*:517-531.
Pitas, R.E., J.K. Boyles, S.H. Lee, D. Foss and R.W. Mahley (1987) Astrocytes synthesize apolipoprotein E and metabolize apolipopropein E-containing lipoproteins. Biochem. Biophys. Acta, *917*:148-161.
Probst, A., D. Langui, S. Ipsen, N. Robakis and J. Ulrich (1991) Deposition of ß/A4 protein along neuronal plasma membranes in diffuse senile plaques. Acta Neuropathol. *83*:21-29.
Probst, A., J. Ulrich and U. Heitz (1982) Senile dementia of Alzheimer type: Astroglial reaction to extracellular neurofibrillary tangles in the hippocampus. Acta Neuropathol.(Berl) *57*:75-79.
Riederer, B.M., A. Guadaño-Ferraz and G.M. Innocenti (1990) Difference in distribution of microtubule associated proteins 5a and 5b during the development of cerebral cortex and corpus callosum of cats: dependence on phosphorylation. Dev. Brain Res. *56*:235-243.
Rogers, J., S. Webster, L.-F. Lue, L. Brachova, W.H. Civin, M. Emmerling, B. Shivers, D. Walker and P. McGeer (1996) Inflammation and Alzheimer's disease pathogenesis. Neurobiol. Aging *17*:681-686.
Schmidt, M.L., R.E. Gur, R.C. Gur and J.Q. Trojanowski (1988) Intraneuronal and extracellular neurofibrillary tangles exhibit mutually exclusive cytoskeletal antigens. Ann. Neurol. *23*:184-189.
Selkoe, D.J. (1989) Molecular pathology of amyloidogenic proteins and the role of vascular amyloidosis in Alzheimer's disease. Neurobiol. Aging *10*:387-395.
Siman, R., J.P. Card, R.B. Nelson and L.G. Davis (1989) Expression of ß-amyloid precursor protein in reactive astrocytes following neuronal damage. Neuron *3*:275-285.
Singhrao, S.K., B.P. Morgan, J.W. Neal and G.R. Newman (1995) A functional role of corpora amylacea based on evidence from complement studies. Neurodegeneration *4*:335-345.
Streit, W. J., M. B. Graeber and G. W. Kreutzberg (1988) Fuctional plasticity of microglia: A review. Glia *1*:301-307.

Terry, R.D. (1992) The pathogenesis of Alzheimer's disease. What causes dementia? In Y.Christen and P. Churchland (eds): Neurophilosophy and Alzheimer's Disease, Research and Perspectives in Alzheimer's Disease. Berlin: Springer-Verlag. pp.123-130.
Thomas, W. E.(1992) Brain macrophages: evaluation of microglia and their functions. Brain Res. Rev. *17*:61-74.
Tomlinson, B.E., G. Blessed and M. Roth (1968) Observations on the brain of non-demented old people. J. Neurol. Sci. *7*:331-356.
Ulloa, L., E. Montejo de Garcini, P. Gómez-Ramos, M. A. Morán and J.Avila (1994) Microtubule-associated protein MAP-1B showing a fetal phosphorylation pattern is present in sites of neurofibrillary degeneration in brains of Alzheimer's disease patients.Mol.Brain Res. *26*:113-122.
Wallin, A., K. Blennow and L.E. Rosengren (1996) Glial fibrillary acidic protein in the cerebrospinal fluid of patients with dementia. Dementia 7:267-272.
Wang, D. and D.G. Muñoz (1995) Qualitative and quantitative differences in senile plaque dystrophic neurites of Alzheimer's disease and normal aged brain. J.Neuropathol. Exp. Neurol. *54*:548-556.
Wegiel, J., H.M. Wisniewski, K.C. Wang, M. Kujawa and B. Lanch (1988) The role of microglia in plaque formation in Alzheimer's disease. J. Neuropathol. Exp. Neurol. (Abst.) *47*:338.
West, M.J., P.D. Coleman, D.G. Flood and J.C. Troncoso (1994) Differences in the pattern of hippocampal neuronal loss in normal ageing and Alzheimer's disease. The Lancet *344*:769-772.
Wilcock, G.K. and M.M. Esiri (1982) Plaque, tangles and dementia: a quantitative study. J.Neurol. Sci. *56*:343-56.
Wisniewski, H.M. and J. Wegiel (1992) Alzheimer's disease neuropathology. Current status of interpretation of lesion development. Physiopathol. Proc.*673*:270-284.
Wisniewski, H.M., J. Wegiel, K.C. Wang, M. Kujawa and B. Lach (1989) Ultrastructural studies of the cells forming amyloid fibers in classical plaques. Can. J. Neurol. Sci. *16*:535-542.
Yamaguchi, H., S. Harai, M. Moritatsu, M. Shoji and Y. Harigaya (1988) Diffuse type of senile plaques in the brains of Alzheimer-type dementia. Acta Neuropathol (Berl) *77*: 113-119.
Yamaguchi, H., Y. Nakazato, S. Harai, M. Shoji, and Y. Harigaya (1989) Electron micrograph in the Alzheimer brain. Am. J. Pathol. *135*:593-597.
Yamaguchi, H., Y. Nakazato, M. Shoji, K. Okamoto, Y. Ihara, M. Morimatsu, and S. Hirai (1991) Secondary deposition of beta amyloid within extracellular neurofibrillary tangles in Alzheimer-type dementia. Am. J. Pathol. *138*:699-705.

20

ADHESION MOLECULES, ANGIOGENESIS AND MALIGNANT GLIOMAS: IMPLICATIONS FOR TUMORIGENESIS

Blanca Conde, Elena Sinués, Ana Gascón,
Angela Alcalá and Mercedes Ruidíaz

A high degree of local invasiveness is a hallmark of malignant glioma. Turmor cell migration and proliferation are highly dependent on the composition of the extracellular matrix. The tumoral invasion seems not to be strictly associated with the grade of malignancy. So, anaplastic astrocytomas frequently show more extensive infiltration of adjacent brain than do glioblastomas. The invasiveness seems to be an intrinsic behaviour of glioma cells rather than a feature of extreme dedifferentiation. The malignant progression of gliomas is associated with the migratory capacity of these cells. A three step process-attachment, degradation of matrix and migration- has been proposed to characterize the profound influence of the extracellular matrix. The altered expression or distribution of adhesion molecules as well as the proteolytic degradation of the matrix by an imbalance between the metalloproteinases and their inhibitors create an intercellular space where the active tumor cell migration leads to dissemination into the surrounding tissue. The molecular characterization of adhesion receptors (integrins, CAMs, cadherins, selectins) provides a tool to study the molecluar basis and biological consequences of these phenomena. The integrins play important roles in tumor angiogenesis or neovascularization by the sprouting of new blood vessels form pre-existing ones. Angiogenesis, essential for the growth of tumors, is prevented by integrins blockade. In fact, their inhibition resulted in apoptosis of the proliferating vascular cell. The endogenous death pathway of apoptosis can be activated by deprivation of cytokines mediators of angiogenesis. The ability to modulate molecules involved in the relations tumor cell-extracellular matrix or endothelial cells may be a method to limit proliferation and local invasion in the gliomas.

The mean survival of patients with malignant gliomas treated with surgery and radiotherapy is calculated to be between 36 and 40 months. This clinical datum, with scarce variations during the last fifteen years, is sufficient reason to search for other strategies than the conventional therapies. Treatment based on conventional and experimental chemotherapy only increase this survival rate slightly. Patients with low grade gliomas either present expectations which are too optimistic, since their most common cause of death is to be found in the dedifferentiation of the original tumors towards malignant gliomas, as evaluated after a five-year survival in 28% of patients with initially low grade gliomas (Vertosick et al., 1991). The design of new therapies needs significant advances in the scientific knowledge of the biological bases of the induction and neoplasic progression of the glia, as a way to identify new molecules that constitute specific targets for the therapy.

CELLULAR DEDIFFERENTIATION IN GLIAL TUMORS

The concept of cancer as the result of an alteration of the differentiation/ dedifferentiation cellular has been accepted and incorporated in clinical practice as shown by the terms "well differentiated tumor" or "poorly differentiated tumor". Up until now, the establishment of the grade of differentiation or malignity of the glial cells has mainly been based on the morphological pattern, which is rather a paradox. Normally, the immature glia tends to present a homogeneous morphology, there being controversy as to the existence of morphological differences between astroblasts, oligodendroblasts, ependimoblasts, glioblasts and meduloblasts. In fact, the prediction and the grade of malignity of the classification systems of glial tumors envisages characteristics of the tumoral architecture such as necrosis or endothelial proliferation more than those of the individual morphology of the cells (Nelson et al., 1982). Clearly, morphology alone is insufficient for the study of glial dedifferentiation and oncogenesis. The development of cell cultures, monoclonal antibodies, molecular biology and techniques of gene transfer has produced powerful tools in the study of glial differentiation.

The normal antigenic phenotype during the glial differentiation in the central nervous system of the rat (CNS) has been studied by means of different antigens and antibodies, which can be seen in table 20-1. The antigens used to identify the different differentiation stages of the glia may be the genetic products of a series of sequential expression genes, whose transcription is controlled by molecules directly related to normal glial progression and differentiation. In this way, the induction of cellular differentiation requires the presence of mediating, diffusible, or transportable molecules, in the cascade of transcriptional regulation. A mutation in a homeobox gene would lead to a new transcription factor with unpredictable effects on gene transcription. Some of the known oncogens encode homeodomains containing proteins, while other encode for other molecules involved in the transcriptional regulation (Fuller and Bigner, 1992). Stronger candidates to play this role are the growth factors, and other candidates include vitamin derivatives such as retinoic acid. Examples of inductive growth factors of cellular differentiation through the induction of the gene expression related to given homeoboxes include FGF, the TGF-ß family of proteins (e.g.: vitellogenin-1, TGF-ß5, activin) and the proteins encoded by the wnt family of genes (Johnson et al., 1993). Even without a categorical demonstration, it seems that the actions of EGF, PDGF, CNTF, basicFGF and insulin on glial differentiation is produced via the induction of homeobox genes (Melton, 1991). Table 20-2 summarizes the growth factors involved in proliferation and differentiation during gliogenesis (Linskey and Gilbert, 1995).

Table 20-1. *Rat glial cell lineage and stage of maduration defined by antigenic phenotype*

Cell identified	GD_3	anti A_2B_5	GFAP	anti O_4	anti Gc	anti NF	anti Fc
Type 1 astrocyte precursor	?	-	-	-	-	-	-
Mature type 1 astrocyte	?	-	+	-	-	-	-
Oligodendrocyte type 2 -astrocyte precursor	+	+	-	-	-	-	-
Immature oligodendrocyte	+	+	-	+	-	-	-
Mature oligodendrocyte	-	-	-	+	+	-	-
Mature type 2 astrocyte	-	+	+	-	-	-	-
Neuron	-	+	-	-	-	+	-
Microglia/macrophage	-	-	-	-	-	-	+

GFAP, glial fibrillary acidic protein; A_2B_5, polysialogangliosides (GQ1c, GQ1b, GT3); O_4, cell surface sulfatides; Gc, galactocerebroside; NF, neurofilament protein; Fc, plasma membrane receptor+, positive; -, negative; ?, unknown (From Linskey and Gilbert, 1995).

Since alterations in the control of proliferation may lead to neoplasic transformations, it has been demonstrated that human gliomas respond to various growth factors (e.g.: EGF, PDGF, IGF-1, acidic FGF and basic FGF) and express both these growth factors and their receptors (e.g.: EGF, TGF-α, EGFR and PDGFR). which suggests possible autocrine and paracrine mechanisms involved in the cellular proliferation of gliomas (Stefanik et al., 1991; Fleming et al., 1992; Hurtt et al., 1992).

Studies based on restriction polymorphisms suggest at least possible two pathways that lead to the malign transformation of the glia. In this way, in gliomas with all the histological grades the loss of heterozygosity affecting chromosomes 13, 17 and 22 has been observed as well as the presence of "double-minutes" chromosomes. While the loss of heterozygosity in the chromosome 9p has been related to anaplastic astrocytomas and multi-form glioblastomas, the affectation of chromosomes 1p and 10 seems be exclusive of multiform glioblastomas (Veter and Thomas, 1991). Although the glioblastomas may derive from low grade gliomas through a series of sequential genetic alterations, it is also possible that this tumoral type emerges directly from the normal glia through isolated alterations of chromosome 10. In the low grade gliomas, trisomies of chromosome 7 and the loss of sexual chromosomes have been detected with relative frequency (Hoang-Xuan et al., 1995). Some suppressive tumor genes are located in delectioned chromosomic regions in the gliomas, such as genes MTS1, Rb, p53 and NF2, located in 9p, 13q, 17p and 22q respectively. In fact, these genes are found frequently altered in glial tumors (Hoang-Xuan et al., 1995). During normal cellular differentiation there is a convergence of characteristics such as the proliferation, migration, cellular invasion and loss of contact inhibition, characteristics that are indicative of malignity in the transformed cells. Since the genes that control the method of cellular differentiation are present in the genome of mature cells, it may be speculated as to whether oncogenesis implies the reactivation and loss of control of the genes involved in proliferation/differentiation, triggered by spontaneous or induced mutations. In a first hypothesis, oncogenesis would commence in a mature cell which, by means of unknown mechanisms, is dedifferentiated toward an immature and potentially malignant glial cell. A second hypothesis postulates the existence of a pool of non-differentiated stem cells that in a latent condition would integrate the central nervous system, precursor cells that would be the sustrate for oncogenesis.

Table 20-2. *Growth Factor dependence of glial proliferation and differentiation during gliogenesis*

Growth Factor	Cell stimulated to proliferate	Cell stimulated to differentiate
Transforming growth factor-α	Multipotential neuronal glial precursor	Reynolds et al.,1992
Epidermal growth factor	Multipotential neuronal glial percursor	Reynolds et al.,1992
Platelet-derived growth factor	O2A progenitor cell	Masters et al.,1991
Basic fibroblast growth factor *	O2A progenitor cell	Benbrook et al.,1988
Insulin / insulin-like growth factor	O2A progenitor cell (type-2 astrocyte)	
	O2A progenitor cell (oligodendrocyte)	Lillien and Raff, 1990
Ciliary neurotrophic factor	O2A progenitor cell (type-2 astrocyte)	Goldman et al., 1986
	O2A progenitor cell (oligodendrocyte)	Mayer et al., 1994

* Only effective in synergistic combination with plateled-derived growth factor
(From Liinskey and Gilbert, 1995)

ONCOGENES AND ANTIONCOGENES IN BRAIN TUMORS

In the vast majority of cases, the genetic changes that cause cancer occur in a somatic cell. Genes that cause cancer can essentially be grouped into three categories: Oncogenes with a dominant action and which generally encode products that stimulate cell growth, inhibit differentiation, or block apoptosis (table 20-3). Suppressor genes are recessive, and in most cases both alleles must be detected or inactivated; their products inhibit cell growth or direct the cells to the apoptotic pathway (table 20-4). A third category is the genes that maintain the integrity of the genome or DNA repair genes (Westermark and Nister, 1995). The oncogenes involved in the gliomas encode proteins involved in the transduction of mitogenic signals from the membrane to the nucleus, which is confirmed by the intense activity of the protein kinase C observed in the tumors derived from the glia (Couldwell et al., 1991; Baltuch et al.,1995). The gene EGFR encodes for the membrane receptor of the epidermic growth factor. This gene is activated in 50-80% of the gliomas and in most cases its overexpression is due to a gene amplification. However, in 20% of the cases, rearrangements have been detected that lead to the synthesis of a superactive truncated protein which is able to transduce signals in a constitutive manner. The grade of amplification of the EGFR gene has been related to the level of malignity of the gliomas (Collins, 1993).

Table 20-3. *Genes involved in gliomas and their mechanism of oncogenic activation*

Genes	Oncogenic activation
Growth Factors	
TGFα (Transforming growth factor alpha)	amplification
PDGFB (Platelet derived growth factor B)	overexpression, amplification
IGF (Insulin growth factor)	overexpression
FGF (Fibroblast growth factor)	overexpression
VEGF (Vascular endothelial growth factor)	overexpression
Growth Factors Receptors (tyrosine-quinase activity)	
EGFR (Epidermal growth factor receptor)	amplification, rearrangement
PDGFRA (PDGF receptor A)	amplification, rearrangement
FGFR (FGF receptor)	overexpression
IGFR (IGF receptor)	overexpression
MET (Hepatocyte growth factor receptor)	amplification
Membrane proteins (tyrosine-quinase activity)	
ros	amplification, rearrangement
abl	rearrangement
src	overexpression
Nuclear proteins, transcripcion factors	
myb	amplification, rearrangement
N-myc	amplification
C-myc	amplification, rearrangement
gli	amplification
fos	overexpression
MDM2 (Murine double minute)	amplification

(From Hoang-Xuan et al., 1995).

Table 20-4. Tumor suppressor genes altered in gliomas

Gene	Protein	Function	% Mutations (gliomas)
p53	p53	Trancription factor Cell cycle control Apoptosis	30-40%
Rb	p110Rb	Transcriptional modulation Cell cycle control	< 10%
NF1	neurofibronime	Ras-GTPase activation	astrocytomas pilocytiques?
MTS1	p16	Protein-quinase Cdk4 inhibition Cell cycle control	< 5%

(From Hoang-Xuan et al., 1995)

Among the tumoral suppressor genes, that of greatest incidence in the gliomas is gene p53. The p53 gene in humans resides in the p13 region of chromosome 17, it codes for a p53 protein of nuclear localization, p53 is subject to post-translational modifications. Thus, $p34^{cdc2}$ phosphorylates human p53, starting at S phase and p53 is also a sustrate for the cyclin B-cdc2 complex, which becomes active during M phase (Bischoff et al., 1990). The binding of wild type p53 to sequence-specific DNA results in transcriptional modulation. Transcriptional activation is mediated through direct interaction with the transcription factor TFIID. Negatively regulated genes include ß-actin, c-jun, c-fos, IL-6 and basic FGF (Bögler et al., 1995). The genes bax (Miyashita and Reed, 1995) and MDM-2 are positively regulated. MDM-2 gene is expressed in 10% of human multiform glioblastoma and grade III astrocytomas. A major cellular function of the p53 protein is to suspend progression through the cell cycle, from G1 to S phase, in response to DNA damage so that it can be repaired before replication. Its function as tumoral suppressor gene is based on its ability to protect the stability of the genome, permitting the repair of injuries or the elimination of tumoral cells through its own suicide by apoptosis (Lane, 1994). The overexpression of wild-type p53 protein leads to growth arrest in many cell types, including glioma-derived cell lines (Bögler et al., 1995). Furthermore, the implication of p53 in senescence allows it to be involved in regulating growth arrest in the absence of DNA damage. Apoptosis, as opposed to G1 arrest in response to DNA damage, is induced by the p53 gene and associated with the ability of p53 to promote transcription of genes that influence cell cycle progression. An important pathway by which p53 acts to arrest the cell cycle is through protein p21; p53 induces the expression of p21, which binds and inactivates cyclin/Cdk complexes and the proliferating cell nuclear antigen PCNA, arresting cells in G1. The overexpression of p21 promotes apoptosis of proliferating cells in vitro (Boudreau et al., 1996).

Gene p53 has been implicated in the malignant transformation of the astrocytomas, considering their altered shapes as markers of a bad tumoral prognosis.

Thus, the tumoral progression of some low grade astrocytomas that reccur as glioblastomas, is accompanied by the clonal expansion of mutant tumoral cells for p53. Equally, the presence of normal p53 favours the expression of physiological inhibitors of angiogenesis (Hoang-Xuan et al., 1995).

Cytogenetic and restriction fragment length polymorphism analyses have revealed interstitial deletions and homozygous delections in the short arm of chromosome 9 in a large number of glioblastoma multiforme. These deletions are localized to a centromeric region of the interferon gene cluster in 9p21. The deletions affect the genes encoding the cyclin-dependent quinase inhibitor p15 and p16 and the genes are denoted MTS2 and MTS1, respectively (Westermark et al., 1995).

TUMORAL PROGRESSION AND EXTRACELLULAR MATRIX

The malignant progression in the gliomas is associated with an extraordinary capacity of migration and local invasion. A process which developes in three stages, adhesion, degradation of the extracellular matrix and migration, it characterizes the profound influences exerted on the tumoral invasion by the extracellular matrix (ECM). The alteration in the expression or in the distribution of the molecules of adhesion, transmembranous receptors such as integrins or lectins of the cellular surface, can induce variations in the adhesion of the tumoral cells to their specific sustrate (Paulus et al., 1993). The proteolytic degradation of the matrix due to an imbalance between the metaloproteinases of the activators of the plasminogene and its inhibitors creates an intercellular space where the migration of the active tumoral cells generates dissemination between the neighbouring tissues (De Clerk et al., 1994).

In the tumoral invasion, the transformed cells interact with proteins of the ECM of the basal membranes (laminin, fibronectin and collagen), a basal membrane which in the central nervous system is restricted to the external limiting glia and vascular endothelia, textures that remain undisrupted for a long time even in the case of malignant tumors. The ECM of the cerebral parenchyma is made up of an amorphous substance that contains fibrous proteins. The main component of the extracellular space is hyaluronic acid and other glycosaminoglycans. A glycoprotein of extracellular matrix, tenascin with adhesive and counteradhesive domains is expressed temporo-spatially during development and in the adult matrix (Erickson and Bourdon, 1989).

Different components of the ECM have been identified in the matrix of the gliomas. Although fibronectin and collagen are only expressed in the hyperplastic blood vessels, laminin, tenascin and vitronectin are synthesized by the tumoral cells of gliomas, being deposited in the ECM. A relationship between the expression of tenascin and the histological grade of malignity has been postulated in these tumors, as well as their role in the promotion of angiogenesis (Higuchi et al., 1993). The front of tumoral invasion is characterized by an increase of hyaluronic acid. These components can facilitate or contribute to glial dissemination. Table 20-5 shows a summary of the ligands of the ECM involved in the invasion of gliomas (Giese at al., 1995).

Table 20-5. *ECM in the central nervous system and human gliomas*

Anatomic structure	Composition	Role in invasion of human gliomas
Glia limitans externa	collagen (types I, III, IV) laminin fibronectin heparin sulfate	Although subpial spread can occur as a route of dissemination , penetration of glioma cells into the subarachnoid space is an infrequent event
Vascular basement membrane	collagen (types IV, V) entactin laminin heparin sulfate vitronectin	Rarely disrupted by glial tumors
Brain parenchyma	hyaluronic acid glycosaminoglycans: -dermatan sulfate -chondoitin sulfate -hyaluronectin tenascin	Most frecuent route of glioma dissemination following white matter fiber tracts (Rosenblum, 1993)
Tumor ECM	tenascin vitronectin laminin hyaluronic acid	Possible supportive substrate for tumor dissemination (Giese et al,1994)

Laminin, fibronectin,and collagen type IVare expressed in hyperplastic tumor blood vessels (From Giese et al. 1995).

The cell-extracellular matrix interactions that modify the migration form part of the normal biology of development. Thus, the migration of the cells of the neural crest in the development of the nervous system is guided not only by the expression of ECM proteins and hyaluronic acid but also by signaling mediated by cell-cell contact. These contacts stimulate the cellular migration as a whole as well as inhibiting the individualized locomotion of the cells (Thomas and Yamada, 1992). The membrane ligands and receptors of the host cells, as the immunoglobulin superfamily, ICAM, NCAM, or CD44, may affect the motility of the invading cells. Thus, the rat glioma cells of transfected with the vector of NCAM expression present less capacity of invasion than the non-transfected ones when they are injected into the adult rat brain. Table 20-6 shows the molecules of adhesion involved in glial tumorigenesis (Couldwell et al., 1992).

Table 20-6. *Adhesion molecules with relevance to glioma biology*

Family adhesion molecule	Natural ligand	Distribution
Integrins		
lymphocyte function-associated antigen-1 (LFA-1)	ICAM-1	eukocytes
very late antigen-4 (VLA-4)	fibronectin, VCAM-1	glioma, lymphocytes, monocytes, neural crest derivatives
Immunoglobulin superfamily		
lymphocyte function-associated antigen-3 (LFA-3)	CD_2	glioma, leukocytes
vascular cell adhesion molecule-1 (VCAM-1)	VLA-4	endothelial cells
intercellular adhesion molecule-1 (ICAM-1)	LFA-1	glioma, leukocytes, fibroblasts, endothelial cells.
neural cell adhesion molecule (NCAM)	NCAM, collagen types I-IV	glioma, neurons
Selectins		
endothelial-leukocyte adhesion molecule-1 (ELAM-1)	sialyl-lex	endothelial cells
Others		
CD44	collagen types I & IV fibronectin, vascular addressin, hyaluronate,	glioma, T lymphocytes, normal brain, fibroblasts

(From Couldwell et al., 1992)

INTEGRINS

The phenomenon of the malignant transformation cannot be explained only as a loss of control of cellular increase, triggered by the activation of oncogenes and the inactivation of tumor suppressor genes, since the transformed cells also present the ability to migrate from the primary tumor and extend over the organism giving rise to metastatic areas (Giancotti and Mainiero, 1994). The neoplasic cells present alterations in the cellular adhesiveness which contribute to their ability to invade locally or distally (Rouslathi, 1984). The integrins are found among the molecules of adhesion altered in cancer (Chan et al. 1991).

Integrins are a large family of homologous cell surface receptors that mediate cell-matrix and cell-matrix and cell-cell adhesion (Hemler, 1990; Springer, 1990; Hynes, 1992). They participate in embryonic development, in the maintenance of tissular organization, in the wound healing of injuries, in inflammation and in homeostasis. Each integrin is composed of a subunit α and a subunit ß. At present at least 15 types of subunits α and 9 ß are known. These can combine amongst themselves to form at least 21 receptors with distinct ligand binding specificities (table 20-7). Most integrins bind to extracellular matrix proteins and promote cell-substratum adhesion. Both the α and the b subunits of integrins are membrane glycoproteins with a large extracellular domain, a single membrane spanning segment and a short cytoplasmatic portion. Because of their structure and domain organization, integrins serve as transmembrane linkers between the extracellular matrix and the cytoskeleton. In addition, recent results indicate that integrins can transmit signals from the extracellular matrix to the cell interior (outside-in signaling) (Juliano and Haskill, 1993). Furthermore the integrins are also the target of regulating signals originating from the cellular interior (inside-out signaling) (Ginsberg et al., 1992).

***Table 20-7.** Subunit integrins*

ß4	α6	ß4: Kal, Mer, Lam. ß1: Lam, Mer
	α1	Coll, Lam
	α2	Coll, Lam
	α3	Lam, Kal, Coll, Fn
ß1	α5	Fn-RGD
	α7	Lam
	α8	Basal lamina
	α9	?
ß7	α4	ß1: Fn-V25, VCAM-1. ß7: Fn-V25, VCAM-1, MAdCAM-1
	αIEL	?
ß5		
ß6	αv	ß1: Fn. ß5: Vn. ß6: Fn. ß8: ?. ß3: Vn, Fb, Fn, VWF, TS, OP
ß8		
ß3	αIIb	Fb, Fn, Vn, VWF, TS
	αL	ICAM-1, ICAM-2, ICAM-3
ß2	αM	iC3b, Fb, FX, ICAM-1
	αx	iC3b, Fb

Schematic representation of the known αß subunit integrin pairs and the ligands they interact with. Many α subunits can associate with only one ß subunit. Thus, subfamilies comprising integrins with the same ß subunit. However, the boundaries between individual subfamilies are poorly defined because some α subunits can combine with several different ß subunits. In this regard, αv is particularly promiscuous. Lam, laminin; Coll, collagen; Kal, kalinin; Mer, merosin; Fn, fibronectin; Fn-RGD denotes binding to the RGD site in fibronectin; Fn-V25 denotes binding to the V25 splice insert of fibroectin. VCAM-1, vascular cell adhesion molecule 1; MadCAM-1, mucosal addressin cell adhesion molecule 1; Vn, vitronectin; Fb, fibrinogen; VWF, Von Willebrand factor; TS, thrombospondin; OP, osteopontin; ICAM-1, 2, 3; iC3b, inactivated C3b component of complement; FX, coagulation factor X (from Giancotti and Mainiero, 1994 and references therein).

The involvement of the integrins in tumorigenesis is twofold. In the first place, the integrins mediate the stable cellular adhesion and the migration over the components of the extracellular matrix. In this way, changes in their level of expression and function can therefore contribute to the tumoral invasion. The steps of the metastatic process in which the integrins are probably be involved would be:

i) Detachment of the primary tumor and penetration in the adjacent cellular matrix and blood vessels. In the same way as changes in the integrins can promote the adquisition of an invasive phenotype of the tumoral cells, there is evidence that suggests that the tumoral invasion is facilitated by the proteolytic degradation of the components of the extracellular matrix (Mignatti and Rifkin, 1993). For example, many neoplasic cells secrete high levels of plasminogen activator. The activation of the plasmin system in the surface of the tumoral cells producing the degradation of fibronectin and other ligands of the extracellular matrix integrins (Quigley et al., 1986; Sullivan and Quigley, 1986; Chen and Chen, 1987). Once the tumoral cells have crossed the basal membrane and the extracellular matrix, they must accede to the blood circulation. This is favoured as the tumoral cells secrete angiogenic factors that stimulate the vascularization of the tumor and the walls of these vessels are discontinuous (Folkman and Klagsbrun, 1987).

ii) Interaction with platelets and leukocytes in the general circulation. The role of platelets in metastasis can be mechanical since they may protect the tumoral cells from shear forces in the bloodstream, and the aggregates formed by platelets and tumoral cells would detained more easily in the capillary beds than the isolated tumoral cells. Some tumoral cells secrete molecules which activate the platelets and lead to the upregulation of $\alpha IIb\beta 3$ fibrinogen receptors and the activation of the coagulation cascade. Many tumoral cells, among which are found the cells of the glioblastomas, express integrin $\alpha V\beta 3$, which can bind to fibrinogen; fibrinogen may act as a bridge between the platelets and the tumoral cells and mediate co-aggregation (Karpatkin et al., 1988; Gawaz et al., 1991). The aggregates of the tumoral cells and platelets can also form co-aggregates with fibrin-coated monocytes: the fibrin is bound to such cells through the $\alpha M\beta 2$ (MAC) integrin (Altieri et al., 1988; Wright et al., 1988).

iii) Detention and extravasation in the target organ. Many tumors metastasize specifically in certain organs. At least two factors have been implicated in this selectivity: one is the expression pattern of adhesion molecules expressed by the metastatic cells, and the other is the existence of suitable growth factors in the parenchyma of the target organ. In diverse studies various integrins have been implicated in such metastatic selectivity, for example the $\alpha 4\beta 1$ integrin may be involved in metastasis to the bone marrow (Giancotti and Mainiero, 1994).

Secondly, the integrins transmit signals from the extracellular matrix to the cellular interior and these signals affect cellular growth and differentiation (Giancotti and Mainiero, 1994).

Normal cells require adhesion to the extracellular matrix to progress throughout the cellular period, a phenomenon called anchorage dependence. Furthermore, certain extracellular matrixes may modulate the gene expression positively or negatively. The effect that the cellular matrix has in cellular growth and in differentiation suggests that the integrins may mediate the intracellular signals. The analysis of the intracellular signals mediated by the integrins can explain the effects that the extracellular matrix has in growth and differentiation and can explain the phenomenon of the invasive tumoral growth (Giancotti and Mainiero, 1994).

When the cells are plated on fibronectin, a ligand of several integrins, the phosphorylation is induced in the tyrosine residues of two components located in focal adhesions: paxillin (80 kDa) and FAK (Focal Adhesion Kinase) (Parsons et al., 1994). This phosphorylation is mediated by the integrins and triggers the expression of immediate-early response genes and cyclins, and consequently the cell cycle. One of the most important functions mediated by FAK phosphorylation is the maintenance of the anchorage dependence. (Dike and Farmer, 1988; Kornberg et al., 1991; Burridge et al., 1992). Another protein involved in integrin signaling is the integrin associated protein (IAP). This plasma membrane molecule is physically and functionally linked to integrins such as b3 (Brown et al., 1990). The molecule is required for integrin-regulated calcium entry in endothelial cells, suggesting that IAP is an integrin associated calcium channel (Schwarz et al., 1993).

The effects that the integrins have in the expression of genes related to differentiation can occur by mechanisms at least partially different of those involved in the regulation of cellular growth. The role of the integrins in the establishment and maintenance of the differentiated phenotype may be relevant for the analysis of the transformed phenotype. The changes in the expression of the integrins that occur during normal differentiation leads to the establishment of a characteristic pattern of integrins for each cellular type (Adams and Watt, 1993). The loss of expression of some of the integrins may contribute to cellular transformation, incapacitating the cell to respond to the differentiation signals originating from the extracellular matrix, and therefore contributing to the loss of differentiation.

Neoplasic cells are characterized by alterations of cellular adhesiveness. Studies into fibroblasts transformed by virus, show in culture a more rounded morphology than non-transformed cells. They often lack a cell surface fibronectin-containing pericellular matrix (Hynes, 1981). This suggests that the transformed cells cannot retain the fibronectin that they produce in their surface, perhaps due to a defect in the integrin receptors. Studies performed on fibronectin expressing cells demonstrate that the loss of integrin $\alpha 5\beta 1$ contributes to the adquisition of a invasive phenotype.

The effects of integrins on cellular behavior are mediated at least in part by their ability to promote the assembly of oligomeric complexes containing cytoskeletal and signaling elements. Tumor cells present a number of cytoskeletal defects (Burridge, 1986). A partial reversion has been observed of the transformed phenotype by overexpressing several integrins or cytoskeletal molecules (Gluck et al., 1993; Fernández et al., 1993) the assembly of which is integrin-dependent.

The expression of given oncogenes (ras, src, met) in transformed cells affects the level of expression of integrins (Plantefaber and Hynes, 1989; Champac 1988; Dedhar and Saulnier, 1990). The comparison between the cells of a primary tumor and the derived metastatic cells shows changes in the expression of various integrins (Mortarini et al., 1991; Klein et al., 1991; Chang et al., 1992). These alterations in the expression of the integrins might contribute to the ability of the neoplasic cells to detach from the original tumor and invade the adjacent tissues.

Studies in tumors have revealed that the loss of the expression of integrins is important for the maintenance of stable adhesion and of tissular organization. For example the reduction of $\alpha5\beta1$ and the $\alpha6\beta4$ in the ductal breast carcinoma and $\alpha2\beta4$ in colon carcinoma (Zutter et al., 1990; Koretz K. et al., 1991; Natali et al., 1992; Zutter et al., 1993). While most integrins are downregulated or otherwise defective in many human tumors in vivo, the $\alpha V\beta3$ integrin, which is capable of binding to multiple ligands, appears to be upregulated in an aggressive class of glioblastoma (Gladson and Cheresh, 1991; Paulus et al., 1993). Vitronectin, which is bound to integrin $\alpha V\beta3$, has been localized to the parenchyma of late-stage glioblastoma but was not found in early stage tumors or normal glia (Gladson and Cheresh, 1991). Tumorigenicity of neuro-ectodermal tumors cells, such as melanoma and glioblastoma, largely depends on the expression ofintegrin $\alpha V\beta3$ (Gladson and Cheresh, 1991; Marshall et al., 1991; Felding-Habermann et al., 1992). $\alpha V\beta3$ co-operates with given growth factors potentializing their effect in the cells. This integrin co-operates with the cytoplasmatic signal transduction of the insulin (and insulin -like growth factor, IGF) receptors (Vuori and Rouslahti, 1994). A protein of 190-kD which is phosphorylated in tyrosine as a result of the activity of the growth factor re-ceptor derived from the platelets (PDGF) also bound to integrin $\alpha V\beta3$, suggests that there may be cooperation between this integrin and the PDGF pathway (Bartfeld et al., 1993).

The $\alpha V\beta3$ upregulation is also observed in invasive melanoma. Interestingly, the upregulation of $\alpha V\beta3$ expression in melanoma coincides with the transition from the radial to the vertical phase of growth and may, therefore, be related to the acquisition of a more malignant phenotype and the penetration of the basal membrane by the tumor cells (Albelda et al., 1990; Feldin-Habermann et al., 1992). The role of $\alpha V\beta3$ has been demonstrated in the angiogenesis of melanomas. Thus, its expression is detected in the new vessels and not in the preexisting ones, and antibodies against it inhibit neovascularization (Brooks et al., 1994). Integrin $\alpha V\beta3$ mediated endothelial cell interaction with fibrin gels resulted in an enhancement of capillary formation (Gamble et al., 1993). Furthermore, the subcutaneous injection of a cyclic peptide antagonist of vitronectin receptor type integrins inhibits retinal neovascularization (Hammes et al., 1996). The expression of integrins in human tumors is to be seen in table 20-8.

Table 20-8. *Integrin expression in human tumors*

	ß1						α6ß4	ß3		αLß2
	α1	α2	α3	α5	α6	α7		αv	αIIb	
Brain tumors										
Astrocytoma			↑							
Glioblastoma			↑					↑		
Breast cancer		↓		↓			↓			
Colon cancer		↓								
Lung adeno cancer							↓			
Lymphoma (Burkitt)										↓
Melanoma			↑		↑			↑		
Pancreatic cancer			↓	↓						
Skin tumors										
Basal cell cancer		↓					↓			
Squamous cancer							↑			

(Giancotti and Mainiero, 1994 and references therein)

Normal astrocytes show the expression of integrin chains α2, α3, α6, ß1 and ß4, but not α4, α5 , αV, αL, αM, αX, ß2 and ß3. The neoplasic astrocytes, both in vivo and in vitro, show an increase in the expression of α3 and ß1, and occasionally of the chains α5, αV, ß3 and ß4. In comparison with the normal vascular endothelia, the neoexpression of α4 and the reduced levels of ß4 observed in the vascular proliferations of the glioblastomas stand out (Paulus et al., 1993).

The understanding of cellular adhesion at the molecular level have led to a number of therapeutic applications in the cure of cancer (Rouslahti et al., 1995). It has been shown that peptides which are able to interfere with the function of the integrins may block experimental metastasis (Humphries et al., 1986; Ruoslahti, 1992).

The expression of the members of the family of transmembrane glycoproteins, CD44, has been related to tumoral metastasis, their inhibition by means of specific antisense oligonucleotide inhibits the cellular invasion of Matrigel matrixes in continuous glial lines, modulating cellular adhesion to fibronectin, laminin, vitronectin and collagen (Merzak et al., 1994). Antibodies against integrin chains α3, α7, αV, ß1 and ß3 induce a decrease of the invasive contents of cell lines derived from glioblastomas, while the blocking of α5 and α6 increases said characteristic (Paulus et al, 1994). By means of a sterotaxic injection in the brain of nude mice of the glioblastoma U-251A1, a glioma cell line with CD44-antisense-gene expression, gliomas developed located in the injection zone with well limited borders between the tumor and the cerebral tissue (Okada et al., 1996).

The cellular determinants of the migration include adhesion to the sustrate, restructuration of the actin of the cytoskeleton as the generating element of motion and secretion of enzymes that remodel the interstitial space, favouring cellular migration. In the context of the local invasion of astrocytoma cells in the brain, the adhesion and contents of migration on different sustrates of seven cell lines of astrocyte origin, NCE, G-22, G-28, G-63, G-112, SF-767 and U-251 have been studied in vitro. Laminin and tenascin constitute the more permissive sustrates for a faster migration as well as for adhesion. The adhesion to collagen, fibronectin and vitronectin is integrin dependent and it can be blocked by means of antibodies against fraction b1. Hyaluronic acid, expressed as a migration sustrate for the normal and transformed astrocytes, is recognized by the receptor CD44 (Giese et al., 1995).

A comparison between adhesion and the ability of cellular migration shows that poor adhesion is associated with minimum levels of migration and that a high cellular adhesiveness is frequently correlated with a rapid migration. Studies which report migration on the autologous ECM of the astrocytomas cited, only attribute to the SF-767 line the ability of promoting the migration of the cells of the other astrocytomas. Thus, migration would be a constitutive characteristic of the gliomas modulated by certain permissive ligands in the environment (Giese et al., 1994).

TUMORAL INVASION AND MATRIX METALLOPROTEINASES

Gliomas present a capacity of invasion which is not strictly related to the grade of cellular malignity or dedifferentiation. Clinically, the low grade gliomas show poorly defined borders and even the anaplastic astrocytomas infiltrate the adjacent structures with greater frequency than the glioblastomas. These observations indicate that the capacity of invasion and infiltration is a characteristic dependent on the intrinsic behaviour of the transformed cells more than on their grade of dedifferentiation. In gliomas the infiltration is mainly produced through the blood vessels or spread along the surface of the brain. However, a predominant pathway of dissemination is the spread of isolated cells throughout the tracts of myelinized fibre of the white matter, a pathway observed in the invasion of the contralateral hemisphere through the callous body. The parenchyma of the white matter practically lacks structural proteins being mainly composed by hyaluronic acid and glycosaminoglycans, molecules that are non-permissive sustrates for the migration and differentiation of different cell types (Oohira et al.,1991). Therefore the dissemination of tumoral cells throughout the myelinized tracts might either be a specific mechanism based on ligands specifically expressed in the white matter or represent a passive method, being facilitated by a pathway of lower resistance.

It has been studied on glial cell lines and cultures of glioblastomas whether crude human central nervous system myelin extracts presented specific ligands involved in the cellular adhesiveness capable of promoting glioma attachment and cell migration. In comparison with the ECM proteins, the studied material showed a variable answer, which suggests that the astrocytomas may be able to use both ligands associated to the membrane of the oligodendrocytes and ECM proteins of the basal membranes to invade the normal brain (Giese et al., 1996).

The proteolytic degradation of the extracellular matrix through the secretion of enzymes which remodel the interstitial space is necessary for the cellular change. The family of structurally related metalloproteinases, namely matrix metalloproteinases (MMPs) plays an important paper in the remodulation of the matrix during development, growth and tissular repair as well as in tumoral invasion and metastasis. In the connective tissue the activity of the metalloproteinases may be modulated by a variety of stimuli, among which are included hormones, interleukines, endocytosis, phorbol esters and agents that affect the cell shape. The effects of these agents on the activity of the proteinases may suppose an increase or decrease of the gene transcription level, the stability of mRNAs, protein secretion, the synthesis of enzymes required to trigger proteolytic activity and the synthesis of endogenous inhibitors (Apodaca et al., 1990).

The MMPs are secreted as proenzymes which require the elimination of a propeptide domain in order to be activated. Eleven members of this family have been identified which may be classified into three groups according to their texture and to the specificity of their sustrate: collagenases which degrade interstitial fibrillar collagen; gelatinases specific to the degradation of type IV collagen; and stromelysins among whose sustrates are included proteoglycans, glycoproteins such as fibronectin and laminin, and type IV and V collagens. The activity of MMPs is strictly regulated by natural inhibitors called tissue inhibitors of metalloproteinases (TIMPS). There are two main members of this unit: TIMP-1 and TIMP-2 (Nakano et al., 1995).

In the medium conditioned by cell line U343, established from a human glioma, and in fetal astrocytes the secretion of interstitial collagenase and 65-kD gelatinase A has been observed (Apodaca et al., 1990) . The glial cell lines secrete small inhibitors of the metalloproteinases called, by analogy with the TIMPs, IMP-1, IMP-2 and IMP-3, the first two of which have also been detected in the medium conditioned by vascular endothelium cells (Herron et al., 1986).

The expression of MMP and TIMP genes in the tumors is regulated not only by the tumoral cells themselves but also by the interactions between the cells of the tumor and those of the host. Thus, the activator of the plasminogen, growth factors, interleukines and ECM components regulate the expression of these genes (Wasylyk et al., 1991).

Studies accomplished in vitro and in vivo on a series of human gliomas show how the concomitant overexpression of gelatinase A, gelatinase B and matrilysin genes is related to glioma malignity and is simultaneous with the overexpression of the TIMP-1 gene. There is a clear difference between the gene expression in the tumoral biopsies and in the cell lines derived from the same. The overexpression of interstitial collagenase, gelatinase B, stromelysin and TIMP-1 genes is induced by EGF, TNF-α, or interleukin-6. TGF-ß triggers the overexpression of gelatinase A and matrilysin genes. Differences in vitro and in vivo support the hypothesis of a biomodulation exercised by various cytokines (Nakano et al. 1995; Yamamoto et al., 1996). This biomodulation can be directly related to the tumoral invasion as is demonstrated in studies in vitro performed on glioblastoma cell lines whose response as compared to TNF-ß was not only inhibition of the proliferation but also of the infiltration ability (Iwasaki et al., 1993).

APOPTOSIS

Programmed cellular death or apoptosis, the most common form of cellular death of the eucariote cells, is produced during embriogenesis, tissular atrophy and tumoral regression. Apoptosis is a process whereby cells actively participate in their own death, defined by biochemical detection of DNA fragmentation and by characteristic morphological features such as nuclear and cytoplasmic condensation (Trauth et al., 1989; Vouri and Rouslahti, 1994). In pluricellular organisms, the imbalance between proliferation and apoptosis may contribute to oncogenesis (Damsky and Werb 1992; Boudreau et al., 1995). The extracellular matrix has been recognized as a regulator of apoptosis, since it has been shown how the normal adherents cells that lose contact with the extracellular matrix (ECM) activate the apoptosis (Lo et al., 1994; Meredith et al., 1993; Frisch and Francis, 1994). In this process, the greater transmittors of the signals derived from the extracellular matrix are the members of the family of the integrins and some of these signals mediated by the integrins seem to promote apoptosis. Most tumoral cells fail to synthesize their matrix or do so to a lesser degree that normal cells. In consequence, cell mobility is not limited by adhesion to the extracellular matrix (Rouslahti, 1991). Thus, tumoral cells would escape from the controls of normal differentiation by means of the loss of function or synthesis of the adhesion molecules or by the modification of the interaction with the ECM (Pignatelli and Bodner, 1990). Alterations in the integrin expression may probably contribute to tumoral development permitting the cellular outlet of apoptosis. During angiogenesis, proliferative endothelial cells become apoptotic in response to antagonists of integrin αVß3. Thus, ligation of endothelial cell αVß3 promotes a critical and specific adhesion-dependent cell survival signal during angiogenesis leading to the inhibition of p53 activity, decreased expression of p21 and suppression of the cell death pathway (Strömbland et al., 1996).

Apoptosis in glioma cells is induced by the application of anti-fas/APO1 antibody, protein kinase C inhibitors and TNF-α (Yin et al., 1995). In the human glioma cell lines U-87MG and U- 251MG, which express high levels of basicFGF and its receptors, the growth inhibitory mechanism triggered by an mouse monoclonal antibody against basicFGF is attributable to apoptosis. Some antibodies cause apoptosis in the same glioma cells that were transplanted into nude mice. In vivo and in vitro, the over-expression of bcl-2 protein by gene transfection prevented induced apoptosis of glioma cells (Murai et al., 1996).

Apoptosis can be induced by DNA damage and growth factor deprivation. Apoptosis caused by agents that damage DNA is mediated by wild-type p53 gene in various cells. The absence of functional p53 is associated with the inability of cells to undergo apoptosis in response to DNA damage and with the development of malignant tumors. Therefore, mutation of the p53 gene is found in 40% of malignant human gliomas (Deen et al.,1993). Ueba et al. (1994) found that the mutant-type p53 gene activates endogenous bFGF gene expression where as wild-type p53 represses bFGF expression in human glioma cells.

The role of p53 in promoting apoptosis is linked to its ability to repress transcription of bcl-2 while activating that of bax (Lloyd et al., 1995; Miyashita and Reed, 1995). Bcl-2 potentiates cell survival based on its ability to dimerize with the death promoting molecule bax (Oltvai et al., 1993). Bcl-2 gene has been shown to block apoptosis and thus may promote tumor development (Wu, 1996). The cell survival is favoured by a high bcl-2/bax ratio. While αVß3 and b1 integrin ligation caused increased bcl-2 expression relative to nonadherent cells, only αVß3 produced a concomitant decrease in bax expression, resulting in a sharp increase in the bcl-2/bax ratio. Accordingly, the antagonists of αVß3 during angio-genesis block bFGF-induced bcl-2 mRNA. These findings support observations in vivo that antagonists of this integrin selectively promote endothelial cells apoptosis during angiogenesis (Strömbland et al., 1996).

Table 20-9. *Frequency of p53 alterations and loss of heterozigosity on chromosome 17p in human gliomas*

Tumor type (grade)	Alterations of p53 No.	%	Loss of heterozigosity on chrom. 17p No.	%
Astrocytoma (I)	6/25	24	1/12	8
Astrocytoma (II)	51/172	30	9/68	13
Astrocytoma (III)	51/165	31	28/81	35
Glioblastoma multiforme	83/250	33	71/246	29

(From Bögler et al., 1995)

Loss or mutation of the p53 suppressor gene has been detected in many types of gliomas (table 20-9) and represents the earliest detectable genetic event in astrocytoma progression. Loss of wild-type p53 may affect two phases in glial tumorigenesis: early in the formation of low-grade disease and later by facilitating progression toward malignancy (Bögler et al., 1995). The overall spectrum of mutations in brain tumors is compared with that of other tumors. However, the frequency of mutations at particular codons differs somewhat among tumors types. For example, in colon cancer, codon 175 is mutated most frequently, while in gliomas it is codon 273. Similarly, codon 249 is commonly mutated in lung cancer, but not as often in gliomas. These comparisons are consistent with the contention that different cell types may be more susceptible to the effects of certain mutations. The types of mutations observed in sporadic and inherited brain tumors are remarkably similar. G:C A:T mutations prevail, and most of these are located at CpG sites. Transition mutations at CpG sites are most likely due to the deamination of 5' methylcytosine residues and are considered to be spontaneous (Bögler et al., 1995). Cortical astrocytes from p53 knock-out mice fail to stop growing on the formation of a confluent monolayer and show no cell death (Bögler et al., 1995).

ANGIOGENESIS IN MALIGNANT NEOPLASIA

Angiogenesis, the process of new blood vessel development, is an essential component of the body's physiology and contributes to the pathogenesis of malignant neoplasia. Angiogenesis represents a crucial stage in the tumoral progression of the gliomas; the proliferation of the vascular endothelia constitutes a histological criterion of the malignity of these tumors. Angiogenesis is controlled by a variety of factors that initiate, control and terminate this complex, multistage process, which includes endothelial cell proliferation, migration, protease elaboration and synthesis of adhesion molecules and components of the basal membrane. Among the active growth factors in the gliomas, VEFG, FGF, TGFα and PDGFß behave as angiogenic factors (Vandenbunder, 1994, Colville and Willoughby, 1997). Also, the angiogenic response may be influenced by many substances, these include extracellular matrix components such as hyaluronic acid or various collagens and fibrin degradation products (Colville and Willoughby, 1997).

The factor of vascular endothelial increment VEFG, also known as vascular permeability factor VPF, is synthesized in different cellular types as a response to intratissular hypoxia or to cytokines such as PDGF and TGFß. VEGF-induced angiogenesis in the rabbit cornea in vivo has also revealed the existence of an alternative integrin pathway using the αVß5 integrin on endothelial cells rather than the previously reported αVß3 integrin pathway induced by FGF-2. It is suggests that this could be used to differentiate angiogenic response in some pathologies such as ischemia (αVß5) from non-ischemia (αVß3) (Friedlander, 1995).

VEGF presents a rate 50 times greater in glioblastomas than in the low grade gliomas. The physiological importance of VEGF and of its receptor, tyrosine kinase Flk1, in the gliomas has been demonstrated by means of the utilization of an monoclonal antibody that specifically blocks the activity of VEGF. The injection of said antibody inhibits the development of tumors obtained through xenografts of glioblastomas in nude mice. Similar results have been reported in the same experimental model after the introduction of a mutant of the VEGF receptor in the endothelia of the vessels located in the proximities of the tumoral cells implanted (Hoang-Xuan et al., 1995).

It has been reported that, in human glioblastomas, VEGF mRNA expression is specifically upregulated in tumor cells which surround necroses. Due to their typical morphology, these cells are known as palisading cells. On the basis of histopathological observations, it has been suggested that they represent dedifferentiated anaplastic glioma cells. Since glioma cells upregulate VEGF in response to hypoxia in vitro, palisading cells could represent the metabolic conditions of these cells, which could be close to cell death. This hypotesis is supported by the fact that no immunoreactivity to the Ki-67 proliferation monoclonal antibody in VEGF mRNA expressing palisading cells could be detected, although tumor cells distant from necrosis showed immunoreactivity to Ki-67 (Plate et al., 1994).

TNFα is an inhibitor of endothelial proliferation in vitro, but can be powerfully angiogenic in vivo due to indirect actions such as the promotion of a inflammatory reaction that then stimulates angiogenesis (Frater-Schroder, 1987).

Brooks et al. (1994a) showed that integrin αVß3 becomes expressed on angiogenic vascular cells where it plays a critical role in angiogenesis induced by basicFGF, TNFα, or fragments of human tumors in the chick embryo. αVß3 antagonists, administered intravenously to chick embryos, caused regression of angiogenic blood vessels due to unscheduled apoptosis (Brooks et al., 1994b). This resulted in the regression of human tumors on the chick chorioallantoic membrane or in human skin transplants on mice (Brooks et al., 1994b; Brooks et al., 1995).

The protein encoded by the wild type p53 gene prevents cells from progressing towards malignancy in three general ways: by retarding the accumulation of the DNA lesions that underlie tumor progression, by restricting the exponential growth of incipient tumor cells, by preventing the developing tumor from attracting new blood vessels essential for its progressive growth in vivo. Growing tumors attract the vessels they need by secreting one or more potent angiogenic factors such as basicFGF, IL-8, VEGF and TGFß. Loss of the wild type p53 tumor suppressor gene can contribute to the development of an angiogenic phenotype. Thus, loss of p53 may stimulate the production of VEGF. The demonstration that thrombospondin in fibroblasts in the same way as a novel inhibitor of angiogenesis, a glioma derived angiogenesis inhibitory factor (GD-AIF), requires wild-type p53 to be expressed, which suggests that the loss of this molecule may shift the balance in favor of angiogenic stimulation (Van Meir et al., 1994; Wu, 1996). Restoring wild type p53 to a human glioblastoma cell line stimulated the secretion of an protein inhibitor of angiogenesis and, as result, the cells evolved from being potently angiogenic to antiangiogenic (Bouck, 1996).

The angiogenic activity in extracts of human tumors was first reported by Folkman et al. (1971). The transformational effects that neoplastic astrocytes may have on the endothelial cells within malignant astrocytomas have been studied by Silbergeld et al. (1991). Endothelial cell cultures were exposed to conditioned medium by glioma derived cell lines, significant changes in the amount and distribution of DNA and RNA were observed in a concentration-dependent manner. It is concluded that human glioblastoma cell lines elaborate a soluble factor/s capable of inducing transformational alterations in target normal endothelial cells, and that said transformation may account for some of the abnormal endothelial cell changes associated with malignant astrocytomas.

Certainly, the strategies based in angiogenesis inhibitors represent a valid component of anticancer therapy by restricting tumor growth. Nevertheless, Parangi et al. (1996) demonstrate the high efficacy and low toxicity associated with antiangiogenic therapy, strengthening the proposition that angiogenesis inhibitors will prove to be effective in combination with conventional cancer therapeutic strategies perhaps to the point of complete remission (Parangi et al., 1996).

REFERENCES

Adams J.C., and Watt F.M. (1993) Regulation of development and differentiation by the extracellular matrix. Development *117*: 1183-1198.

Albelda S.M., S.A. Mette, D.E. Elder, R.M. Elder, R.M. Stewart, L. Damjanovich, M. Herlyn, and C.A. Buck (1990) Integrin distribution in malignant melanoma: association of the beta 3 subunit with tumor progression. Cancer Res. *5*: 6757-6764.

Altieri D.C., R. Bader, P.M. Mannucci, and T.S. Edginton (1988) Oligospecificity of the cellular adhesion receptor Mac-1 encompasses an inducible recognition specificity for fibrinogen. J. Cell Biol. *107*: 1893-1900.

Apodaca G., J.T.Rutka, K.Bouhana, M.E.Berens, J.R.Giblin, M.L.Rosenblum, J.H.McKerrow and M.J.Banda (1990) Expression of metalloproteinases and metalloproteinase inhibitors by fetal astrocytes and glioma cells. Cancer-Res, *50*: 2322-2329.

Bartfeld N.S., E.B. Pasquali, J.E. Geltosky, and L.L. Languino (1993). The αvß3 integrin associates with a 190-kDa protein that is phosphorylated on tyrosine in response to platelet-derived growth factor. J. Biol. Chem. *268*: 17270-17276.

Benbrook D., E. Lernhardt, and M. Pfahl (1988) A new retinoic acid receptor identified from a hepatocellular carcinoma. Nature. *333*: 669-672.

Bischoff J.R., P.N.Fiedman, D.R.Marshak, C. Prives and D. Beach (1990) Human p53 is phosphorylated by p60-cdc2 and cyclin B-cdc2. Proc. Natl. Acad. Sci. U.S.A. *87*:4766-4770.

Bögler O., H.J. Su Huang, P. Kleihues, and W.K. Cavenee (1995) The p53 gene ands its role in human brain tumors. Glia. *15*:308-327.

Bouck N.(1996) P53 and angiogenesis. Biochim. Biophys. Acta. *1287*: 63-66.

Boudreau N., C.J. Sympson, Z. Werb, and M.J. Bissell (1995) Suppression of ICE and apoptosis in mammary epithelial cells by extracellular matrix. Science *267*: 891-893.

Boudreau N., Z.Werb, and M.J. Bissell (1996) Supression of apoptosis by basement membrane requires three-dimensional tissue organization and withdrawal from the cell cycle. Proc. Natl. Acad. Sci. USA. *93*:3509-3513.

Brooks P.C., R.F. Clark, and D.A. Cheresh (1994a) Requirement of vascular integrin αVß3 αVß3 αVß3 αVß3 for angiogenesis. Science *264*:569-571.

Brooks P.C., A.M. Montgomery, M. Rosenfeld, R.A. Reisfeld, T.Hu, G. Klier, and D.A. Cheresh (1994b) Integrin αVß3 antagonists promotes further tumor regression by inducing apoptosis of angiogenic blood vessels. Cell. *79*:1157-1164.

Brooks P.C., S. Stròmbald, R. Klemke, D. Visscher, F.H. Sarkar, and D.A. Cheresh (1995) Antiintegrin αVß3 blocks human breast cancer growth and angiogenesis in human skin. J.Clin.Invest. *96*:1815-1822.

Brown E., L. Hooper, T. Ho, and H. Gresham (1990) Integrin-associated protein: a 50-kD plasma membrane antigen physically and functionally associated with integrins. J. Cell Biol. *111*: 2785-2794.

Burridge K., C.E. Turner, and L.H. Romer (1992) Tyrosine phosphorylation of paxillin and pp125FAK accompanies cell adhesion to extracellular matrix: a role in cytoskeletal assembly. J. Cell Bio. *119*: 893-903.

Collins V.P. (1993) Amplified genes in human gliomas. Semin. Cancer Biol. *4*: 27-33.
Couldwell W.T., J.H.Uhm, J.P. Antel and V.W. Yong (1991)Enhanced protein kinase C activity correlates with the growth rate of malignant gliomas in vitro. Neurosurgery. *29*:119-126.
Couldwell W.T., N.de Tribolet., J.P. Antel, T.Gauthier, and M.C.Kuppner (1992). Adhesion molecules and malignant gliomas: implications for tumorigenesis. J Neurosurg, *76*: 782-791.
Chan B.M., N. Matsuura, Y. Takada, B.R. Zetter, and M.E. Hemler (1991) In vitro and in vivo consequences of VLA-2 expression on rhabdomyosarcoma cells. Science *251*: 1600-1602.
Chang Y.S., Y.Q. Chen, J .Timar, K.K. Nielsen, I.M.Grossi, L.A. Fitzgerald, C.A. Diglio, and K.V. Honn (1992) Increased expression of alpha II beta 3 integrin in subpopulations of murine melanoma cells with high lung-colonizing ability. Int. J. Cancer *51*: 445-451.
Chen J.M., and W.T. Chen (1987) Fibronectin-degrading proteases from membranes of transformed cells. Cell. *48:* 193-203.
Damsky C.H., and Z Werb. (1992) Signal transduction by integrin receptors for extracellular matrix: cooperative processing of extracellular information. Curr. Opin. Cell. Biol. *4*: 772-781.
De Clerk Y.A.,H. Shimada, I. González-Gómez, and C. Raffel (1994) Tumoral invasion in the central nervous system J. Neurooncol. *18*:11-121.
Dedhar S, and R. Saulnier. (1990) Alterations in integrin receptor expression on chemically transformed human cells: specific enhancement of laminin and collagen receptor complexes. J. Cell Biol. *110*: 481-489.
Deen D.F., A. Chiarodo, and E.A. Grimm (1993) Brain tumoor working group report on the 9th international conference on brain tumor research and therapy. J. Neurooncol. *16*:243-272.
Dike L.E., and S.R. Farmer (1988) Cell adhesion induces expression of growth-associated genes in suspension-arrested fibroblasts.Proc. Natl. Acad. Sci. USA *85*: 6792-6796.
Erickson H.P. and M.A. Bourdon (1989) An extracellular matrix protein prominent in specialized embryonic tissues and tumors. Annu Rev Cell Biol, *5*:71-92.
Felding-Habermann B., B.M. Mueller, C.A. Romerdahl, and D.A. Cheresh (1992) Involvement of integrin alpha v gene expression in human melanoma tumorigenicity. J. Clin. Invest *89*: 2018-2022.
Fernandez J.L., B. Geiger, D. Salomon, and A. Ben-Ze'ev. (1993) Suppression of vinculin expression by antisense transfection confers changes in cell morphology, motility and anchorage-dependent growth of 3T3 cells. J. Cell Biol. *122*: 1285-1294.
Fleming T.P., A. Savena, W.C. Clark, J.T. Robertson, E.H. Oldfield, S.A. Aaronson, and I.V. Ali (1992) Amplification and/or overexpression of platelet derived growth factor receptors and epidermal growth factor receptor in human glial tumors. Cancer Res. *52*:4550-4553.
Folkman J., and M. Klagsbrun (1986) Angiogenic. Science *235*: 442-447.
Frisch S.M., and H. Francis (1994) Disruption of epithelial cell-matrix interactions induces apoptosis. J. Cell Biol. *124*: 619-626.
Fuller G.N. and S.H. Bigner (1992) Amplified cellular oncogene in neoplasms of the human central nervous system. Mutat Res. *276*:299-306.
Gamble J.R., L.J. Matthias, G. Meyer, P Kaur, G. Russ, R. Faull, M.C. Berndt, and M.A. Vadas (1993). Regulation of in vitro capillary tube formation by anti-integrin antibodies. J. Cell Biol. *121*:931-943.
Gawaz M.P., J. Loftus, M.L. Bajt, M.M. Frojmovic, E.F. Plow, and M.H. Ginsberg (1991) Ligand bridging mediates integrin alpha Iib beta 3 (platelet GPIIB-IIA) dependent homotypic and heterotypic. J. Clin. Invest. *88*: 1128-1134.
Giese A., M.D. Rief, M.A. Loo, and M.E. Berens (1994). Determinants of human glioma migration. Cancer Res, *54*: 3897-3904.
Giese A., A.M. Loo, M.D. Rief, N. Tran and M.E. Berens (1995) Substrayes for astrocytoma invasion. Neurosurgery, *37*: 294-301.
Giese, A., L. Kluwe, B. Laube, H. Meissner, M.E. Berens, and M. Westphal (1996) Migration of human glioma cells on myelin. Neurosurgery. *38*(4): 755-64.
Ginsberg M.H., X. Du, and E.F. Plow (1992) Inside-out integrin signalling. Curr. Op. Cell Biol. *4*: 766-771.
Gladson C.L., and D.A. Cheresh (1991) Glioblastoma expression of vitronectin and the alpha v beta 3 integrin. Adhesion mechanism for transformed glial cells. J. Clin. Invest. *88*: 1924-1932.
Gluck U., D.J. Kwiatkowski, and A. Ben-Ze'ev.(1993) Suppression of tumorigenicity in Simian virus 40 transformed 3T3 cells transfected with alpha-actinin cDNA. Proc. Natl. Acad. Sci. USA. *90*: 383-387.

Goldman J.E., S. Geiser and M. Hirano (1986) Differentiation of astrocytes and oligodendrocytes from germinal matrix cells in primary culture. J Neurosci, *6*: 52-60.
Hammes H-P, M. Brownlee, A. Jonczyk, A. Sutter, and K.T. Preissner (1996) Subcutaneous injection of a cyclic peptide antagonist of vitronectin receptor-type integrins inhibits retinal neovascularization. Nature Medicine *8*: 529-533.
Hemler M.E. (1990) VLA proteins in the integrin family: structures, functions and their role on leukocytes. Annu. Rev. Immunol. *8*: 365-400.
Herron G.S., Z.Werb and K. Dwyer (1986). Secretion of metalloproteinases by stimulated capillary endothelial cells. Production of procollagenase and prostromelysin exceeds expression of proteolytic activity. J Biol Chem, *261*:2810-2813.
Hoang-Xuan K., Y.J. Li, R. Hamelin and D.Delattre (1995) Molecular biology in cerebral gliomas: recent advances. Rev. Neurol. *151*: 608-618.
Humphries M.J., S.K. Akiyama, A. Komoyiya, K, Olden, and K.M. Yamada. (1986) Identification of an alternatively spliced site in human plasma fibronectin that mediates cell type-specific adhesion. J. Cell Biol. *103*: 2637-2647.
Hurtt M.R., J. Moossy, M. Donovan-Peluso, and J. Locker (1992) Amplification of epidermal growth factor receptor gene in gliomas: Histopathology and prognosis. J. Neuropathol. Exp. Neurol., *51*:84-92.
Hynes R.O. (1981) in Hay, E.D. (ed.), The Cell Biology of the Extracellular Matrix, Plenum Plublishing, New York, pp 295-231.
Hynes R.O. (1992) Integrins: versatility, modulation, and signaling in cell adhesion. Cell *69*: 11-25.
Iwasaki K, L.R.Rogers, G.H.Barnett, M.L.Estes and B.P.Barna (1993) Effect of recombinant tumor necrosis factor-alpha on three-dimensional growth, morphology, and invasiveness of human glioblastoma cells in vitro. J Neurosurg, *8*: 952-958.
Johnson M,D., M.T.Jennings, L.I.Gold and H.L.Moses (1993). Transforming growth factor-beta in neural embryogenesis and neoplasia. Human Pathol. *24*: 457-462.
Juliano R.L., and S. Haskill (1993) Signal transduction from extracellular matrix. J. Cell Biol. *120*: 577-585.
Karpatkin S., E. Pearlstein , C. Ambrogio, and B.S. Coller (1988) Role of adhesive proteins in platelet tumor interaction in vitro and metastasis formation in vivo. J. Clin. Invest. *81*: 1012-1019.
Klein C.E., D. Dressel T. Steinmayer, C. Mauch, B. Eckes, T. Krieg, R.B. Bankert, and L. Weber (1991) Integrin alpha 2 beta 1 is upregulated in fibroblasts and highly aggresive melanoma cells in three-dimensional collagen lattices and mediates the reorganization of collagen Y fibrils. J. Cell Biol. *115*: 1427-1436.
Koretz K., P. Schalg, L. Boumsell, and P. Moller (1991) Expression of VLA-alpha 2, VLA-alpha 6, and VLA-beta 1 chains in normal mucosa and adenomas of the colon, and in colon carcinoma and their liver matastases. Am. J. Pathol. *138*: 741-750.
Lane D.P. (19994) Guardian of the genome. Nature. *3*:15-16.
Lillien L.E. and M.C. Raff (1990) Analysis of cell-cell interactions that control type-2 astrocyte development in vitro. Neuron. *4*:525-534.
Lo S.H., E. Weisberg, and L.B. Chen (1994) Tensin: A potencial link between the cytoskeleton and signal transduction. Bioessays *16*: 817-823.
Marshall J.F., S.A. Nesbitt, M.H. Helfrich, M.A. Horton, K. Polakova, and I.R. Hart (1991). Integrin expression in human melanoma cell lines: Heterogeneity of vitronectin receptor composition and function. Int. J. Cancer *49*: 924-931.
Masterts B.A, H.Werner , C.T. Jr. Roberts, D.Le Roith and M.K.Raizada (1991) Insulin-like growth factor I (IGF-I) receptors and IGF-I action in oligodebdrocytes from rat brains. Reg. Pept. *33*:117- 131.
Mayer M., K. Bhakoo and M. Noble (1994) ciliary neurotrophic factor and leukemis inhibitory factor promote the genertion, maturation, and survival of oligodendrocytes in vitro. Development. *120*: 143-153.
Melton D.A. (1991) Pattern formation during animal development. Science. *252*: 234-241.
Meredith J.E., B. Fazeli, and M.A. Schwartz. (1993) The extracellular matrix as a cell survival factor. Mol. Biol. Cell *4*: 953-961.
Merzak A, S. Koocheckpour, and G.J.Pilkington.(1994) CD44 mediates human glioma cell adhesion and invasion in vitro. Cancer Res. *54*: 3988-3992.

Mignatti P., and D.B. Rifkin (1993) Biology and biochemistry of proteinases in tumor invasion. Physiol. Rev. *73*:161-195.

Miyashita T., and J.C. Reed (1995) Tumor suppressor p53 is a direct transcripcional activator of the human bax gene. Cell. *80*: 293-299.

Mortarini R., A Anichini, and G. Parmiani. (1991) Heterogeneity for integrin expression and cytokine-mediated VLA modulation can influence the adhesion of human melanoma cells to extracellular matrix proteins. Int. J. Cancer *47*: 551-557.

Murai N.:, T. Ueba, J.A. Takahashi, H.Q. Yang, H. Kikuchi, H. Hiai, M. Hatanaka and M. Fukumoto (1996) Apoptosis of human glioma cells in vitro and in vivo induced by a neutralizing antibody against human basic fibroblast growth factor. J. Neurosurg. *85*:1072-1077.

Nakano A., E.Tani, K.Miyazaki, Y.Yamamoto and J.Furuyama.(1995) Matrix metalloproteinases and tissue inhibitors of metalloproteinases in human gliomas. J. Neurosurg. *83*: 298-307.

Natali P.G., M.R. Nicotra, C. Botti, M. Mottolese, A. Bigotti, and O. Segatto (1992) Changes in expression of alpha 6/beta 4 integrin heterodimer in primary and metastatic breast cancer. Br. J. Cancer *66*: 318-322.

Nelson J.S., D. Schoenfeld, Y. Tsukada, and F.Fulling (1982).Histologic criteria with prognostic sifnificance for malignant gliomas, in Chang CH, Housepian EM (eds): Tumor of the Central Nervous System: Modern Radiotherapy in Multidisciplinary management. New York, Masson Publishing USA, Inc. pp 1-4.

Okada H., J.Yoshida, M. Sokabe, T. Wakabayashi and M. Hagiwara (1996) Suppression of CD44 expression decreases migration and invasion of human glioma cells. Int. J. Cancer *66*: 255-260.

Oltvai Z.N., C.L. Millman, and S.J. Korsmeyer (1993) Bcl-2 heterodimerizers in vivo with a conserved homologue, bax, that accelerates programmed cell death. Cell. *74*:609-619.

Oohira A., F. Matsui and R. Katoh-Semba (1991) Effects of brain chondroitin sulfate proteoglycans on neurite ouygrowth from PC12 cells. J Neurosci. *11*: 822-827.

Parsons J.T., M.D. Schaller, J. Hildebrand, T.Z. Leu, A. Richardson, and C. Otey (1994). Focal adhesion kinase: Structure and signalling. J. Cell Sci. *18*: 109-113.

Paulus W., Y. Baur, D. Schuppan, and W. Roggendorf (1993) Characterization of integrin receptors in normal and neoplastic human brain. Am. J. Pathol. *143*: 154-163.

Paulus, W., I.Baur, D.Schuppan and W.Roggendorf (1993). Characterization of integrin receptors in normal and neoplastic human brain. Am. J. Pathol. *143*: 154-163.

Paulus, W., and J.N.Tonn (1994) Basement membrane invasion of glioma cells mediated by integrin receptors.J. Neurosurg.*80*: 515-519

Pignatelli M., and W.F. Bodmer (1990) Integrin cell adhesion molecules and colorectal cancer. J. Path. *162*: 95-97.

Plantefaber L., and R.O. Hynes (1987) Changes in integrin receptors on oncogenically transformed cells. Cell *56*: 181-190.

Quigley J.P., L.I. Gold, R. Schwimmer, and Sullivan L.L. (1986) Limited cleavage of cellular fibronectin by plasminogen activator purified from transformed cells. Proc. Natl. Acad. Sci. USA *84*: 2776-2780.

Reynolds B.A., W.Tetzlaff, and S Weiss (1992) A multipotent EGF-responsive striatal embryonic progenitor cell produces neurons and astrocytes. J Neurosci. *12*: 4565-4574

Rosenblum M.I.(1993) factors influencing tumors cell trafic in the central nervous system. J. Neurooncol. *69*:155-177.

Rouslathi E (1984). Fibronectin in cell adhesion and invasion. Cancer Metastasis Rev. *3*: 43-51.

Rouslahti E. (1991) Integrins. J. Clin. Invest. *87*: 1-5.

Rouslahti E. (1992) Control of cell motility and tumor invasion by extracellular matrix interactions. Br. J. Cancer *86*: 239-242.

Rouslahti E., M.D. Pierschbacher, V.L. Woods (1995) The α5ß1 integrin in tumor supression. Bull Inst. Pasteur *92*: 242-247.

Saga S., W. T. Chen, and Yamada K.M. (1988) Enhanced fibronectin receptor expression in Rous sarcoma virus-induced tumors. Cancer Res. *48*: 5510-5513.

Schwarz M.A., E. Brown, and B. Fazeli (1993) A 50-kD integrin-associated protein is required for integrin regulated calcium entry in endothelial cells. J. Biol. Chem. *27*:19931-19934.

Springer T.A. (1990) Adhesion receptors of the immune system. Nature *346*: 425-434.

Strömblad S., J.C. Becker, M. Yebra, P.C. Brooks, and D.A. Cheresh (1996) Suppresion of p53 activity and p21 expression by vascular cell integrin αVß3 during angiogenesis. J.Clin. Invest. *98*:426-433.

Sullivan L.M. and J.P. Quigley (1986) An anticatalytic monoclonal antibody to avian plasminogen activator: its effect on behaviour of RSV-transformed chick fibroblasts. Cell *45*:905-915.

Thomas L.A. and K.M. Yamada (1992) Contact stimulation of cell migration. J Cell Sci, *103*:1211-1214

Trauth B., C. Klas, A.M. Peters, S. Matzku, P. Möller, W. Falk, K.M. Debatin, and P.H. KKrammer (1989) Monoclonal antibody-mediated tumor regression by induction of apoptosis. Science, *245*:301-305.

Ueba y., T. Nosaka and J.A. Takahashi (1994) Transcriptional regulation of basic fibroblastic growth factor gene by p53 in human glioblastoma and hepatocellular carcinoma cells. Proc.Natl.Acad.Sci USA *91*:9009-9013.

Vandenbunder B.,V. Fafeur, N. Wernert and D.Steheln (1994) Analyse moleculaire de l'angiogenese tumorale. Med. Science *5*:516-527.

Van Meir E.G., P.J. Polverini, V.R.Chazin, H. Huang, J.S.de Tribolet and W.K.Cavenee (1994) Release of a inhibitor of angiogenesis upon induction of wild-type p53 expression in glioblastoma cells. Nature Genet. *8*:171-176.

Vertosick FT, R.G. Selker, and V.C. Arena (1991) Survival of patients with well-differentiatedm astrocytomas diagnosed in the era of computed tomography. Neurosurgery. *28*:496-501.

Veter D.J.and D.G. Thomas (1991) Multiple sequential abnormalities in the evolution of human gliomas.Br J Cancer Res. *63*:445-448.

Vuori K., and E. Rouslahti. (1994) Association of insulin receptor substrate-1 with integrins. Science. *266*: 1576-1578.

Wasylyk C., A.Gutman and R. Nicholson (1991). The c-Ets oncoprotein activates the stromelysin promoter trough the same elements as several nonnuclear oncoproteins. EMBO J. *10*:1127-1134.

Westermark B., and M. Nistér (1995) Molecular genetics of human glioma. Curr.Opin.Oncol. *7*:220-225.

Wright S.D., J.I. Weitz, A.J. Huang, S.M. Levin, S.C. Silverstein, and J.D. Loike (1988) Complement receptor type three (CD11b/CD18) of human polymorphonuclear leukocytes recognizes fibrinogen. Proc. Natl. Acad. Sci. USA *85*: 7734-7738.

Yamamoto M., S.Mohanam, R.Sawaya, G.N.Fuller, M.Seiki, H.Sato, Z.L.Gokaslan, L.A.Liotta, G.L.Nicolson and J.S.Rao (1996). Differential expression of membrane-type matrix metalloproteinase and its correlation with gelatinase A activation in human malignant brain tumors in vivo and in vitro. Cancer Res. *56*: 384-392.

Zutter M.M., G. Mazoujian, and S.A Santoro (1990) Decreased expression of integrin adhesive protein receptors in adenocarcinoma of the breast. Am. J. Pathol. *137*: 863-870.

Zutter M.M., H.R. Krigman, and S.A. Santoro (1993) Altered integrin expression in adenocarcinoma of the breast. Analysis by in situ hybridization. Am. J. Pathol. *142*:1439-1448.

INDEX

GPSR Compliance

The European Union's (EU) General Product Safety Regulation (GPSR) is a set of rules that requires consumer products to be safe and our obligations to ensure this.

If you have any concerns about our products, you can contact us on ProductSafety@springernature.com

In case Publisher is established outside the EU, the EU authorized representative is:

Springer Nature Customer Service Center GmbH
Europaplatz 3
69115 Heidelberg, Germany

Batch number: 10755284

Printed by Printforce, the Netherlands